Helicobacter pylori:
Molecular and Cellular Biology

Edited by

Mark Achtman
Berlin, Germany

and

Sebastian Suerbaum
Würzburg, Germany

Copyright © 2001
Horizon Scientific Press
P.O. Box 1
Wymondham
Norfolk NR18 0EH
England

www.horizonpress.com

British Library Cataloguing-in-Publication Data

A catalogue record for this book is available from the British Library

ISBN: 1-898486-25-5

Printed and bound in Great Britain
by Biddles Ltd, Guildford and King's Lynn

Contents

List of Contributors   vi

Preface   x

Section 1 - History, Taxonomy and Epidemiology

1. *Helicobacter pylori*: A Historical Overview   1
Drahoslava Pantoflickova and André L. Blum

2. Taxonomy and Phylogeny of *Helicobacter*   15
James Versalovic and James G. Fox

3. Epidemiologic Observations and Open Questions About Disease and Infection   29
Caused by *Helicobacter pylori*
Roger A. Feldman

Section 2 - Host Responses

4. Antigastric Autoimmunity and Pathology in *Helicobacter pylori* Gastritis   53
Gerhard Faller and Thomas Kirchner

5. Cytokine Responses in *Helicobacter pylori* Infection   63
Jean E. Crabtree

6. Role of Th1/Th2 Cells in *H. pylori*–induced Gastric Disease   85
Mario M. D'Elios and Gianfranco Del Prete

7. *Helicobacter pylori* induced Apoptosis   99
Siegfried Wagner and Winfried Beil

8. Animal Models for *Helicobacter pylori* Gastritis   113
Kathryn A. Eaton

9. Transgenic Mouse Models for Studying the Relationship Between *Helicobacter pylori*   133
Infection and Gastric Cancer
Per G. Falk, Andrew J. Syder, Janaki L.Guruge, Lars G. Engstrand, and Jeffrey I. Gordon

Section 3 -Bacterial Virulence

10. *Helicobacter pylori* Urease   155
Harry L.T. Mobley

11. *Helicobacter* Motility and Chemotaxis   171
Christine Josenhans and Sebastian Suerbaum

12. *Helicobacter pylori*, an Adherent Pain in the Stomach 185
    Markus Gerhard, Siiri Hirmo, Torkel Wadström, Halina Miller-Podraza, Susann Teneberg,
    Karl-Anders Karlsson, Ben Appelmelk, Stefan Odenbreit, Rainer Haas,
    Anna Arnqvist and Thomas Borén

13. *Helicobacter pylori* Lipopolysaccharides 207
    Anthony P. Moran

14. The *cag* Type IV Secretion system of *Helicobacter pylori* and CagA Intracellular Translocation 227
    Markus Stein, Rino Rappuoli and Antonello Covacci

15. *Helicobacter pylori* VacA Vacuolating Cytotoxin and HP-Nap Neutrophil Activating Protein 245
    Cesare Montecucco, Emanuele Papini, Marina de Bernard, William G. Dundon, Mario Zoratti,
    Jean-Marc Reyrat, John L. Telford, Giuseppe del Giudice and Rino Rappuoli

16. Development of a *Helicobacter pylori* Vaccine 263
    Subhas Banerjee and Pierre Michetti

Section 4 - Genetic Relationships

17. Comparative Analysis of the *Helicobacter pylori* Genomes 275
    Richard A. Alm and Trevor J. Trust

18. Molecular Typing of *Helicobacter pylori* 293
    Leen-Jan van Doorn, Céu Figueiredo, and Wim Quint

19. Population Structure of *Helicobacter pylori* and Other Pathogenic Bacterial Species 311
    Mark Achtman

Colour Plates 323

Index 325

Books of Related Interest

Development of Novel Antimicrobial Agents: Emerging Strategies 2001
Editor: Karl Lohner

Prokaryotic Nitrogen Fixation: A Model System for the Analysis of 2000
a Biological Process
Editor: Eric W. Triplett

NMR in Microbiology: Theory and Applications 2000
Editors: Jean-Noël Barbotin and Jean-Charles Portais

Oral Bacterial Ecology: The Molecular Basis 2000
Editors: Howard K. Kuramitsu and Richard P. Ellen

Molecular Marine Microbiology 2000
Editor: Douglas H. Bartlett

Cold Shock Response and Adaptation 2000
Editor: Masayori Inouye

Prions: Molecular and Cellular Biology 1999
Editor: David A. Harris

Probiotics: A Critical Review 1999
Editor: Gerald W. Tannock

Peptide Nucleic Acids: Protocols and Applications 1999
Editors: Peter E. Nielsen and Michael Egholm

Current Issues in Molecular Biology 1999

Intracellular Ribozyme Applications:Protocols and Applications 1999
Editors: John J. Rossi and Larry Couture

For further information on these books contact:

Horizon Scientific Press
P.O. Box 1, Wymondham
Norfolk
NR18 0EH England

Tel: +44(0)1953-601106
Fax: +44(0)1953-603068
Email: mail@horizonpress.com
Internet: www.horizonpress.com

Our Web site has details of all our books including full chapter abstracts, book reviews, and ordering information:

www.horizonpress.com

Contributors

Mark Achtman,
Max-Planck Institut für Infektionsbiologie,
Schumannstr. 21/22,
D-10117 Berlin,
Germany

Richard A Alm,
Infection Discovery,
AstraZeneca R&D Boston,
35 Gatehouse Drive,
Waltham,
Massachusetts 02451,
U.S.A.

Ben J. Appelmelk,
Department Med. Microbiology,
Vrije Universiteit,
Medical School,
1081 BT Amsterdam,
The Netherlands

Anna Arnqvist,
Department of Odontology,
Umeå University,
SE-901 87 Umeå,
Sweden

Subhas Banerjee,
Division of Gastroenterology,
Beth Israel Deaconess Medical Center,
Harvard Medical School,
Boston,
Massachusetts,
U.S.A.

Winfried Beil,
Department of Pharmacology,
Medizinische Hochschule Hannover,
Carl-Neuberg-Str. 1,
D-30625 Hannover,
Germany

André L. Blum,
Gastroenterology Department,
CHUV, anc. CI-MAT,
CH-1011, Lausanne,
Switzerland

Thomas Borén,
Department of Odontology,
Umeå University,
SE-901 87 Umeå,
Sweden

Antonello Covacci,
Immunobiol. Research Inst. Siena (IRIS),
Chiron S.p.A.,
53100 Siena,
Italy

Jean E. Crabtree,
Molecular Medicine Unit,
St. James's University Hospital,
Leeds LS9 7TF,
England

Marina de Bernard ,
Centro CNR Biomembrane,
and Dipartimento di Scienze Biomediche,
Università di Padova,
Via G. Colombo 3,
35121 Padua,
Italy

Giuseppe del Giudice,
Centro Ricerche IRIS,
CHIRON-Vaccines,
53100 Siena,
Italy

Mario M. D'Elios,
Department Internal Med. Immunoallergology,
University of Florence,
Viale Morgagni 85,
50134 Florence,
Italy

Gianfranco Del Prete,
Department Internal Med. Immunoallergology,
University of Florence,
Viale Morgagni 85,
50134 Florence,
Italy

William G. Dundon,
Centro CNR Biomembrane,
and Dipartimento di Scienze Biomediche,
Università di Padova,
Via G. Colombo 3,
35121 Padua,
Italy

Kathryn A. Eaton,
Ohio State University,
1925 Coffey Road,
Columbus,
OH 43210,
U.S.A.

Lars G. Engstrand,
Swedish Inst. Infectious Disease Control,
SE-171 82 Solna,
Sweden

Per G. Falk,
GI Pharmacology,
AstraZeneca R & D Mölndal,
SE-431 83 Mölndal,
Sweden

Gerhard Faller,
Institute of Pathology,
University of Erlangen-Nuremberg,
Krankenhaustrasse 8-10,
D-91054 Erlangen,
Germany

Roger Feldman,
The Royal London Hospital,
Whitechapel Road,
London E1 1BB,
England

Céu Figueiredo,
Delft Diagnostic Laboratory,
R. de Graafweg 7,
2625 AD Delft,
the Netherlands

James G. Fox,
Division of Comparative Medicine,
Massachusetts Institute of Technology,
77 Massachusetts Ave.,
Cambridge MA 02139,
U.S.A.

Markus Gerhard,
Department of Medicine II,
Technical University of Munich,
81675 Munich,
Germany

Jeffrey I. Gordon,
Department Mol. Biol. Pharmacology,
660 S. Euclid Ave., Box 8103,
Washington Univ. Sch. Medicine,
St. Louis,
MO. 63110,
U.S.A.

Janaki L.Guruge,
Department Mol. Biol. Pharmacology,
660 S. Euclid Ave., Box 8103,
Washington Univ. Sch. Medicine,
St. Louis,
MO. 63110,
U.S.A.

Rainer Haas,
Max von Pettenkofer-Institut für Hyg.,
und Medizinische Mikrobiologie,
Lehrstuhl Bakteriologie,
Pettenkofer Str. 9a,
D-80336 Munich,
Germany

Siiri Hirmo,
Department Infect. Dis. Med. Microbiol.,
University of Lund,
SE-223 62 Lund,
Sweden

Christine Josenhans,
Institut für Hygiene und Mikrobiologie,
Universität Würzburg,
Josef-Schneider-Strasse 2,
D-97080 Würzburg,
Germany

Karl-Anders Karlsson,
Inst. Medical Biochemistry,
Göteborg University,
SE-405 30 Göteborg,
Sweden

Thomas Kirchner,
Institute of Pathology,
University of Erlangen-Nuremberg,
Krankenhaustrasse 8-10,
D-91054 Erlangen,
Germany

Pierre Michetti
Division de Gastroentérologie, BH-10N,
Centre Hospitalier Universitaire Vaudois,
CH-1011 Lausanne,
Switzerland

Harry L.T. Mobley,
Department Microbiol. Immunol.,
Univ. Maryland School of Medicine,
Baltimore,
Maryland 21201,
U.S.A.

Cesare Montecucco,
Centro CNR Biomembrane
and Dipartimento di Scienze Biomediche,
Università di Padova,
Via G. Colombo 3,
35121 Padua,
Italy

Anthony P. Moran,
Department of Microbiology,
National University of Ireland Galway,
University Road,
Galway,
Ireland

Stefan Odenbreit,
Max von Pettenkofer-Institut für Hyg.,
und Medizinische Mikrobiologie,
Lehrstuhl Bakteriologie,
Pettenkofer Str. 9a,
D-80336 Munich,
Germany

Drahoslava Pantoflickova,
Gastroenterology Department,
CHUV anc. CI-MAT,
CH-1011, Lausanne,
Switzerland

Emanuele Papini,
Dipt. Sci. Biomed. Oncologia Umana,
Università di Bari,
70124 Bari,
Italy

Halina Miller-Podraza,
Institute of Medical Biochemistry,
Göteborg University,
SE-405 30 Göteborg,
Sweden

Wim Quint,
Delft Diagnostic Laboratory,
R. de Graafweg 7,
2625 AD Delft,
the Netherlands

Rino Rappuoli ,
Centro Ricerche IRIS,
CHIRON-Vaccines,
53100 Siena,
Italy

Jean-Marc Reyrat,
Unité de Génétique Mycobactérienne,
Institut Pasteur,
75724 Paris,
France

Markus Stein,
Immunobiol. Res. Inst. Siena (IRIS),
Chiron S.p.A.,
53100 Siena,
Italy

Sebastian Suerbaum,
Institut für Hygiene und Mikrobiologie,
Universität Würzburg,
Josef-Schneider-Strasse 2,
D-97080 Würzburg,
Germany

Andrew J. Syder,
Department Mol. Biol. Pharmacology,
660 S. Euclid Ave., Box 8103,
Washington University School of Medicine,
St. Louis, MO. 63110,
U.S.A.

John L. Telford,
Centro Ricerche IRIS,
CHIRON-Vaccines,
53100 Siena,
Italy

Susann Teneberg,
Institute of Medical Biochemistry,
Göteborg University,
SE-405 30 Göteborg,
Sweden

Trevor J. Trust,
Infection Discovery,
AstraZeneca R&D Boston,
35 Gatehouse Drive,
Waltham,
Massachusetts 02451,
U.S.A.

Leen-Jan van Doorn,
Delft Diagnostic Laboratory,
R. de Graafweg 7,
2625 AD Delft,
the Netherlands

James Versalovic,
Division of Comparative Medicine,
Massachusetts Institute of Technology,
77 Massachusetts Ave.,
Cambridge MA 02139,
U.S.A.

Torkel Wadström,
Department Infect. Dis. Med. Microbiol.,
University of Lund,
SE-223 62 Lund,
Sweden

Siegfried Wagner,
Dept. Gastroenterology & Hepatology,
Medizinische Hochschule Hannover,
Carl-Neuberg-Str. 1,
D-30625 Hannover,
Germany

Mario Zoratti,
Centro CNR Biomembrane
and Dipartimento di Scienze Biomediche,
Università di Padova,
Via G. Colombo 3,
35121 Padua,
Italy

Preface

An awe-inspiring amount of knowledge has been acquired about *Helicobacter pylori* biology since Robin Warren's and Barry Marshall's seminal publication in 1983. As of June 2000, the Pubmed database contained 11882 entries for "pylori" (Salmonella, 45036; Shigella, 8249; Yersinia, 6976; Legionella, 3659; Bartonella, 710). The quality of research in the *H. pylori* field has also increased continuously over the last decade, especially through the recruitment of numerous scientists with special expertise from diverse fields. The diversity of the approaches and the sheer numbers of publications make it difficult for even the most diligent reader to stay abreast of progress on this fascinating pathogen. This book distills exciting findings from this vast body of literature for scientists and clinicians from other fields as well as providing updates within various disciplines for *H. pylori* experts. It contains 19 brief but thorough review chapters that summarize the state of knowledge in early 2000 on the molecular and cellular biology of infection with *H. pylori*. Due to its particular focus, it does not deal with patient diagnosis and treatment and contains very little concerning pathology.

The book begins with background chapters covering the history, taxonomy, and epidemiology of *H. pylori* infection, because we believe that these topics are paramount to an integrated understanding of *H. pylori* infection and associated diseases. It continues with chapters on host responses including immunity, apoptosis and animal models followed by chapters on virulence factors and vaccine development. The final section summarizes genome analyses, molecular typing and population genetics. Chapters are independent and can be read in any desired sequence.

Given that funding is currently so dependent on numerous primary publications in journals with high impact factors, we anticipated difficulties in convincing the leading figures in this area to write review chapters for our book. To our great pleasure, essentially all of the authors we approached enthusiastically provided a contribution. We are extremely grateful to all the authors for their input.

We thank Ms Annette Griffin of Horizon Scientific Press for initiating this book. She has followed the project thoughout its course, and has been extremely supportive.

Mark Achtman, *Berlin*,
and Sebastian Suerbaum, *Würzburg*,
20.6.2000

References

Warren, J.R., and Marshall, B.J. 1983. Unidentified curved bacilli on gastric epithelium in active gastritis. Lancet i: 1273-1275.

From: *Helicobacter pylori: Molecular and Cellular Biology*
ISBN 1-898486-25-5 © 2001 Horizon Scientific Press, Wymondham, UK.

1

Helicobacter pylori: A Historical Overview

Drahoslava Pantoflickova[*] and André L. Blum

Abstract

For more than 200 years, generations of pathologists, clinicians and epidemiologists have tried to explain the cause of peptic ulcer disease. Beginning with Robert Koch and Louis Pasteur in the 19th century, one century was necessary to establish the connection between *Helicobacter pylori* and peptic ulcer disease. The isolation of *H. pylori* by Marshall and Warren 15 years ago initiated one of the major medical revolutions of the 20th century (Marshall and Warren, 1984). However, it is becoming clear that the relationship between *H. pylori* and human disease is highly complex. We all seek simple answers but, as in the case of *H. pylori*, "truth is rarely pure and never simple" (Oscar Wilde, The Importance of Being Earnest, act.1). With this in mind, we present a historical overview of *H. pylori* infection.

History of Peptic Ulcer

The first recorded case of gastric ulcer was published in 1586 (Donati, 1586). Duodenal ulcer was firstly described by Hamberger in 1757 (Hamberger, 1757). Detailed descriptions of the symptoms and morbid anatomy of gastric ulcer appeared in the 17th century (Littré, 1872). The disease was, however, so rare that some pathologists did not believe their eyes when confronted with a case of perforated gastric ulcer (Littré, 1872). Peptic ulcer disease and its complications were rare in the 18th century (Rawlinson, 1727; Morgani, 1756), but at the beginning of the 19th century, physicians began to observe a rapidly increasing number of ulcers (Brinton, 1857) which first decreased in the late 1960s (Tytgat, 1999). The age and sex distribution as well as the occurrence of two different ulcer types have also changed. In the beginning of the 19th century, gastric ulcer was typically a disease of young women (Jennings, 1940). Later, duodenal ulcer started to appear and became more frequent in men (Jennings, 1940; Sonnenberg, 1995). By the end of the 19th century, the incidence of duodenal ulcer had surpassed that of gastric ulcer.

For a long time, the origins of digestive complaints remained speculative. In the 17th century, gastric illness was thought to come from "the atmosphere" or from external mechanical irritants and was accordingly treated with a gastric brush (Blum, 1997). In the 18th century, it appeared difficult to understand why food in the lumen of the stomach would be liquefied whilst the wall of the stomach remained intact (Hunter, 1772). Further reflections on the matter resulted in the recognition that the stomach itself must harbour an intrinsic

[*]corresponding author, email: dpantofl@hola.hospvd.ch

Table 1.

Timetable of curved bacteria in the stomach

Author	Observation	Year
Bottcher	Spiral bacteria in animal stomach	1874
Rappin	Spiral bacteria in canine stomach	1881
Bizzozero	Spiral bacteria in dog stomach	1893
Salomon	Complete description of spiral bacteria in animal stomach	1896
Jaworski	Spiral bacteria in human gastric washings	1889
Pel	Spiral bacteria in human gastric juice	1899
Krienitz	Spiral bacteria in human cancerous stomach	1906
Konjetzny	Spiral bacteria in human ulcerous stomach	1928
Doenges	Spiral bacteria cause inflammation of human stomach	1939
Freedberg	Spiral bacteria present more frequently in ulcerating stomach than in normal stomach	1940
Ivy	Spiral bacteria does not cause duodenal ulcer	1950
Palmer	No spiral bacteria is present in human stomach	1954
Steer	Spiral bacteria in active superficial gastritis	1975
Warren and Marshall	Identified *H. pylori* in human gastritis	1983

Timetable of bacteria in human diseases

Year	Observation	Author
1683	First observation of bacteria	van Leeuwenhoek
1865	*Bacillus anthracis* - anthrax	Davaine
1874	*Mycobacterium leprae* - lepra	Hansen
1879	*Neisseria gonorrhoeae* - gonorrhoea	Neisser
1880	*Plasmodium malariae* - malaria	Laveran
1881	*Staphylococcus pyrogenes aureus* - abscess	Ogston
1882	*Mycobacterium tuberculosis* - tuberculosis	Koch
1883	*Vibrio cholerae* - cholera	Koch
1884	*Corynebacterium diphtheriae* - diphtheria	Loeffler
1889	*Bacillus tetani* - tetanus	Kitasato
1892	*Clostridium perfringens* -gangrene	Welch and Nuttall
1894	*Yersinia pestis* - plague	Yersin
1905	*Treponema pallidum* - syphilis	Schaudinn and Hoffman
1906	*Bordetella pertussis*-whooping cough	Bordet
1909	*Rickettsia prowazecki* - typhus	Ricketts
1977	*Legionella pneumoniae* - pneumonia	Mc Dade
1982	*Borrelia burgdorferi* -Lyme disease	Burgdorfer
1990	*Bartonella henselae*-cat scratch fever	Relman

"vital force" capable of resisting digestion. Issues such as the existence of "a locus minoris resistentiae" backed the concept that ulcer is a local digestion of the gastric mucosa. The fact that the stomach produced hydrochloric acid became accepted at the beginning of 19th century (Prout, 1824). The paradigm which would dominate gastric speculation until the late 1980's originated in studies on the regulation of gastric secretion (Beaumont, 1833; Pavlov, 1902). It was thought that excessive nervousness and other pathologic nervous stimuli led either to acid overproduction or to weakening of the local defence or to both; deregulation of the balance between aggressive and defensive factors was responsible for peptic ulceration (Broussais, 1835; Riegel, 1905; Schindler, 1947; Weiner *et al.*, 1957; Feldman *et al.*, 1986). Defensive factors often invoked were blood flow, mucus and alkali secretion. At the end of 19th century, when bacteria were proven to cause many human diseases (Table 1), some microbiologists attempted to identify the bacteria that cause peptic ulcer. However, current opinion did not permit a bacterial etymology. The views of those who pointed towards bacterial theory of peptic ulcer were brushed away by the keepers of the prevailing paradigm "no acid-no ulcer" (Schwartz, 1910) until the early 1980's (Table 1).

Discovery of Bacteria in the Stomach

At the beginning of the 19th century, it became clear that acid gastric juice exerts a strong bactericidal action. It still holds true that only particularly adapted organisms can resist acid gastric juice (Goodwin and Zeikus, 1987). It was probably a German gastroenterologist, Boas, who observed for the first time a correlation between acid and microorganisms in gastric juice: acid gastric juice contained only few bacteria while gastric juice whose pH was nearly neutral contained many (Boas, 1890). The presence of microorganisms in the stomach was thus associated with pathological conditions resulting in low acidity such as mucosal atrophy and gastric cancer (Boas, 1894). The dogma that a normal stomach is sterile except for transiently present oral flora persisted for the next 100 years.

Discovery of Curved Bacteria in the Stomach

Histologists first described spiral microorganisms in the stomach of animals in the latter quarter of the 19th century (Bottcher, 1874). During a meeting in 1881, Rappin mentioned that he had observed spiral bacteria in canine stomach which he «modestly» named *Spirocheta rappini* (Blum, 1997). In 1893, the Italian anatomist G. Bizzozero published beautiful drawings of spirochetal organisms in the dog's stomach (Bizzozero, 1893). Seen with modern eyes, they look like *Helicobacter heilmannii*. Three years later, Salomon gave a complete bacteriological account of spiral organisms, probably *H. felis*, in the animal stomach, including flagellar structure, motility, animal hosts, and handling in the laboratory (Salomon, 1896).

The first report on spiral bacteria in the human stomach appeared in 1889. A Polish gastroenterologist, Jaworski, noted spiral structures in the sediment of gastric washings obtained from humans which he named *Vibrio rugula* (Jaworski, 1889). He suggested that they might play a role in gastric disease. However, Boas claimed that these structures were generated by simple chemical reactions and suggested that Jaworski's cells were the results of an interaction of gastric mucus and acid (Boas, 1907). In 1899, a Dutch gastroenterologist described similar structures and suggested that they may cause gastric disease (Pel, 1899). After additional sporadic reports (Krienitz, 1906; Luger and Neuberger, 1921), Doenges published an impressive series of observations (Doenges, 1939). Upon autopsying a large series of young adults, he found spiral organisms in 40% of the stomachs. Most likely Doenges saw *H. pylori* but he chose *H. heilmanni* -like bacteria with a more pronounced spiral structure for his published microphotographs.

Gastric Ammonia and Urease

Ammonia was already noted in gastric juice in the Middle Ages (Spallazani 1780). Later it was even suggested that ammonia was a normal constituent of gastric juice (Leuret and Lassaigne, 1825). In 1823, a brilliant English physician Prout, while demonstrating that the stomach secretes hydrochloric acid, found ammonia in the stomach of one subject (Prout, 1824). Prout's explanation of this anomaly was that the patient must have ingested ammonia-containing syrup. At the turn of the century, German investigators observed ammonia crystals in gastric juice and correctly speculated that it was of bacterial origin (Boas, 1890). Gastric mucosal urease was first characterised in 1924 (Luck and Seth, 1924). In the early 1950's, a clinical role for urea in gastric juice was postulated by Fitzgerald, who claimed that gastric urease functioned as a protective agent for the mucosa by providing ions to neutralise acid (Fitzgerald and Murphy, 1950). Five years later, histochemical studies demonstrated that tissues surrounding gastric ulcers were particularly rich in urease, whilst cancerous or achlorhydric stomach was devoid of urease activity (Kornberg and Davies, 1955). Investigations of gastric urease-containing tissue suspensions showed the presence of urea-splitting organisms. This led to the suggestion that gastric urease might be of bacterial origin (Kornberg and Davies, 1955). Subsequently, the availability of the ^{14}C urea breath test allowed, for the first time, the study of urea metabolism *in vivo* (Walser and Bodenlos, 1959). Studies of the effect of antibiotics on ammonia production in animals and men clearly showed that gastrointestinal urease-containing bacteria transformed urea into ammonia (Delluva *et al.*, 1968). These experiments and observations, while establishing that gastric urease was of bacterial origin, unfortunately failed to stimulate an investigation of the relationship between urease-containing bacteria and ulcer disease. One curious anomaly was that *H. pylori*, at the time of its isolation, was described as urease negative (Marshall and Warren, 1984). In the same year, Langenberg *et al.* noted the discrepancy between Marshall's report and their own observations on rapid urea hydrolysis by gastric campylobacters (Langenberg *et al.*, 1984).

Gastritis as an Infectious Disease

The term "gastritis" was first used in 1728 by Stahl. In his *Collegium practicum* he postulated that some episodes of fever might be related to superficial irritation of the gastric mucosa, especially in those cases which had a tendency to develop ulcer (Stahl, 1728). One hundred and fifty years later, an etiological role of bacteria for gastritis was discussed, albeit vaguely, by Konjetzny (Konjetzny, 1928). He observed bacteria and gastritis in resected ulcerated stomachs and thought that the bacteria might be involved in the disease process but he did not go beyond this point. A statement made in 1954 (Palmer, 1954) that "no structure which could be reasonably considered to be spirochetal in nature can be identified in the stomach" set back gastric bacterial research by a further 20 years.

In the 1970's, Steer (1975) and then Fung *et al.*, (1979) described the full entity of gastritis: their electron microscopic images show *H. pylori* attached to the gastric epithelial cells, and active superficial gastritis was characterised. However, Steer's carefully formulated suggestions on a possible etiopathological role of the bacteria in gastritis and gastric ulceration (Steer, 1975) were ignored by the scientific community which was, at that time, concentrating on the development of H_2-receptor antagonists as potent anti-ulcer agents (Black *et al.*, 1972). Thus, the credit for the rediscovery of *H. pylori* is given to Warren and Marshall, whose letter in 1983 drew attention to a close correlation between the organism and gastritis (Warren and Marshall, 1983). The authors claimed also, with much less evidence, that infected patients may develop ulcer and cancer (Warren and Marshall, 1983). A few years later Morris, after a successful attempt to repeat Marshall's experiment of self-inoculation, showed than *H. pylori* gastritis was not a self-limited disorder (Morris and Nicholson, 1987). Long-lasting antral-predominant active gastritis is now recognised as a typical pathology associated with long term *H. pylori* infection (Dixon, 1991).

Gastritis and Ulcer

The inflammatory origin of peptic ulcer was proposed by several authors at the beginning of the 19th century (Broussais, 1823; Abercrombie, 1824). A French pathologist, J. Cruveilhier, gave a complete gross pathologic description of peptic ulcers and described their inflammatory appearance (Cruveilhier, 1842). In 1928, Konjetzny described the relationship between gastritis and ulcer disease (Konjetzny, 1928). However, his 300-page report was based on a string of isolated observations and did not convince the scientific community. It was an easy target for criticisms by powerful leaders of opinion, who published solid experimental data on the relationship between acid and ulcer (Büchner, 1927, 1931). In addition, pioneers of gastroscopy claimed than in patients with duodenal ulcer, they never saw gastritis (Schindler, 1947). Apparently, dogma dulls the senses: even astute observers are blindfolded by preconceived ideas.

Ulcer as an Infectious Disease

Early Theories

At the end of the 19th century, the evocation of bacteria as a cause of disease had become almost routine in clinical medicine (Table 1). Therefore, Letulle who postulated in 1888 (Letulle, 1888) a bacterial origin of gastric ulcer, was, for his contemporaries, one of the bacteriomaniacs who tried to explain every disease by invoking a bacterial cause. Microorganisms that were frequently incriminated were *Treponema pallidum*, *Staphylococcus*, *Streptococcus*, *Lactobacillus*, *Pneumococcus*, and *Escherichia coli* (Letulle, 1888; Chiari, 1891; Wurtz and Leudet, 1891; Besançon and Griffon, 1899). At the beginning of this century, some authors thought that a hematogenous invasion of bacteria from carious teeth would result in the formation of an ulcer (Rosenow, 1913, 1923; Saunders, 1930). In 1950, the "bacterial theory" was, as it appeared, irrevocably refuted by three leading American gastroenterologists, Ivy, Grossman and Barach (1950). They critically reviewed the previous data and published their own negative experience with bacteria.

Herpes Virus and Ulcer

In the 1970's, another attempt to show a link between infection and ulcer disease was stimulated by the finding of latent Herpes virus infection in the vagal ganglia. Since both Herpes virus diseases and peptic ulcer are recurrent conditions confined to specific and circumscribed anatomical locations (Warren *et al.*, 1978), it was suggested that Herpes virus might be a cause of peptic ulcer (Vestergaard and Rune, 1980). However, when it turned out that anti-herpetic treatment did not heal ulcer (Borg and Andrén, 1980), the Herpes hypothesis was abandoned. In the late 1990's, the Herpes theory was reactivated and it was claimed that some *H. pylori* negative ulcers are in fact due to Herpes virus. Today, the putative role of Herpes virus in gastric pathology is still unclear (Gesser *et al.*, 1995; Al-Samman *et al.*, 1994; Nelson and Crippin, 1997).

H. pylori and Ulcer

In 1940, Freedberg and Baron investigated the presence of spirochetes in the gastric mucosa of patients who had undergone partial resection surgery (Freedberg and Baron, 1940). Since they failed to detect spirochetes in 50% of ulcerated stomachs but did find them in 10% of non-ulcerated stomach, these authors did not arrive at a firm conclusion and were unable to confirm the hypothesis that ulcer was caused by spirochetal organisms.

At the time of isolation of *H. pylori*, there was still reluctance to believe it responsible for peptic ulceration (Brooks, 1985; Jones *et al.*, 1986; Hui *et al.*, 1987). The firm belief that the primary cause of peptic ulcer was gastric acid did not leave room for an alternative opinion (Halter and Eigenmann, 1987). When Marshall suggested that *H. pylori* infection may lead to the development of duodenal ulcer, the link between bacteria and ulcer still remained to be proven (Marshall *et al.*, 1985). Reports from other investigators who found the curved bacteria in the stomach of patients with gastritis and ulcer were consistent with Marshall's hypothesis (Kasper and Dickgiesser, 1984; Steer and Newell, 1985). The connection between *H. pylori* and ulcers was further strengthened by epidemiological studies showing an increased incidence of ulcers in persons infected with bacteria (Hirschl *et al.*, 1986; Johnston *et al.*, 1986). A causal role of *H. pylori* in duodenal ulcer disease was finally accepted by the scientific community in 1987, when Coghlan showed that the suppression of the organism by bismuth therapy led to rapid healing and reduced relapse of duodenal ulcer (Coghlan *et al.*, 1987). It is ironic that bismuth was already used in the 18th and 19th century for the treatment of gastrointestinal symptoms (Hochradel, 1928) and was described as an effective anti-ulcer agent in the 20th century (Kidd and Modlin, 1998).

Gastric Malignancy: An Infectious Disease?

Gastric Cancer

As early as 1771, Boerhaave suggested that gastric carcinoma was due to chronic inflammation (Boerhaave, 1771). The death of Napoleon Bonaparte from gastric cancer in 1821 and the preoccupation of his physicians with gastritis (Broussais, 1835; Corvisart, 1802) set the stage for a flurry of publications on gastric disorders. In 1821, Nepveu published a book stating his belief that chronic gastritis evolved to gastric cancer (Nepveu, 1821). The first link between bacteria and gastric cancer was made in 1894 by Boas. He identified a bacteria - the "Oppler Boas bacillus"- present in anacidic stomach and linked it to the development of cancer (Boas, 1894). In 1906, Kreinitz identified spirochetes in the gastric contents of a patient with gastric cancer but did not consider them to be pathogens. Fifteen years later, Luger reported spirochetal organisms in the gastric juice of patients with ulcerating carcinomas of the stomach (Luger and Neuberger, 1920). In 1926, J. Fibiger won the Nobel Prize for demonstrating (albeit erroneously) that an infection - *Spiroptera carcinoma* - caused stomach cancer in rats (Wernstedt, 1965).

It is now well recognized that most gastric cancers arise as the result of the sequence: chronic gastritis - mucosal atrophy - intestinal metaplasia (Correa *et al.,* 1975). Identification of *H. pylori* infection as a cause of chronic gastritis placed this infection at the top of this pathway (Marshall *et al* ., 1985; Fox *et al*., 1989). The first well documented - albeit disputed - reports linking *H. pylori* infection with gastric cancer appeared in 1991 (Forman *et al*., 1991; Nomura *et al.,* 1991; Parsonnet *et al*., 1991). Subsequent studies associating *H. pylori* with the development of precancerous lesions such as atrophic gastritis, intestinal metaplasia and dysplasia seemed to consolidate the *H. pylori* link to gastric cancer (Craanen *et al*., 1992; Kuipers *et al.*, 1995). In 1994, a Working Group of the International Agency for Research on Cancer considered the evidence for a causal role of *H. pylori* in gastric cancer conclusive enough to designate this organism as a class I carcinogen (IARC, 1994). However, with increasing insight into the pathogenesis and epidemiology of gastric cancer, it became more difficult to accept the close link between *H. pylori* and gastric cancer (Muñoz, 1994; Crespi and Citarda, 1996). Curing of *H. pylori* infection did not prevent the development of gastric cancer or its precursors (Satoh *et al*., 1998; Suriani *et al*., 1998; van der Hulst *et al*., 1998), the reliable diagnosis of atrophy has been disputed (Schlemper *et al*., 1997; Genta, 1998), and the association of intestinal metaplasia and cancer is uncertain (Cheli *et al.* 1998; van der Hulst *et al*., 1998). Today, the role of *H. pylori* in gastric cancer is still a matter of considerable dispute (Kidd, 1999; Genta, 1998).

Gastric Lymphoma

Early studies already pointed to an association between the appearance of lymphoid tissue in gastric mucosa, the so called "pseudolymphomas", and the occurrence of peptic ulcer (Watson and O'Brien, 1970; Chiles and Platz, 1975; Saraga *et al*., 1981). A malignant potential of gastric pseudolymphomas to progress to MALT-lymphoma was also suggested (Stroehlen *et al*., 1977, Brooks and Enterline, 1983). The causal connection between bacteria and malignancy was, however, yet to be made. The first histopathological studies of *H. pylori* showed that lymphoid proliferation of gastric mucosa is a common feature of the infection (Stolte and Eidt, 1989). Early epidemiological studies linked *H. pylori* with a risk for the development of malignant gastric lymphoma of the MALT-type (Wotherspoon *et al*., 1991). The recognition that low-grade gastric MALT-lymphoma often regresses following anti-*H. pylori* therapy has been particularly exciting and seemed to transform management of the disease (Wotherspoon *et al*., 1993). The causative role of *H. pylori* in gastric lymphoma, today established beyond doubt, should, however, be seen in perspective. It is often difficult to differentiate true malignancy from monoclonal lymphoid proliferation (Collins, 1997). Thus, it is not clear how many patients with essentially benign lymphoid proliferation were in the past labeled as having malignant disease (Blaser, 1999). In addition, not all patients with a low-grade stage of disease benefit from the eradication of *H. pylori* (Muller *et al*., 1995). Finally, anti-*H. pylori* treatment is not effective in a high-grade MALT lymphoma (Muller *et al*., 1995).

How Dangerous is *H. pylori*?

The isolation and culturing of *H. pylori* 17 years ago launched a medical revolution in which, like many other revolutions, three successive stages can be recognized. At the time of the discovery of *H. pylori* infection, the medical community adhered to the acid paradigm and refuted the suggestion that the organism might be responsible for gastrointestinal disease. In a second stage, a landslide change of medical opinion occurred. This culminated in the enthusiastic recommendation for universal elimination of *H. pylori* from humans (Graham, 1997; Axon, 1999).

In the third stage, the new paradigm "only a dead *Helicobacter pylori* is a good *Helicobacter pylori*" has again been challenged (Blaser, 1997; Sprung, 1998). It has begun to become clear that the relationship between *H. pylori* infection and gastrointestinal diseases is more complex than previously thought. Firstly, the presence of *H. pylori* does not customarily imply the presence of disease, except, of course, for those who consider chronic gastritis to be a disease. Not all *H. pylori* infected individuals develop peptic ulcer or gastric cancer and serious disease may also occur in *H. pylori* negative individuals (Axon, 1999). Successful cure of *H. pylori* heals gastritis, many duodenal ulcers and a substantial proportion of low-grade gastric MALT-lymphomas. However, treatment of *H. pylori* does not provide symptomatic relief in non-ulcer dyspeptic patients (Blum *et al.*, 1998; Talley *et al.*, 1999a; Talley *et al.*, 1999b; Pantoflickova *et al.*, 2000) and its role in the prevention of gastric cancer is unclear (Cheli *et al.*, 1998; Farinati *et al.*, 1998). This implies that other factors besides *H. pylori*, for example host or environmental factors, are important in determining the course of the infection. Furthermore, the epidemiology of *H. pylori* and its associated diseases did not develop in parallel throughout the last three centuries. Colonization with *H. pylori* was nearly universal before 1800 and then began to diminish until the present (Sonnenberg, 1995; Jyotheeswaran *et al.*, 1998). In contrast, peptic ulcer was rare before 1800, then rose sharply throughout the 19th and the 20th century. By the end of the 20th century, the prevalence of peptic ulcer started to fall (Tytgat, 1999; Jyotheeswaran *et al.*, 1998). Thus, it is possible that *H. pylori* per se is not central to the pathogenesis of peptic ulceration. Finally, a protective role of *H. pylori* has been suggested. The decline in *H. pylori* incidence during the modern era seems to correlate with a rise in reflux esophagitis and its complications such as Barrett's esophagus and adenocarcinomas of the gastric cardia and esophagus (Labenz *et al.*, 1997; Chow *et al.*, 1998; Werdmuller and Loffeld, 1997). In view of these findings, some authors have even suggested that *H. pylori* carriage may be of benefit to the host and that the elimination of the organism would be disadvantageous (Blaser, 1997; Blaser, 1999). In this context, *H. pylori* is considered as a part of the indigenous or "normal" flora, which was, for millennia, well adapted to cohabit with the host without giving the rise to clinical disease until the industrial era started (Blaser, 1997; Blaser, 1999). Today, further research is needed to determine in whom, when and how to eradicate *H. pylori* infection.

References

Abercombie, 1824. Inflammatory affections and ulcerations of the stomach. Edinburg M. and S. J. 21: 1.

Al-Samman, M., Zuckerman, M.J., Verghese, A. and Boman, D. 1994. Gastric ulcers associated with herpes simplex esophagitis in a nonimmunocompromised patient. J. Clin. Gastroenterol. 18(2): 160.

Axon, A. 1999. *Helicobacter pylori* is not a commensal. Curr. Opin. Gastroenterol. 15 (Suppl. 1): S1-S4.

Beaumont, W. 1833. Experiments and observations on the gastric juice, and the physiology of digestion. Plattsburg. 280 p.

Besançon, F., and Griffon, V. 1899. Ulcérations gastriques au cours de la septicémie pneumococcique chez le cobaye. Bull. Soc. Anat. de Paris. 74: 409-416.

Bizzozero, G. 1893. Über die schlauchförmigen Drüsen des Magendarmkanals und die Beziehungen ihres Epithels zu dem Oberflächenepithel der Schleimhaut. Arch. für Mikrosk. Anat. 42: 82-152.

Black, J.W., Duncan, W.A., Durant, C.J., Ganellin, C.R., and Parsons, E.M. 1972. Definition and antagonism of histamine H 2 -receptors. Nature 236: 385-90.

Blaser, M.J. 1997. Not all *Helicobacter pylori* strains are created equal: should all be eliminated? Lancet 349: 1020-1022.

Blaser, M.J. 1999. Where does *Helicobacter pylori* come from and why is it going away? JAMA 282: 2260-2262.

Blum, A.L. 1997. An historical overview of Helicobacter-associated disorders. In: The Immunobiology of *H. pylori*. Ernst, P.B., Michetti, P and Smith, P.D. New York. xiv-xix.

Blum, A.L., Talley, N.J., O'Morain, C., et al. 1998. Lack of effect of treating *Helicobacter pylori* infection in patients with nonulcer dyspepsia. N. Engl. J. Med. 339: 1875-1881.

Boas, I. 1890. Allgemeine Diagnostik und Therapie der Magenkrankheiten nach dem heutigen Stande der Wissenschaft. Thieme, Leipzig.

Boas, I. 1894. Über das Vorkommen von Milchsäure im gesunden und kranken Magen nebst Bemerkungen zur Klinik des Magenkarzinoms. Z. Klin. Med. 25: 285-302.

Boas, I. 1907. Diseases of the stomach. Davis, Philadelphia.

Boerhaave, H. 1771. Ventriculi inflammatio. In: Gerardi van Swieten: Commentaria. Ed.Parisiis.

Bordet, J.J. 1906. Le microbe de la coqueluche. Ann. Inst. Pasteur. 20: 731-741.

Borg, I., and Andrén, L. 1980. Herpes simplex virus as a cause of peptic ulcer. Scand. J. Gastroenterol. 15 (Suppl. 63): 56-61.

Bottcher, G. 1874. Dorpater Med. Z. 5: 184.

Brinton, W. 1857. On the pathology, symptoms, and treatment of ulcers of the stomach, with an appendix of cases. London: J. Churchill: 1-227.

Brooks J.J., and Enterline, H.T. 1983. Gastric pseudolymphoma. Its three subtypes and relation to lymphoma. Cancer Res. 51(3): 476-486.

Brooks, F.P. 1985. The pathophysiology of peptic ulcer disease. Dig. Dis. Sci. 30 (Suppl. 11): 15S-29S.

Broussais, F. 1835. Histoire des phlegmons ou inflammations chroniques. In: Recherches sur les Effets de la Saignée dans quelques Maladies Inflammatoires, et sur l'action de l'émétique et des Vésicatoires dans la Pneumonie. Louis, P.C.A., ed. Baillière, Paris.

Büchner, F., and Molloy, P.J. 1927. Das echte peptische Geschwür der Ratte. Klin. Wschr. 6: 2193-2194.

Büchner, F., and Schneider, E. 1931. Neue Experimente zur Entstehung des peptischen Geschwürs. Klin. Wschr. 10: 522-523.

Burgdorfer, W., Barbour, A.G., Hayes, S.F., Benach, J.L., Grunwaldt, E., and Davis, J.D. 1982. Lyme disease – a tick borne spirochetosis?. Science. 216:1317-19.

Cheli, R., Crespi, M., and Testino, G. 1998. Gastric cancer and *Helicobacter pylori*: biologic and epidemiologic inconsistencies. J. Clin. Gastroenterol. 26(1):3-6.

Chiari, H. 1891. Über Magensyphilis. Int. Beitr. Wiss. Med. 2: 295-321.

Chiles, J.T., and Platz, C.E. 1975. The radiographic manifestations of pseudolymphoma of the stomach. Radiology. 116: 551-6.

Chow, W.H., Blaser, M.J., Blot, W.J., Gammon M.D., et al. 1998. An inverse relation between cagA+ strains of *Helicobacter pylori* infection and risk of esophageal and gastric cardia adenocarcinoma. Cancer Res. 58(4): 588-590.

Coghlan, J.G., Gilligan, D., and Humphries, H. 1987. *Campylobacter pylori* and recurrence of duodenal ulcers- a 12 month follow-up study. Lancet 2(8568): 1109-1111.

Collins, R.D. 1997. Is clonality equivalent to malignancy: specifically, is immunoglobulin gene rearrangement diagnostic of malignant lymphoma? Hum. Pathol. 28(7): 757-759.

Correa, P., Haenszel, W., and Cuello, C. 1975. A model for gastric cancer epidemiology. Lancet 2(7924): 58-60.

Corvisart, J.N., and Leroux, J.J. 1802. Suite d'observations sur les ulcères fistuleux à l'estomac. J. Méd. Chir. Pharm. (Paris). 3: 503-527.

Craanen, M.E., Dekker, W., Blok, P., Ferwerda, J., and Tytgat, G.N. 1992. Intestinal metaplasia and *Helicobacter pylori*: an endoscopic bioptic study of the gastric antrum. Gut 33(1): 16-20.

Crespi, M., and Citarda, F. 1996. *Helicobacter pylori* and gastric cancer: an overrated risk? Scand. J. Gastroenterol. 31(11): 1041-1046.

Cruveilhier, J. 1842. Considérations générales sur les ulcérations folliculaires de l'estomac. In: Atlas d'Anatomie pathologique. Baillières, Paris.

Davaine, C.J. 1863. Recherches sur les infusories du sang dans la maladie connue sous le nom de sang de rate. C. R. Acad. Sci. (Paris). 57: 220-223.

Delluva, A.M., Markley, K., and Davies, R.E. 1968. The absence of gastric urease in germ-free animals. Biochim. Biophys. Acta 151: 646-650.

Delluva, A.M., Markley, K., Davies, R.E. 1968. The absence of gastric urease in germ-free animals. Biochim. Biophys. Acta. 151: 646-650.

Dixon, M.F. 1991. *Helicobacter pylori* and peptic ulceration: histopathological aspects. J. Gastroenterol. Hepatol. 6: 125-130.

Doenges, J.L. 1939. Spirochetes in the gastric glands of Macacus rhesus and of man without related disease. Arch. Pathol. Lab. Med. 27: 469-477.

Donati, M. 1586. De medica historia mirabili. Mantuae, Fr. Osanam.

Farinati, F., Foschia, F., Di Mario, F., et al. 1998. *H. pylori* eradication and gastric precancerous lesions. Gastroenterology 115: 512-514.

Feldman, M., Walker, P., Green, J.L., and Weingarden, K. 1986. Life events stress and psychosocial factors in men with peptic ulcer disease. A multidimensional case-controlled study. Gastroenterology 91(6): 1370-1379.

Fitzgerald, O., and Murphy, P. 1950. Studies on the physiological chemistry and clinical significance of urease and urea with special reference to the human stomach. Ir. J. Med. Sci. 292: 97-159.

Forman, D., Newell, D.G., and Fullerton, F. 1991. Association between infection with *Helicobacter pylori* and risk of gastric cancer: evidence from a prospective investigation. BMJ 302(6788): 1302-1325

Fox, J.G., Correa, P., Taylor, N.S., Zavala, D., et al. 1989. *Campylobacter pylori*-associated gastritis and immune response in a population at increased risk of gastric carcinoma. Am. J. Gastroenterol. 84(7): 775-781.

Freedberg, A.S., and Barron, L.E. 1940. The presence of Spirochetes of the human. Am. J. Dig. Dis. 38: 443-445.

Fung, W.P., Papadimitriou, J.M., and Marz, L.R. 1979. Endoscopic, histologic and ultrastructural correlations in chronic gastritis. Am. J. Gastroenterol. 71: 269-279.

Genta, R.M. 1998. Atrophy, metaplasia and dysplasia: are they reversible? Ital. J. Gastroenterol. Hepatol. 30 (Suppl. 3): S324-S325.

Gesser, R.M., Valyi-Nagy, T., Fraser, N.W., and Altschuler, S.M. 1995. Oral inoculation of SCID mice with an attenuated herpes simplex virus- 1 strain causes persistent enteric nervous system infection and gastric ulcers without direct mucosal infection. Lab. Invest. 73(6): 880-889.

Goodwin, S., and Zeikus, J.G. 1987. Physiological adaptations of anaerobic bacteria to low pH: metabolic control of proton motive force in *Sarcina ventriculi*. J. Bacteriol. 169: 2150-2157.

Graham, D.Y. 1997. The only good *Helicobacter pylori* is a dead *Helicobacter pylori*. Lancet 350(9070): 70-71.

Halter, F., and Eigenmann, F. 1987. Is it really more difficult to treat prepyloric ulcers? Aliment. Pharmacol. Ther. 1 (Suppl. 1): 433S-438S.

Hamberger, G.E. De ruptura intestini duodeni. In: Disputationes ad Morborum Historiam et Curationem Pacientes.Quas Collegit, Edidit et recensuit. Albertus Hallerus. Tomus Tertius. Ad Morbos Abdominis. Bousquet, Lausanne, 1757.

Hansen, G.H. 1874. Indberetning til det Norske medicinske Selskab i Christiania om en med understøttelse af selskabet foretagen reise for at anstille undersøgelser agående spedalskhedens årsager, tildels udførte sammen med forstander. Hartwig. Norsk. Mag. f. Laegevidensk. 3R. : 49Heft, 1-88.

Hirschl, A., Potzi, R., Stanek, G., et al. 1986. Occurrence of *Campylobacter pylori*dis in patients from Vienna with gastritis and peptic ulcers. Infection 14(6): 275-278.

Hochradel, J. 1928. Historisches zur Wismutfrage. Münch. Med. Wschr. 177 p.

Hui, W.M., Lam, S.K., Chau, P.Y., et al. 1987. Persistence of *Campylobacter pylori*dis despite healing of duodenal ulcer and improvement of accompanying duodenitis and gastritis. Dig. Dis. Sci. 32(11): 1255-1260.

Hunter, J. 1772. On the digestion of the stomach after death. Philos. Trans. R. Soc. Lond. 62: 447-454.

IARC Working Group on the Evaluation of Carcinogenic Risks to Humans. 1994. IARC Monogr. Eval. Carcinog. Risks Hum. Lyon 61: 1-241.

Ivy, A.C., Grossman, M.I., and Bachrach, W.H. 1950. Peptic Ulcer. Ed. Blakiston, Philadelphia.

Jaworski, W. 1889. Podrecznik Chorob zoladka. Wydawnictwa Dziel Lakarskich Polskich. 32.

Jennings, D. 1940. Perforated peptic ulcer. Changes in age-incidence and sex- distribution in the last 150 years. Lancet i: 395-398.

Johnston, B.J., Reed, P.I., and Ali, M.H. 1986. Campylobacter like organisms in duodenal and antral endoscopic biopsies: relationship to inflammation. Gut 27(10): 1132-1137.

Jones, D.M., Eldridge, J., Fox, A.J., et al. 1986. Antibody to the gastric campylobacter-like organism ("*Campylobacter pylori*dis")- clinical correlations and distribution in the normal population. J. Med. Microbiol. 22(1): 57-62.

Jyotheeswaran, S., Shah, A.N., and Jin, H.O. 1998. Prevalence of *Helicobacter pylori* in peptic ulcer patients in greater Rochester, NY: is empirical triple therapy justified? Am. J. Gastroenterol. 93(4): 574-5478.

Kasper, G., and Dickgiesser, N. 1984. Isolation of campylobacter-like bacteria from gastric epithelium. Infection 12(3): 179-180.

Kidd, M., and Modlin, I.M. 1998. A century of *Helicobacter pylori*: paradigms lost-paradigms regained. Digestion 59(1): 1-15.

Kidd, M., Louw. J.A., and Marks, I.N. 1999. *Helicobacter pylori* in Africa: observations on an 'enigma within an enigma'. J. Gastroenterol. Hepatol. 14: 851-8.

Kitasato, S. 1889. "Ueber den Tetanusbacillus." Ztschr. Hyg. u. Infektionskrank. 7:225-234.

Koch, R. 1882. Die Aetiologie der Tuberculose. Berl. Klin. Wchnschr., xix: 221-230.

Koch, R. 1884. Uber die Cholerbacterien. Dtsch. Med. Wschr.; 10: 725-728.

Konjetzny, G.E. 1928. Die Entzündungen des Magens. In: Handbuch Pathologie, Anatomie und Histologie IV. 2: 768-1116.

Kornberg, H.L., and Davies, R.E. 1955. Gastric urease. Phyl. Rev. 35: 169-177.

Krienitz, U. 1906. Über das Auftreten von Spirochaeten verschiedener Form im Mageninhalt bei Carcinoma ventriculi. Dtsch Med. Wschr. 32: 872.

Kuipers, E.J., Uyterlinde, A.M., and Pena, A.S. 1995. Long-term sequelae of *Helicobacter pylori* gastritis. Lancet 345(8964): 1525-1528.

Labenz, J., Blum, A.L., Bayerdörffer, E. et al. 1997. Curing *Helicobacter pylori* infection in patients with duodenal ulcer may provoke reflux esophagitis. Gastroenterology 112: 1442-1447.

Langenberg, M.L., Tytgat, G.N., Schipper, M.E.I., Rietra, P.J.G.M., and Zanen, H.C. 1984. Campylobacter-like organisms in the stomach of patients and healty individuals: letter. Lancet 1: 1348.

Laveran, A. 1880. A new parasite found in the blood of malarial patients. Parasitic origin of malarial attacks. Bull. mem. soc. med. hosp. Paris. 17: 158-164.

Leeuwenhoek, A. 1693-1718. Ontledingen en ontedekkingen, *etc.* 6 vols. Leiden, Delft.

Letulle, M. 1888. Origine infectieuse de certains ulcères simples de l'estomac ou du duodénum. Soc. méd. hôp. de Paris 5: 360-385.

Leuret, F.L., and Lassaigne, M.L. 1825. Recherches physiologiques et chimiques pour servir à l'histoire de la digestion. Huzard, Paris.

Littré, E. 1872. Henriette d'Angleterre est-elle morte empoisonnée? In: Médecine et Médecins. Didier, Paris.

Loeffler, F. 1884. Utersuchung uber die Bedeutung der Mikroorganismen fir die Entstehung der Diptherie beim Menschen, bei der taube und beim Kalbe. Mitth. a. d. kaiserl. Gesundheitsampte. Ii: 421-499.

Luck, J.M., and Seth, T.N. 1924. The physiology of gastric urease. Biochem. J. 18: 357-365.

Luger, A., and Neuberger, H. 1920. Über Spirochätenbefunde im Magensaft und deren diagnostische Bedeutung für das Carcinoma ventriculi. Wien Med. Wschr. 70: 54-75.

Marshall, B.J., and Warren, J.R. 1984. Unidentified curved bacilli in the stomach of patients with gastric and peptic ulceration. Lancet i: 1311-1315.

Marshall, B.J., McGechie, D.B., Rogers, P.A., and Glancy, R.J. 1985. Pyloric Campylobacter infection and gastroduodenal disease. Med. J. Aust. 142(8): 439-44.

McDade, J. E., Sheperd, C. C., Fraser, D. W.,. Tsai, T. R, Redus, M. A., Dowdle, W. R., and Laboratory Investigation Team. 1977. Legionaire's disease: Isolation of a bacterium and demonstration of its role in other respiratory disease. N. Engl. J. Med. 297: 1197-1203.

Morgani, G.B. 1756. De sedibus et causis morborum per anatomen indigatis libri quinque. Typographia Remondiniana, Venetiis.

Morris, A., and Nicholson, G. 1987. Ingestion of *Campylobacter pylori*dis causes gastritis and raised fasting gastric pH. Am. J. Gastroenterol. 82(3): 192-199.

Muller, A.F., Maloney, A., Jenkins D., et al. 1995. Primary gastric lymphoma in clinical practice 1973-1992. Gut 36(5): 679-83.

Muñoz, N. 1994. Is *Helicobacter pylori* a cause of gastric cancer? An appraisal of the sero-epidemiological evidence. Cancer Epidemiol. Biomarkers Prev. 3(5): 445-451.

Neisser, A. 1879. Ueber eine der Gonorrhoe eigenthumliche Micrococcusform. Vorlaufige Mitteilung. Cbl. F. d. Med. Wiss. 28: 497-500.

Nelson, A.C., and Crippin, J.S. 1997. Gastritis secondary to herpes simplex virus. Am. J. Gastroenterol. 92(11): 2116-2117.

Nepveu, P.F. 1821. Sur le cancer de l'estomac, considéré comme l'une des terminaisons de la gastrite chronique. Paris.

Nomura, A., Stemmermann, G.N., and Chyou, P.H. 1991. *Helicobacter pylori* infection and gastric carcinoma among Japanese Americans in Hawaii. N. Engl. J. Med. 325(16): 1132-1136.

Ogston, A. 1881. Report upon microorganism in surgical diseases. Br. Med. J.; i: 369-375.

Palmer, E.D. 1954. Investigation of the gastric Spirochetes of the human. Gastroenterology 27: 218-220.

Pantoflickova, D. Talley, N.J., Koelz, H.R., Blum A.L. 2000. *Helicobacter pylori* and non-ulcer dyspepsia: comments on a flawed meta-analysis. Letter. BMJ 320: 1208.

Parsonnet, J., Friedman, G.D., and Vandersteen, D.P. 1991. *Helicobacter pylori* infection and the risk of gastric carcinoma. N. Engl. J. Med. 325(16): 1127-1131.

Pavlov, J.P. 1902. The work of digestive glands. Charles Griffin and Cie. London. p. 196.

Pel, P.K. 1899. De Ziekten van de Maag met het oog op de behoeften der geneeskundige praktijk geschetst. Bohn ed., Harlem.

Prout, W. 1824. On the nature of the acid and saline matters usually existing in the stomachs of animals. Philos. Trans. R. Soc. Lond. part I: 45-49.

Rawlinson, C. 1727. A prenatural perforation found in the upper part of the stomach, with the symptoms it produc'd. Phil. Trans. 1727;35: 361-362

Relman, L.W., Loutit, J.S., Schmidt, T.M., Falkow, S., and Tompkins, L.S. 1990. The agents of bacillary angiomatosis. An approach to the identification of uncultured pathogens. N. Engl. J. Med. 323:1573-80.

Ricketts, H. 1909. A micro-organism which apparently has a specific relationship to Rocky Mountain spotted fever. J. Am. Med. Soc. 52: 379-380.

Riegel, F. 1905. Diseases of the stomach. In: Szockton CG Eds. Nothnagel's Encyclopedia of Practical Medecine. Saunders, Philadelphia.

Rosenow, E.C. 1913. The production of ulcer of the stomach by injection streptococci. JAMA 61: 1947.

Rosenow, E.C. 1923. Etiology of spontaneous ulcer of stomach in domestic animals. J. Infect. Dis. 32: 384-399.

Salomon, H. 1896. Über das Spirillium des Säugetiermagens und sein Verhalten zu den Belegzellen. Zentralbl. für Bakteriol. Parasitenkd. Infektionskr. 19: 434-442.

Saraga, P., Hurlimann, J., and Ozzello, L. 1981. Lymphomas and pseudolymphomas of the alimentary tract. An immunohistochemical study with clinicopathologic correlations. Hum. Pathol. 12(8): 713-723.

Satoh, K., Kimura, K., Takimoto, T., et al. 1998. A follow-up study of atrophic gastritis and intestinal metaplasia after eradication of *Helicobacter pylori*. Helicobacter 3: 236-240.

Saunders, E.W. 1930. The serologic and etiologic specificity of the alpha streptococcus of gastric ulcer. Arch. Int. Med. 45: 347-382.

Schaudinn, F.R., and Hoffman E. 1905. Uber Spirochatenbefunde in Lymphdrusensaft Syphilitischer. Dtsch. med. Wschr. 31: 711.

Schindler, R. 1947. Gastritis. Waverly Press, Grune and Stratton, New York.

Schlemper, R.J., Itabashi, M., and Kato, Y. 1997. Differences in diagnostic criteria for gastric carcinoma between Japanese and western pathologists. Lancet 349(9067): 1725-1729.

Schwartz, K. 1910. über penetrierende Magen- und Jejunal-Geschwüre. Beitr. Klin. Chir. 67: 96-128.

Sonnenberg, A. 1995. Temporal trends and geographical variations of peptic ulcer disease. Aliment. Pharmacol. Ther. 9 (Suppl. 2): 3-12.

Spallanzani L. 1780. Dissertazioni di fisica animale e vegetabile. Modena, Società Tipografica. vol 1.

Sprung, D.J. 1998. *H. pylori*: is it really such a bad guy after all? Am. J. Gastroenterol. 93(9): 1596.

Stahl, G.E. 1728. Collegium praticum. Caspar Jacob Eyssell: Leipzig.

Steer, H.W. 1975. Ultrastructure of cell migration through the gastric epithelium and its relationship to bacteria. J. Clin. Pathol. 28: 639-646.

Steer, H.W., and Newell, D.G. 1985. Immunological identification of *Campylobacter pylori*dis in gastric biopsy tissue. Lancet 2(8445): 38.

Stolte, M., and Eidt, S. 1989. Lymphoid follicles in antral mucosa: immune response to *Campylobacter pylori*? J. Clin. Pathol. 42(12): 1269-1271.

Stroehlen, J.R., Weiland, L.H., Hoffman, H.N., and Judd, E.S. 1977. Untreated gastric pseudolymphoma. Am. J. Dig. Dis. 22: 465-70.

Suriani, R., Mazzuco, D., Cerutti, E., et al. 1998. Duodenitis, gastric duodenal metaplasia, and body chronic gastritis correlations in duodenal ulcer patients after 5 years of *H. pylori* eradication. Gut 43 (Suppl. 2): A101.

Talley, N.J., Janssens, J., Lauritsen, K., and Racz, I. 1999a. Bolling-Sternevald E. Eradication of *Helicobacter pylori* in functional dyspepsia: randomised double blind placebo controlled trial with 12 months follow up. BMJ 27; 318(7187): 833-837.

Talley, N.J., Vakil, N., and Ballard, E.D. 2nd, and Fennerty, M.B. 1999b. Absence of benefit of eradicating *Helicobacter pylori* in patients with nonulcer dyspepsia. N. Engl. J. Med. 341(15): 1106-1111.

Tytgat, G.N. 1999. *Helicobacter pylori*-reflections for the next millennium. Gut 45 (Suppl. 1): I45-147.

van der Hulst, R.W.M., Kate, F.J.W., Rauws, E.A.J., *et al.* 1998. The relation of cagA and the long-term sequele of gastritis after successful cure of *H. pylori* infection: A long-term follow-up study. Gastroenterology 114: A318.

Vestergaard, B.F., and Rune, S.J. 1980. Type-specific Herpes simplex-virus antibodies in patients with recurrent duodenal ulcer. Lancet 1(8181): 1273-1274 .

Walser, M., and Bodenlos, L.J. 1959. Urea metabolism in man. J. Clin. Invest. 38: 1617-1626.

Warren, J.R., and Marshall, B.J. 1983. Unidentified curved bacilli on gastric epithelium in active gastritis. Lancet i: 1273-1275.

Warren, K.G., Brown, S.M., Wroblewska, Z., *et al.* 1978. Isolation of latent herpes simplex virus from the superior cervical and vagus ganglions of human beings. N. Engl. J. Med. 298(19): 1068-1069.

Watson, R.J., and O'Brien, M.T. 1970. Gastric pseudolymphoma (lymphofollicular gastritis). Ann. Surg. 171(1): 98-106.

Weiner, H., Thaler, M., Reiser, M.F., and Mirsky, I.A. 1957. Etiology of duodenal ulcer. I. Relation of specific psychological caracteristics to rate of gastric secretion (serum pepsinogen). Psychosom. Med. 19: 1.

Welch, W. H., and Nuttall G. 1892. A gas-producing bacillus *(B. aerogenes capsulatus*, nov. spec.), capable of rapid development in the blood vessels after death. Johns Hopkins Hosp. Bull. 3: 81-91.

Werdmuller, B.F., and Loffeld, R.J. 1997. *Helicobacter pylori* infection has no role in the pathogenesis of reflux esophagitis. Dig. Dis. Sci. 42(1): 103-105.

Wernstedt, W. 1965. Presentation speech for the Nobel Prize in Physiology or Medecine, 1926. In: Nobel Foundation Eds. Nobel Lectures: Physiology and Medecine, 1922-1941. Elsevier, Amsterdam. 119-121.

Wotherspoon, A.C., Doglioni, C., Diss, T.C., et al. 1993. Regression of primary low-grade B-cell gastric lymphoma of mucosa- associated lymphoid tissue type after eradication of *Helicobacter pylori*. Lancet 342(8871): 575-577.

Wotherspoon, A.C., Ortiz-Hidalgo, C., Falzon, M.R., and Isaacson, P.G. 1991. *Helicobacter pylori*-associated gastritis and primary B-cell gastric lymphoma. Lancet 338 (8776): 1175-1176.

Wurtz, R., and Leudet, R. 1891. Recherches sur l'action pathogène du bacille lactique. Arch. Méd. Exp. Anat. Pathol. 3: 485.

Yersin, A. 1894. La peste bubonique à Hong Kong. C.R. Acad. Sci. 119: 356.

From: *Helicobacter pylori: Molecular and Cellular Biology*
ISBN 1-898486-25-5 © 2001 Horizon Scientific Press, Wymondham, UK.

2

Taxonomy and Phylogeny of *Helicobacter*

James Versalovic and James G. Fox[*]

Abstract

The genus *Helicobacter* was formally proposed in 1989 and represents a rapidly expanding taxon with considerable medical interest. The importance of *Helicobacter pylori* stimulated investigations of multiple related species isolated from the gastrointestinal tracts of animals and humans. Currently, twenty formally named species have been designated and proposed names such as *"Helicobacter rappini"* likely include multiple taxa. Helicobacters have been grouped as gastric or enteric (enterohepatic), depending on their preferred site of colonization. These organisms are uniformly Gram-negative non-sporeforming bacilli which are motile and usually possess multiple or single bipolar sheathed flagella. Biochemical, genetic, and physiologic features distinguish Helicobacters from the related genera, *Campylobacter* and *Arcobacter*, and emphasize their unique contributions to the gastrointestinal microbiota.

Introduction

The first recorded observations of gastric spiral-shaped bacteria in animals were made by Rappin in 1881 and Bizzozero in 1893 (see Chapter 1). In 1898, Salomon noted spiral bacteria in the stomachs of dogs, cats, and Norway rats (Salomon, 1898). Despite these reports, the role of curved Gram-negative organisms in human gastric diseases was not appreciated until the 1980s. In 1982, *Campylobacter pyloridis* (later known as *Helicobacter pylori*) was successfully cultured from stomach biopsies of human patients with gastritis (Warren *et al.*, 1983). Subsequently other spiral Gram-negative bacteria have been observed and isolated from the gastrointestinal tracts of mammals such as cats, dogs, ferrets and rodents (Fox *et al.*, 1997). Initially, many spiral Gram-negative bacteria isolated from the mammalian gastrointestinal tract were grouped as campylobacters. This classification was based on similar microscopic and ultrastructural morphologies, common microaerobic growth requirements, and a similar ecologic niche. However, partial sequencing of 16S rRNA genes provided evidence that *Campylobacter pylori* belonged in a different genus from other campylobacters (Romaniuk *et al.*, 1987). The genus *Helicobacter* was formally distinguished from other Gram-negative curved rods (e.g. *Campylobacter*) following extensive analysis of enzymatic activities, fatty acid profiles, growth characteristics, nucleic acid hybridization profiles, and 16S rRNA sequence analysis (Goodwin *et al.*, 1989; Paster *et al.*, 1991; Vandamme *et al.*, 1991).

[*]corresponding author, email: jgfox@mit.edu

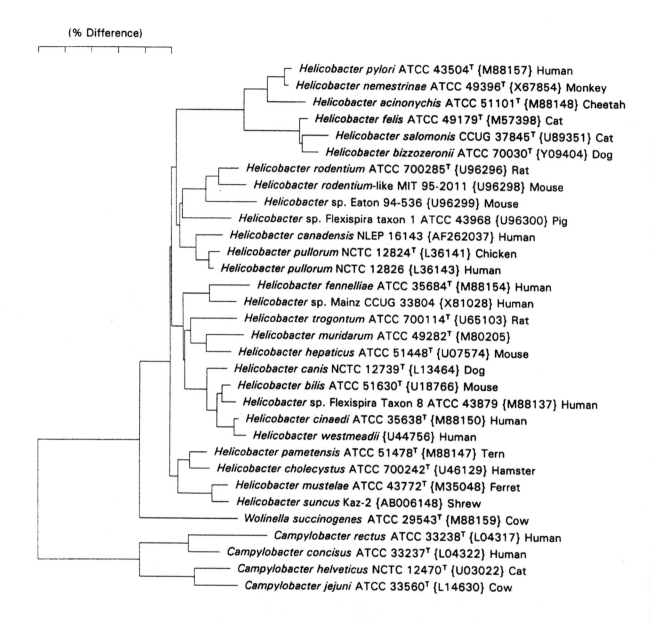

Figure 1. Phylogenetic tree constructed on the basis of 16S rRNA sequence similarity values using the neighbor-joining method. Scale bar equals 5% difference in nucleotide sequences as determined by measuring the lengths of the horizontal lines connecting the two species. (Courtesy of Floyd Dewhirst).

Approaches to Identification

The genus *Helicobacter* (twenty formally named species) includes spiral or curved bacilli ranging from 0.3-1.0 mm in width and 1.5-10.0 mm in length (Figure 1). Helicobacters are Gram-negative, non-sporeforming rods that may form spheroid or coccoid bodies upon prolonged culture. These bacteria are motile and usually possess multiple or single bipolar sheathed flagella (Table 1). *H. pylori* isolates usually have multiple unipolar sheathed flagella. Only three known helicobacters, *Helicobacter pullorum, H. canadensis,* and *Helicobacter rodentium,* have unsheathed flagella similar to those of the campylobacters.

These organisms are microaerobic and possess a respiratory type of metabolism. Successful cultivation of helicobacters requires a humid atmosphere enriched for carbon dioxide (5-12%). The optimal growth temperature is between 37 and 42°C. Atmospheric hydrogen (as much as 5-10%) is either required or stimulates the growth of these organisms. Most *Helicobacter* species grow poorly, if at all, in routine aerobic or anaerobic atmospheres. An exception is *Helicobacter westmeadii* (Trivett-Moore *et al.*, 1997), which requires anaerobic conditions to grow on solid media. *Helicobacter hepaticus* isolates often grow well in anaerobic and microaerobic conditions.

Several biochemical and genetic criteria distinguish this genus (Table 1, Figure 1), although significant intragenus variation has been described. All helicobacters are oxidase positive and do not appear to ferment or oxidize carbohydrates by conventional biochemical tests. However, nuclear magnetic resonance (NMR) spectroscopy studies (Hazell *et al.*, 1997) and functional genomics (Tomb *et al.*, 1997) have revealed the existence of carbohydrate catabolic activities in *H. pylori*. Members of the genus *Helicobacter* have G+C contents ranging from 30-48 mol% (Fox *et al.*, 1997), similar to that of the campylobacters which range from 29-45 mol% (Vandamme *et al.*, 1991). Comparative analysis of 16S rDNA/rRNA sequences established rRNA superfamily VI including curved Gram-negative rods (ε class of proteobacteria) such as *Arcobacter, Campylobacter, Helicobacter,* and *Wolinella* (Vandamme *et al.*, 1991). By phylogenetic analysis using 23S rDNA sequences, separate "rRNA clusters" could be distinguished, among members of rRNA superfamily VI consisting of campylobacters (rRNA cluster I), arcobacters (rRNA cluster II) and helicobacters (rRNA cluster III) (Vandamme *et al.*, 1991). Phylogenetic analysis of 16S rDNA sequencing data (Goodwin *et al.*, 1989; Romaniuk *et al.*, 1987; Vandamme *et al.*, 1991) also clearly distinguished *H. pylori* from campylobacters and justified the formation of a new genus.

Helicobacter species have been isolated from the gastrointestinal tracts of mammals and birds. In this chapter, we will refer to gastric (stomach) and enteric (intestinal) helicobacters according to their preferred gastrointestinal niche. Division by ecologic niche generally corresponds to phylogenetic divisions as most gastric helicobacters cluster separately from enteric organisms. The gastric helicobacters primarily inhabit the stomach either within or beneath the mucus gel layer adjacent to the epithelium and rarely invade the bloodstream. In contrast to enteric helicobacters, gastric helicobacters are bile-susceptible, tend to have multiple polar flagella, and are uniformly urease-positive.

Gastric Helicobacters

Different gastric helicobacters have been characterized from various mammals and are associated with gastric inflammation (gastritis). *H. pylori* colonizes the corpus (body) and antrum (distal portion) of the human stomach. *H. pylori* naturally infects other mammals such as cats (Fox *et al.*, 1995) and nonhuman primates (Handt *et al.*, 1997) and has caused experimental infections of laboratory mice (Marchetti *et al.*, 1995). The larger, corkscrew-shaped

Table 1. Characteristics which differentiate *Helicobacter* species[a]

Taxon	Primary site	Catalase production	Nitrate reduction	Alkaline phosphatase hydrolysis	Urease	Indoxyl acetate hydrolysis	γ-glutamyl transpeptidase activity	Growth at 42°C	Growth with 1% glycine	Susceptibility to: Nalidixic acid (30-μg disc)	Cephalothin (30-μg disc)	Peri-plasmic fibers	No. of flagella	Distribution of flagella
H. rodentium	Intestine	+	+	-	-	-	-	+	+	R	R	-	2	Bipolar
H. pullorum	Intestine	+	+	-	-	-	ND	+	ND	R	S	-	1	Monopolar
H. pylori	Stomach	+	-	+	+	-	+	-	-	R	S	-	4-8	Monopolar
H. nemestrinae	Stomach	+	-	+	+	-	ND	+	-	R	S	-	4-8	Bipolar
H. acinonyx	Stomach	+	+	+	+	-	+	-	-	R	S	-	2-5	Bipolar
H. felis	Stomach	+	+	+	+	+	+	+	-	R	S	+	14-20	Bipolar
H. fennelliae	Intestine	+	-	+	-	+	-	-	+	S	S	-	2	Bipolar
H. trogontum	Intestine	+	+	+	+	ND	+	+	ND	R	R	+	5-7	Bipolar
H. muridarum	Intestine	+	-	+	+	+	+	-	-	R	R	+	10-14	Bipolar
H. hepaticus	Intestine	+	+	ND	+	+	ND	-	+	R	R	-	2	Bipolar
H. canis	Intestine	-	-	+	-	+	ND	+	ND	S	I	-	2	Bipolar
H. bilis	Intestine	+	+	ND	+	-	ND	+	+	R	R	+	3-14	Bipolar
'H. rappini'	Intestine	+	-	-	-	ND	+	-	-	S	I	+	10-20	Bipolar
H. cinaedi	Intestine	+	+	-	-	-	-	-	+	S	S	-	1-2	Bipolar
H. pametensis	Intestine	+	+	+	-	-	-	+	+	S	S	-	2	Bipolar
H. cholecystus	Gallbladder	+	+	+	-	+	-	+	+	I	R	-	1	Monopolar
H. canadensis	Intestine	+	+	-	-	+	-	+	+	R	R	-	1-2	Bipolar
H. mustelae	Stomach	+	+	+	+	+	+	+	-	S	R	-	4-8	Peritrichous
H. suncus	Stomach	+	+	+	+	-	-	ND	ND	R	R	-	2	Bipolar
H. typhlonicus	Intestine	+	+	-	-	-	-	-	+	ND	ND	-	2	Bipolar
H. westmeadii	Unknown	+	+	+	-	ND	ND	-	ND	S	R	-	1	Monopolar

[a] Data were obtained from references (Fox *et al.*, 2000, 1999 and 1997; Franklin *et al.*, 1999; Goto *et al.*, 1998; Versalovic *et al.*, 1999). Symbols and abbreviations: +, positive reaction; -, negative reaction; S, susceptible; R, resistant; I, intermediate; ND, not determined.

Helicobacter heilmannii resides in the human stomach. *H. heilmannii*-like organisms (Fox *et al.*, 1997; Lee *et al.*, 1993; Queiroz *et al.*, 1996), *Helicobacter bizzozeronii* (Hanninen *et al.*, 1996), *Helicobacter felis* (Lee *et al.*, 1988; Paster *et al.*, 1991; Eaton *et al.*, 1996), and *Helicobacter salomonis* (Jalava *et al.*, 1997) have been isolated from the stomachs of cats and dogs.

Phylogenetic and fatty acid analyses distinguish gastric helicobacters from enteric (enterohepatic) helicobacters. Gastric helicobacters have been subdivided phylogenetically into two groups, the "*H. pylori-H. felis-H. heilmannii*" and the "*H. mustelae-H. suncus*" clusters (Figure 1). Most gastric helicobacters cluster together in the "*H. pylori-H. felis-H. heilmannii*" group by phylogenetic analysis based on comparative 16S rDNA studies. *H. pylori* (Goodwin *et al.*, 1989; Marshall *et al.*, 1983; Romaniuk *et al.*, 1987) and other gastric helicobacters such as *Helicobacter acinonychis* from cheetahs (Eaton *et al.*, 1993), *Helicobacter felis* from cats, dogs, and humans (Lee *et al.*, 1988), *H. heilmanni* (humans) (Heilmann *et al.*, 1991; Hilzenrat *et al.*, 1995), and *Helicobacter nemestrinae* (nonhuman primates) (Bronsdon *et al.*, 1991) are closely related and form a distinctive phyletic grouping within the genus. Canine gastric helicobacters such as *Helicobacter salomonis* (Jalava *et al.*, 1997) and *Helicobacter bizzozeroni* (Hanninen *et al.*, 1996) also cluster tightly with the related canine/feline helicobacter, *H. felis*. The gastric pathogen, *H. mustelae* (Fox *et al.*, 1988; Paster *et al.*, 1991), is quite different by phylogenetic analysis and clusters with the shrew helicobacter *H. suncus* (Goto *et al.*, 1998) and the enteric avian species, *Helicobacter pametensis* (Dewhirst *et al.*, 1994).

Phenotypic diversity is present in gastric helicobacters and justifies the application of phylogenetic approaches for identification. Gastric helicobacters may have strikingly different morphologies by light and electron microscopy. The corkscrew-shaped, larger gastric helicobacters appear very different from the slightly curved, smaller rods of *H. pylori*. Importantly, the "corkscrew" phenotype may be induced during bacteriologic culture of *H. pylori*. A morphology typical of *H. pylori* in gastric biopsies was observed after culture on plated media, whereas corkscrew phenotypes reminiscent of *H. heilmannii* were apparent following culture in liquid media (Fawcett *et al.*, 1999). Therefore, phylogenetic studies are necessary for accurate classification of the multiple gastric helicobacters. Additional morphologic features such as the presence of multiple monopolar or bipolar flagella distinguish gastric helicobacters from most enteric helicobacters.

H. pylori and *H. mustelae* (ferret gastric organism) were the cardinal members of the new genus *Helicobacter* following its transition from the genus *Campylobacter* (Goodwin *et al.*, 1989). In contrast to campylobacters, *H. pylori* is hippurate-negative and urease-positive. Robust urease activity represents a biochemical hallmark of gastric helicobacters, indicating the importance of urease for gastric colonization. Urea appears important for the survival of urease-positive organisms in acidic environments (Clyne *et al.*, 1995), and isogenic urease-negative mutants of *H. mustelae* fail to colonize the ferret stomach (Andrutis *et al.*, 1995). It is noteworthy that the urease-negative helicobacters colonize the lower gastrointestinal tract and only urease-positive organisms have been isolated from the stomach. Like other helicobacters, *H. pylori* isolates possess catalase, oxidase, nitrate reducing, and γ-glutamyl-transpeptidase activities (Table 1).

"*H. heilmannii*" (formerly "*Gastrospirillum hominis*") has been observed in human gastric biopsies (Hilzenrat *et al.*, 1995; Solnick *et al.*, 1993) and cultured from human stomach tissue (Andersen *et al.*, 1996). The species name has not been formally recognized since significant nucleic acid sequence variation was found among various human isolates. Infection with "*H. heilmannii*" or "*H. heilmannii*"-like organisms has been associated with gastritis and peptic ulcer disease in swine (Queiroz *et al.*, 1996; Cantet *et al.*, 1999) and abdominal pain, gastritis, and peptic ulcer disease in humans (Heilmann *et al.*, 1991; Hilzenrat *et al.*, 1995). "*H. heilmanni*" differs morphologically from its human counterpart, *H. pylori*, but retains common biochemical activities such as oxidase, catalase, urease, and nitrate reduction. Its colonial morphology (small, translucent colonies) was indistinguishable from that of *H. pylori* (Andersen *et al.*, 1999). By phylogenetic cluster analysis, it is closely related with *H. pylori*, *H. felis*, and other gastric helicobacters. This identification of human *H. felis* was based on 16S rDNA sequence analysis of PCR-generated amplicons. *H. heilmannii* type I shares 99.5% rDNA sequence identity with Candidatus *Helicobacter suis* (formerly *Gastrospirillum suis*) (see below) and may represent the same species. (Candidatus status is granted by the International Committee on Systematic Bacteriology (Murray *et al.*, 1994) for uncultivable bacteria based on genomic and antigenic data.)

Gastric helicobacters (known as CLOs, campylobacter-like organisms) were first visualized in canine and feline stomachs during the late nineteenth century. Several species have been cultured successfully from gastric tissues of dogs and cats. *Helicobacter felis* was the first gastric helicobacter to be identified and characterized extensively from feline sources (Lee *et al.*, 1988) and has also been identified in a human gastric biopsy of a Melanesian patient (Germani *et al.*, 1997). Both phenotypic and genotypic analyses supported the designation of *H. felis* as a member of the genus *Helicobacter*. Subsequently other canine species with larger, corkscrew morphologies were identified such as *H. bizzozeronii* and *H. salomonis*. These large, spiral gastric helicobacters are closely related to one another and cluster together by 16S rDNA analysis (Hanninen *et al.*, 1996; Jalava *et al.*, 1997). Comparative analyses of 16S rDNA sequences reveal differences of 1% or less for *H. bizzozeronii*, *H. felis*, and *H. salomonis* (De Groote *et al.*, 1999; Hanninen *et al.*, 1996; Jalava *et al.*, 1997). Nearly identical 16S rDNA sequences among canine gastric helicobacters and the "*H. heilmannii*" (formerly *Gastrospirillum*) group highlight the difficulties of species designation by nucleotide sequence comparisons of closely related organisms.

Gastric helicobacters from other mammals have expanded our appreciation of the importance of these organisms in the gastrointestinal microbiota. *H. nemestrinae* (Bronsdon *et al.*, 1991) is very closely related to *H. pylori* phylogenetically and phenotypically and has been isolated from nonhuman primates. Whether *H. nemestrinae* represents a variant of *H. pylori* represents a fair question as 16S rDNA sequence similarities exceed 99% and both organisms have the rather unique feature (within *Helicobacter*) of multiple monopolar flagella. Additionally, *H. pylori* infects nonhuman primates. New Candidatus designations for bovine (De Groote *et al.*, 1999) and porcine (De Groote *et al.*, 1999) helicobacters have been proposed. Candidatus *Helicobacter suis* (formerly *G. suis*) represents a spiral, urease-positive helicobacter from the stomachs of slaughterhouse pigs (De Groote *et al.*, 1999). Detailed analysis of 16S rDNA sequences place this organism firmly in the gastric helicobacter cluster and related to "*H. heilmanni*" and the canine helicobacters. Candidatus *Helicobacter bovis* (De Groote *et al.*, 1999) is a urease-positive helicobacter from gastric (abomasum) specimens of European cattle. It resembles *H. pylori* morphologically and is distinct from the corkscrew gastric helicobacters, clustering loosely with the *H. pylori*- *H. felis* cluster.

The *H. mustelae*- *H. suncus* cluster of gastric helicobacters form a distinct phyletic group with unusual phenotypic features and host ranges. Like other gastric helicobacters, *H. mustelae* (Fox *et al.*, 1988) and *H. suncus* (Goto *et al.*, 1998) express robust urease activities and are associated with chronic gastritis in ferrets and musk shrews, respectively. *H. mustelae* and *H. suncus* possess unique flagellar configurations for gastric helicobacters. The presence of peritrichous flagella represents a unique feature of *H. mustelae*, whereas the presence of two bipolar (one at each end) flagella in *H. suncus* is a feature usually found in enteric helicobacters. Additionally *H. suncus* appears predominantly coccoid in culture but spiral-shaped *in vivo*.

Enteric and Enterohepatic Helicobacters

In contrast to the gastric helicobacters, the enteric helicobacters inhabit the lower gastrointestinal tract (small intestine, colon, rectum, and hepatobiliary tract) of mammals and birds. Enteric helicobacters form phylogenetic clusters distinct from most gastric helicobacters and constitute a diverse group of organisms (Figure 1). As with gastric helicobacters, isolates have been obtained from both animal and human sources. Several species have been found in both animal and human hosts, whereas no species have been strictly found in the human alimentary tract. As a group and in contrast to gastric organisms, enteric helicobacters often possess fusiform morphologies, single polar flagella at each pole (bipolar) (exception is *H. rappini* group), may be urease-negative, are bile-tolerant, and often have periplasmic fibers (Table 1).

In humans, enteric helicobacters (*Helicobacter canis, Helicobacter cinaedi, Helicobacter fennelliae, H. pullorum*) have been isolated from rectal swabs and feces (Burnens *et al.*, 1993; Grayson *et al.*, 1989; Laughon *et al.*, 1988; Quinn *et al.*, 1984; Stanley *et al.*, 1994; Stanley *et al.*, 1993). *H. cinaedi* also comprises part of the normal intestinal flora of hamsters (Gebhart *et al.*, 1989). *H. fennelliae* organisms have been isolated from macaque primates and dogs (Kiehlbauch *et al.*, 1995). *H. canis* has been cultured from the bile, intestines, and feces of diarrheic dogs (Stanley *et al.*, 1993) and bengal cats (Foley *et al.*, 1999). The avian helicobacter, *H. pullorum*, has been cultured from the intestines and feces of asymptomatic chickens and hens with hepatitis (Fox *et al.*, 1997; Stanley *et al.*, 1994). Animal enteric and enterohepatic helicobacters also include *H. bilis* from several animals (rodents, dogs, cats) (Fox *et al.*, 1995), *H. hepaticus* from mice (Fox *et al.*, 1994), *H. muridarum* from rodents (Lee *et al.*, 1992), *H. pametensis* from birds (Dewhirst *et al.*, 1994), *H. rodentium* from mice (Shen *et al.*, 1997), and *H. trongontum* from rats (Mendes *et al.*, 1996). The *H. rappini* group (Schauer *et al.*, 1993) includes diverse organisms from animal and human sources. Additional enteric helicobacters have been recently identified from mice (Fox *et al.*, 1999; Franklin *et al.*, 1999) and nonhuman primates (Saunders *et al.*, 1999).

In addition to the "*H. rappini*" group, rodent enteric helicobacters have been identified from the lower intestinal tract. *H. cinaedi* is considered part of the normal flora of hamsters and isolates have been cultured from fecal specimens. Animal and human isolates of *H. cinaedi* were distinguished by host source based on ribotyping (Kiehlbauch *et al.*, 1995), indicating the possible existence of host-specific subspecies. *H. muridarum* was isolated from the intestinal mucosa of rats and mice and possess a unique ultrastructural morphology (Lee *et al.*, 1992). Multiple (9-11) periplasmic fibers appearing as concentric helical ridges and bipolar tufts of 10-14 sheathed flagella were visualized. These organisms are phylogenetically related to enterohepatic species such as *H. bilis* and *H. hepaticus*, although *H. muridarum* has only been identified in the intestine. *H. rodentium* was isolated from the intestines of laboratory mice (Shen *et al.*, 1997) and has several unusual features such as unsheathed flagella and a urease-negative phenotype. Multiple urease-negative enteric helicobacters have been identified whereas all gastric helicobacters possess detectable urease activity.

Novel urease-negative enteric helicobacters have recently been identified from laboratory mice and nonhuman primates (e.g. cotton-top tamarins), extending the phylogenetic diversity of this organism class. The first tamarin helicobacter was isolated from colonic biopsies and feces of cotton-top tamarins afflicted with chronic colitis (Saunders *et al.*, 1999). This organism is closely related to an urease-negative enteric helicobacter found in humans, *H. fennelliae*.

A urease-negative enteric helicobacter named "*H. typhlonicus*" from mice was associated with colitis and typhlitis in BALB/c and interleukin-10-deficient laboratory mice and experimentally reproduced the disease in IL-10-/- as well as SCID mice (Fox *et al.*, 1999; Franklin *et al.*, 1999). Infection of SCID mice with this organism caused typhlocolitis but not liver disease. "*H. typhlonicus*" is closely related by phylogenetic analysis to the urease-positive murine species *H. muridarum* and is distinguished from *H. rodentium* by the presence of sheathed flagella.

"*Helicobacter rappini*" (formerly *Flexispira rappini*) represents a diverse set of enteric helicobacters from different animal hosts (Schauer *et al.*, 1993). *F. rappini* has reportedly induced abortion and caused fetal hepatic necrosis in sheep (Bryner *et al.*, 1987; Kirkbride *et al.*, 1985; On *et al.*, 1995). Experimentally it causes abortions in guinea pigs (Bryner *et al.*, 1987; Kirkbride *et al.*, 1985). It also has been isolated from diarrheic humans (Archer *et al.*, 1988) and has been found in children and immunocompromised bacteremic patients (Sorlin *et al.*, 1999; Tee *et al.*, 1998; Weir *et al.*, 1999). "*H. rappini*" are characterized as spindle-shaped organisms with spiral periplasmic fibers and bipolar tufts of sheathed flagella (Dewhirst *et al.*, in press). The classic *Flexispira* phenotype includes distinctive periplasmic fibers with a crisscross appearance in negatively stained electron micrographs and multiple bipolar sheathed flagella. The flagellar organization of the "*H. rappini*" group contrasts with the single flagella at each pole found in many enteric helicobacters. The species *H. bilis* and *H. trongontum* also are included in the *H. rappini* group. Additional isolates of the "*H. rappini*" taxa were recently studied by 16S rDNA sequence analysis (Dewhirst *et al.*, 2000), resulting in the identification of ten distinct taxa among 36 strains.

Enterohepatic helicobacters (Table 1, Figure 1) include the species, *H. hepaticus*, *H. bilis*, *H. pullorum*, *H. canis*, and *H. cholecystus* that not only colonize the intestinal tract but also have been isolated from the hepatobiliary tract. Enterohepatic helicobacters do not cluster separately from other enteric helicobacters by phylogenetic analysis. During 1991-1993, an outbreak of chronic, active hepatitis in colonies of inbred mice at the National Cancer Institute (NCI) in Frederick, Maryland stimulated the search for a causative agent. Evidence of chronic, active hepatitis was found in at least 15 different strains of mice at the facility suggesting that an exogenous agent and not something intrinsic to a particular inbred strain was responsible for the disease outbreak. Fox *et al.* (1994) isolated a microaerobic, curved Gram-negative rod from the intestines and livers of mice inhabiting affected colonies. Following isolation, ultrastructural analysis and 16S rRNA gene sequencing, this novel *Helicobacter* organism was named *H. hepaticus*.

H. bilis possess a *Flexispira*-like morphology and was originally isolated from bile, livers, and intestines of aged, inbred mice with chronic hepatitis (Fox *et al.*, 1995). While investigating possible infection with the known hepatic pathogen, *H. hepaticus*, spiral organisms with a distinct ultrastructural morphology were isolated. The organisms were fusiform, slightly curved rods with periplasmic fibers and multiple bipolar sheathed flagella. This distinct morphology resembles that of "*Flexispira rappini*" (see above) from animal species such as sheep, mice, and dogs. Subsequent 16S rRNA gene sequencing and phylogenetic analysis led to its designation as a novel species, closely related to "*Helicobacter rappini*" isolates. Known 16S rDNA sequences have been useful for the identification of *H. bilis* sequences in human bile specimens from patients with chronic cholecystitis (Fox *et al.*, 1998).

H. cholecystus was isolated from the gallbladders of Syrian hamsters with cholangiofibrosis and centrilobular pancreatitis (Franklin *et al.*, 1996). This disease entity of cholangiofibrosis or chronic hepatitis and cirrhosis was described in hamsters three decades ago and is associated with elevated liver enzyme levels. Three-to-eight week old hamsters were examined by liver and gallbladder histopathology and microbial culture. Whereas liver cultures did not yield organisms, pinpoint bacterial colonies were isolated from anaerobic gallbladder cultures of 22/32 hamsters. These organisms were identified as a novel species, *H. cholecystus*, based on 16S rDNA gene sequencing.

H. pullorum was isolated initially from livers, duodena, and ceca of broiler chickens and from stool specimens of human patients with gastroenteritis (Stanley *et al.*, 1994). *H. pullorum* represents yet another bile-tolerant, urease-negative *Helicobacter* colonizing the lower gastrointestinal tract. This organism has been isolated from the livers of chickens with vibrionic hepatitis. *H. pullorum* isolates possess distinctive features such as unsheathed flagella (like *H. rodentium*) and the absence of the membrane component, cholesteryl glucoside (Steinbrueckner *et al.*, 1998). As described earlier for *H. cinaedi*, *H. pullorum* isolates from poultry and human sources have distinct genotypes (Gibson *et al.*, 1999).

Other Helicobacters

Recent reports have documented the isolation of novel helicobacters from human blood without established gastric or intestinal sources. One organism labeled *Helicobacter westmeadii* (named for Westmead Hospital, Australia) was identified from blood specimens of patients with AIDS (Trivett-Moore *et al.*, 1997). Isolates from blood cultures were cultivated in anaerobic conditions on horse blood-containing media. Their cellular morphology (rod or spiral shaped with a single, sheathed, unipolar flagellum) resembles that of other urease-negative, human enteric helicobacters such as *H. cinaedi* and *H. fennelliae* (Table 1). Phylogenetic studies confirmed their close relationship with *H. cinaedi* and raise the possibility that these organisms, like *H. cinaedi*, constitute normal human flora. Interestingly, *H. cinaedi* also causes bloodstream infections in patients with AIDS (Burman *et al.*, 1995). A *Helicobacter* isolate from the blood of a patient with AIDS has been recently-described in the USA (Weir *et al.*, 1999). This isolate possesses single bipolar flagella and also clusters phylogenetically with *H. westmeadii* and *H. cinaedi*. Possibly these organisms represent emerging infections in patients with AIDS and must be monitored with the advent of widespread antiviral resistance in HIV-I associated with treatment failures. Finally, a human helicobacter from joint fluid of a patient with AIDS and septic arthritis has been reported (Husmann *et al.*, 1994). This human helicobacter clustered phylogenetically with *H. fennelliae* and emphasizes again the probable enteric origin of helicobacters isolated from blood or sterile body fluids.

Summary and Conclusions

Helicobacter species comprise a robust genus with many different species characterized by phenotypic and phylogenetic approaches. These organisms possess unusual morphologic features and occupy characteristic ecologic niches in the gastrointestinal tract. Organisms can be grouped phylogenetically by their gastric or enteric niche and, in some cases, by host range. Gastric helicobacters such as *H. pylori* have been associated with important human and animal diseases (ulcer disease and cancer). *H. pylori* represents a model for comparative genomics and our understanding of subspecies evolution at the molecular level. Enteric helicobacters have been associated with mucosal inflammation and represent key components of intestinal flora in many different mammalian and avian species. *Helicobacter* organisms represent an important link in our understanding of microbial-host interactions in the gastrointestinal tract, and accurate phylogenetic data will facilitate our understanding of this fascinating genus.

References

Andersen, L.P., Boye, K., Blom, J., Holck, S., Norgaard, A., and Elsborg, L. 1999. Characterization of a culturable *"Gastrospirillum hominis"* (*Helicobacter heilmannii*) strain isolated from human gastric mucosa. J. Clin. Microbiol. 37: 1069-1076.

Andersen, L.P., Norgaard, A., Holck, S., Blom, J., and Elsborg, L. 1996. Isolation of *Helicobacter heilmanni*-like organism from the human stomach (letter). Eur. J. Clin. Microbiol. Infect. Dis. 15: 95-96.

Andrutis, K.A., Fox, J.G., Schauer, D.B., Marini, R.P., Murphy, J.C., Yan, L., and Solnick, J.V. 1995. Inability of an isogenic urease-negative mutant strain of *Helicobacter mustelae* to colonize the ferret stomach. Infect. Immun. 63: 3722-3725.

Archer, J.R., Romero, S., Ritchie, A.E., Hamacher, M.E., Steiner, B.M., Bryner, J.H., and Schell, R.F. 1988. Characterization of an unclassified microaerophilic bacterium associated with gastroenteritis. J. Clin. Microbiol. 26: 101-105.

Bronsdon, M.A., Goodwin, C.S., Sly, L.I., Chilvers, T., and Schoenknecht, F.D. 1991. *Helicobacter nemestrinae* sp. nov., a spiral bacterium found in the stomach of a pigtailed macaque (*Macaca nemestrina*). Int. J. Syst. Bacteriol. 41: 148-153.

Bryner, J.H., Ritchie, A.E., Pollet, L., Kirkbride, C.A., and Collins, J.E. 1987. Experimental infection and abortion of pregnant guinea pigs with a unique spirillum-like bacterium isolated from aborted ovine fetuses. Am. J. Vet. Res. 48: 91-97.

Burman, W.J., Cohn, D.L., Reves, R.R., and Wilson, M.L. 1995. Multifocal cellulitis and monoarticular arthritis as manifestations of *H. cinaedi* bacteremia. Clin. Infect. Dis. 20: 564-570.

Burnens, A.P., Stanley, J., Schaad, U.B., and Nicolet, J. 1993. Novel Campylobacter-like organism resembling *Helicobacter fennelliae* isolated from a boy with gastroenteritis and from dogs. J. Clin. Microbiol. 31: 1916-1917.

Cantet, F., Magras, C., Marais, A., Federighi, M., and Megraud, F. 1999. *Helicobacter* species colonizing pig stomach: molecular characterization and determination of prevalence. Appl. Environ. Microbiol. 65: 4672-4676.

Clyne, M., Labigne, A., and Drumm, B. 1995. *Helicobacter pylori* requires an acidic environment to survive in the presence of urea. Infect. Immun. 63: 1669-1673.

De Groote, D., van Doorn, L.J., Ducatelle, R., Verschuuren, A., Tilmant, K., Quint, W.G., Haesebrouck, F., and Vandamme, P. 1999. Phylogenetic characterization of 'Candidatus *Helicobacter bovis*', a new gastric helicobacter in cattle. Int. J. Syst. Bacteriol. 49: 1707-1715.

De Groote, D., van Doorn, L.J., Verschuuren, A., Haesebrouck, F., Quint, W.G., Jalava, K., and Vandamme, P. 1999. 'Candidatus *Helicobacter suis*', a gastric helicobacter from pigs, and its phylogenetic relatedness to other gastrospirilla. Int. J. Syst. Bacteriol. 49: 1769-1777.

Dewhirst, F.E., Fox, J.G., Mendes, E.N., Paster, B.J., Gates, C.E., Kirkbride, C.A., and Eaton, K.A. 2000. *Flexispira rappini* strains represent at least ten *Helicobacter* taxa. Int. J. Syst. Bacteriol. 50: 1781-1787.

Dewhirst, F.E., Seymour, C., Fraser, G.J., Paster, B.J., and Fox, J.G. 1994. Phylogeny of Helicobacter isolates from bird and swine feces and description of *Helicobacter pametensis* sp. nov. Int. J. Syst. Bacteriol. 44: 553-560.

Eaton, K.A., Dewhirst, F.E., Paster, B.J., Tzellas, N., Coleman, B.E., Paola, J., and Sherding, R. 1996. Prevalence and varieties of *Helicobacter* species in dogs from random sources and pet dogs: Animal and public health implications. J. Clin. Microbiol. 34: 3165-3170.

Eaton, K.A., Dewhirst, F.E., Radin, M.J., Fox, J.G., Paster, B.J., Krakowka, S., and Morgan,

D.R. 1993. *Helicobacter acinonyx* sp. nov., isolated from cheetahs with gastritis. Int. J. Syst. Bacteriol. 43: 99-106.

Fawcett, P.T., Gibney, K.M., and Vinette, K. 1999. *Helicobacter pylori* can be induced to assume the morphology of *Helicobacter heilmannii*. J. Clin. Microbiol. 37: 1045-1048.

Foley, J.E., Marks, S., Munson, L., Melli, A., Dewhirst, D.E., Yu, S., Shen, Z., and Fox, J.G. 1999. Isolation of *Helicobacter canis* from a colony of Bengal cats with endemic diarrhea. J. Clin. Microbiol. 37: 3271-3275.

Fox, J.G., Batchelder, M., Marini, R.P., Yan, L., Handt, L., Li, X., Shames, B., Hayward, A., Campbell, J., and Murphy, J.C. 1995. *Helicobacter pylori* induced gastritis in the domestic cat. Infect. Immun. 63:(7) 2674-2681.

Fox, J.G., Cabot, E.B., Taylor, N.S., and Laraway, R. 1988. Gastric colonization of *Campylobacter pylori* subsp. mustelae in ferrets. Infect. Immun. 56: 2994-2996.

Fox, J.G., Chien, C.C., Dewhirst, F.E., Paster, B.J., Shen, Z., Melito, P.L., Woodward, D.L., and Rodgers, F.G. 2000. *Helicobacter canadensis* sp. nov. isolated from humans with diarrhea: an example of an emerging pathogen. J. Clin. Microbiol. 38: 2546-2549.

Fox, J.G., Dewhirst, F.E., Shen, Z., Fen, Y., Taylor, N.S., Paster, B., Erickson, R.L., Lau, C.N., Correa, P., Araya, J.C., and Roa, I. 1998. Hepatic *Helicobacter* species identified in bile and gallbladder tissue from Chileans with chronic cholecystitis. Gastroenterology 114: 755-763.

Fox, J.G., Dewhirst, F.E., Tully, J.G., Paster, B.J., Yan, L., Taylor, N.S., Collins, M.J., Jr, Gorelick, P.L., and Ward, J.M. 1994. *Helicobacter hepaticus* sp. nov, a microaerophilic bacterium isolated from livers and intestinal mucosal scrapings from mice. J. Clin. Microbiol. 32: 1238-1245.

Fox, J.G., Gorelick, P.L., Kullberg, M.C., Ge, Z., Dewhirst, F.E., and Ward, J.M. 1999. A novel urease-negative *Helicobacter* species associated with colitis and typhlitis in IL-10-deficient mice. Infect. Immun. 67: 1757-1762.

Fox, J.G., and Lee, A. 1997. The role of *Helicobacter* species in newly recognized gastrointestinal tract diseases of animals. Lab. Anim. Sci. 47: 222-255.

Fox, J.G., Yan, L.L., Dewhirst, F.E., Paster, B.J., Shames, B., Murphy, J.C., Hayward, A., Belcher, J.C., and Mendes, E.N. 1995. *Helicobacter bilis* sp. nov., a novel Helicobacter isolated from bile, livers, and intestines of aged, inbred mice. J. Clin. Microbiol. 33: 445-454.

Franklin, C.L., Beckwith, C.S., Livingston, R.S., Riley, L.K., Gibson, S.V., Besch-Williford, C.L., and Hook, R.R., Jr. 1996. Isolation of a novel *Helicobacter* species, *Helicobacter cholecystus* sp. nov., from the gallbladders of Syrian hamsters with cholangiofibrosis and centrilobular pancreatitis. J. Clin. Microbiol. 34: 2952-2958.

Franklin, C.L., Riley, L.K., Livingston, R.S., Beckwith, C.S., Hook, R.R., Jr, Besch-Williford, C., Huziker, R., and Gorelick, P.L. 1999. Enteric lesions in SCID mice infected with "*Helicobacter typhlonicus*", a novel urease-negative *Helicobacter* species. Lab. Anim. Sci. 49: 496-505.

Gebhart, C.J., Fennell, C.L., Murtaugh, M.P., and Stamm, W.E. 1989. *Campylobacter cinaedi* is normal intestinal flora in hamsters. J. Clin. Microbiol. 27: 1692-1694.

Germani, Y., Dauga, C., Duval, P., Huerre, M., Levy, M., Pialoux, G., Sansonetti, P., and Grimont, P.A. 1997. Strategy for the detection of *Helicobacter* species by amplification of 16S rRNA genes and identification of *H. felis* in a human gastric biopsy. Res. Microbiol. 148: 315-326.

Gibson, J.R., Ferrus, M.A., Woodward, D., Xerry, J., and Owen, R.J. 1999. Genetic diversity in *Helicobacter pullorum* from human and poultry sources identified by an amplified fragment length polymorphism technique and pulsed-field gel electrophoresis. J. Appl. Microbiol. 87: 602-610.

Goodwin, C.S., Armstrong, J.A., Chilvers, T., Peters, M., Collins, M.D., Sly, L., McConnell, W., and Harper, W.E.S. 1989. Transfer of *Campylobacter pylori* and *Campylobacter mustelae* to *Helicobacter pylori* gen. nov. and *Helicobacter mustelae* comb. nov. and *Helicobacter mustelae* comb. nov., respectively. Int. J. Syst. Bacteriol. 39: 397-405.

Goto, K., Ohashi, H., Ebukuro, S., Itoh, K., Tohma, Y., Takakura, A., Wakana, S., Ito, M., and Itoh, T. 1998. Isolation and characterization of *Helicobacter* species from the stomach of the house musk shrew (*Suncus murinus*) with chronic gastritis. Curr. Microbiol. 37: 44-51.

Grayson, M.L., Tee, W., and Dwyer, B. 1989. Gastroenteritis associated with *Campylobacter cinaedi*. Med. J. Aust. 150: 214-215.

Handt, L.K., Fox, J.G., Yan, L., Shen, Z., Pouch, W.J., Ngai, D., Motzel, S.L., Nolan, T.E., and Klein, H.J. 1997. Diagnosis of *Helicobacter pylori* infection in a colony of rhesus monkeys (*Macaca mulatta*). J. Clin. Microbiol. 35: 165-168.

Hanninen, M.L., Happonen, I., Saari, S., and Jalava, K. 1996. Culture and characteristics of *Helicobacter bizzozeronii*, a new canine gastric *Helicobacter* sp. Int. J. Syst. Bacteriol. 46: 160-166.

Hazell, S.L., and Mendz, G.L. 1997. How *Helicobacter pylori* works: An overview of the metabolism of *Helicobacter pylori*. Helicobacter 2: 1-12.

Heilmann, K.L., and Borchard, F. 1991. Gastritis due to spiral shaped bacteria other than *Helicobacter pylori*: Clinical, histological, and ultrastructural findings. Gut 32: 137-140.

Hilzenrat, N., Lamoureux, E., Weintrub, I., Alpert, E., Lichter, M., and Alpert, L. 1995. *Helicobacter heilmannii*-like spiral bacteria in gastric mucosal biopsies. Arch. Pathol. Lab. Med. 119: 1149-153.

Husmann, M., Gries, C., Jehnichen, P., Woelfel, T., Gerken, G., Ludwig, W., and Bhakdi, S. 1994. *Helicobacter* sp. strain *Mainz* isolated from an AIDS patient with septic arthritis: case report and nonradioactive analysis of 16S rRNA sequence. J. Clin. Microbiol. 32: 3037-3039.

Jalava, K., Kaartinen, M., Utrianinen, M., Happonen, I., and Hanninen, M.L. 1997. *Helicobacter salomonis* sp. nov., a canine gastric *Helicobacter* sp. related to *Helicobacter felis* and *Helicobacter bizzozeronii*. Int. J. Syst. Bacteriol. 47: 975-982.

Kiehlbauch, J.A., Brenner, D.J., Cameron, D.N., Steigerwalt, A.G., Makowski, J.M., Baker, C.N., Patton, C.M., and Wachsmuth, I.K. 1995. Genotypic and phenotypic characterization of *H. cinaedi* and *H. fennelliae* strains isolated from humans and animals. J. Clin. Microbiol. 22: 2940-2947.

Kirkbride, C.A., Gates, C.E., and Collins, J.E. 1985. Ovine abortion associated with an anaerobic bacterium. J. Am. Vet. Med. Assoc. 186: 789-791.

Laughon, B.E., Vernon, A.A., Druckman, D.A., Fox, R., Quinn, T.C., Polk, F., and Bartlett, J. 1988. Recovery of *Campylobacter* species from homosexual men. J. Infect. Dis. 158: 464-467.

Lee, A., Hazell, S.L., and O'Rourke, J. 1988. Isolation of a spiral-shaped bacterium from the cat stomach. Infect. Immun. 56: 2843-2850.

Lee, A., and O'Rourke, J. 1993. Gastric bacteria other than *Helicobacter pylori*. Gastroenterol. Clin. North Am. 22: 21-42.

Lee, A., Phillips, M.W., O'Rourke, J.L., Paster, B.J., Dewhirst, F.E., Fraser, G.J., Fox, J.G., Sly, L.I., Romaniuk, P.J., Trust, T.J., and Kouprach, S. 1992. *Helicobacter muridarum* sp. nov., a microaerophilic helical bacterium with a novel ultrastructure isolated from the intestinal mucosa of rodents. Int. J. Syst. Bacteriol. 42: 27-36.

Marchetti, M., Arico, B., Burroni, D., Figura, N., Rappuoli, R., and Ghiara, P. 1995. Development of a mouse model of *Helicobacter pylori* infection that mimics human disease. Science 267: 1655-1658.

Marshall, B.J., and Warren, J.R. 1983. Unidentified curved bacillus on gastric epithelium in active chronic gastritis. Lancet i: 1273-1275.

Mendes, E.N., Queiroz, D.M.M., Dewhirst, F.E., Paster, B.J., Moura, S.B., and Fox, J.G. 1996. *Helicobacter trogontum* sp. nov., isolated from the rat intestine. Int. J. Syst. Bacteriol. 46: 916-921.

Murray, R.G., and Schleifer, K.H. 1994. Taxonomic notes: a proposal for recording the properties of putative taxa of procaryotes. Int. J. Syst. Bacteriol. 44: 174-176.

On, S.L., and Holmes, B. 1995. Classification and identification of Campylobacters and Helicobacters and allied taxanumerical analysis of phenotypic characters. Syst. Appl. Microbiol. 18: 374-390.

Paster, B.J., Lee, A., Fox, J.G., Dewhirst, F.E., Tordoff, L.A., Fraser, G.J., O'Rourke, J.L., Taylor, N.S., and Ferrero, R. 1991. Phylogeny of *Helicobacter felis* sp. nov., *Helicobacter mustelae*, and related bacteria. Int. J. Syst. Bacteriol. 41: 31-38.

Queiroz, D.M.M., Rocha, G.A., Mendes, E.N., de Moura, S.B., Oliveira, A.M.R., and Miranda, D. 1996. Association between Helicobacter and gastric ulcer disease of the pars esophagea in swine. Gastroenterology 111: 19-27.

Quinn, T.C., Goodell, S.E., Fennell, C.L., Wang, S.P., Schuffler, M.D., Holmes, K.K., and Stamm, W.E. 1984. Infections with *Campylobacter jejuni* and Campylobacter-like organisms in homosexual men. Ann. Intern. Med. 101: 187-192.

Romaniuk, P.J., Zolfowska, B., Trust, T.J., Lane, T.J., Olsen, D.J., Pace, N.R., and Stahl, D.A. 1987. *Campylobacter pylori*, the spiral bacterium associated with human gastritis, is not a true *Campylobacter sp.* J. Bacteriol. 169: 2137-2141.

Salomon, H. 1898. Über das Spirillum des Säugetiermagens und sein Verhalten zu den Belegzellen. Zentralbl. Baketeriol. Parasitenkd. Infektionskr. Hyg. Abt. 1 19: 422-441.

Saunders, K.E., Shen, Z., Dewhirst, F.E., Paster, B.J., Dangler, C.A., and Fox, J.G. 1999. A novel intestinal *Helicobacter* species isolated from cotton-top tamarins (*Saguinus oedipus*) with chronic colitis. J. Clin. Microbiol. 37: 146-151.

Schauer, D.B., Ghori, N., and Falkow, S. 1993. Isolation and characterization of '*Flexispira rappini*' from laboratory mice. J. Clin. Microbiol. 31: 2709-2714.

Shen, Z., Fox, J.G., Dewhirst, F.E., Paster, B.J., Foltz, C.J., Yan, L., Shames, B., and Perry, L. 1997. *Helicobacter rodentium* sp. nov., a urease negative *Helicobacter* species isolated from laboratory mice. Int. J. Syst. Bacteriol. 47: 627-634.

Solnick, J.V., O'Rourke, J., Lee, A., Paster, B.J., Dewhirst, F.E., and Tompkins, L.S. 1993. An uncultured gastric spiral organism is a newly identified Helicobacter in humans. J. Infect. Dis. 168: 379-385.

Sorlin, P., Vandamme, P., Nortier, J., Hoste, B., Rossi, C., Pavlof, S., and Struelens, M.J. 1999. Recurrent "Flexispira rappini" bacteremia in an adult patient undergoing hemodialysis: case report. J. Clin. Microbiol. 37: 1319-1323.

Stanley, J., Linton, D., Burens, A.P., Dewhirst, F.E., On, S.L.W., Porter, A., Owen, R.J., and Costas, M. 1994. *Helicobacter pullorum* sp. nov. - genotype and phenotype of a new species isolated from poultry and from human patients with gastroenteritis. Microbiology 140: 3441-3449.

Stanley, J., Linton, D., Burens, A.P., Dewhirst, F.E., Owen, R.J., Porter, A., On, S.L.W., and Costas, M. 1993. *Helicobacter canis* sp. nov., a new species from dogs: an integrated study of phenotype and genotype. J. Gen. Microbiol. 139: 2495-2504.

Steinbrueckner, B., Haerter, G., Pelz, L., Burnens, A., and Kist, M. 1998. Discrimination of *Helicobacter pullorum* and *Campylobacter lari* by analysis of whole cell fatty acid extracts. FEMS Microbiol. Lett. 168: 209-212.

Tee, W., Leder, K., Karroum, E., and Dyall-Smith, M. 1998. *Flexispira rappini* bacteremia in a child with pneumonia. J. Clin. Microbiol. 36: 1679-1682.

Tomb, J.F., White, O., Kerlavage, A.R., Clayton, R.A., Sutton, G.G., Fleischmann, R.D., Ketchum, K.A., Klenk, H.P., Gill, S., Dougherty, B.A., Nelson, K., Quackenbush, J., Zhou, L., Kirkness, E.F., Peterson, S., Loftus, B., Richardson, D., Dodson, R., Khalak, H.G., Glodek, A., McKenney, K., Fitzgerald, L.M., Lee, N., Adams, M.D., and *et al.* 1997. The complete genome sequence of the gastric pathogen *Helicobacter pylori*. Nature 388: 539-547.

Trivett-Moore, N.L., Rawlinson, W.D., Yuen, M., and Gilbert, G.L. 1997. *Helicobacter westmeadii* sp. nov., a new species isolated from blood cultures of two AIDS patients. J. Clin. Microbiol. 35: 1144-1150.

Vandamme, P., Falsen, E., Rossau, R., Hoste, B., Segers, P., Tytgat, R., and De, L.J. 1991. Revision of *Campylobacter*, *Helicobacter*, and *Wolinella* taxonomy: Emendation of generic description and proposal of *Arcobacter* gen. nov. Int. J. Syst. Bacteriol. 41: 88-103.

Versalovic, J., and Fox, J.G. 1999. Helicobacter. In: Manual of Clinical Microbiology 7th ed., P. R. Murray, E. J. Baron, M. A. Pfaller, F. C. Tenover and R. H. Yolken (eds.). ASM Press, Washington. p. 727-738.

Warren, J.D., and Marshall, B.J. 1983. Unidentified curved bacilli on gastric epithelium in active chronic gastritis. Lancet 1: 1273-1275.

Weir, S., Cuccherini, B., Whitney, A.M., Ray, M.L., MacGregor, J.P., Steigerwalt, A., Daneshvar, M.I., Weyant, R., Wray, B., Steele, J., Strober, W., and Gill, V.J. 1999. Recurrent bacteremia caused by a "Flexispira"-like organism in a patient with X-linked (Bruton's) agammaglobulinemia. J. Clin. Microbiol. 37: 2439-2445.

Weir, S.C., Gibert, C.L., Gordin, F.M., Fischer, S.H., and Gill, V.J. 1999. An uncommon Helicobacter isolate from blood: evidence of a group of *Helicobacter* spp. pathogenic in AIDS patients. J. Clin. Microbiol. 37: 2729-2733.

3

Epidemiologic Observations and Open Questions about Disease and Infection Caused by *Helicobacter pylori*

Roger A. Feldman[*]

Abstract

This book documents the variety of virulence factors, geographic variations, and variation in response to environmental factors that exist within *H. pylori*. As a result, earlier epidemiologic analyses of prevalence rates and disease associations, which generally considered *H. pylori* as a single entity, benefit from reinterpretation and newer studies will deal separately and comparatively with the various genetic clusters, geographic variants and environmental factors identified in *H. pylori*. This chapter reviews the most recently available data dealing with several specific epidemiologic questions. Data are reviewed concerning mode of spread, risk factors, and acquisition of infection, both in children and in adults. Studies in children are analyzed concerning the frequency of both stable and transient infection. With adults, the data summarized relate both to the frequency and source of adult infection and to rates of reinfection after successful treatment. The questions addressed also relate to multiple infections and the association of persistent *H. pylori* infection with any of a multitude of diseases, ranging from iron deficiency anemia, heart disease and other non-gastroenterologic problems to gastritis, peptic ulcer, gastric cancer, and MALT lymphoma.

Introduction

The enteric bacterium *Helicobacter pylori* has been of great interest to gastroenterologists as a bacterium that colonizes the stomach and is associated with gastric inflammation and peptic ulcer disease. This book is a result of the burgeoning interest in molecular studies of *H. pylori*. It demonstrates the variety of virulence factors, geographic variations, and variation in response to environmental factors that exist within the species (Blaser, 1997, 1999a). As a result of the new data, earlier epidemiologic analyses of prevalence rates and disease associations may benefit from reinterpretation, and much new epidemiologic information can be anticipated.

Earlier studies of the disappearance from Europe of *Shigella dysenteriae* by the 1930's and of *Shigella flexneri* by the 1960's, despite the persistence of *Shigella sonnei,* suggests differences in the major modes of spread of species within the genus *Shigella*. Changes of a similar character and magnitude may exist within *H. pylori*. Some genetic clusters of *H.*

[*]email: r.a.feldman@qmw.ac.uk

pylori may have different modes of spread, perhaps as a result of differences in ability to withstand heat, drying, ability to take up iron, and other important agent characteristics. As strain variations within *H. pylori* are better defined, there may be strain variants that are more likely to be found in children, or in environments with poor water supplies. There may be variants more likely to persist in the stomach, having out-competed other strains, or strains that are more likely to persist in nutritionally deprived individuals. Nutritional circumstances, particularly related to iron availability, may have effects on expression of virulence factors (Keenan and Allardyce, 1999; Keenan *et al.*, 2000). What, at present, is a description of *H. pylori* will possibly become more like the descriptions of other enteric bacterial genetic clusters and serologic investigation will have to become subtler to keep up with the advances in descriptions of the various clusters identified in *H. pylori*.

This chapter deals with several distinct epidemiologic questions. One question concerns mode of spread and risk factors. Another concerns the acquisition of infection, both in children and in adults. Studies in children are analyzed for information on the frequency of both stable and transient infection. With adults, the data summarized also relate to the frequency and source of infection and reinfection after successful treatment. The questions addressed also relate to multiple infections and the association of persistent *H. pylori* infection with any of a multitude of diseases, ranging from iron deficiency anaemia, heart disease and other non gastroenterologic problems to gastritis, peptic ulcer, gastric cancer, and MALT lymphoma. And finally I speculate on the future prevalence of *H. pylori* due to the observations that infections are becoming rarer.

Mode of Spread and Risks Factors for Infection

Mode of Spread

Observations made with *H. pylori* resemble those for other enteric bacteria, which are frequently transmitted in families by children. Most enteric bacteria are transmitted in association with food and water, often from person-to-person, and are uncommonly transmitted by saliva or vomit. The major vehicles of infection for any particular bacterial species are occasionally different in different geographic environments. Water is more commonly a vehicle in developing countries. It is easy to reason by analogy, and to anticipate that newer epidemiologic studies of *H. pylori* will follow paths common with other enteric bacteria. One example comes from studies to determine the mode of spread of enteric bacteria in households. Using serologic methods to analyse differences in the disease associations and frequencies of different serogroups of *Vibrio cholerae*, there was found to be one major epidemic strain, while other strains were less frequently associated with illness. And although clustering of infection with *V. cholerae* is found within homes, the clustering reflects the necessity for a high infectious dose. Food contaminated from water is often a major vehicle for infection within the home because the organisms can attain the high levels needed for infection through multiplication in the food. Thus, despite multiple infections within homes and among family members, person-to-person transmission within the home is uncommon.

Published data suggest that food, and possibly water (Klein *et al.*, 1991; Teh *et al.*, 1994; Ozturk *et al.*, 1996; Malaty *et al.*, 1996) are vehicles of transmission for *H. pylori* (Goodman *et al.*, 1996; Begue *et al.*, 1998; McKeown *et al.*, 1999). However, the patterns of infection in those studies suggest that the presence of bacteria in food and water does not directly reflect the risk of infection. Genetic signals from *H. pylori* (adhesin and 16SrRNA) have been found in water (Hulten *et al.*, 1998; Hegarty *et al.*, 1999) in areas where infections are infrequent. Although meat has been suggested to be a possible vehicle, adult vegetarians and populations that abstain from eating either pork or beef do not differ in seropositivity from those with less restricted diets (Megraud *et al.*, 1989; Webberley *et al.*, 1992). In contrast, in Chile, Helicobacter seropositivity did increase with consumption of uncooked vegetables (Hopkins *et al.*, 1993) and children in rural Columbia who ate raw vegetables were at increased risk of infection (Goodman *et al.*, 1996).

Other controversies related to the mode of spread concern whether spread results from faecal-oral, oral-oral (Lee *et al.*, 1991) or gastro-oral transmission (Goodman and Correa, 1995; Megraud, 1995; Cammarota *et al.*, 1996; Axon, 1996). Often, the age-specific prevalence of *H. pylori* has been compared with the age-specific serologic profiles of other infections that are known to be spread by oral-oral or faecal-oral transmission (Webb *et al.*, 1996; Luzza *et al.*, 1997, 1998; Fujisawa *et al.*, 1999). However, if the infectiousness of the two agents being compared differs significantly, their patterns of age-specific acquisition may be different even if the mode of spread is similar.

Some oral bacteria are commonly spread from mother to child and within families (Saarela *et al.*, 1993; van Steenbergen *et al.*, 1993; Petit *et al.*, 1993; Preus *et al.*, 1994; Kononen *et al.*, 1994; Li *et al.*, 1995; Matto *et al.*, 1996; Tuite-McDonnell *et al.*, 1997). Many people who are culture positive for *H. pylori* in the stomach are also PCR positive in the mouth, and the oral cavity has been proposed as a reservoir of infection (Oshowo *et al.*, 1998). Some studies have reported frequent isolation of *H. pylori* from oral specimens, particularly dental plaque (Pytko-Polonczyk *et al.*, 1996), while others reported only occasional isolations (Namavar *et al.*, 1995; Cheng *et al.*, 1996) or have consistently failed to isolate the organism from the mouth. Premastication of food was commoner in Burkino Faso families with both mother and child seropositive for *H. pylori* compared to the frequency in families with a seropositive mother and a seronegative child (Albenque *et al.*, 1990). A physician became infected with *H. pylori* after giving mouth-to-mouth resuscitation to an *H. pylori* positive patient who had recently vomited (Figura, 1996). The primary argument against oral-oral transmission is the frequent lack of success in culturing these bacteria from the mouth. Those who consider that vomitus may be a significant element in transmission have cultured vomitus (Leung *et al.*, 1999). And most recently, positive cultures have been obtained uniformly from vomitus, and occasionally from saliva (Parsonnet *et al.*, 1999).

The availability of new tests for *H. pylori* in the stool has suggested that perhaps stools are important in transmission. The organism has been successfully cultured from faeces only infrequently (Thomas *et al.*, 1992; Kelly *et al.*, 1994) and most of the evidence is based on PCR. The paucity of successful culture data lends little support to the role of faeces as a primary mode of spread.

At one time, the major modes of spread of polioviruses were controversial because polioviruses could be cultured both from the mouth and from the stool. The quantities were greater and the duration of excretion was longer in the stool, but it was difficult to be certain whether transmission was predominantly or entirely from stool. Although never totally resolved scientifically, current opinion assumes that poliovirus transmission is predominantly from stool, and the issue no longer seems important. The recent ability to culture *H. pylori* from vomitus supports vomitus playing a prominent role in the spread of *H. pylori* but studies in children are needed to confirm this conclusion.

Risk Factors for Infection

Low socio-economic status and/or a low level of education correlate with increased prevalence of *H. pylori* infection (Anonymous, 1993; Veldhuyzen van Zanten, 1995; Malaty *et al.*, 1996) and indirect measures of community circumstances, such as religion or ethnicity, also correlate with risk (Lindkvist *et al.*, 1998; Boey *et al.*, 1999; El-Serag and Sonnenberg, 1999).

In some studies, the use of alcohol correlated to seropositivity, but this may reflect an indirect measure of socioeconomic and cultural variables. There does not appear to be any relationship between *H. pylori* seropositivity and ABO blood group (Henriksson *et al.*, 1993; Umlauft *et al.*, 1996) nor with secretor status (Dickey *et al.*, 1993). Human milk IgA protected infants from early infection with *H. pylori*. The protection was demonstrable at 9 months, but by 12 months 9/12 breast-fed children were seropositive for *H. pylori*, including the two children whose mothers had the highest IgA levels of antibody in breast milk (Thomas *et al.*, 1993).

Country of Exposure

In addition to geographic variations in the strains of *H. pylori* (Achtman *et al.*, 1999; Achtman and Suerbaum, 2000), geographic areas also differ in the degree of exposure. High rates of seropositivity in children are found in many developing countries. Some studies (Blecker and Vandenplas, 1993; Banatvala *et al.*, 1995) showed high seropositivity rates in children from developing countries, even when they were born in countries with low rates of infection. It is possible that these individuals became infected while visiting relatives in the developing countries or that they were infected more frequently from their parents than were other children. An increased rate of acquisition was also found in persons moving from an area of lower seropositivity to one of high seropositivity (Ramirez-Ramos *et al.*, 1994; Becker *et al.*, 1999) and in persons travelling as part of military actions (Smoak *et al.*, 1994). Other environmental influences might also influence the frequency of infection-associated disease (Leon-Barua *et al.*, 1997).

Gender

The seroprevalence of *H. pylori* infection is often similar in males and females (Anonymous, 1993; Gasbarrini *et al.*, 1995). However, higher infection rates in males were found in rural Columbian Andes communities (Goodman *et al.*, 1996), New Zealand, (Fawcett *et al.*, 1996) and a variety of ethnic groups in California (Replogle *et al.*, 1995). If gender-specific differences exist in infection rates, it may be that, as in the Columbian study, the differences will be most dramatic in childhood.

Occupation

Increased *H. pylori* seropositivity has been seen in gastroenterologists (Lin *et al.*, 1994; Liu *et al.*, 1996) and endoscopists (Goh *et al.*, 1996; Su *et al.*, 1996), but not in dentists (Malaty *et al.*, 1992; Luzza *et al.*, 1995; Lin *et al.*, 1998). Submariners showed a higher rate of *H. pylori* seropositivity than did other military groups (Hammermeister *et al.*, 1992).

Age and Acquisition of Infection

A question frequently studied with other enteric bacteria concerns the importance of the age at which infection occurs for subsequent spread. Salmonellae have many different animal and food reservoirs, but once a *Salmonella* infection occurs in a family, subsequent person-to-person transmission depends predominantly on the age of the infected person(s) in the household and the infectious dose. With *Salmonella* and *Shigella*, infections are most frequently transmitted by children, more often by pre-schoolers than by older children.

Preschool and School Children

Studies of *H. pylori* infections from many countries and continents suggest that children, particularly pre-school children, have higher infection rates than older children from the same geographic location and socio-economic situation (Goodman and Correa, 1995; Replogle *et al.*, 1996; Clemens *et al.*, 1996; Ashorn *et al.*, 1996; Gold *et al.*, 1997; Rothenbacher *et al.*, 1998; Lindkvist *et al.*, 1999b; Bassily *et al.*, 1999). In the developing world, seropositivity of over 50% is not unusual by age 10. The number of siblings per household increased the rate of infection in pre-school children (Lin *et al.*, 1999). A Swedish study (Granstrom *et al.*, 1997) found higher prevalence rates in younger children than might have been expected from the frequency of infection of young adults in the same area. A study in the southern Columbian Andes (Goodman and Correa, 2000) showed that birth order, birth spacing, and the infection status of sibling all influenced the odds of *H. pylori* infection, independently of the number of children in the home. The data suggested transmission from older to younger children and are compatible with the older children becoming infected outside the home.

The rate of acquisition of *H. pylori* in infants and young and older children seems to have steadily decreased over the past decades, which may partially explain the age-specific differences in cross sectional data concerning seropositivity (Banatvala *et al.*, 1993). In Japanese studies, both Kumagai *et al.* (1998) and Replogle *et al.* (1996) found cohort effects, with Replogle describing increased seropositivity from 1900-1959 followed subsequently by decreases.

Environmental variables correlated with infection rates in several studies. In Peru, children whose homes had external water sources were three times more likely to be infected than were children whose homes had internal water sources (Klein *et al*, 1991). In addition, children from high-income families, whose homes were supplied with municipal water, were 12 times more likely to be infected than were those from low-income families whose water supply came from community wells (Klein *et al*, 1991). It was possible to PCR amplify genes from the municipal water (Hulten *et al.*, 1996) and the use of municipal water correlated with *H. pylori* infection among Lima children from both low and high socio-economic families.

Sero-reversion in Children

In an Estonian cohort of 100 infants, followed for three years from birth, *H. pylori* seroconversion rates in those initially seronegative were 27% in the first year, 25% in the second year, and 12% in the third year (Lindkvist *et al.*, 1999). In Ethiopian children (Lindkvist *et al.*, 1998), similar high rates of acquisition were found in the first three years of life. However, 25% of the Estonian birth cohort who were seropositive at age one were seronegative at age two. And of those newly seropositive at age 2, 80% were seronegative at age 3. Similarly, breath tests suggested that, in Peru, seroreversion was frequent in children (Klein *et al*, 1991). Seroreversion has also been reported from the developed world: In the cohort of Swedish children studied initially as infants and then to age 11 years by Granstrom (1997), 10% were seropositive for *H. pylori* at age 2 and only 3% were positive at age 11. 32 of the 40 originally seropositive children had seroreverted. In a Japanese cohort, Kumagai (1998) found 13% seropositivity in the age group 6-12 and 48% seropositivity in the age group 13-19 years. In this group followed from 1986 to 1994, there was a 9% (2/22) seroreversion in children. An American cross sectional study (Redlinger *et al.*, 1999) found a decreasing seroprevalence by age in children 4-7 years of age, and interpreted the changes as due to seroreversion. A French study (Ganga-Zandzou *et al.*, 1999) found seroreversion of one of 18

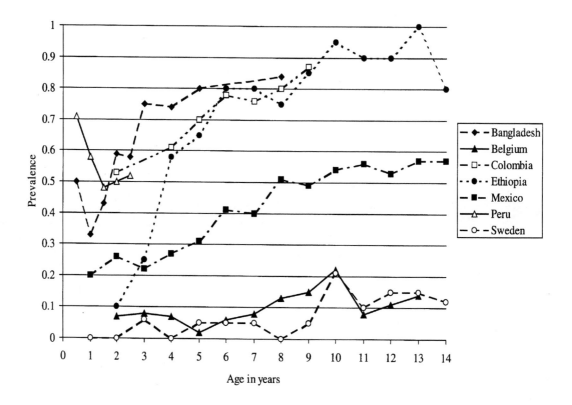

Figure 1. Prevalence of *H. pylori* infection in selected populations of asymptomatic children. Prevalence was approximated from graphs for Bangladesh, Belgium, Ethiopia, and Sweden. Plotted points are category midpoints. *H. pylori* was detected by urea breath test in Bangladesh, Colombia, and Peru and by serology in all other countries. Population details: Bangladesh - 406 periurban village residents aged 0-9 years, near Dhaka (Mahalanabis *et al.*, 1996); Belgium - 466 Academic Children's Hospital elective surgery outpatients aged 2-14 years, Brussels (Blecker *et al.*, 1993); Colombia- 684 residents of a rural Andean village in Narino aged 2-9 years (Goodman *et al.*, 1996); Ethiopia - 242 children of hospital employees who participated in vaccine trials in Addis Ababa (Lindkvist *et al.*, 1998); Mexico - 4,190 children aged 1-14 years sampled at random from the population of Mexico as part of a national serologic survey conducted in 1987-8 to form the National Serum Bank (Torres *et al.*, 1998); Peru - 56 residents of a periurban community near Lima followed at six-month intervals from 6 to 30 months of age (Klein *et al.*, 1994); Sweden - 295 non-gastric surgical patients of St. Goran Hospital, Stockholm (Lindkvist *et al*, 1996)

children around 7 years of age in a 2-year follow-up. However, a Finnish study (Rehnberg-Laiho *et al.*, 1998) found no sero reversions in children.

Frequent seroreversion in children, especially pre-school children, raises questions about the immunologic capacity of children to eliminate an early infection that are important for vaccine development, and for subsequent disease occurrence. None of the published studies report anything distinctive in the initial immune response or present data on the genotypes of the infecting strains. In the Swedish cohort (Tindberg *et al.*, 1999), antimicrobial use was more frequent in the children who seroreverted. It is possible that environmental factors may be of great importance in the development of persistent infection in children. Perhaps the absence or presence of iron, or some other environmental variable, is associated with persistence. Regardless of the ultimate explanations, it is probable that more children, particulaarly preschool children, are infected than is measured by cross sectional seropositivity in prevalence studies, both in developed and developing countries.

Table 1. Rates of "reinfection" of adults after "successful" treatment.

Citation	Range of overall treatment efficacy	Duration of time used to define successful treatment	Estimate of rate of "reinfection"
Coelho *et al.*, 1992	60%	2 months	20.7%
Bell and Powell, 1996	60-79%	6 months	4.6%
Bell and Powell, 1996	80%	6 months	1.7%
Berstad *et al.*, 1995	82-95%	12 months	1.7%

Prevalence in Asymptomatic Infants and Children

The prevalence of infection in developed and undeveloped countries has been recently compared (Goodman *et al.*, 1999) (Figure 1).The global patterns are similar, but the rates are higher in the developing world.

Infections and Reinfections in Adults

An Infectious Dose in Adults

The adult infectious dose can be estimated as being relatively high from experiments where one volunteer became infected after ingesting a dose of 10^9 organisms of a field strain and a second was not infected with a dose of 10^7 (Morris *et al.*, 1991). The second volunteer was thereupon successfully infected with 10^5 organisms of a different field strain. Outbreaks of achlorhydria in the United States, and acute mucosal lesion syndrome in Japan represent examples of iatrogenic *H. pylori* infection transmitted by endoscopes that are contaminated with gastric fluid (Tytgat, 1995). It is unclear what dose was involved in these incidents. Although doses of 10^3/ml (Young *et al*, 2000) have been found in gastric juice, it is uncertain whether such a dose can be present in a washed endoscope. These infections were introduced directly, and may have little relevance to the actual infectious dose needed to transmit an infection to an adult. Further studies to estimate the infectious dose in adults will need to precede vaccine trials, but they will probably attempt to measure the ID_{50}, that is, the dose that infects 50% of the subjects on average. In nature, it is possible that, similarly to Salmonella, the normal infecting dose is one that is large enough only to infect one in 100, an ID_1 (Blaser and Newman, 1982). And it is possible that the size of the dose is related to persistence, with smaller doses leading to spontaneously resolving infections and larger doses resulting in persistent infections.

Rates of Acquisition in Adults

The preferred method to measure the rate of acquisition in adults requires study of the same individuals over long periods. Acquisition can also be estimated from cross sectional seroprevalence data from independent groups with sera taken several decades apart but the power of this method is lower. The common observation from cross sectional studies has been that seroprevalence increases with each decade of age, suggesting a rather steady acquisition rate for all adult ages. An alternative explanation is that there has been a steady decrease in acquisition in the last 100 years, and the differences in seroprevalence by age are predominantly a result of a birth cohort effect. Several different studies of adults in the developed world showed an acquisition rate of 3 to 4 percent per decade (Kuipers *et al.*, 1993; Cullen *et al.*, 1993; Kosunen *et al.*, 1997; Sipponen *et al.*, 1996).

Sero-reversion in Adults

Cohort studies in adults have measured seroreversion rates ranging from 0.7 to 16% per decade (Xia and Talley, 1997; Kumagai *et al.*, 1998). In an isolated rural area of Japan where seropositivity among 20-80 year old adults was 84%, 15% seroreversion was observed over 10 years. In 173 persons over age 100 in Finland, Rehnberg-Laiho (1999) interpreted seropositivity of only 66% as seroreversions in association with frequent gastric atrophy, loss of infection, and of measurable antibody.

Risk of Reinfection, after Successful Treatment, in Adults

It is difficult to distinguish reinfection with *H. pylori* after "successful" treatment from recrudescence of an incompletely treated *H. pylori* infection with *H. pylori*. The important criteria used to differentiate reinfection and recrudescence are the time after apparently successful curing, which is often 4 weeks after treatment, and the overall success of the treatment regime used (Bell and Powell 1996). The percentage of reinfection varied strongly with the time interval used to define success as well as the effectiveness of the treatment protocol (Table 1).

Without reference to the overall effectiveness of the therapeutic regimen, low reinfection rates (1-2.5%) were found one year after therapy (Cutler and Schubert, 1993; Borody *et al.*, 1994; Mitchell *et al*, 1998). Reinfections were mostly observed before the end of the first year after therapy (Figueroa *et al*, 1996; Archimandritis *et al*, 1999; Rollan *et al*, 2000).

H. pylori isolates from before and after therapy are often but not always indistinguishable by molecular methods. Isolates that are distinguishable may still be compatible with recrudescence because some patients are infected with more than one strain (Fraser *et al.*, 1992; Owen *et al*, 1993; Bamford *et al.*, 1993). Alternatively, different strains may reflect reinfection, especially because in some instances, isolates from different family members are not identical (Nwokolo *et al.*, 1992; Suerbaum *et al.*, 1998, Kersulyte *et al.*, 1999). In a study of six members of three generations of a family with a high incidence of duodenal ulcer disease, three family members (one from each generation) harboured clonal variants of a single strain of *H. pylori* and three other members harboured three different strains (Nwokolo *et al.*, 1992). In one study documented "treatment failures" most likely reflected recurrences, since they were overwhelmingly due to the persistence of the same strain (Xia *et al.*, 1995).

Acquisition within the Family

Many serologic and molecular studies have identified households with multiple infected members, some of whom carry similar or identical strains (Drumm *et al.*, 1990; Bamford *et al.*, 1993, Mitchell *et al.*, 1993; Wang *et al.*, 1993; Vincent *et al.*, 1994; Schultz *et al.*, 1995; Georgopulos, *et al.*, 1996; Suerbaum *et al.*, 1998; Dominici *et al.*, 1998). Often, studies have shown no clear relationship between the Helicobacters found in family members. And in those homes in which concordance has been found, it is unclear whether the infection is transmitted from the children to the adults, or from an adult to other adults and/or from adults to children.

In studies of transmission of Shigella within families, where transmission was apparently person-to-person, preschool children were more infectious than older children, and older children were more infectious than adults. Similar characteristics of spread of poliovirus in families were found during studies of the introduction of oral vaccine into families, and in addition to the effects of age, mothers were more infectious than fathers. Reasoning by analogy with these enteric agents, it seems likely that transmission of *H. pylori* may occur from any individual, but that child-to-child, and child-to-adult is more frequent than adult to child or adult to adult.

Some data from Helicobacter studies support these analogies. The seroprevalence among children under age 16 related positively to both birth order and the number of children in the family (Teh *et al.*, 1994). In rural Columbia, the strongest predictor of *H. pylori* seroprevalence was the number of children in the family.

In 52 children who were followed for an average of 2 years after successful eradication of *H. pylori* infection, 6 children (11.5%) became reinfected (Rowland *et al.*, 1999). The mean age of those who became reinfected was 6 years compared with 12 years for those who remained clear of infection (P = 0.00001). Living with infected parents and siblings and low socioeconomic status were not risk factors for reinfection.

Preferential clustering of infections in identified families compared to families in the general population can be inferred from many studies, in both developed (Dominici *et al.*, 1999; Rothenbacher *et al.*, 1999; Suzuki *et al*, 1999; Brenner *et al.*, 2000) and undeveloped countries (Ma *et al.*, 1998). But it is unclear in which direction the infection flows when there is a relation between isolates from a child and the parents (Elitsur *et al.*, 1999).

Frequency of Multiple Strains

There is little reason to believe, especially in developing countries with high infection rates in children, that individuals are infected only once with *H. pylori*. As a result, it might be expected that gastric mucosal biopsies from different sites would reveal multiple strains. The frequency of such multiple colonization will clearly be affected by differences between wild strains in growth rate, ability to evade the immune system and response to environmental factors. van der Ende found, although all family members in one family cluster had similar isolates, seven of the eight persons studied had more than one subtype of *H. pylori* in their stomach (van der Ende *et al*, 1996). Two or more different vacA genotypes were isolated from gastric biopsy specimens from 17 of 20 patients in Mexico (Morales-Espinosa *et al*, 1999). Enroth *et al.* (1999) described strains with variable resistance to Clarithromycin in one patient and anti-*H. pylori* treatment including Clarithromycin resulted in overgrowth of a strain that was uncommon before treatment. In Poland, Hennig *et al.*(1999) found multiple *H. pylori* strains in more than 50% of duodenal or gastric ulcer patients. Multiple strains present a problem for relating virulence determinants to disease because in many laboratories, single colony isolation is not a general practice resulting in experiments with mixed cultures. When single colonies are isolated, normally only one bacterium, the most common one, is preserved.

Association of *H. pylori* with Illness

Risk of Heart Disease

The possibility that *H. pylori* might be associated with an increased risk of coronary heart disease has been investigated with case control studies and with cohorts followed for variable periods. The published studies have slightly different end points, case ages, and socioeconomic backgrounds. Overall, the results indicate no increased risk or at most a small increased risk.

Wald *et al.*(1997) performed a nested case-control study of 648 men who died from ischaemic heart disease plus 1296 age matched controls. The subjects were from a group of

21,520 professional men of similar socioeconomic status who attended a medical centre in London between 1975 and 1982 at the age of 35-64 years. The average follow-up was 15.6 years. The odds ratio of death from ischaemic heart disease for those with a positive serology for *H. pylori* was 1.06 (with 95% CI [confidence interval] of 0.86 to 1.31). In the same cohort, over the same time interval of observation, the odds ratio for the association of *H. pylori* and gastric cancer was 4.0 (95%CI 1.9 to 8.2)

Strachan *et al.*(1998) completed a prospective study of 1796 men (the Caerphilly, South Wales cohort), from whom blood was collected from 1979 to 1983, and who were followed for an average of 13.7 years. The odds ratio for overall incident ischaemic heart disease, including mortality, in those seropositive for *H. pylori* was 1.05 (95% CI of 0.8 to 1.39).

Koenig *et al.* (1999) tested whether the presence of CagA in *H. pylori* might be important in affecting the risk of angiographically-defined stable coronary heart disease. The 312 individuals in the case group were mostly male, with an average age about 57 years. After correcting for age, gender, and other known cardiovascular risk factors, no residual association was found, and no effect of CagA status.

Folsom *et al.* (1998), in a case-control study of American men and women from 45-64 years of age, with a three-year period of observation, found no association with an increased risk of acute coronary heart disease in those seropositive for *H. pylori* (OR 1.03, 95% CI .68 to 1.57) after correcting for other cardiovascular risk factors.

Danesh *et al.* (1999) reported results of two studies using different end points and comparison groups. They compared 1122 survivors of suspected acute myocardial infarction at ages 30-49 with age and sex matched controls with no history of coronary disease, in an attempt to maximise the strength of an association. After correcting for smoking, socioeconomic indicator, serum lipids and obesity, the odds ratio for those that were seropositive was 1.75 (99% CI 1.29 to 2.36). Danesh *et al.* also studied 510 sibling pairs (mean age 59 years) in which one sibling survived a myocardial infarction and the other had no history of coronary heart disease, in an attempt to reduce confounding. After the same corrections, the odds ratio for those that were seropositive was 1.29 (99% CI .78 to 2.13).

Risk of MALT Lymphoma

Gastric lymphomas are the commonest of the extranodal lymphomas. In an American study of gastric lymphomas, the rates of new diagnoses in the period from 1973 to 1986 in the age group 60-79 years of age was about 2 per 100,000 per year (Severson and Davis, 1990). The tumors have clinicopathologic features more closely related to the structure and function of so-called mucosa-associated lymphoid tissue (MALT) than to peripheral lymph nodes. In contrast to peripheral lymph nodes, which are adapted to deal with antigens carried to the node in afferent lymphatics, MALT appears to have evolved to protect mucosal tissue, which is directly in contact with antigens in the external environment. Normally there is no lymphoid tissue in the stomach, nor in the salivary gland or thyroid, the common sites for MALT lymphomas, and the development of MALT lymphoma is preceded in these organs by a chronic inflammatory process. Certain histologic features of the gastric MALT lymphoma suggest that its growth is subject to immunologic stimuli, in particular those related to *H. pylori*.

The presence of lymphoid follicles in the gastric mucosa is virtually pathognomonic of *H. pylori* infection. Five independent sources of evidence link prior *H. pylori* infection and MALT lymphoma (Wotherspoon, 1998; Isaacson, 1999). These include: a) the presence of *H. pylori* in more than 90% of cases of gastric MALT lymphoma; b) a high incidence of gastric MALT lymphoma in north-east Italy, in association with a high seroprevalence of *H. pylori*; c) case-control data showing an association between previous *H. pylori* infection and the development of primary gastric lymphoma; d) studies showing that treatment of the *H. pylori* infection leads to resolution of low grade MALT lymphoma, with approximately 70% of cases in stage IE of gastric MALT lymphoma regressing following successful treatment; e) in the absence of T cells in the gastric lymphoma, there was no in vitro response to heat-killed whole-cell preparations of *H. pylori*. Additional evidence concerns the frequency of CagA positive bacteria in association with MALT lymphoma. Analysis of biopsies from 123 *H. pylori*-related gastric biopsies from Italy (Peng *et al.*, 1998) showed that 30% (17/56) of gastritis cases and 38% (14/37) of the low-grade MALT lymphoma were CagA+, while 77% (23/30) of the high-grade MALT lymphomas were CagA positive (P < 0.05).

Risk of Iron Deficiency Anaemia

Chronic gastritis is associated with iron deficiency anaemia. Serum ferritin, haemoglobin, and immunoglobulin G (IgG) antibodies against *H. pylori* were assessed in a population survey of 2794 Danish adults (Milman *et al.*, 1998). Although the seroprevalence of *H. pylori* infection did not relate to hemoglobin levels, serum ferritin levels were significantly lower in men (114 vs. 120 µg/l; P = 0.01) and in postmenopausal women (63 vs.77 µg/l; P = 0.02) who were IgG positive than in seronegative individuals. IgG-positive people more often had reduced serum ferritin levels (<=30 µg/l) than seronegative people. This association persisted in multivariate analysis after adjusting for possible confounding factors (odds ratio, 1.4; 95% CI1.1-1.8).

With this background, a trial of treatment of *H. pylori* was initiated in 30 patients with a long history of iron deficiency anaemia (Annibale *et al.*, 1999). After 12 months in the absence of iron therapy, 91.7% of the patients cured of *H. pylori* had recovered from anaemia, and had increased serum ferritin levels.

A double-blind, placebo-controlled therapeutic trial was conducted in paediatric patients (mean age, 15.4 years) with iron-deficiency anaemia (Choe *et al.*, 1999). Patients were assigned randomly to three groups; those given oral ferrous sulphate and antimicrobials to treat *H. pylori*; those given placebo for iron and anti-Helicobacter therapy; and those given oral ferrous sulphate and a placebo instead of anti-Helicobacter therapy. In the groups that received anti-Helicobacter therapy, there was a significant increase in haemoglobin level as compared with the group on placebo (p = 0.009).

Risk of Gastric Cancer

Gastric cancer rates increase with age, and rise sharply after age 65 to over 250/100,000 year at age 80. Males are more frequently affected than females. *H. pylori* is a chronic infection, associated with inflammation of the gastric mucosa, and progresses over time to precancerous conditions, including gastric mucosal atrophy, metaplasia and dysplasia. As a result, there was the possibility of a variable association with gastric cancer. It has been hypothesised that the age of acquisition may be a significant factor associated with disease occurrence, but it is difficult to obtain direct data on the age of infection. If infection occurs predominantly in children, then those of higher birth orders, and in large families, should have early acquisition of infection, and perhaps be at higher risk of illness. A nested case-control study (Blaser *et al.*, 1995) of a cohort of Japanese-American men in Hawaii matched controls with cases for sibship size and birth order. *H. pylori*-infected but not *H. pylori*-uninfected men from larger sibships (odds ratio, 2.06) and of higher birth order (odds ratio, 1.67) were at increased risk for developing gastric cancer. There was no relation between birth order or sibship size and the occurrence of duodenal ulcer.

Table 2. Disease-to-infection (D/I) ratios and point prevalence calculated from the number of infected individuals with stigmata of peptic ulcer disease and the number of infected individuals screened

Study area and citation	Point prevalence (%)	Lifetime D/I ratio (%)
India, Katelaris *et al.*, 1992	7.8	13-16
India, Khuroo *et al.*, 1989	4.7	11-13
Norway, Bernersen *et al*, 1990	15	
Hawaii, Nomura *et al*, 1994	6.0	12.5
Italy, Vaira *et al.*, 1994	20	
Japan, Schlemper *et al,*, 1996		25
Holland, Schlemper *et al.*, 1996		14
USA, Dooley *et al.*, 1989		3
Holland, Kuipers *et al.*, 1995		3

In measuring the association, retrospective studies have compared the rates of *H. pylori* infection in gastric cancer cases with the infection rate in controls, identified at the time of diagnosis of the cases. Prospective studies have followed cohorts of individuals, whose blood was drawn at the time of entry into the study, and, as a nested case-control study, compared rates of gastric cancer in seropositive individuals with age and sex matched seronegative controls drawn from the cohort.

In the retrospective studies, which predominantly involved an elderly group of patients, the occurrence of gastric atrophy leads to loss of seropositivity, and can be a cause of bias. However, this bias is minimised among gastric cancer cases at an early age and in such a study in Japan (Kikuchi *et al.*, 1995), with all cases under 40 years of age, the odds ratio was 13 (95% CI 5.3-36.0).

The pooled odds ratio for association in three early retrospective studies was 3.8 (95% CI 2.3-6.2). However, when the data were stratified by duration of follow-up, the odds ratio increased to 8.7 (95% CI 2.7-44.7) for the group with 15 or more years between the assessment of infection and a diagnosis of gastric cancer (Forman *et al.*, 1994). In a Finnish study, the odds ratio for association increased from 0.3 for 2 years of follow-up to 3.3 for at least 7 years of follow-up. Clearly, a more valid measurement of risk comes from a longer period between the assessment of infection and diagnosis of cancer. A recently published Norwegian prospective study (Hansen *et al.*, 1999) with over 200 cases showed an association of *H. pylori* seropositivity and non-cardia gastric cancer, with an odds ratio of 5.1 (95% CI 2.8-9.4).

There are conflicting data concerning the significance of the association of gastric cancer with the CagA status of *H. pylori*. The risk of gastric cancer in young adult Japanese (Kikuchi *et al.*, 1999) was equal for bacteria that were CagA- (odds ratio 15, CI 6.4-35.2) and CagA+ (odds ratio 14.6, CI 6.7-31.9). However, an increased association of gastric cancer with the *H. pylori* CagA genotype was found in young Italian patients (Rugge *et al.*, 1999).

Disease-to-Infection ratio for Helicobacter-associated Peptic Ulcer Disease

H. pylori infection is a major risk factor for peptic ulcer disease (Axon and Forman, 1997). Although most people infected by *H. pylori* do not report any clinical symptoms, chronic infection modestly raises the likelihood of disease at some future time. The ratios of peptic ulcer disease-to *H. pylori*- infection (Feldman *et al.*, 1998) are summarized in Table 2 using data from India, where there is a high *H. pylori* seroprevalence (Graham *et al.*, 1991; Prasad *et al.*, 1994; Gill *et al.*, 1994), and Europe, where peptic ulcer disease is common (Anonymous, 1993).

The prevalence of peptic ulcer disease varies between populations and over time (Susser *et al.*, 1962; Holcombe *et al.*, 1992; Sonnenberg, 1995). Although peptic ulcer disease began to decline before antimicrobials or anti-acid treatments became commonplace (Coggon *et al.*, 1981; Gustavsson and Nyren, 1989), the available data suggests that between 6 and 20 percent of untreated *H. pylori* infections will lead to peptic ulcer disease (Table 2).

Anticipated Future Prevalence of *H. pylori*

Data from various sources suggests that *H. pylori* prevalence is decreasing globally (Parsonnet, 1995; Kosunen *et al.*, 1997), and concomitantly there has been a decrease in the age-specific rates of peptic ulcer (Sonnenberg, 1995), gastric cancer, and in the acquisition of infection in the younger age groups (Blaser, 1999b). These observations imply that peptic ulcers caused by *H. pylori* will also decrease. It is not clear that any change occurred in the pattern of human enteric disease after *Shigella dysenteriae* almost vanished as a human pathogen. However, as *H. pylori* vanishes, there may be changes in the patterns of gastro-oesophageal disease (Chow *et al.*, 1998; Vicari *et al.*, 1998; Blaser, 1999b, 1999c).

References

Achtman, M., Azuma, T., Berg, D.E., Ito, Y., Morelli, G., Pan, Z.J., Suerbaum, S., Thompson, S.A., and van der Ende, A., van Doorn, L.J. 1999. Recombination and clonal groupings within *Helicobacter pylori* from different geographical regions. Mol. Microbiol. 32: 459-470.

Achtman, M., and Suerbaum, S. 2000. Sequence variation in *Helicobacter pylori*. Trends Microbiol. 8: 57-58.

Albenque, M., Tall, F., Dabis, F., and Megraud, F. 1990. Epidemiological study of *Helicobacter pylori* transmission from mother to child in Africa. Rev. Esp. Enferm. Dig. 78: Suppla. 1: 48.

Annibale, B., Marignani, M., Monarca, B., Antonelli, G., Marcheggiano, A., Martino, G., Mandelli, F., Caprilli, R., and Delle Fave, G. 1999. Reversal of iron deficiency anemia after *Helicobacter pylori* eradication in patients with asymptomatic gastritis. Ann. Intern. Med. 131: 668-672.

Anonymous. 1993. Epidemiology of, and risk factors for, *Helicobacter pylori* infection among 3194 asymptomatic subjects in 17 populations. The EUROGAST Study Group. Gut. 34: 1672-1676.

Archimandritis, A., Balatsos, V., Delis, V., Manika, Z., and Skandalis, N. 1999. "Reappearance" of *Helicobacter pylori* after eradication: implications on duodenal ulcer recurrence: a prospective 6 year study. J. Clin. Gastroenterol. 28: 345-347.

Ashorn, M., Maki, M., Hallstrom, M., Uhari, M., Akerblom, H.K., Viikari, J., and Miettinen, A. 1995. *Helicobacter pylori* infection in Finnish children and adolescents. A serologic cross-sectional and follow-up study. Scand. J. Gastroenterol. 30: 876-879.

Ashorn, M., Miettinen, A., Ruuska, T., Laippala, P., and Maki, M. 1996. Seroepidemiological study of *Helicobacter pylori* infection in infancy. Arch. Dis. Child. Fetal Neonatal Ed. 74: F141-142.

Axon, A., and Forman, D. 1997. Helicobacter gastroduodenitis: a serious infectious disease. B.M.J. 314: 1429-1430.

Axon, A.T.R. 1996. The transmission of *Helicobacter pylori*: which theory fits the facts? Eur. J. Gastroenterol. Hepatol. 8: 1-2.

Bamford, K.B., Bickley, J., Collins, J.S., Johnston, B.T., Potts, S., Boston, V., *et al.* 1993. *Helicobacter pylori*: comparison of DNA fingerprints provides evidence for intrafamilial infection. Gut. 34: 1348-1350.

Banatvala, N., Clements, L., Abdi, Y., Graham, J.Y., Hardie, J.M., and Feldman, R.A. 1995. Migration and *Helicobacter pylori* seroprevalence: Bangladeshi migrants in the U.K. J. Infect. 31: 133-135.

Banatvala, N., Mayo, K., Megraud, F., Jennings, R., Deeks, J.J., and Feldman, R.A. 1993. The cohort effect and *Helicobacter pylori*. J. Infect. Dis. 168: 219-221.

Bassily, S., Frenck, R.W., Mohareb, E.W., Wierzba, T., Savarino, S., Hall, E., Kotkat, A., Naficy, A., Hyams, K.C., and Clemens, J. 1999. Seroprevalence of *Helicobacter pylori* among Egyptian newborns and their mothers: a preliminary report. Am. J. Trop. Med. Hyg. 61: 37-40.

Becker, S.I., Smalligan, R.D., Frame, J.D., Kleanthous, H., Tibbitts, T.J., Monath, T.P., and Hyams, K.C. 1999. Risk of *Helicobacter pylori* infection among long-term residents in developing countries. Am. J. Trop. Med. Hyg. 60: 267-270.

Begue, R.E., Gonzales, J.L., Correa-Gracian, H., and Tang, S.C. 1998. Dietary risk factors associated with the transmission of *Helicobacter pylori* in Lima, Peru. Am. J. Trop. Med. Hyg. 59: 637-640

Bell, G.D., and Powell, K.U. 1996. *Helicobacter pylori* reinfection after apparent eradication—the Ipswich experience. Scand. J. Gastroenterol. Suppl. 215: 96-104.

Bernersen, B., Johnsen, R., Straume, B., Burhol, P.G., Jenssen, T.G., and Stakkevold, P.A. 1990. Towards a true prevalence of peptic ulcer: the Sorreisa gastrointestinal disorder study. Gut. 31: 989-992

Berstad, A., Hatlebakk, J.G., Wilhelmsen, I., Berstad, K., Bang, C.J., Nysaeter, G., Hausken, T., Weberg, R., Nesje, L.B., and Hjartholm, A.S. 1995. Follow-up on 242 patients with peptic ulcer disease one year after eradication of *Helicobacter pylori* infection. Hepatogastroenterol. 42: 655-659.

Blaser, M.J., and Newman, L.S. 1982. A review of human salmonellosis: I. Infective dose. Rev. Infect. Dis. 4: 1096-1106.

Blaser, M.J., Chyou, P.H., and Nomura, A. 1995. Age at establishment of *Helicobacter pylori* infection and gastric carcinoma, gastric ulcer, and duodenal ulcer risk. Cancer Res. 55: 562-565.

Blaser, M.J. 1997. Heterogeneity of *Helicobacter pylori*. Eur. J. Gastroenterol. Hepatol. 9 (Suppl 1): S3-S7.

Blaser, M.J. 1999a. Allelic variation in *Helicobacter pylori*: progress but no panacea. Gut. 45: 477.

Blaser, M.J. 1999b. Where does *Helicobacter pylori* come from and why is it going away? J.Amer. Med. Assoc. 282: 2260-2262.

Blaser, M.J., 1999c. Hypothesis: the changing relationships of *Helicobacter pylori* and humans: implications for health and disease. J. Infect. Dis.179: 1523-1530.

Blecker, U., Hauser, B., Lanciers S., Peeters, S., Suys, B., and Vandenplas, Y. 1993. The prevalence of *Helicobacter pylori*-positive serology in asymptomatic children. J. Pediatr. Gastroenterol. Nutr. 16: 252-256.

Blecker, U., and Vandenplas, Y. 1993. Ethnic differences in *Helicobacter pylori* infection. Eur. J. Pediatr. 152: 176.

Boey, C.C., Goh, K.L., Lee, W.S., and Parasakthi, N. 1999. Seroprevalence of *Helicobacter pylori* infection in Malaysian children: evidence for ethnic differences in childhood. J. Paediatr. Child Health. 35: 151-152.

Borody, T.J., Andrews, P., Mancuso, N., McCauley, D., Jankiewicz, E., Ferch, N., *et al*. 1994. *Helicobacter pylori* reinfection rate, in patients with cured duodenal ulcer. Am. J. Gastroenterol. 89: 529-532.

Brenner, H., Bode, G., and Boeing, H. 2000. *Helicobacter pylori* infection among offspring of patients with stomach cancer. Gastroenterology. 118: 31-35.

Cammarota, G., Tursi, A., Montalto, M., Papa, A., Veneto, G., Bernadi, S., Boari, A., Colizzi, V., Fedeli, G., and Gasbarrini, G. 1996. Role of dental plaque in the transmission of *Helicobacter pylori* infection. J. Clin. Gastroenterol. 22: 174-177.

Cheng, L.H., Webberley, M., Evans, M., Hanson, N., and Brown, R. 1996. *Helicobacter pylori* in dental plaque and gastric mucosa. Oral Surg. Oral Med. Oral Path. Oral Radiol. Endodont. 81: 421-423.

Choe, Y.H., Kim, S.K., Son, B.K., Lee, D.H., Hong, Y.C., and Pai, S.H. 1999. Randomized placebo-controlled trial of *Helicobacter pylori* eradication for iron-deficiency anemia in preadolescent children and adolescents. Helicobacter. 4: 135-139.

Chow, W.H., Blaser, M.J., Blot, W.J., Gammon, M.D., Vaughan, T.L., Risch, H.A., Perez-Perez, G.I., Schoenberg, J.B., Stanford, J.L., Rotterdam, H., West, A.B., and Fraumeni, J.F. Jr. 1998. An inverse relation between cagA+ strains of *Helicobacter pylori* infection and risk of esophageal and gastric cardia adenocarcinoma. Cancer Res. 58: 588-590.

Clemens, J., Albert, M.J., Rao, M., Huda, S., Qadri, F., Van Loon, F.P., Pradhan, B., Naficy, A., and Banik, A. 1996. Sociodemographic, hygienic and nutritional correlates of *Helicobacter pylori* infection of young Bangladeshi children. Pediatr. Infect. Dis. J. 15: 1113-1118.

Coelho, L.G., Passos, M.C., Chausson, Y., Costa, E.L., Maia, A.F., Brandao, M.J., *et al*. 1992. Duodenal ulcer and eradication of *Helicobacter pylori* in a developing country. An 18-month follow-up study. Scand. J. Gastroenterol. 27: 362-366.

Coggon, D., Lambert, P., and Langman, M.J. 1981. 20 years of hospital admissions for peptic ulcer. Lancet: 1: 1302-1304.

Cullen, D.J., Collins, B.J., Christiansen, K.J., Epis, J., Warren, J.R., Surveyor, I., *et al*. 1993. When is *Helicobacter pylori* infection acquired? Gut. 34: 1681-1682.

Cutler, A.F., and Schubert, T.T. 1993. Long-term *Helicobacter pylori* recurrence after successful eradication with triple therapy. Am. J. Gastroenterol. 88: 1359-1361.

Danesh, J., Youngman, L., Clark, S., Parish, S., Peto, R., and Collins, R. 1999. *Helicobacter pylori* infection and early onset myocardial infarction: case-control and sibling pairs study. B.M.J. 319: 1157-1162.

Dickey, W., Collins, J.S., Watson, R.G., Sloan, J.M., and Porter, K.G. 1993. Secretor status and *Helicobacter pylori* infection are independent risk factors for gastroduodenal disease. Gut. 34: 351-353.

Dominici, P., Bellentani, S., Di Biase, A.R., Saccoccio, G., Le Rose, A., Masutti, F., Viola, L., Balli, F., Tiribelli, C., Grilli, R., Fusillo, M., and Grossi, E. 1999. Familial clustering of *Helicobacter pylori* infection: population based study. B.M.J. 319: 537-40

Dooley, C.P., Cohen, H., Fitzgibbons, P.L., Bauer, M., Appleman, M.D., Perez-Perez, G.I., and Blaser, M.J. 1989. Prevalence of *Helicobacter pylori* infection and histologic gastritis in asymptomatic persons. N. Engl. J. Med 321: 1562-1566.

Drumm, B., Perez-Perez, G.I., Blaser, M.J., and Sherman, P.M. 1990. Intrafamilial clustering of *Helicobacter pylori* infection. N. Engl. J. Med. 322: 359-363.

Elitsur, Y., Adkins, L., Saeed, D., and Neace, C. 1999. *Helicobacter pylori* antibody profile in household members of children with *Helicobacter pylori* infection. J. Clin. Gastroenterol. 29: 178-182.

El-Serag, H.B., and Sonnenberg, A. 1999. Ethnic variations in the occurrence of gastroesophageal cancers. J. Clin. Gastroenterol. 28: 135-139.

Enroth, H., Bjorkholm, B., and Engstrand, L. 1999. Occurence of resistance mutation and clonal expansion in *Helicobacter pylori* multiple-strain infection: a potential risk in clarithromycin-based therapy. Clin. Infect. Dis. 28: 1305-1307.

Fawcett, J.P., Shaw, J.P., Cockburn, M., Brooke, M., and Barbezat, G.O. 1996. Seroprevalence of *Helicobacter pylori* in a birth cohort of 21-year-old New Zealanders. Eur. J. Gastroenterol. Hepatol. 8: 365-369.

Feldman, R.A., Eccersley, A.J., and Hardie, J.M. 1998. Epidemiology of *Helicobacter pylori*: acquisition, transmission, population prevalence and disease-to-infection ratio. Br. Med. Bull. 54: 39-53.

Figueroa, G., Acuna, R., Troncoso, M., Portell, D.P., Toledo, M.S., Albornoz, V., *et al.* 1996. Low *Helicobacter pylori* reinfection rate after triple therapy in Chilean duodenal ulcer patients. Am. J. Gastroenterol. 91: 1395-1399.

Figura, N. 1996. Mouth-to-mouth resuscitation and *Helicobacter pylori* infection. Lancet 347:1342.

Folsom, A.R., Nieto, F.J., Sorlie, P., Chambless, L.E., and Graham, D.Y. 1998. *Helicobacter pylori* seropositivity and coronary heart disease incidence. Atherosclerosis Risk In Communities (ARIC) Study Investigators. Circulation. 98: 845-850.

Forman, D., Webb, P., and Parsonnet, J. 1994. *Helicobacter pylori* and gastric cancer. Lancet. 343: 243-244.

Fraser, A.G., Bickley, J., Owen, R.J., and Pounder, R.E. 1992. DNA fingerprints of *Helicobacter pylori* before and after treatment with omeprazole. Am. J. Clin. Pathol. 45: 1062-1065.

Fujisawa, T., Kumagai, T., Akamatsu, T., Kiyosawa, K., and Matsunaga, Y. 1999. Changes in seroepidemiological pattern of *Helicobacter pylori* and hepatitis A virus over the last 20 years in Japan. Am. J. Gastroenterol. 94: 2094-2099.

Ganga-Zandzou, P.S., Michaud, L., Vincent, P., Husson, M.O., Wizla-Derambure, N., Delassalle, E.M., Turck, D., and Gottrand, F. 1999. Natural outcome of *Helicobacter pylori* infection in asymptomatic children: a two-year follow-up study. Pediatrics. 104: 216-221.

Gasbarrini, G., Pretolani, S., Bonvicini, F., Gatto, M.R., Tonelli, E., Megraud, F, *et al.* 1995. A population based study of *Helicobacter pylori* infection in a European country: the San Marino Study. Relations with gastrointestinal diseases. Gut. 36: 838-844.

Georgopoulos, S.D., Mentis, A.F., Spiliadis, C.A., Tzouvelekis, L.S., Tzelepi, E., Moshopoulos, A., *et al.* 1996. *Helicobacter pylori* infection in spouses of patients with duodenal ulcers and comparison of ribosomal RNA gene patterns. Gut. 39: 634-638.

Gill, H.H., Majmudar, P. 1994. Age related prevalence of *Helicobacter pylori* antibodies in Indian subjects Indian J. Gastroenterology. 13: 92-94.

Goh, K.L., Parasakthi, N., and Ong, K.K. 1996. Prevalence of *Helicobacter pylori* infection in endoscopy and non-endoscopy personnel: results of field survey with serology and 14C-urea breath test. Am. J. Gastroenterol. 91: 268-270.

Gold, B.D., Khanna, B., Huang, L.M., Lee, C-Y., and Banatvala, N. 1997. *Helicobacter pylori* acquisition in infancy after decline of maternal passive immunity. Ped. Research. 41: 641-645.

Goodman, K.J., Correa, P., Tengana Aux, H.J., Ramirez, H., DeLany, J.P., Guerrero Pepinosa, O., *et al.* 1996. *Helicobacter pylori* infection in the Colombian Andes: a population-based study of transmission pathways. Am. J. Epidemiol. 144: 290-299.

Goodman, K.J., and Correa, P. 1995. The transmission of *Helicobacter pylori*. A critical review of the evidence. Int. J. Epidemiol. 24: 875-887.

Goodman, K.J., and Correa, P. 2000. Transmission of *Helicobacter pylori* among siblings. Lancet 355: 358-362.

Goodman, K.J., Redlinger, T., and O'Rourke, K. 1999. The authors respond to Dr. Taylor. Am. J. Epidemiol. 150: 233-234.

Graham, D.Y., Adam, E., Reddy, G.T., Agarwal, J.P., Agarwal, R., Evans, D.J. Jr, Malaty, H.M., and Evans, D.G. 1991. Seroepidemiology of *Helicobacter pylori* infection in India. Comparison of developing and developed countries. Dig. Dis. Sci. 36: 1084-1088.

Granstrom, M,. Tindberg, Y., and Blennow, M. 1997. Seroepidemiology of *Helicobacter pylori* infection in a cohort of children monitored from 6 months to 11 years of age. J. Clin. Microbiol. 35: 468-470.

Gustavsson, S., and Nyren, O. 1989. Time trends in peptic ulcer. Ann. Surg. 210: 704-709.

Hammermeister, I., Janus, G., Schamarowski, F., Rudolf, M., Jacobs, E., Kist, M. 1992. Elevated risk of *Helicobacter pylori* infection in submarine crews. Eur J. Clin. Microbiol. Infect. Dis. 11: 9-14.

Hansen, S., Melby, K.K., Aase, S., Jellum, E., and Vollset, S.E. 1999. *Helicobacter pylori* infection and risk of cardia cancer and non-cardia gastric cancer. A nested case-control study. Scand. J. Gastroenterol. 34: 353-360.

Hegarty, J.P., Dowd, M.T., and Baker, K.H. 1999. Occurrence of *Helicobacter pylori* in surface water in the United States. J. Appl. Microbiol. 87: 697-701.

Hennig, E.E., Trzeciak, L., Regula, J., Butruk, E., and Ostrowski, J. 1999. VacA genotyping directly from gastric biopsy specimens and estimation of mixed *Helicobacter pylori* infections in patients with duodenal ulcer and gastritis. Scand. J. Gastroenterol. 34: 743-749.

Henriksson, K., Uribe, A., Sandstedt, B., and Nord, C.E. 1993. *Helicobacter pylori* infection, ABO blood group, and effect of misoprostol on gastroduodenal mucosa in NSAID-treated patients with rheumatoid arthritis. Dig. Dis. Sci. 38: 1688-1696.

Holcombe, C. 1992. *Helicobacter pylori*: the African enigma. Gut. 33: 429-431.

Hopkins, R.J., Vial, P.A., Ferreccio, C., Ovalle, J., Prado, P., Sotomayor, V., *et al.* 1993. Seroprevalence of *Helicobacter pylori* in Chile: vegetables may serve as one route of transmission. J. Infect. Dis. 168: 222-226.

Hulten, K., Enroth, H., Nystrom, T., and Engstrand, L. 1998. Presence of Helicobacter species DNA in Swedish water. J. Appl. Microbiol. 85: 282-286.

Hulten, K., Han, S.W., Enroth, H., Klein, P.D., Opekun, A.R., Gilman, R.H., *et al.* 1996. *Helicobacter pylori* in the drinking water in Peru. Gastroenterology. 110: 1031-1035.

Isaacson, P.G., 1999. Mucosa-associated lymphoid tissue lymphoma. Semin. Hematol. 36:139-147.

Katelaris, P.H., Tippett, G.H.K., Norbu, P., Lowe, D.G., Brennan, R., and Farthing, M.J.G. 1992. Dyspepsia, *Helicobacter pylori* and peptic ulcer in a randomly selected population in India. Gut. 33: 1462-466.

Keenan, J.I., and Allardyce, R.A. 1999. Iron levels during growth influence the expression of *Helicobacter pylori* virulence factors. 10th International workshop on *Campylobacter*, *Helicobacter* and Related Organisms, Baltimore, USA. 154.

Keenan, J.I., Day, T., Neale, S., Cook, B., Perez-Perez, G., Allardyce, R.A., and Bagshaw, P.F. 2000. A role for the bacterial outer membrane in the pathogenesis of *Helicobacter pylori* infection. F.E.M.S. Microbiol. Lett. 182: 259-264.

Kelly, S.M., Pitcher, M.C., Farmery, S.M., and Gibson, G.R. 1994. Isolation of *Helicobacter pylori* from feces of patients with dyspepsia in the United Kingdom. Gastroenterology. 107: 1671-1674.

Kersulyte, D., Chalkauskas, H., and Berg, D.E. 1999. Emergence of recombinant strains of *Helicobacter pylori* during human infection. Mol. Microbiol. 31: 31-43.

Khuroo, M.S., Mahajan,R., Zargar, S.A., Javid, G., and Munshi, S. 1989. Prevalence of Peptic Ulcer in India: an endoscopic and epidemiological study in Urban Kashmir. Gut. 30: 930-934.

Kikuchi, S., Crabtree, J.E., Forman, D., and Kurosawa, M. 1999. Association between infections with CagA-positive or -negative strains of *Helicobacter pylori* and risk for gastric cancer in young adults. Am. J. Gastroenterol. 94: 3455-3459.

Kikuchi, S., Wada, O., Nakajima, T. *et al.* 1995. Serum anti-*Helicobacter pylori* antibody and gastric carcinoma among young adults. Cancer. 75: 2789-2793.

Klein, P.D., Gilman, R.H., Leon-Barua, R., Diaz, F., Smith, E.O., and Graham, D.Y. 1994. The epidemiology of *Helicobacter pylori* in Peruvian children between 6 and 30 months of age. Am. J. Gastroenterol. 89: 2196-2200.

Klein, P.D., Graham, D.Y., Gaillour, A., Opekun, A.R., and Smith, E.O. 1991. Water source as risk factor for *Helicobacter pylori* infection in Peruvian children. Gastrointestinal Physiology Working Group. Lancet. 337: 1503-1506.

Koenig, W., Rothenbacher, D., Hoffmeister, A., Miller, M., Bode, G., Adler, G., Hombach, V., Marz, W., Pepys, M.B., and Brenner, H. 1999. Infection with *Helicobacter pylori* is not a major independent risk factor for stable coronary heart disease: lack of a role of cytotoxin-associated protein A-positive strains and absence of a systemic inflammatory response. Circulation. 100:23 26-31.

Kononen, E., Saarela, M., Karjalainen, J., Jousimies-Somer, H., Alaluusua, S., and Asikainen, S. 1994. Transmission of oral Prevotella melaninogenica between a mother and her young child. Oral Microbiol Immunol. 9: 310-314..

Kosunen, T.U., Aromaa, A., Knekt, P., Salomaa, A., Rautelin, H., Lohi, P., and Heinonen, O.P. 1997. *Helicobacter* antibodies in 1973 and 1994 in the adult population of Vammala, Finland. Epidemiol. Infect. 119: 29-34.

Kuipers, E.J., Pena, A.S., van Kamp, G., Uyterlinde, A.M., Pals, G., Pels, N.F., *et al.* 1993. Seroconversion for *Helicobacter pylori*. Lancet. 342: 328-331.

Kuipers, E.J., Thijs, J.C., and Festen, H.P.M 1995. The prevalence of *Helicobacter* in peptic ulcer disease. Aliment. Pharmacol. Ther. 9 (Supp. 2): 59-69.

Kumagai, T., Malaty, H.M., Graham, D.Y., Hosogaya, S., Misawa, K., Furihata, K., Ota, H., Sei, C., Tanaka, E., Akamatsu, T., Shimizu, T., Kiyosawa, K., and Katsuyama, T. 1998. Acquisition versus loss of *Helicobacter pylori* infection in Japan: results from an 8-year birth cohort study. J. Infect. Dis. 178: 717-721.

Lee, A., Fox, J.G., Otto, G., Dick, E.H., and Krakowka, S. 1991. Transmission of *Helicobacter* spp. A challenge to the dogma of faecal-oral spread. Epidemiol. Infect. 107: 99-109.

Leon-Barua, R., Berendson-Seminario,R., Recavarren-Arce, S., and Gilman, R.H. 1997. Geographic factors probably modulating alternative pathways in *Helicobacter pylori*-associated gastroduodenal pathology: a hypothesis. Clin. Infect. Dis. 25: 1013-1016.

Leung, W.K., Siu, K.L., Kwok, C.K., Chan, S.Y., Sung, R., and Sung, J.J. 1999. Isolation of *Helicobacter pylori* from vomitus in children and its implication in gastro-oral transmission. Am. J. Gastroenterol. 94: 2881-2884.

Li, Y., and Caufield, P.W. 1995. The fidelity of initial acquisition of mutans streptococci by infants from their mothers. J. Dent. Res. 74: 681-685.

Lin, D.B., Nieh, W.T., Wang, H.M., Hsiao, M.W., Ling, U.P., Changlai, S.P., Ho, M.S., You, S.L., and Chen, C.J. 1999. Seroepidemiology of *Helicobacter pylori* infection among preschool children in Taiwan. Am. J. Trop. Med. Hyg. 61: 554-558.

Lin, S.K., Lambert, J.R., Schembri, M.A., Nicholson, L., and Johnson, I.H. 1998. The prevalence of *Helicobacter pylori* in practising dental staff and dental students. Aust. Dent. J. 43:35-39.

Lin, S.K., Lambert, J.R., Schembri, M.A., Nicholson, L., and Korman, M.G. 1994. *Helicobacter pylori* prevalence in endoscopy and medical staff. J. Gastroenterol. Hepatol. 9: 319-324.

Lindkvist, P., Asrat, D., Nilsson, I., Tsega, E., Olsson, G.L., Wretlind, B., and Giesecke, J. Age at acquisition of *Helicobacter pylori* infection: comparison of a high and a low prevalence country. Scand. J. Infect. Dis. 1996: 28:181-184.

Lindkvist, P., Enquselassie, F., Asrat, D., Muhe, L., Nilsson, I., and Giesecke, J. 1998. Risk factors for infection with *Helicobacter pylori*—a study of children in rural Ethiopia. Scand J. Infect. Dis. 30: 371-376.

Lindkvist, P., Enquselassie, F., Asrat, D., Nilsson, I., Muhe, L., and Giesecke, J. 1999a. *Helicobacter pylori* infection in Ethiopian children: a cohort study. Scand. J. Infect. Dis.31: 475-480.

Lindkvist, P., Tammur, R., Nilsson, I., Wretlind, B., and Giesecke, J. 1999b. *Helicobacter pylori* infection in Estonia – a three year follow-up of children from birth. Gut. (suppl. III), 45: A 45.

Liu, W.Z., Xiao, S.D., Jiang, S.J., Li, R.R., and Pang, Z.J. 1996. Seroprevalence of *Helicobacter pylori* infection in medical staff in Shanghai. Scand. J. Gastroenterol. 31: 749-752.

Luzza, F., Imeneo, M., Maletta, M., Paluccio, G., Giancotti, A., Perticone, F., and Foca, A., Pallone, F. 1997. Seroepidemiology of *Helicobacter pylori* infection and hepatitis A in a rural area: evidence against a common mode of transmission. Gut. 41: 164-168.

Luzza, F., Imeneo, M., Maletta, M., Paluccio, G., Nistico, S., Perticone, F., Foca, A., and Pallone, F. 1998. Suggestion against an oral-oral route of transmission for *Helicobacter pylori* infection: a seroepidemiological study in a rural area. Dig. Dis. Sci. 43: 1488-1492.

Luzza, F., Maletta, M., Imeneo, M., Fabiano, E., Doldo, P., Biancone, L., *et al.* 1995. Evidence against an increased risk of *Helicobacter pylori* infection in dentists: a serological and salivary study. Eur. J. Gastroenterol. Hepatol. 7: 773-776.

Ma, J.L., You, W.C., Gail, M.H., Zhang, L., Blot, W.J., Chang, Y.S., Jiang, J., Liu, W.D., Hu, Y.R., Brown, L.M., Xu, G.W., and Fraumeni, J.F. Jr. 1998. *Helicobacter pylori* infection and mode of transmission in a population at high risk of stomach cancer. Int. J. Epidemiol. 27: 570-573.

Mahalanabis, D., Rahman, M.M., Sarker, S.A., Bardhan, P.K., Hildebrand, P., Beglinger, C., and Gyr. K. 1996. *Helicobacter pylori* infection in the young in Bangladesh: prevalence, socioeconomic and nutritional aspects. Int. J. Epidemiol. 25: 894-898.

Malaty, H.M., Evans, D.J. Jr., Abramovitch, K., Evans, D.G., Graham, D.Y. 1992. *Helicobacter pylori* infection in dental workers: a seroepidemiology study. Am. J. Gastroenterol. 87: 1728-1731.

Malaty, H.M., Kim, J.G., Kim, S.D., and Graham, D.Y. 1996. Prevalence of *Helicobacter pylori* infection in Korean children: inverse relation to socioeconomic status despite a uniformly high prevalence in adults. Am. J. Epidemiol. 143: 257-262.

Matto, J., Saarela, M., von Troil-Linden, B., Kononen, E., Jousimies-Somer, H., Torkko, H., Alaluusua, S., and Asikainen, S. 1996. Distribution and genetic analysis of *oral Prevotella intermedia* and *Prevotella nigrescens*. Oral Microbio. Immunol. 11: 96-102.

McKeown, I., Orr, P., Macdonald, S., Kabani, A., Brown, R., Coghlan, G., Dawood, M., Embil, J., Sargent, M., Smart, G., and Bernstein, C.N. 1999 *Helicobacter pylori* in the Canadian arctic: seroprevalence and detection in community water samples. Am. J. Gastroenterol. 94: 1823-1829.

Megraud, F., Brassens-Rabbe, M.P., Denis, F., Belbouri, A., and Hoa, D.Q. 1989. Seroepidemiology of *Campylobacter pylori* infection in various populations. J. Clin. Microbiol. 27: 1870-1873.

Megraud, F. Transmission of *Helicobacter pylori*: faecal-oral versus oral-oral route. 1995. Aliment. Pharmacol. Ther. 9 Supplement 2: 85-91.

Milman, N., Rosenstock, S., Andersen, L., Jorgensen, T., and Bonnevie, O. 1998. Serum ferritin, hemoglobin, and *Helicobacter pylori* infection: a seroepidemiologic survey comprising 2794 Danish adults. Gastroenterology 115: 268-274.

Mitchell, H.M., Bohane, T., Hawkes, R.A., and Lee, A. 1993. *Helicobacter pylori* infection within families. Zentralbl. Bakteriol. 280: 128-136.

Mitchell, H.M., Li, Y.Y., Hu, P.J., Liu, Q., Chen, M., Du, G.G., *et al*. 1992. Epidemiology of *Helicobacter pylori* in southern China: identification of early childhood as the critical period for acquisition. J. Infect. Dis. 166: 149-153.

Mitchell, H.M., Hu, P., Chi, Y., Chen, M.H., Li, Y.Y., and Hazell, S.L. 1998. A low rate of reinfection following effective therapy against *Helicobacter pylori* in a developing nation (China). Gastroenterology. 114: 256-261.

Morales-Espinosa, R., Castillo-Rojas, G., Gonzalez-Valencia, G., Ponce de Leon, S., Cravioto, A., Atherton, J.C., and Lopez-Vidal, Y. 1999. Colonization of Mexican patients by multiple *Helicobacter pylori* strains with different vacA and cagA genotypes. J. Clin. Microbiol. 37: 3001-3004.

Morris, A.J., Ali, M.R., Nicholson, G.I., Perez-Perez, G.I., and Blaser, M.J. 1991. Long-term follow-up of voluntary ingestion of *Helicobacter pylori*. Ann. Intern. Med. 114: 662-663.

Namavar, F., Roosendaal, R., Kuipers, E.J., de Groot, P., van der Bijl, M.W., Pena, A.S., de and Graaff, J.1995. Presence of *Helicobacter pylori* in the oral cavity, oesophagus, stomach and faeces of patients with gastritis. European J. Clin. Microbiol. Infect. Dis. 14: 234-237.

Nomura, A., Stemmermann, G.N., Chyou, P.H., Perez-Perez, G.I., and Blaser, M.J 1994. *Helicobacter pylori* infection and the risk for duodenal and gastric ulceration. Ann. Intern. Med. 120: 977-981.

Nwokolo, C.U., Bickley. J., Attard, A.R., Owen, R.J., Costas, M., and Fraser, I.A. 1992. Evidence of clonal variants of *Helicobacter pylori* in three generations of a duodenal ulcer disease family. Gut. 33: 1323-1327.

Oshowo, A., Gillam, D., Botha, A., Tunio, M., Holton, J., Boulos, P., and Hobsley, M. 1998. *Helicobacter pylori*: the mouth, stomach, and gut axis. Ann. Periodontol. 3: 276-280.

Owen, R.J., Desai, M., Figura, N., Bayeli, P.F., Di Gregorio, L., Russi, M., *et al*. 1993. Comparisons between degree of histological gastritis and DNA fingerprints, cytotoxicity and adhesivity of *Helicobacter pylori* from different gastric sites. Eur. J. Epidemiol. 9: 315-321.

Ozturk, H., Senocak, M.E., Uzunalimoglu, B., Hascelik, G., Buyukpamukcu, N., and Hicsonmez, A. 1996. *Helicobacter pylori* infection in symptomatic and asymptomatic children: a prospective clinical study. Eur. J. Pediatr. Surg. 6: 265-269.

Parsonnet, J., Shmuely, H., and Haggerty, T. 1999. Fecal and oral shedding of *Helicobacter pylori* from healthy infected adults. J. Amer. Med. Assoc. 282: 2240-2245.

Parsonnet, J. 1995. The incidence of *Helicobacter pylori* infection. Aliment. Pharmacol. Ther. 9 Suppl 2: 45-51.

Peng, H., Ranaldi, R., Diss, T.C., Isaacson, P.G., Bearzi, I., and Pan, L. 1998. High frequency of CagA+ *Helicobacter pylori* infection in high-grade gastric MALT B-cell lymphomas. J. Pathol. 185: 409-412.

Petit, M.D., van Steenbergen, T.J., de Graaff, J., and van der Velden, U. 1993. Transmission of *Actinobacillus actinomycetemcomitans* in families of adult periodontitis patients. J. Periodontal Res. 28: 335-345.

Prasad, S., Mathan, M., *et al.* 1994. Prevalence of *Helicobacter pylori*. J. Gastroenterol. Hepatol. 9: 501-506.

Preus, H.R., Zambon, J.J., Dunford, R.G., and Genco, R.J. 1994. The distribution and transmission of *Actinobacillus actinomycetemcomitans* in families with established adult periodontitis. J. Periodontol. 65: 2-7.

Pytko-Polonczyk, J., Konturek, S.J., Karczewska, E., Bielanski, W., Kaczmarczyck-and Stachowska, A. 1996. Oral cavity as permanent reservoir of *Helicobacter pylori* and potential source of infection. J. Physiol. Pharmacol. 47: 121-129.

Ramirez-Ramos, A., Gilman, R.H., Watanabe, J., Recavarren, A.S., Spira, W., Miyagui, J. *et al.* 1994. *Helicobacter pylori* infection in long-term and short-term Japanese visitors to Peru. Lancet. 344:1017.

Redlinger, T., O'Rourke, K., and Goodman, K.J. 1999. Age distribution of *Helicobacter pylori* seroprevalence among young children in a United States/Mexico border community: evidence for transitory infection. Am. J. Epidemiol. 150: 225-230.

Rehnberg-Laiho, L., Louhija, J., Rautelin, H., Jusufovic, J., Tilvis, R., Miettinen, A., and Kosunen, T.U.1999. *Helicobacter* antibodies in Finnish centenarians. J. Gerontol. A. Biol. Sci. Med. Sci. 54: M400-403.

Rehnberg-Laiho, L., Rautelin, H., Valle, M., and Kosunen, T.U. 1998. Persisting *Helicobacter* antibodies in Finnish children and adolescents between two and twenty years of age. Pediatr. Infect. Dis. J., 17: 796-799.

Replogle, M.L., Glaser, S.L., Hiatt, R.A., and Parsonnet, J. 1995. Biologic sex as a risk factor for *Helicobacter pylori* infection in healthy young adults. Am. J. Epidemiol. 142: 856-863.

Replogle, M.L., Kasumi, W., Ishikawa, K.B., Yang,, S.F., Juji, T., Miki, K., Kabat, G.C., and Parsonnet, J. 1996. Increased risk of *Helicobacter pylori* associated with birth in wartime and post-war Japan. Int. J. Epidemiol.. 25: 210-214.

Rollan,A., Giancaspero, R., Fusterm F., Acevedom C., Figueroam C., Holam K., Schulz, M., and Duarte, I. 2000. The long-term reinfection rate and the course of duodenal ulcer disease after eradication of *Helicobacter pylori* in a developing country. Am. J. Gastroenterol. 95:50-56.

Rothenbacher, D., Bode, G., Berg, G., Gommel, R., Gonser, T., Adler, G., and Brenner, H. 1998. Prevalence and determinants of *Helicobacter pylori* infection in preschool children: a population-based study from Germany. Int. J. Epidemiol. 27: 135-141.

Rothenbacher, D., Bode, G., Berg, G., Knayer, U., Gonser, T., Adler, G., and Brenner, H. 1999. *Helicobacter pylori* among preschool children and their parents: evidence of parent-child transmission. J. Infect. Dis. 179: 398-402.

Rowland, M., Kumar, D., Daly, L., O'Connor, P., Vaughan, D., and Drumm, B. 1999. Low rates of *Helicobacter pylori* reinfection in children. Gastroenterology. 117: 336-341.

Rugge, M., Busatto, G., Cassaro, M., Shiao, Y.H., Russo, V., Leandro, G., Avellini, C., Fabiano, A., Sidoni, A., and Covacci, A. 1999. Patients younger than 40 years with gastric carcinoma: *Helicobacter pylori* genotype and associated gastritis phenotype. Cancer. 85: 2506-2511.

Saarela, M., von Troil-linden, B., Torkko, H., Alaluusua, S., Jousimies-Somer, H., and Asikainen, S. 1993. Transmission of oral bacterial species between spouses. Oral microbiol. Immunol. 8: 349-354.

Schlemper, R.J., Van der Werf, S., Biemond, I., and Lamers, C.B.H.W. 1996. Seroepidemiology of gastritis in Japanese and Dutch male employees with and without ulcer disease. Eur. J. Gastro. Hepatol. 8:33-39.

Schutze, K., Hentschel, E., Dragosics, B., and Hirschl, A.M. 1995. *Helicobacter pylori* reinfection with identical organisms: transmission by the patients' spouses. Gut. 36: 831-833.

Severson RK, and Davis S. 1990. Increasing incidence of primary gastric lymphoma. Cancer. 66: 1283-1287.

Sipponen, P., Kosunen, T.U., Samloff, I.M., Heinonen, O.P., and Siurala, M. 1996. Rate of *Helicobacter pylori* acquisition among Finnish adults: a fifteen year follow-up. Scand. J. Gastroenterol. 31: 229-232.

Smoak, B.L., Kelley, P.W., and Taylor, D.N. 1994. Seroprevalence of *Helicobacter pylori* infections in a cohort of US Army recruits. Am. J. Epidemiol. 139: 513-519.

Sonnenberg, A. 1995. Temporal trends and geographical variations in peptic ulcer disease. Aliment. Pharmacol. Ther. 9 supp 2: 3-12

Strachan, D.P., Mendall, M.A., Carrington, D., Butland, B.K., Yarnell, J.W., Sweetnam, P.M., and Elwood, P.C. 1998. Relation of *Helicobacter pylori* infection to 13-year mortality and incident ischaemic heart disease in the Caerphilly prospective heart disease study. Circulation. 98: 1286-1290.

Su, Y-C., Wang, W-M., Chen, L-T., Chiang, W., Chen, C-Y., Lu, S-N. *et al*. 1996. High seroprevalence of IgG against *Helicobacter pylori* among endoscopists in Taiwan. Dig. Dis. Sci. 41: 1571-1576.

Suerbaum, S., Maynard Smith, J., Bapumia, K., Morelli, G., Smith, N.H., Kunstmann, E. *et al*. 1998. Free recombination within *Helicobacter pylori*. Proc. Natl. Acad. Sci. USA 95: 12619-12624.

Susser, M., and Stein, Z. 1962. Civilisation and peptic ulcer. Lancet: 155-158.

Suzuki, J., Muraoka, H., Kobayasi, I., Fujita, T., and Mine, T. 1999. Rare incidence of interspousal transmission of *Helicobacter pylori* in asymptomatic individuals in Japan. J. Clin. Microbiol. 37: 4174-4176.

Teh, B.H., Lin, J.T., Pan, W.H., Lin, S.H., Wang, L.Y., Lee, T.K., *et al*. 1994. Seroprevalence and associated risk factors of *Helicobacter pylori* infection in Taiwan. Anticancer Res. 14: 1389-1392.

Thomas, J.E., Austin, S., Dale, A., McClean, P., Harding, M., Coward, W.A., *et al*. 1993. Protection by human milk IgA against *Helicobacter pylori* infection in infancy. Lancet. 342: 121.

Thomas, J.E., Gibson, G.R., Darboe, M.K., Dale, A., and Weaver, L.T. 1992. Isolation of *Helicobacter pylori* from human faeces. Lancet. 340: 1194-1195.

Tindberg, Y., Blennow, M., and Granstrom, M. 1999. Clinical symptoms and social factors in a cohort of children spontaneously clearing *Helicobacter pylori* infection. Acta Paediatr. 88: 631-635.

Torres, J., Leal-Herrera, Y., Perez-Perez, G., Gomez, A., Camorlinga-Ponce, M., Cedillo-Rivera, R., Tapia-Conyer, R., and Munoz, O. 1998. A community-based seroepidemiologic study of *Helicobacter pylori* infection in Mexico. J. Infect. Dis. 178: 1089-1094.

Tuite-McDonnell, M., Griffen, A.L., Moeschberger, M.L., Dalton, R.E., Fuerst, P.A., and Leys, E.J. 1997. Concordance of *Porphyromonas gingivalis* colonization in families. J. Clin. Microbiol. 35: 455-461.

Tytgat, G.N.J..1995. Endoscopic transmission of *Helicobacter pylori*. Aliment. Pharmacol. Ther. 9 (Suppl.2): 105-110.

Umlauft, F., Keeffe, E.B., Offner, F., Weiss, G., Feichtinger, H., Lehmann, E., Kilga-Nogler, S., Schwab, G., Propst, A., Grussnewald, K., and Judmaier, G. 1996. *Helicobacter pylori* infection and blood group antigens: lack of clinical association. Am. J. Gastroenterol. 91: 2135-2138.

Vaira, D., Miglioli, M., Mule, P., Holton, J., Menegatti, M. *et al.* 1994. Prevalence of peptic ulcer in *Helicobacter pylori* positive blood donors. Gut. 35: 309-312.

van der Ende, A., Rauws, E.A., Feller, M., Mulder, C.J., Tytgat, G.N., and Dankert, J. 1996. Heterogeneous *Helicobacter pylori* isolates from members of a family with a history of peptic ulcer disease. Gastroenterology. 111: 638-647.

van Steenbergen, T.J., Petit, M.D., Scholte, L.H., van der Velden, U., and de Graaff, J. 1993. Transmission of *Porphyromonas gingivalis* between spouses. J. Clin. Periodont. 20: 340-345.

Veldhuyzen van Zanten, S.J. 1995. Do socio-economic status, marital status and occupation influence the prevalence of *Helicobacter pylori* infection? Aliment. Pharmacol. Ther. 9 Suppl 2: 41-44.

Vicari, J.J., Peek, R.M., Falk, G.W., Goldblum, J.R., Easley, K.A., Schnell, J., Perez-Perez, G.I., Halter, S.A., Rice, T.W., Blaser, M.J., and Richter, J.E. 1998. The seroprevalence of cagA-positive *Helicobacter pylori* strains in the spectrum of gastroesophageal reflux disease. Gastroenterology. 115: 50-57.

Vincent, P., Gottrand, F., Pernes, P., Husson, M.O., Lecomte-Houcke, M., Turck, D., *et al.* 1994. High prevalence of *Helicobacter pylori* infection in cohabiting children. Epidemiology of a cluster, with special emphasis on molecular typing. Gut. 35: 313-316.

Wald, N.J., Law, M.R., Morris, J.K., and Bagnall, A.M. 1997. *Helicobacter pylori* infection and mortality from ischaemic heart disease: negative result from a large, prospective study. B.M.J. 315: 1199-1201.

Wang, J.T., Sheu, J.C., Lin, J.T., Wang, T.H., and Wu, M.S. 1993. Direct DNA amplification and restriction pattern analysis of *Helicobacter pylori* in patients with duodenal ulcer and their families. J. Infect. Dis. 168: 1544-1548.

Webb, P.M., Knight, T., Newell, D.G., Elder, J.B., and Forman, D. 1996. *Helicobacter pylori* transmission: evidence from a comparison with hepatitis A virus. Eur. J. Gastroenterol. Hepatol.. 8: 439-441.

Webberley, M. J., Webberley, J.M., Newell, D.G., Lowe, P., and Melikian, V. 1992. Seroepidemiology of *Helicobacter pylori* infection in vegans and meat-eaters. Epidemiol. Infect. 108: 457-462.

Wotherspoon, A.C.. 1998. *Helicobacter pylori* infection and gastric lymphoma. Br. Med. Bull. 54: 79-85.

Xia, H.H., and Talley, N.J. 1997. Natural acquisition and spontaneous elimination of *Helicobacter pylori* infection: clinical implications. Am. J. Gastroenterol. 92:1780-1787.

Xia, H.X., Windle, H.J., Marshall, D.G., Smyth, C.J., Keane, C.T., and O'Morain, C.A. 1995. Recrudescence of *Helicobacter pylori* after apparently successful eradication: novel application of randomly amplified polymorphic DNA fingerprinting. Gut. 37:30-34.

Young, K.A., Akyon, Y., Rampton, D.S., Barton, S.G.R.G., Allaker, R.P., Hardie. J.M., and Feldman, R.A. 2000. Quantitative culture of *Helicobacter pylori* from gastric juice: the potential for transmission. J. Med. Microbiol. 49: 343-347.

From: *Helicobacter pylori: Molecular and Cellular Biology*
ISBN 1-898486-25-5 © 2001 Horizon Scientific Press, Wymondham, UK.

4

Antigastric Autoimmunity and Pathology in *Helicobacter pylori* Gastritis

Gerhard Faller[*] and Thomas Kirchner

Abstract

Chronic and active *Helicobacter pylori* gastritis represents the common platform for several gastroduodenal complications, such as duodenal or gastric ulcer, gastric mucosa atrophy, gastric carcinoma or gastric MALT-type lymphoma. These clinical outcomes are reflected by the histological pattern and topographical distribution of the inflammatory infiltrates in the gastric mucosa. Classification of chronic gastritis describes the different patterns of mucosa alterations and can also provide information on the probable clinical course and the aetiology of the disease. So far, autoimmune gastritis and *H. pylori* gastritis were considered strictly distinct. However, autoimmune gastritis and corpus predominant, atrophic *H. pylori* gastritis show many histological similarities. Furthermore, recent immunological studies revealed a close pathogenic relationship between autoimmune gastritis and one subset of *H. pylori* gastritis. These new findings are summarized in this chapter and lead to the speculation whether the strict distinction between autoimmune gastritis and *H. pylori* gastritis should be maintained in the future.

Topography and Classification of Chronic Gastritis

The most obvious histopathological consequence of the gastric mucosa colonisation by *Helicobacter pylori* is the acquisition of a mucosa associated lymphoid tissue (MALT) which is accompanied by the accumulation of polymorphonuclear leucocytes. The development of such chronic active *H. pylori* induced gastritis is the common platform for the various clinical outcomes of H. pylori infection in different subjects. In most individuals the result is asymptomatic. However, complicated forms of disease also occur and can progress to duodenal ulcer, gastric ulcer, gastric mucosa atrophy, gastric carcinoma or gastric lymphoma (Warren *et al.,* 1983; Parsonnet *et al.,* 1991; Wotherspoon *et al.,* 1991; Kuipers *et al.,* 1995; Kuipers *et al.,* 1995). An important task for histopathologists is to characterise distinct histological phenotypes and topographical patterns of chronic active *H. pylori* gastritis as prognostic indicators for the clinical outcome (Kirchner *et al.,* 1998).

Studies on the topographical pattern of gastritis have shown that antrum predominant gastritis, with only minor pathological alterations in the corpus mucosa increases the risk of duodenal ulcerations. In contrast, corpus predominant gastritis and particularly atrophic cor-

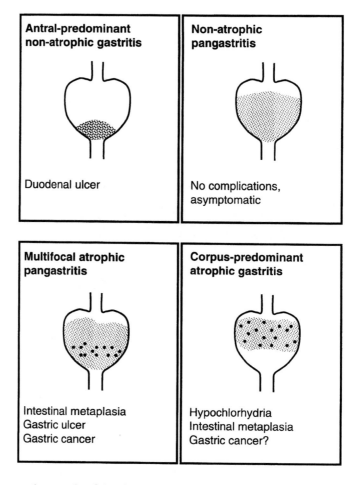

Figure 1. Phenotypes and topography of chronic *H. pylori* gastritis and their clinical relevance.

pus gastritis, i.e. gastritis with a loss of acid producing glands in the corpus mucosa, is a characteristic of those patients with an increased risk of gastric ulcer and gastric cancer (Dixon 1991; Meining *et al.,* 1997; Meining *et al.,* 1998; Schultze *et al.,* 1998). Finally, non-atrophic pangastritis is thought to have only little tendency to progress to gastric or duodenal complications.

Based on the proposals by Correa (1995) and increasing knowledge about infections with *H. pylori*, four clinically relevant phenotypes of *H. pylori* gastritis can be differentiated (Rubin 1997) (Figure 1):

1. Predominantly antral, non-atrophic *H. pylori* gastritis, which predisposes to hyperacidity and the development of duodenal ulcer.
2. Non-atrophic *H. pylori* pangastritis, generally low grade, with no apparent change in acid secretion, no tendency to develop ulcers, atrophy or intestinal atrophy, which seems to persist without complications or symptoms in the majority of *H. pylori* infected patients.
3. Multifocal atrophic *H. pylori* pangastritis, accompanied by a risk of gastric ulcer with residual acid secretion and a risk of gastric carcinoma with increasing hypoacidity.
4. Atrophic *H. pylori* gastritis located predominantly in the corpus, with the antrum mostly unaffected and an increased risk of gastric cancer.

The current updated Sydney classification of chronic gastritis (Dixon *et al.*, 1996) distinguishes gastritis hierarchically based on the aetiological cause, namely *H. pylori* gastritis or autoimmune gastritis. Autoimmune gastritis typically shows high titers of serum antibodies against the gastric H,K-ATPase in the parietal cells of the corpus mucosa (Toh *et al.*, 1997). However, autoimmune gastritis, and atrophic, corpus predominant *H. pylori* gastritis, show significant similarities. Autoimmune gastritis and its presumed precursor lesion, active autoimmune gastritis, are histologically characterised by severe inflammatory infiltrates, gland destruction, mucosa atrophy and hyperplasia of endocrine cells in the corpus mucosa, while the antrum shows only minor morphological alterations (Stolte *et al.*, 1992; Eidt *et al.*, 1996). Atrophic, corpus predominant *H. pylori* gastritis with gland destruction develops in up to 30% of all *H. pylori* infected patients over a period of 10 to 32 years (Kuipers *et al.*, 1995; Valle *et al.*, 1996; Maaroos *et al.*, 1998). Hyperplasia of endocrine cells in the body mucosa can also be observed in atrophic *H. pylori* gastritis (Maaroos *et al.*, 1998). Thus, it is doubtful whether a clear etiological classification into autoimmune or *H. pylori* induced gastritis is feasible based only on morphology.

In addition to these morphological resemblances, there is increasing evidence for a close etiological relationship between *H. pylori* gastritis and autoimmune gastritis and that autoimmune host factors contribute to the type and outcome of *H. pylori* gastritis.

Evidence for an Aetiological Relationship between Autoimmune Gastritis and *H. pylori* Gastritis

Several studies indicate that a certain proportion of patients with autoimmune gastritis currently are or have been infected by *H. pylori*. Antibodies against *H. pylori* were detected in up to 83% of patients with autoimmune gastritis, although the colonisation of the gastric mucosa with *H. pylori* was rarely detected (Faisal *et al.*, 1990; Cariani *et al.*, 1991; Karnes *et al.*, 1991; Ma *et al.*, 1994; Eidt *et al.*, 1996). On the other hand, *H. pylori* infection and the development of antiparietal cell antibodies were significantly associated with increasing age (Uibo *et al.*, 1995) and a proportion of *H. pylori* infected patients observed for 32 years developed anti-parietal cell antibodies as well as chronic atrophic gastritis and eventually became *H. pylori* negative (Valle *et al.*, 1996). The sera of up to 84% of *H. pylori* gastritis patients also contained antigastric autoantibodies (Negrini *et al.*, 1991).

Based on these findings, we attempted to determine the prevalence, specificity and clinical relevance of antigastric autoimmunity in *H. pylori* gastritis using our own analyses.

Detection of Antigastric Antibodies in *H. pylori* Gastritis

Patient sera were tested on normal human gastric mucosa by immunohistochemical methods which permitted reliable determination of the presence of antigastric autoantibodies. It was not only possible to demonstrate the presence of antibodies, but also to differentiate between two binding patterns of these autoantibodies (Faller *et al.*, 1996):
- antiluminal autoantibodies, which marked the luminal cell membrane of foveolar epithelia of antrum and corpus and
- anticanalicular autoantibodies, which bound to the canaliculi of the parietal cells in gastric corpus mucosa (Figure 2).

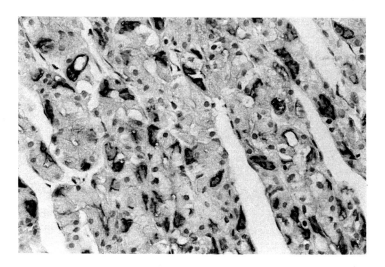

Figure 2. Detection of anticanalicular serum-autoantibodies by immunostaining of canaliculi within human parietal cells. A colour version of this figure is printed on page 323 of this volume.

The prevalence of antigastric autoimmunity is high and both types of autoantibodies, in particular anticanalicular autoantibodies, are significantly associated with *H. pylori* gastritis. According to a prospective study of 126 patients with epigastric discomfort and diagnostic gastroscopy, 50% of *H. pylori* infected patients exhibited antigastric autoantibodies. Of these patients, one third showed antiluminal antibodies, one third possessed anticanalicular antibodies and one third showed both types of autoantibody (Faller *et al.*, 1997). Subsequent analyses confirmed the consistency of this phenomenon (Negrini *et al.*, 1996; Claeys *et al.*, 1998; Faller *et al.*, 1998; Faller *et al.*, 1999, 2000; Vorobjova *et al.*, 2000).

Clinical Relevance of Antigastric Autoimmunity in *H. pylori* Gastritis

The high prevalence and particular specificity of autoreactivity in *H. pylori* gastritis poses questions regarding its clinical relevance. In several studies, the clinical and histopathological parameters of chronic gastritis were evaluated and compared to the presence or absence of antigastric autoantibodies. The most interesting and consistent findings in Hp gastritis were observed with anticanalicular autoantibodies, which correlated with:
- a higher grade corpus gastritis (Faller *et al.*, 1997),
- an increased rate of apoptosis in corpus glands (Steininger *et al.*, 1998),
- morphological atrophy of gastric mucosa (Faller *et al.*, 1996, 1997; Negrini *et al.*, 1996; Claeys *et al.*, 1998; Vorobjova *et al.*, 2000) ,
- low pepsinogen I/II ratio as a sign of functional atrophy (Faller *et al.*, 1997),
- reduced basal secretion of gastric acid in patients with non ulcer dyspepsia and in duodenal ulcer patients (Faller *et al.*, 1998; Parente *et al.*, 1999) and
- raised fasting levels of gastrin (Faller *et al.*, 1997, 1998).

It is obvious that these parameters are not independent of each other. However the results show that the presence of anticanalicular autoantibodies in Hp gastritis correlates with atrophic body gastritis and that this subtype of gastritis therefore resembles classic autoimmune gastritis.

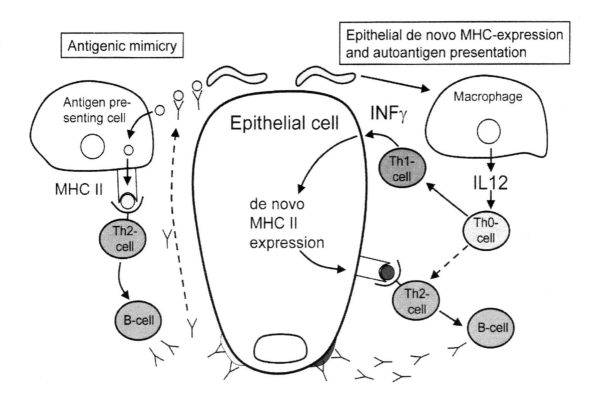

Figure 3. Models for the pathogenesis of antigastric autoimmunity in *H. pylori* infection. A colour version of this figure is printed on page 323 of this volume.

Models for Autoimmune Pathogenesis in *H. pylori* Gastritis

Two basic models exist which may account for antigastric autoimmunity in *H. pylori* gastritis (Figure 3) (Appelmelk *et al.*, 1997; Appelmelk *et al.*, 1998; Kirchner *et al.*, 1998; Faller *et al.*, 2000). In the first model, it is assumed that there is molecular mimicry between *H. pylori* and the host and that antibacterial antibodies induced in the course of the infection cross-react with antigens of the host gastric mucosa. This pathogenic model was proposed because Lewis X and Y blood group antigens are present in the lipopolysaccharides of most *H. pylori* strains and also on gastric epithelial cells and in the highly glycosylated β-subunit of gastric H,K-ATPase of the parietal cells (Negrini *et al.*, 1991; Appelmelk *et al.*, 1996; Aspinall *et al.*, 1996; Negrini *et al.*, 1996). Indeed it has been demonstrated that molecular mimicry plays a role in the development of *H. pylori*-associated antigastric autoimmunity in different animal models (Negrini *et al.*, 1991; Appelmelk *et al.*, 1998; Guruge *et al.*, 1998). However, this does not seem to be the case in humans, where the immunological reaction against Lewis X and Lewis Y is independent from infection by *H. pylori* (Amano *et al.*, 1997; Claeys *et al.*, 1998). Moreover, human antigastric autoantibodies are not absorbed onto Lewis X or Y positive *H. pylori* strains (Faller *et al.*, 1998). Finally, analyses of the fine specificity of antigastric autoantibodies (see below) argue against the mimicry hypothesis (Claeys *et al.*, 1998).

The second model proposes that the lymphoid infiltrate in *H. pylori* gastritis gives rise via cytokines to the *de novo* expression of MHC class II molecules on gastric epithelial cells.

This permits the immunogenic presentation of autoantigens and the activation of potentially autoreactive T cells during the course of gastritis.

Immunogenic presentation may be caused by Th1 type lymphocytes which are generated in chronic *H. pylori* gastritis through the release of interleukin 12. Th1 cells are predominant over Th2 cells in chronic *H. pylori* gastritis and, via interferon γ secretion, induce MHC class II expression in gastric epithelium (Engstrand *et al.*, 1989; Kirchner *et al.*, 1990; Haeberle *et al.*, 1997; Bamford *et al.*, 1998; Sommer *et al.*, 1998). At the same time, interferon γ gives rise to coexpression of B7-1 (CD80) and B7-2 (CD86) as costimulatory factors in gastric mucosa (Ye *et al.*, 1997). Thus, the essential requirements for immunogenic presentation of autoantigens by gastric epithelium cells are fulfilled, enabling the initiation and stimulation of autoreactive T and B cells. It is highly interesting that the pathogenic process for the initiation of antigastric autoimmunity in *H. pylori* gastritis requires time after infection. In a twelve year follow-up study, a significant increase of anticanalicular autoantibodies was observed in subjects who were continuously infected with *H. pylori* (Vorobjova *et al.*, 2000). Additionally, three independent groups found that the prevalence of this type of autoantibodies is significantly lower in the pediatric population, where the duration of *H. pylori* infection is much shorter (Ierardi *et al.*, 1998; Faller *et al.*, 1999; Kolho *et al.*, 2000).

Fine Specificity of Anticanalicular Antibodies

More evidence for the close relationship between *H. pylori* gastritis and antigastric autoantibodies/autoimmune gastritis came from analysis of the fine specificity of anticanalicular autoantibodies. Immunohistochemical studies revealed anticanalicular autoantibodies in 100% of patients with classical autoimmune gastritis and in about 40% of cases with *H. pylori* gastritis. In immunoprecipitation experiments using the recombinant murine gastric H,K-ATPase, 50% of *H. pylori* infected patients with anticanalicular autoantibodies also had autoantibodies against the α and/or β subunits of this enzyme, which are known to be the key targets in classical autoimmune gastritis (Claeys *et al.*, 1998). A significant association between anticanalicular/anti-H,K-ATPase autoantibodies and gastric mucosa atrophy was again confirmed in these studies. Thus, the anti-H,K-ATPase reactivity is not only a marker for classic autoimmune gastritis but also for a certain proportion of atrophic *H. pylori* gastritis.

Conclusions for the Assesment of Gastritis and Tasks for Future Research

The detection of anticanalicular autoimmunity in *H. pylori* gastritis could represent a tool which helps to differentiate between the various phenotypes of gastritis. It might be possible to distinguish between two of the complicated progressions of *H. pylori* gastritis, depending on the presence or absence of antigastric autoimmunity:
- Non-atrophic, predominantly antral gastritis with the risk of developing duodenal ulcers in *H. pylori* infections without anticanalicular autoantibodies and
- Multifocal atrophic pangastritis or predominantly corpus gastritis with the risk of gastric ulcer or gastric carcinoma in *H. pylori* infections with anticanalicular autoantibodies.

However, the precise role of antigastric autoantibodies for the development of atrophy needs to be evaluated further in prospective follow-up studies. Also the importance of a decrease of antigastric autoantibodies after successful eradication for reducing gastric atrophy needs to be examined (Faller *et al.,* 1999).

Extensive research should also be done on the role of T-cells in chronic gastritis, particularly in human autoimmune gastritis, atrophic *H. pylori* gastritis and/or in *H. pylori* gastritis with antigastric autoantibodies. It is quite possible that different phenotypic T-cell populations accumulate in different types of chronic gastritis and play a pivotal role for the various clinical complications of chronic gastritis (Fox *et al.,* 2000). Autoreactive T-cells are expected to occur in the various clinical entities just mentioned. The loss of glands in the mucosa, i.e. the development of gastric atrophy, could be caused by a cellular attack against epithelial cells rather than through autoantibodies. Consequently, antigastric autoantibodies might be only a marker for ongoing antigastric autoimmunity carried out primarily by autoreactive T cells. The construction and characterisation of human monoclonal antigastric autoantibodies might enable the comparison of humoral antigastric autoimmunity in autoimmune gastritis with autoantibodies in *H. pylori* gastritis on a molecular level and could then contribute to a better understanding of the relationship between the various histological appearances of chronic gastritis.

References

Amano, K., Hayashi, S., Kubota, T., Fujii, N., and Yokota, S. 1997. Reactivities of Lewis antigen monoclonal antibodies with the lipopolysaccharides of *Helicobacter pylori* strains isolated from patients with gastroduodenal diseases in Japan. Clin Diagn Lab Immunol. 4: 540-544.

Appelmelk, B.J., Faller, G., Claeys, D., Kirchner, T., and Vandenbroucke-Grauls, C.M. 1998. Bugs on trial: the case of *Helicobacter pylori* and autoimmunity. Immunol. Today. 19: 296-299.

Appelmelk, B.J., Negrini, R., Moran, A.P., and Kuipers, E.J. 1997. Molecular mimicry between *Helicobacter pylori* and the host. Trends Microbiol. 5: 70-73.

Appelmelk, B.J., Simoons-Smit, I., Negrini, R., Moran, A.P., Aspinall, G.O., Forte, J.G., De Vries, T., Quan, H., Verboom, T., Maaskant, J.J., Ghiara, P., Kuipers, E.J., Bloemena, E., Tadema, T.M., Townsend, R.R., Tyagarajan, K., Crothers, J.M., Jr., Monteiro, M.A., Savio, A., and De Graaff, J. 1996. Potential role of molecular mimicry between *Helicobacter pylori* lipopolysaccharide and host Lewis blood group antigens in autoimmunity. Infect. Immun. 64: 2031-2040.

Appelmelk, B.J., Straver, S., Verboom, T., Kuipers, E.J., Claeys, D., Faller, G., Kirchner, T., Negrini, R., Krakowka, S., De Pont, J.J.H.H.M., Simoons-Smit, I., Maaskant, J.J., and Vandenbroucke-Grauls, C.M.J.E. 1998. Molecular mimicry between *Helicobacter pylori* and the host. In: *Helicobacter pylori*. Basic mechanisms to clinical cure 1998. R.H. Hunt and G.N.J. Tytgat, eds. Kluwer Academic Publishers, Dordrecht. p. 33 - 42.

Aspinall, G.O., and Monteiro, M.A. 1996. Lipopolysaccharides of *Helicobacter pylori* strains P466 and MO19: structures of the O antigen and core oligosaccharide regions. Biochemistry 35: 2498-2504.

Bamford, K.B., Fan, X., Crowe, S.E., Leary, J.F., Gourley, W.K., Luthra, G.K., Brooks, E.G., Graham, D.Y., Reyes, V.E., and Ernst, P.B. 1998. Lymphocytes in the human gastric mucosa during *Helicobacter pylori* have a T helper cell 1 phenotype. Gastroenterology 114: 482-492.

Cariani, G., Bonora, G., Vandelli, A., Mazzoleni, G., and Fontana, G. 1991. *Helicobacter pylori* in autoimmune gastritis. Gastroenterology 101: 759.

Claeys, D., Faller, G., Appelmelk, B.J., Negrini, R., and Kirchner, T. 1998. The gastric H+,K+-ATPase is a major autoantigen in chronic *Helicobacter pylori* gastritis with body mucosa atrophy. Gastroenterology 115: 340-347.

Correa, P. 1995. Chronic gastritis. In: Gastrointestinal and oesophageal pathology. R. Whitehead, eds. Churchill Livingstone, New York. p. 485-502.

Dixon, M.F. 1991. *Helicobacter pylori* and peptic ulceration: histopathological aspects. J. Gastroenterol. Hepatol. 6: 125-130.

Dixon, M.F., Genta, R.M., Yardley, J.H., and Correa, P. 1996. Classification and grading of gastritis. The updated Sydney System. International Workshop on the Histopathology of Gastritis, Houston 1994. Am J Surg Pathol 20: 1161-1181.

Eidt, S., Oberhuber, G., Schneider, A., and Stolte, M. 1996. The histopathological spectrum of type A gastritis. Pathol Res Pract 192: 101-106.

Engstrand, L., Scheynius, A., Pahlson, C., Grimelius, L., Schwan, A., and Gustavsson, S. 1989. Association of Campylobacter pylori with induced expression of class II transplantation antigens on gastric epithelial cells. Infect. Immun. 57: 827-832.

Faisal, M.A., Russell, R.M., Samloff, I.M., and Holt, P.R. 1990. *Helicobacter pylori* infection and atrophic gastritis in the elderly. Gastroenterology 99: 1543-1544.

Faller, G., and Kirchner, T. 2000. Role of antigastric autoantibodies in chronic *Helicobacter pylori* infection. Microsc. Res. Tech. 48: 321-326.

Faller, G., Ruff, S., Reiche, N., Hochberger, J., Hahn, E.G., and Kirchner, T. 2000. Mucosal production of antigastric autoantibodies in *Helicobacter pylori* gastritis. Helicobacter. 5: 123-134.

Faller, G., Steininger, H., Appelmelk, B., and Kirchner, T. 1998. Evidence of novel pathogenic pathways for the formation of antigastric autoantibodies in *Helicobacter pylori* gastritis. J. Clin. Pathol. 51: 244-245.

Faller, G., Steininger, H., Eck, M., Hensen, J., Hann, E.G., and Kirchner, T. 1996. Antigastric autoantibodies in *Helicobacter pylori* gastritis: prevalence, in-situ binding sites and clues for clinical relevance. Virchows Arch. 427: 483-486.

Faller, G., Steininger, H., Keller, K.M., Kühlwein, D., and Kirchner, T. 1999. *Helicobacter pylori* gastritis and antigastric autoantibodies in children and adolescents. Pathol. Res. Pract. 195: 302.

Faller, G., Steininger, H., Kranzlein, J., Maul, H., Kerkau, T., Hensen, J., Hahn, E.G., and Kirchner, T. 1997. Antigastric autoantibodies in *Helicobacter pylori* infection: implications of histological and clinical parameters of gastritis. Gut 41: 619-623.

Faller, G., Winter, M., Steininger, H., Konturek, P., Konturek, S.J., and Kirchner, T. 1998. Antigastric autoantibodies and gastric secretory function in *Helicobacter pylori*-infected patients with duodenal ulcer and non-ulcer dyspepsia. Scand. J. Gastroenterol 33: 276-282.

Faller, G., Winter, M., Steininger, H., Lehn, N., Meining, A., Bayerdorffer, E., and Kirchner, T. 1999. Decrease of antigastric autoantibodies in *Helicobacter pylori* gastritis after cure of infection. Pathol. Res. Pract. 195: 243-246.

Fox, J.G., Beck, P., Dangler, C.A., Whary, M.T., Wang, T.C., Shi, H.N., and Nagler-Anderson, C. 2000. Concurrent enteric helminth infection modulates inflammation and gastric immune responses and reduces helicobacter-induced gastric atrophy [In Process Citation]. Nature Med. 6: 536-542.

Guruge, J.L., Falk, P.G., Lorenz, R.G., Dans, M., Wirth, H.P., Blaser, M.J., Berg, D.E., and Gordon, J.I. 1998. Epithelial attachment alters the outcome of *Helicobacter pylori* infection. Proc. Natl. Acad. Sci. USA. 95: 3925-3930.

Haeberle, H.A., Kubin, M., Bamford, K.B., Garofalo, R., Graham, D.Y., El-Zaatari, F., Karttunen, R., Crowe, S.E., Reyes, V.E., and Ernst, P.B. 1997. Differential stimulation of interleukin-12 (IL-12) and IL-10 by live and killed *Helicobacter pylori in vitro* and association of IL-12 production with gamma interferon-producing T cells in the human gastric mucosa. Infect. Immun. 65: 4229-4235.

Ierardi, E., Francavilla, R., Balzano, T., Negrini, R., and Francavilla, A. 1998. Autoantibodies reacting with gastric antigens in *Helicobacter pylori* associated body gastritis of dyspeptic children. Ital. J. Gastroenterol. Hepatol. 30: 478-480.

Karnes, W.J., Samloff, I.M., Siurala, M., Kekki, M., Sipponen, P., Kim, S.W., and Walsh, J.H. 1991. Positive serum antibody and negative tissue staining for *Helicobacter pylori* in subjects with atrophic body gastritis [see comments]. Gastroenterology 101: 167-174.

Kirchner, T., Faller, G., and Price, A. 1998. The year in *Helicobacter pylori* 1998: Pathology and autoimmunity. Curr. Opin. Gastroenterol. 14 (suppl 1): S35-S39.

Kirchner, T., Melber, A., Fischbach, W., Heilmann, K.L., and Müller-Hermelink, H.K. 1990. Immunohistological patterns of the local immune response in *Helicobacter pylori* gastritis. In: *Helicobacter pylori*, gastritis and peptic ulcer. P. Malfertheiner and H. Ditschuneit, eds. Springer-Verlag, Berlin, Heidelberg. p. 213-222.

Kolho, K.L., Jusufovic, J., Miettinen, A., Savilahti, E., and Rautelin, H. 2000. Parietal cell antibodies and *Helicobacter pylori* in children. J Pediatr Gastroenterol Nutr 30: 265-268.

Kuipers, E.J., Thijs, J.C., and Festen, H.P. 1995. The prevalence of *Helicobacter pylori* in peptic ulcer disease. Aliment Pharmacol Ther 2: 59-69.

Kuipers, E.J., Uyterlinde, A.M., Pena, A.S., Roosendaal, R., Pals, G., Nelis, G.F., Festen, H.P., and Meuwissen, S.G. 1995. Long-term sequelae of *Helicobacter pylori* gastritis. Lancet 345: 1525-1528.

Ma, J.Y., Borch, K., Sjostrand, S.E., Janzon, L., and Mardh, S. 1994. Positive correlation between H,K-adenosine triphosphatase autoantibodies and *Helicobacter pylori* antibodies in patients with pernicious anemia. Scand J Gastroenterol 29: 961-965.

Maaroos, H.I., Havu, N., and Sipponen, P. 1998. Follow-up of *Helicobacter pylori* positive gastritis and argyrophil cells pattern during the natural course of gastric ulcer [see comments]. Helicobacter 3: 39-44.

Meining, A., Bayerdorffer, E., Muller, P., Miehlke, S., Lehn, N., Holzel, D., Hatz, R., and Stolte, M. 1998. Gastric carcinoma risk index in patients infected with *Helicobacter pylori* [see comments]. Virchows Arch 432: 311-314.

Meining, A., Stolte, M., Hatz, R., Lehn, N., Miehlke, S., Morgner, A., and Bayerdorffer, E. 1997. Differing degree and distribution of gastritis in *Helicobacter pylori*- associated diseases. Virchows Arch 431: 11-15.

Negrini, R., Lisato, L., Zanella, I., Cavazzini, L., Gullini, S., Villanacci, V., Poiesi, C., Albertini, A., and Ghielmi, S. 1991. *Helicobacter pylori* infection induces antibodies cross-reacting with human gastric mucosa. Gastroenterology 101: 437-445.

Negrini, R., Savio, A., Poiesi, C., Appelmelk, B.J., Buffoli, F., Paterlini, A., Cesari, P., Graffeo, M., Vaira, D., and Franzin, G. 1996. Antigenic mimicry between *Helicobacter pylori* and gastric mucosa in the pathogenesis of body atrophic gastritis. Gastroenterology 111: 655-665.

Parente, F., Negrini, R., Imbesi, V., Maconi, G., Cucino, C., and Bianchi Porro, G. 1999. The presence of gastric autoantibodies impairs gastric secretory function in patients with *H. pylori*-positive duodenal ulcer. Gut 45 Suppl3: A40.

Parsonnet, J., Friedman, G.D., Vandersteen, D.P., Chang, Y., Vogelman, J.H., Orentreich, H., and Sibley, R. 1991. *Helicobacter pylori* infection and the risk of gastric carcinoma. N Engl J Med 325: 1127-1131.

Rubin, C.E. 1997. Are there three types of *Helicobacter pylori* gastritis? Gastroenterology 112: 2108-2110.

Schultze, V., Hackelsberger, A., Gunther, T., Miehlke, S., Roessner, A., and Malfertheiner, P. 1998. Differing patterns of *Helicobacter pylori* gastritis in patients with duodenal, pre-pyloric, and gastric ulcer disease. Scand J Gastroenterol 33: 137-142.

Sommer, F., Faller, G., Konturek, P., Kirchner, T., Hahn, E.G., Zeus, J., M, R.l., and Lohoff, M. 1998. Antrum- and Corpus Mucosa-Infiltrating CD4(+) Lymphocytes in *Helicobacter pylori* Gastritis Display a Th1 Phenotype. Infect Immun 66: 5543-5546.

Steininger, H., Faller, G., Dewald, E., Brabletz, T., Jung, A., and Kirchner, T. 1998. Apoptosis in chronic gastritis and its correlation with antigastric autoantibodies. Virchows Arch 433: 13-18.

Stolte, M., Baumann, K., Bethke, B., Ritter, M., Lauer, E., and Eidt, H. 1992. Active autoimmune gastritis without total atrophy of the glands. Z Gastroenterol 30: 729-735.

Toh, B.H., van Driel, I.R., and Gleeson, P.A. 1997. Pernicious anemia. N Engl J Med 337: 1441-1448.

Uibo, R., Vorobjova, T., Metskula, K., Kisand, K., Wadstrom, T., and Kivik, T. 1995. Association of *Helicobacter pylori* and gastric autoimmunity: a population-based study. FEMS Immunol Med. Microbiol. 11: 65-68.

Valle, J., Kekki, M., Sipponen, P., Ihamaki, T., and Siurala, M. 1996. Long-term course and consequences of *Helicobacter pylori* gastritis. Results of a 32-year follow-up study. Scand. J. Gastroenterol. 31: 546-550.

Vorobjova, T., Faller, G., Maaroos, H.I., Sipponen, P., Villako, K., Uibo, R., and Kirchner, T. 2000. Significant increase in antigastric autoantibodies in a long-term follow-up study of *H. pylori* gastritis. Virchows Arch. 437: 37-45.

Warren, J.R., and Marshall, B.J. 1983. Unidentified curved bacilli on gastric epithelium in active chronic gastritis. Lancet. 1: 1273-1275.

Wotherspoon, A.C., Ortiz-Hidalgo, C., Falzon, M.R., and Isaacson, P.G. 1991. *Helicobacter pylori*-associated gastritis and primary B-cell gastric lymphoma. Lancet 338: 1175-1176.

Ye, G., Barrera, C., Fan, X., Gourley, W.K., Crowe, S.E., Ernst, P.B., and Reyes, V.E. 1997. Expression of B7-1 and B7-2 costimulatory molecules by human gastric epithelial cells. J. Clin. Invest. 99: 1628 - 1636.

From: *Helicobacter pylori: Molecular and Cellular Biology*
ISBN 1-898486-25-5 © 2001 Horizon Scientific Press, Wymondham, UK.

5

Cytokine Responses In *Helicobacter pylori* Infection

Jean E. Crabtree[*]

Abstract

Cytokines play a critical role in the initiation and modulation of gastric mucosal inflammatory responses to *Helicobacter pylori*. The gastric epithelium which secretes chemokines in response to *H. pylori* has an important role in the initiation of the acute inflammatory response. NF-κB activation and secretion of C-X-C chemokines such as IL-8 and GRO-α are induced by strains with the *cag* pathogenicity island (PAI). Mutational studies show multiple genes in the *cag* PAI are required for induction of this signalling response. *In vivo cag* PAI positive infection is associated with increased mucosal C-X-C chemokine expression and enhanced inflammatory responses. Strain heterogeneity may contribute to the broad spectrum of clinical outcome of infection. Antigen-specific chronic inflammation is a predominant Th1 response characterized by gamma interferon secreting effector cells. The mucosal production of gamma interferon and Th1 inducing cytokines such as IL-12 and IL-18 may be important in the induction of mucosal damage and development of autoimmune responses. Cytokines also play an important role in the changes in host gastric physiological responses associated with infection.

Introduction

Helicobacter pylori infection is one of the best opportunities we have to investigate human mucosal immune responses to a pathogenic agent. Infection is not only very common but occurs, in the most part, in the absence of other microorganisms allowing the mucosal inflammatory and immune responses to be related to the phenotypic characteristics of infecting *H. pylori* strains. The availability of the complete genome sequence of *H. pylori* (Tomb *et al.*, 1997) allows analysis of immune responses to defined antigens and virulence determinants. Understanding the mechanisms by which *H. pylori* induces acute and chronic inflammation in the gastric mucosa is important in elucidating the pathogenic role of this organism in gastroduodenal disease.

Given the clinical importance of *H. pylori* infection considerable attention has focused on investigation of host-pathogen interactions and the role of cytokines in generation of the chronic inflammatory response (Bodger and Crabtree, 1998). The gastric cellular response to *H. pylori*, which in part reflects the bacterial density of infection (Khulusi *et al.*, 1995), is characterised by mononuclear and polymorphonuclear cell infiltration (Dixon *et al.*, 1996).

[*]email: MSJJC@stjames.leeds.ac.uk

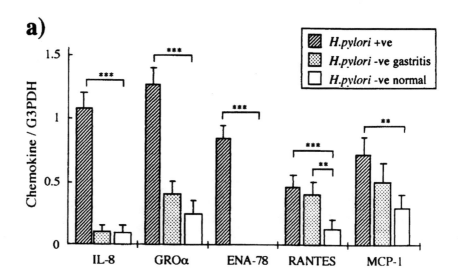

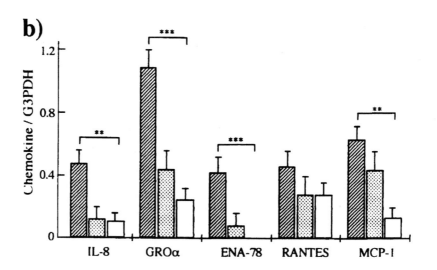

Figure 1. Chemokine mRNA expression in the antral (A) and corpus (B) mucosa in patients with or without *H. pylori* infection. ** p < 0.001, *** p < 0.001 versus control *H. pylori* negative group. mRNA for chemokines and glyceraldehyde 3-phosphate dehydrogenase (G3PDH) has been assessed by semi-quantitative RT-PCR. IL-8, GRO-α and ENA-78 are members of the C-X-C chemokine family and RANTES and MCP-1 are C-C chemokines. (Reproduced with permission from Shimoyama *et al.*, J. Clinical Pathology,1998.)

Cytokines play a critical role in regulating both the extent and characteristics of both innate and specific mucosal inflammatory cellular response and they also contribute to the pertubations in gastric physiological responses associated with infection (Crabtree, 1998). Cytokines, directly or indirectly, have an important role in promoting gastric epithelial cell proliferation (Yasunaga *et al.*, 1996; Berg *et al.*, 1998) and apoptosis (Wagner *et al.*, 1997; Fan *et al.*, 1998; Shibata *et al.* 1999; Houghton *et al.*, 1999) in *H. pylori* infection. In addition, gastric cytokines are likely to be involved in the regulation of autoimmune responses (Falcone and Sarvetnick, 1999) which can result as a consequence of long term chronic *H. pylori* infection.

This chapter focuses on the role of cytokines in induction of *H. pylori* associated inflammation in the human gastroduodenal mucosa, the evidence that inflammatory response may relate to specific bacterial virulence factors and the multiple functional roles of cytokines in disease pathogenesis.

Mucosal Chemokines and Acute Inflammatory Responses in *H. pylori* Infection

Initial infection with *H. pylori* in humans is associated with a prominent neutrophilic infiltration (Sobala *et al.*, 1991). Although in chronic infection the acute neutrophilic response resolves (Sobala *et al.*, 1991), there is a significant neutrophil infiltration into the epithelial surface layer which is defined histologically as active chronic gastritis (Dixon *et al.*, 1996). Many proinflammatory and immunoregulatory cytokines are increased in the human gastroduodenal mucosa in *H. pylori* infection and the relation with histological parameters has been investigated in some detail. Particular interest has focused on the chemokine family of cytokines which have been identified as key molecules in the recruitment and activation of immune cells (Baggiolini *et al.*, 1997). The family is divided into four groups, the two major groups being the C-X-C subfamily, which has the two amino terminal cysteines separated by a non-conserved amino acid residue, and the C-C subfamily which lack an intervening amino acid at this position. The two subfamilies differ in their target cell specificity with the ERL (Glu-Leu-Arg) containing C-X-C chemokines having chemotactic activity for neutrophils, but not monocytes, and the C-C family having effects primarily on monocytes and lymphocytes. Chemokines are early mediators of inflammatory responses and are induced by a variety of bacterial, viral and parasitic pathogens.

The most studied gastric chemokine is the neutrophil chemoattractant IL-8, a member of the C-X-C subfamily. Infection with *H. pylori* is associated with elevated gastric mucosal IL-8 protein and IL-8 mRNA expression (Crabtree *et al.*, 1993, 1994a; Peek *et al.*, 1995; Yamaoka *et al.*, 1996; Ando *et al.* 1996; Shimoyama *et al.*, 1998). Early studies demonstrated that in active chronic gastritis (characterised by the intraepithelial infiltration of polymorphonuclear cells) higher levels of IL-8 are secreted from mucosal biopsies during short term *in vitro* culture than from histologically normal gastric biopsies or biopsies with inactive gastritis, indicating a direct association between mucosal IL-8 and neutrophil infiltration (Crabtree *et al.*, 1993; Ando *et al.*, 1996). Assessment of IL-8 protein concentrations in homogenates of gastric biopsies (Peek *et al.*, 1995; Yamaoka *et al.*, 1998) and IL-8 mRNA expression in gastric mucosa (Shimoyama *et al.*, 1998) has also confirmed this direct association between mucosal IL-8 and neutrophil infiltration. Gastric IL-8 transcript and protein levels and neutrophil infiltration decline rapidly after *H. pylori* eradication demonstrating the direct effect of *H. pylori* infection on IL-8 gene expression (Moss *et al.*, 1994; Ando *et al.*, 1998). Apart from IL-8, other C-X-C chemokines GRO-α and ENA-78 and C-C chemokines such as MCP-1 and RANTES are also increased in the gastric mucosa in *H. pylori* infection (Shimoyama *et al.*, 1998) (Figure 1). As with IL-8, there is a strong correlation between GRO-α and ENA-78 mRNA expression in the antral mucosa and the extent of polymorphonuclear cell infiltration. In contrast, no correlation is found between C-C chemokine mRNA and neutrophil infiltration (Shimoyama *et al.*, 1998).

Interestingly in infected patients mRNA expression of IL-8 and ENA-78 is less in the corpus mucosa than in the antrum which does not relate to differences in bacterial density at the two sites (Shimoyama *et al.*, 1998). Environmental influences such as high acid in the

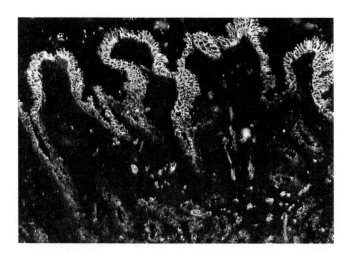

Figure 2. Immunofluorescent localization of interleukin-8 in the antral mucosa of an *H. pylori*-positive patient with chronic gastritis. (Reproduced with permission of Journal of Clinical Pathology from Crabtree *et al.*, 1994).

corpus lumen may influence the interacton between bacteria and host and thus chemokine induction. pH may also influence the transcription of bacterial genes in the *cag* pathogenicity island of relevance to chemokine induction (Karita *et al.*, 1996).

Epithelial Chemokines

Early studies on the gastric mucosa identified the epithelium as a major source of C-X-C chemokines such as IL-8 (Crabtree *et al.*, 1994a; Ando *et al.*, 1996) but other cells such as monocytes (Harris *et al.*, 1996) and neutrophils (Kim *et al.*, 1998) are also a source of chemokines after stimulation with *H. pylori* products. *H. pylori* infection is associated with increased IL-8 immunoreactivity in the epithelium of both the corpus and antral mucosa (Crabtree *et al.*, 1994a) (Figure 2). The presence of other C-X-C chemokines such as GRO-α and ENA-78 in gastric epithelial cells has not been fully investigated *in vivo*. As GRO-α is secreted by colonic epithelial cells (Yang *et al.*, 1997) and ENA-78 is expressed in human

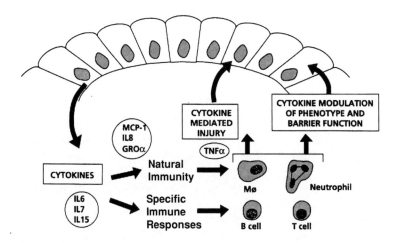

Figure 3. The role of the gastrointestinal epithelium in mucosal defence. Cytokines secreted by the epithelium influence both innate and specific immune responses in the gastrointestinal mucosa (Reproduced with permission of Digestive Disease and Sciences from Crabtree, 1998a)

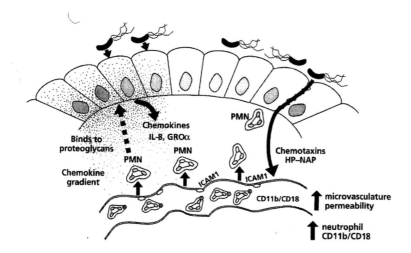

Figure 4. Pathways of induction of neutrophil infiltration by host chemokines and bacterial factors following *Helicobacter pylori* infection. (Reproduced with permission of European J. Gastroenterology and Hepatology from Crabtree 1998b)

colonic epithelium and upregulated in inflammatory states (Keates *et al.*, 1997a), it is likely these C-X-C chemokines are also expressed in gastric epithelial cells in *H. pylori* infection.

There is increasing evidence that epithelial cells at mucosal sites have a central role in host defence because they release a range of inflammatory and immunoregulatory molecules (Kagnoff and Eckmann, 1997). In addition, epithelial cells express numerous cytokine receptors and, as a consequence, cytokines released from activated immune cells within the mucosa will, in turn, modify epithelial phenotype and barrier function (Figure 3) (Crabtree, 1998a). The increase in epithelial C-X-C chemokines in *H. pylori* infection is thought to play an important role in the recruitment of neutrophils. *H. pylori* will directly stimulate the secretion of chemokines such as IL-8 from gastric epithelial cells (Crabtree *et al.*, 1994b, 1995a; Sharma *et al.*, 1995). From analogous studies with *Salmonella typhimurium* (McCormick *et al.*, 1993, 1995), it is appears that the basal secretion of chemokines from gastric epithelial cells and the binding of chemokines to proteoglycans within the matrix of the lamina propria (Webb *et al.*, 1993) generate a chemotactic gradient to promote directional migration of neutrophils towards the epithelium (Figure 4). Recent studies with *S. typhimurium* suggest that the final transepithelial migration of neutrophils is not dependent on IL-8, but on a low molecular weight apically secreted chemoattractant peptide (McCormick *et al.*, 1998) which is secreted by a different epithelial signalling pathway than IL-8 (Gerwirtz *et al.*, 1999).

It is pertinent to ask whether the gastric chemokine response has a role to play in innate defences against *H. pylori*. In experimental models of enteric *Shigella flexneri* infection blocking intestinal IL-8 responses markedly reduces intestinal pathology, however, it results in increased bacterial transepithelial translocation, suggesting inhibition of C-X-C chemokines may impair mucosal defence (Sansonetti *et al.*, 1999). The functional importance of C-X-C chemokines in *H. pylori* infection is unclear. Whilst the initial immunological and histopathological events following gastric colonisation have rarely been documented, isolated studies on subjects with acute *H. pylori* infection suggest that there is a marked neutrophilic component (Sobala *et al.*, 1991; Marshall *et al.*, 1985). The persistence of a neutrophilic response in chronic *H. pylori* infection suggests that these cells play a major role in inflammatory reactions to the organism. Identification of the bacterial virulence factors inducing mucosal inflammatory responses is an important goal of in delineating the mechanisms by which *H. pylori* produces clinical disease.

Molecular Analysis of Epithelial Chemokine Induction by *H. pylori*

Early studies demonstrated that the induction of chemokines such as IL-8 in gastric epithelial cells is dependent on viable *H. pylori* (Crabtree *et al.*, 1994b, 1995a; Sharma *et al.*, 1995), suggesting that complex bacterial-epithelial interactions are involved. The bacteria lose the ability to induce IL-8 in gastric epithelial cells once converted to the coccoid form (Cole *et al.*, 1997). Colonic (Crabtree *et al.*, 1994b) and bronchial epithelial (Sharma *et al.*, 1995) cells can secrete IL-8 in response to other stimuli, but not to *H. pylori*, indicating target cell specificity for gastric epithelial cells. The molecular mechanisms by which *H. pylori* induces epithelial chemokines has been investigated in some detail. Tyrosine kinase inhibitors such as herbimycin will block *H. pylori* induced IL-8 transcription (Li *et al.*, 1999) and IL-8 protein secretion (Aihara *et al.*, 1997; Beales and Calam, 1997; Li *et al.*, 1999), but protein kinase C and G inhibitors have no effect on *H. pylori* induced IL-8 transcription in L5F11 gastric epithelial cells (Li *et al.*, 1999).

The induction of chemokines following interaction of *H. pylori* with epithelial cells is dependent on activation of NF-κB (Aihara *et al.*, 1995; Keates *et al.*, 1997b; Muntzenmaier *et al.*, 1997; Sharma *et al.*, 1998). Following proteolytic degradation of the inhibitory protein IkB, the nuclear translocation of NF-κB p50/p65 heterodimers and p50 homodimers will stimulate IL-8 transcription (Aihara *et al.*, 1995; Keates *et al.*, 1997b). Activated NF-κB has also been identified immunohistochemically in the gastric epithelial cells of *H. pylori* infected patients (Keates *et al.*, 1997b). Mutational studies in gastric epithelial cell lines have shown that transcriptional regulation of IL-8 by *H. pylori* involves primarily the NF-κB site in the promoting region of the IL-8 gene and to a lesser extent the AP-1 site (Aihara *et al.*, 1995; Masamune *et al.*, 1999). Recent studies show activation of the sphingomyelin-ceramide pathway may be involved in *H. pylori* induced signal transduction (Masamune *et al.*, 1999). *H. pylori* increases ceramide in gastric epithelial cells and C2-ceramide alone will directly induce IL-8 expression in gastric epithelial cells (Masamune *et al.*, 1999). Ceramide is considered to induce IL-8 predominately through activation of NF-κB and partly via AP-1 (Masamune *et al.*, 1999). AP-1 activation in gastric epithelial cells by *H. pylori* has also recently been shown to involve c-Jun terminal kinase, MAP kinase kinase 4, p21 activated kinase and small Rho-GTPases (Naumann *et al.*, 1999).

NF-κB is also activated in primary guinea pig gastric epithelial cells (Rokatun *et al.*, 1997) and gastric epithelial cell lines (Shimada *et al.*, 1999) following stimulation with oxidants. *H. pylori* infection is associated with increased gastric mucosal reactive oxygen metabolites (Davies *et al.*, 1994), the main origin of which is likely to be neutrophils (Jones-Blackett *et al.*, 1999) but recent studies also indicate that gastric epithelial cells themselves produce superoxide in response to *H. pylori* (Teshima *et al.*, 1998). *In vitro* thiol agents such as pyrrolidine dithiocarbamate (Keates *et al.*, 1997b), antioxidants such as curcumin (Muntzenmaier *et al.*, 1997) and N-Acetylcysteine (Shimada *et al.*, 1999) have been shown to decrease activation of IL-8 induced by *H. pylori* or oxidants in human gastric epithelial cells. The antioxidant status of dietary components could therefore potentially have an effect on the induction of gastric epithelial chemokine responses and other host epithelial genes activated by *H. pylori* via redox sensitive mechanisms.

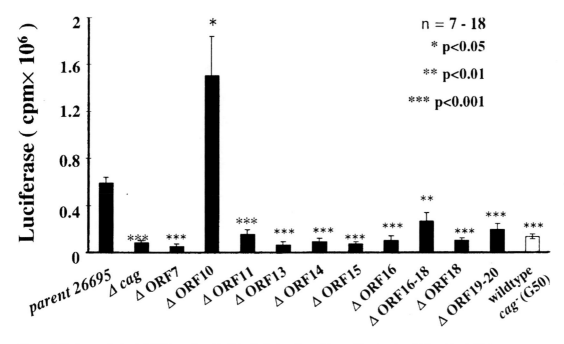

Figure 5. IL-8 transcription in L5F11 gastric epithelial cells induced by wild-type *H. pylori* (*cag* PAI positive 26695 and *cag* PAI negative G50) and *cag* locus isogenic mutant strains of genes in the left half of the *cag* PAI constructed in strain 26695. ORFs as detailed in Akopypants *et al.*, 1998. *cag* is a deletion of the entire *cag* PAI in strain 26695. The figure is based on results in Li *et al.*, 1999 and Crabtree *et al.*, 1999. ORF10 is the *Agrobacterium virD4* homologue.

The Role of the *cag* Pathogenicity Island in Epithelial Chemokine Responses

There is a marked dichotomy in the ability of *H. pylori* to induce epithelial chemokines. Epithelial IL-8 (Crabtree *et al.*, 1994b, 1995a; Sharma *et al.*, 1995) and associated NF-κB activation (Sharma *et al.*, 1998; Glocker *et al.*, 1998) are only induced by strains carrying the 40 kb *cag* pathogenicity island (PAI) (Censini *et al.*, 1996; Akopyants *et al.*, 1998). The genes encoded by the *cag* PAI are part of a complex secretory system involved in host-pathogen interactions comparable to those previously documented in *Agrobacterium tumefaciens* and *Bordetella pertussis* (Christie, 1997; Stein *et al.*, chapter 14). Wildtype *cag* PAI negative strains induce minimal IL-8 in gastric epithelial cells (Crabtree *et al.*, 1994b, 1995a; Sharma *et al.*, 1995). Strains with partial deletions within the *cag* PAI, which are found very infrequently (Censini *et al.*, 1996), also induce much lower levels of IL-8 in gastric epithelial cells than strains with a complete *cag* PAI (Maeda *et al.*,1999).

Mutational studies have shown that multiple genes throughout the whole *cag* PAI are essential for the induction of IL-8 transcription or protein secretion in gastric epithelial cells *in vitro* (Tummuru *et al.*, 1995; Censini *et al.*, 1996; Akopyants *et al.*, 1998; Li *et al.*, 1999) and NF-kB activation (Glocker *et al.*, 1998). Only four genes have so far been shown to be non-essential in inducing either NF-kB activation and/or IL-8 in epithelial cells, *cagA* (Crabtree *et al.*, 1995b; Sharma *et al.*, 1995), *cagF* (Glocker *et al.* 1998), *cagN* (Censini *et al.*, 1996) and ORF10, the *Agrobacterium virD4* homologue (Crabtree *et al.*, 1999a). Interestingly in-activation of the *virD4* homologue in strain 26695 results in a three-fold increase in IL-8 gene transcription (Crabtree *et al.*, 1999a), which markedly contrasts with the essential role of the four *virB* homologues in induction of IL-8 in gastric epithelial cells (Tummuru *et al.*, 1995; Censini *et al.*, 1996; Li *et al.*, 1999) (Figure 5). VirD4 in *Agrobacterium* is essential for protein and nucleoprotein (T DNA) transfer from bacterial to target cells (Christie, 1997). It is possible that the VirD4 homologue in the *cag* PAI of *H. pylori* helps modulate inflammation by down regulating the induction of IL-8 in epithelial cells (Crabtree *et al.*, 1999a).

The interaction of *cag* PAI positive *H. pylori* with gastric epithelial cells has also been associated with tyrosine phosphorylation of a protein of 145 kDa in gastric epithelial cells (Segal *et al.*, 1996). Whilst initially considered to be a host protein (Segal *et al.*, 1996), more recent *in vitro* studies have identified the tyrosine-phosphorylated protein to be the immunodominant *H. pylori* CagA protein in the *cag* PAI (Segal *et al.*, 1999; Asahi *et al.*, 2000). Size variation in the tyrosine-phosphorylated protein relates to the variable molecular mass of CagA (Asahi *et al.*, 2000). As with the induction of chemokines (Censini *et al.*, 1996; Li *et al.*, 1999), tyrosine phosphorylation of the 145kDa protein is dependent on many genes in the *cag* PAI (Segal *et al.*, 1997). The tyrosine phosphorylation of CagA is likely to be important in the actin cytoskeletal rearrangements in epithelial cells observed *in vitro* after *H. pylori* interaction (Segal *et al.*, 1999), rather than in epithelial cytokine induction. Whilst both phosphorylation of epithelial translocated CagA (Segal *et al.*, 1999; Asahi *et al.*, 2000) and *H. pylori*-induced IL-8 transcription and IL-8 protein secretion (Aihara *et al.*, 1995; Li *et al.*, 1999) can be blocked by tyrosine kinase inhibitors, mutational inactivation of *cag*A in *H. pylori* has no effect on the ability of *H. pylori* to induce IL-8 in gastric epithelial cells (Crabtree *et al.*, 1995b; Sharma *et al.*, 1995). In some instances deletion of *cag*A can even result in enhanced epithelial IL-8 secretion induced by *H. pylori* (J.E.Crabtree, unpublished data), suggesting a possible inhibitory effect of CagA on the epithelial translocation of the IL-8 inducing bacterial components. These collective studies suggest that the epithelial translocation and phosphorylation of CagA is not essential for chemokine induction in epithelial cells.

Examination of gastric chemokine mRNA expression in patients infected with *cag* positive and *cag* negative strains of *H. pylori* has substantiated that *cag* PAI positive *H. pylori* strains induce greater C-X-C chemokine responses (Peek *et al.*, 1995; Yamaoka *et al.*, 1996; Shimoyama *et al.*, 1998). Semi-quantitative RT-PCR analysis of IL-8, GRO-α and ENA 78 shows that antral mRNA expression of these C-X-C chemokines is significantly greater in those infected with *cag* positive strains than in those with *cag* negative infection (Shimoyama *et al.*, 1998). Enhanced mucosal C-X-C chemokines which correlate with the extent of neutrophil infiltration (Shimoyama *et al.*, 1998) probably account for the enhanced neutrophilic infiltration observed in *cag* positive infection (Crabtree *et al.*, 1991). Such *in vivo* evidence linking enhanced gastric C-X-C chemokines with *cag* positive infection implies that the epithelial chemokine response has a significant role in neutrophil infiltration.

The dichotomy in the ability of *H. pylori* strains to induce epithelial chemokine responses similar to those observed with enteritis inducing pathogens is of great interest. Even within wildtype strains possessing the *cag* PAI marked variation in the ability to activate epithelial chemokine responses *in vitro* is observed (Crabtree *et al.*, 1994b, 1995a; Li *et al.*, 1999; Maeda *et al.*, 1999). Although current studies have focused on investigating induction of chemokine genes, cDNA array screening has recently identified numerous other known host genes (Crabtree *et al.*, 1999b) and novel genes (Crabtree *et al.*, 2000) which are transcriptionally activated following the interaction of *cag* positive *H. pylori* with the gastric epithelial cells.

Leukocyte Chemokine/Cytokine Responses

Epithelial C-X-C chemokines such as IL-8 can be induced not only by direct bacterial stimulation, but also following exposure to the pro-inflammatory mediators IL-1 and TNF-α (Yashimoto *et al.*, 1992). Both mucosal TNF-α (Crabtree *et al.*, 1991a; Yamaoka *et al.*, 1996; Shibata *et al.*, 1999) and IL-1 (Peek *et al.*, 1995) mRNA expression and protein are increased

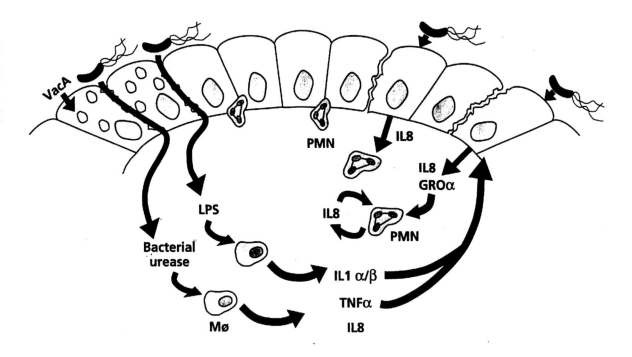

Figure 6. Pathways of *Helicobacter pylori* induction of mucosal chemokines involving both direct bacterial stimulation of chemokines such as IL-8 and GRO-α and activation of lamina propria macrophages. The release of proinflammatory cytokines such as IL-1a/ß and TNF-α from activated macrophages will also increase C-X-C chemokines in the gastric epithelium.

in the gastric mucosa in *H. pylori* infection (Figure 6). Mucosal macrophages are likely to be the main the source of IL-1 and TNF-α, but these proinflammatory cytokines may also be derived from activated neutrophils (Lord *et al.*, 1991) and *H. pylori* antigen-specific gastric T cells (D' Elios *et al.*, 1997).

The bacterial components inducing the pro-inflammatory cytokines such as TNF-α, IL-1 and IL-8 and also immunoregulatory cytokines such as IL-10 from lamina propria mononuclear cells are likely to be multiple. Both *H. pylori* lipopolysaccharide (LPS) (Bliss *et al.*,1998) and LPS-free bacterial components such as the urease (Harris *et al.*, 1996, 1998) will induce mononuclear cell cytokine and chemokine secretion. The uptake of bacterial components such as urease (Mai *et al.*, 1992) will be facilitated by disruption of epithelial barrier function associated with infection (Terres *et al.*, 1998), VacA related mucosal damage (Telford *et al.*, 1994) and neutrophil mediated disruption of the epithelial layer.

Although to date only IL-8 secretion has been documented from neutrophils following *H. pylori* stimulation (Kim *et al.*, 1998), activated neutrophils are known to secrete several C-X-C and C-C chemokines, including MIP-1α and MIP-1β (Cassatella, 1995) and thus further amplify the cellular response to infection (Figure 6). Whilst many putative neutrophil activating factors have been described in *H. pylori* (Kurose *et al.*, 1994; Noraard *et al.*, 1995), the components stimulating chemokine and cytokine secretion by neutrophils await detailed investigation. As with macrophages, cytokine secretion is likely to be mediated in part by LPS. Purified *H. pylori* LPS, however has a low biological activity relative to LPS of other bacterial species (Bliss *et al.*, 1998; Moran, chapter 13).

Chemokine secretion from both neutrophils and mononuclear cells is regulated by IL-10 (Kasama *et al.*, 1994), probably via the inhibition of the nuclear localisation of the transcription factor NF-κB (Wang *et al.*, 1995). Both IL-10 protein (Bodger *et al.*, 1997) and IL-10 mRNA expression (Peek *et al.*, 1995; Hida *et al.*, 1999) are increased in the gastric mucosa in *H. pylori* infection. IL-10 will inhibit neutrophil IL-8 secretion induced by *H. pylori in vitro* (Crabtree *et al.*, 1996). mRNA for IL-10 is increased in those with *cag* PAI positive infection (Peek *et al.*, 1995; Hida *et al.*, 1999). In addition, transcript expression levels of IL-12, a cytokine known to induce T cells to secrete IL-10 (Meyaard *et al.*, 1996), correlate strongly with IL-10 mRNA in the gastric mucosa (Hida *et al.*, 1999). These studies suggest the down regulatory cytokine IL-10 is increased in relation to the extent of mucosal damage. Studies with IL-10 deficient mice have confirmed the essential role of IL-10 in down-regulating *Helicobacter* induced gastric inflammatory responses. *Helicobacter felis* infection results in rapid development of severe hyperplastic gastritis and loss of corpus physiological function (Berg *et al.*, 1998). Thus the balance between proinflammatory cytokines and down-regulatory cytokines, such as IL-10, in the gastric mucosa may influence the extent of mucosal damage and modify epithelial cell proliferation. However, there is no evidence to suggest that impairment in IL-10 relates to ulceration, similar gastric IL-10 transcript levels are observed in ulcer and non-ulcer *H. pylori* infected patients (Hida *et al.*, 1999). In addition, to down regulating proinflammatory cytokine responses, IL-10 may also promote the generation of T regulatory 1 (Tr1) cells (Groux *et al.*, 1997), which may regulate immune responses to both enteric pathogens and to self antigens (Groux and Powrie, 1999).

Another important cell in mucosal defence against infectious agents is the natural killer (NK) cell (Scott and Trinchieri, 1995). Stimulation of peripheral blood lymphocytes with *H. pylori* induces non-MHC restricted NK activity and gamma interferon secretion (Tarkkanen *et al.*, 1993) and natural killer cells are increased in the gastric mucosa in *H. pylori* infection (Agnihotri *et al.*, 1998). The increase in mucosal IL-12 (Karttunen *et al.*, 1997; Hida *et al.*,1999; Buaditz *et al.*, 1999) and IL-18 (Tomita *et al.*, 1999) which occurs in *H. pylori* infection will promote NK cell activity and gamma interferon secretion. NK activity is defective in both IL-18 and IL-12 deficient mice (Takeda *et al.*, 1998) and these two cytokines have a synergistic role in activating NK cell gamma interferon secretion and promoting Th1 responses (Dinarello, 1999).

It is currently unclear whether the acute inflammatory response is in some instances effective in bacterial clearance. Although acute inflammatory responses in humans have been rarely documented, clearance of infection after acute gastritis has been described (Marshall *et al.*, 1985). The genetic host background of the individual may be important and gene polymorphisms, particularly in proinflammatory or immunoregulatory cytokines, may affect the risk of infection and clinical outcome (Kunstmann *et al.*, 1999). There is currently little data available in this area.

Cytokines and Mucosal Damage in Chronic Gastric Inflammation

Apart from regulating non-specific responses to *H. pylori* infection, cytokines will play a critical role in the polarisation of T cell responses (Trinchieri *et al.*, 1995) and the generation of Th1 mediated mucosal damage. Both human and murine studies have shown that the T helper response in chronic *Helicobacter* infection has a Th1 profile characterised by gamma interferon secreting effector cells (Karttunen *et al.*, 1995; Mohammadi *et al.*, 1996; D'Elios *et al.*, 1997; Mohammadi *et al.*, 1997; Bamford *et al.*, 1998). The importance of T cell responses in the generation of *Helicobacter*-induced gastric pathology has recently been demonstrated in RAG-1 -/- mice (Roth *et al.*, 1999) and by passive transfer of T-lymphocytes to SCID mice (Eaton *et al.*, 1999).

The high levels of IL-12 (Karttunen *et al.*, 1997; Hida *et al.*, 1999; Bauditz *et al.*, 1999) and IL-18 (Tomita *et al.*, 1999), both Th1 stimulating cytokines, in the gastric mucosa will

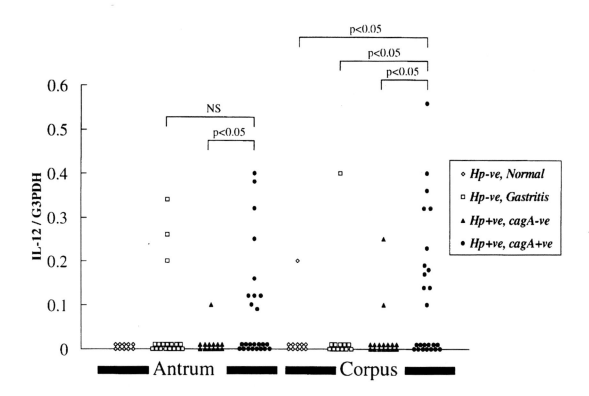

Figure 7. Expression of interleukin-12 (p40) mRNA in the antral and corpus mucosa in *H. pylori* negative and positive patients. mRNA for IL-12 (p40) and glyceraldehyde 3-phosphate dehydrogenase (G3PDH) has been assessed by semi-quantitative RT-PCR. In the *H. pylori* positive patients *cag*A expression in the gastric mucosa has been determined by RT-PCR. Reproduced with permission of Journal of Clinical Pathology from Hida *et al.*, 1999.

promote polarisation of Th1 responses. Recent studies have shown that IL-12 (p40) mRNA expression is increased in the gastric mucosa in those with *cag* PAI positive infections but not in those infected with *cag* negative strains (Figure 7) (Hida *et al.*, 1999). Additionally, the presence of gastric IL-12 mRNA is more frequent in *H. pylori* infected patients with duodenal ulcers than in those with only chronic gastritis (Hida *et al.*, 1999), suggesting that expression of this important Th1 stimulating cytokine may relate to mucosal damage. In contrast, IL-18 mRNA is elevated in both ulcer and non-ulcer patients (Tomita and Crabtree, unpublished data). Interestingly IL-12, but not IL-18, has recently been shown to significantly increase T-cell mediated tissue injury in human fetal gut explants (Monteleone *et al.*, 1999). In this experimental system IL-12 induced mucosal damage was associated with increases in TNF-α, interstitial collagenase and stromelysin-1 (Montelone *et al.*, 1999). Generation of active matrix metalloproteinases by IL-12 activated Th1 cells may thus potentially contribute to gastric mucosal damage in *H. pylori* infection.

The gamma interferon produced by Th1 cells will upregulate epithelial expression of HLA-DR (Valnes *et al.*, 1990; Fan *et al.*, 1998) and B7-2 involved in T cell activation (Ye *et al.*, 1997) and also mediate changes in epithelial barrier function (Madara and Stafford, 1989). Cytokine induced changes in mucus glycoprotein secretion may similarly account for alterations in gastric permeability (Takahashi *et al.*, 1999). These epithelial phenotypic and functional changes are likely to contribute to both enhanced antigen uptake and exacerbation of chronic inflammatory responses. Cytokines are also likely to be important in stimulating epithelial apoptosis which is increased in *H. pylori* infection (Moss *et al.*, 1996; Wagner and Beil, chapter 7). In epithelial cell culture TNF-α and gamma interferon potentiate *H. pylori* induced apoptosis (Wagner *et al.*, 1997; Fan *et al.*, 1998) and the pro-apoptotic effects of TNF-α may be regulated by the release of soluble TNF-receptors from gastric epithelial cells (Shibata *et al.*, 1999). The enhanced expression of the Fas antigen on gastric epithelial cells in *H. pylori* infection (Rudi *et al.*, 1998) is also postulated to be mediated by local cytokines (Houghton *et al.*, 1999).

Infection with *H. pylori* is accompanied by marked changes in gastric physiology, including increased basal and post-prandial plasma gastrin and decreased mucosal somatostatin (Calam, 1997). Mucosal cytokines have been implicated in these physiological changes. Th1 cytokines such as gamma interferon (Lehmann *et al.*, 1996) and proinflammatory cytokines IL1β, TNF-α (Weigert *et al.*, 1996) and IL-8 (Beales *et al.*, 1997a) induce gastrin secretion from cultured G cells. The effects of IL-8 are markedly potentiated by *H. pylori* extracts (Beales *et al.*, 1997a). In contrast to the stimulatory effects of cytokines on gastrin release, long term exposure of isolated D cells to TNF-α results in reduced cellular somatostatin (Beales *et al.*, 1997b). IL-1β also decreases histamine release from enterochromaffin cells (Prinz *et al.*, 1997) and inhibits histamine stimulated pepsinogen secretion by human peptic cells (Serrano *et al.*, 1997). The effect of mucosal cytokines on endocrine and exocrine cell function in the gastric mucosa will relate strongly to the colonisation patterns of *H. pylori* and the distribution of gastritis.

The long term outcome of *H. pylori* infection is the development of chronic atrophic gastritis and intestinal metaplasia (Dixon *et al.*, 1996). There is increasing evidence that *cag* PAI positive *H. pylori* infection increases the risk of developing this gastric cancer precursor condition (Kuipers *et al.*, 1995; Beales *et al.*, 1996; Webb *et al.*, 1999). Infection with *cag* PAI positive *H. pylori* is also a risk factor for developing intestinal type gastric cancer (Blaser *et al.*, 1995). The enhanced cytokine and cellular responses associated with *cag* PAI positive infections are considered to be an important factor in the clinical outcome. Autoantibody responses to the parietal cell canaliculi targeted at the H+K+ ATPase α and β chains occur frequently in infected subjects who develop atrophic gastritis (Claeys *et al.*, 1998; Faller and Kirchner, chapter 4). Whilst little is currently known about autoreactive gastric T cells in *H. pylori* infection, the local expression of cytokines is likely to be involved in the generation of autoimmune responses (Falcone and Sarvetnick, 1999). IL-12, which is preferentially induced by *cag* PAI positive strains (Hida *et al.*, 1999), is thought to have an important role in autoimmune responses by its potential effects on antigen presenting cells (Falcone and Sarvetnick, 1999). Interestingly, the disease promoting effects of IL-12 in induction of experimental autoimmunity can be antagonised by IL-10 in an antigen non-specific manner (Segal *et al.*, 1998). The rapid induction of corpus gastritis and loss of parietal function in *Helicobacter* infected IL-10 -/- mice (Berg *et al.*, 1998) concurs with the hypothesis that the IL-12/IL-10 balance may regulate the development of autoimmune responses and associated atrophic gastritis.

Conclusions

In conclusion, gastric cytokines induced by *H. pylori* infection are important not only in regulating the inflammatory responses to the bacterium, but also in modulating gastric physiological responses, altering gastric permeability, epithelial differentiation and potentially the induction of autoimmune responses. The acquisition of the *cag* pathogenicity island by *H. pylori* has resulted in more severe clinical outcome of infection, which may in part, be explained by the variation in the magnitude and characteristics of the mucosal cytokine responses. Whether particular cytokine profiles and cellular responses have a protective role in *H. pylori* eradication is still under active investigation. Initial studies suggested that the effector mechanism of therapeutic vaccination in mice involved a switch from Th1 to Th2 responses (Ghiara *et al.*, 1997). However more recent murine studies have linked enhanced cellular responses to bacterial clearance (Eaton *et al.*, 1999) and emphasised the importance of both IL-12 and gamma interferon in reducing gastric *Helicobacter* colonisation (Jiang *et al.*, 1999). Direct extrapolation from murine studies to the human situation is however difficult. It is clear that the discovery of *H. pylori* has promoted a great interest in gastric mucosal immune and inflammatory responses. The knowledge generated on host-pathogen interactions should hopefully improve our understanding of "idiopathic" enteric inflammatory conditions such as Crohn's disease or ulcerative colitis.

Acknowledgements

This work was supported by Yorkshire Cancer Research and the European Commission (contract number ICA4-CT-1999-10010).

References

Agnihotri, N., Bhasin, D.K., Vohra, H., Ray, P., Singh, K., and Ganguly, N.K. 1998. Characterization of lymphocyte subsets and cytokine production in gastric biopsy samples from *Helicobacter pylori* patients. Scand. J. Gastroenterology 33: 704- 709.

Aihara, M., Tsuchimoto, D., Takizawa, H., Azuma, A., Wakebe, H., Ohmoto, Y., Imagawa, K., Kikuchi, M., Mukaida, N., and Matsushima, K.1997. Mechanisms involved in *Helicobacter pylori*-induced interleukin-8 production by a gastric cancer cell line, MKN 45. Infect. Immun. 65: 3218-3224.

Akopyants, N.S., Clifton, S.W., Kersulyte, D., Crabtree, J.E., Youree, B.E., Reece, C.A., Bukanov, N.O., Drazek, S.E., Roe, B.A., and Berg, D.E. 1998. Analyses of the *cag* pathogenicity island of *Helicobacter pylori*. Mol. Microbiol. 28: 37-54.

Ando, T., Kusugami, K., Ohsuga, M., Shinoda, M., Sakakibara, M., Saito, H., Fukatsu, A., Ichiyama, S., and Ohta, M. 1996. Interleukin-8 activity correlates with histological severity in *Helicobacter pylori*-associated antral gastritis. Am. J. Gastroenterol. 91: 1150 - 1156.

Ando, T., Kusugami, K., Ohsuga, M., Ina, K., Shinoda, M., Konagaya, T., Sakai, T., Imada, A., Kasuga,N., Nada, T., Ichiyama, S., and Blaser, M.J. 1998. Differential normalization of mucosal interleukin-8 and interleukin-6 activity after *Helicobacter pylori* eradication. Infect. Immun. 66: 4742-4747.

Asahi, M., Azuma, T., Ito, S., Ito, Y., Soyo, H., Nagai, Y., Tsubokawa, M., Tohyama, Y., Maeda, S., Omata, M., Suzuki, T., and Sasakawa, C. 2000. *Helicobacter pylori* CagA protein can be tyrosine phosphorylated in gastric epithelial cells. J. Exp. Med. 191: 593-602.

Baggiolini, M., Dewald, B., and Moser, B. 1997. Human Chemokines: an update. Annu. Rev. Immunol. 15: 675-705.

Bamford, K.B., Fan, X., Crowe, S.E., Leary, J.F., Gourley, W.K., Luthra G.K., Brooks, E.G., Graham, D.Y., Reyes, V.E., and Ernst, P.B. 1998. Lymphocytes in the human gastric mucosa during *Helicobacter pylori* have a T helper cell 1 phenotype. Gastroenterology 114: 482-92.

Bauditz, J., Ortner, M., Bierbaum, M., Niedobitek, G., Lochs, H., and Schrieber, S. 1999. Production of IL-12 in gastritis relates to infection with *Helicobacter pylori*. Clin. Exp. Immunol. 117: 316-323.

Beales I.L.P., Crabtree, J.E., Scunes, D., Covacci, A., and Calam, J. 1996. Antibodies to CagA are associated with gastric atrophy in *Helicobacter pylori* infection. Eur. J. Gastroenterol. Hepatol. 8: 645-649.

Beales, I.L.P., and Calam, J. 1997. Stimulation of IL-8 production in human gastric epithelial cells by *Helicobacter pylori*, IL-1β and TNF-α requires tyrosine kinase activity, but not protein kinase C. Cytokine 9: 514-520.

Beales I., Blaser, M.J., Srinivasin, S., Calam, J., Perez-Perez, G.I., Yamada, T., Scheiman, J., Post, L., and del Valle, J. 1997a. Effect of *Helicobacter pylori* products and recombinant cytokines on gastrin release from cultured canine G cells. Gastroenterology 113: 465 - 471.

Beales, I., Calam, J., Post, L., Srinivasan, S., Yamada, T., and del Valle, J. 1997b. Effect of tumour necrosis factor-α on somatostatin release from canine fundic G cells. Gastroenterology 112: 136-143.

Berg, D.J., Lynch, N.A., Lynch, R.G., Lauricella, D.M. 1998. Rapid development of severe hyperplastic gastritis with gastric epithelial dedifferentiation in *Helicobacter felis*-infected IL-10-/- mice. Am. J. Pathol. 152: 1377-1386.

Blaser, M.J., Perez-Perez, G.I., Kleanthous,H., Cover, T.L., Peek, R.M., Chyou, P.H., Stemmermann, G.N., and Nomura, A. 1995. Infection with *Helicobacter pylori* strains possessing *cag*A associated with an increased risk of developing adenocarcinoma of the stomach. Cancer Res. 55: 2111-2115.

Bliss, C.M., Golenbock, D.T., Keates, S., Linevsky, J.K., Kelly, C.P. 1998. *Helicobacter pylori* lipopolysaccharide binds to CD14 and stimulates release of interleukin-8, epithelial neutrophil-activating peptide 78, and monocyte chemotactic protein 1 by human monocytes. Infect. Immun. 66: 5357-5363.

Bodger, K., Wyatt, J.I., and Heatley, R.V. 1997. Gastric mucosal secretion of interleukin-10: relations to histopathology, *H. pylori* status and TNF-a secretion. Gut 40: 739-744.

Bodger, K., and Crabtree, J.E. 1998. *Helicobacter pylori* and gastric inflammation. Brit. Med. Bull. 54: 139-150.

Calam, J. 1997. Host mechanisms: are they the key to the various clinical outcomes of *Helicobacter pylori* infection? Ital. J. Gastroenterol. Hepatol. 29: 375-382.

Cassatella, M.A. 1995. The production of cytokines by polymorphonuclear neutrophils. Immunol. Today 16: 21-26.

Censini, S., Lange, C., Xiang, Z., Crabtree, J.E., Ghiara, P., Borodovsky, M., Rappuoli, R., and Covacci, A. 1996. *cag*, a pathogenicity island of *Helicobacter pylori*, encodes Type I-specific and disease-associated virulence factors. Proc. Natl. Acad. Sci. USA. 93: 14648-14653.

Christie, P.J. 1997. *Agrobacterium tumefaciens* T-complex transport apparatus: a paradigm for a new family of multifunctional transporters in eubacteria. J. Bacteriol. 179: 3085-3094.

Claeys, D., Faller, G., Appelmelk, B.J., Negrini, R., and Kirchner, T.1998. The gastric H+ K+- ATPase is major autoantigen in chronic *Helicobacter pylori* gastritis with body mucosal atrophy. Gastroenterology 115: 340-347.

Cole, S., Cirillo, D., Kagnoff, M.F., Guiney, D.G. and Eckmann L.1997. Coccoid and spiral *Helicobacter pylori* differ in their abilities to adhere to gastric epithelial cells and induce interleukin-8 secretion. Infect. Immun. 65: 843-846.

Crabtree, J.E., Shallcross, T.M., Heatley, R.V., and Wyatt, J.I. 1991a. Mucosal tumour necrosis factor-alpha and interleukin-6 in patients with *Helicobacter pylori*-associated gastritis. Gut 44: 768-771.

Crabtree J.E., Taylor, J.E., Wyatt, J.I., Heatley, R.V., Shallcross, T.M., Tompkins, D.S., and Rathbone, B.J. 1991b. Mucosal IgA recognition of *Helicobacter pylori* 120 kDa protein, peptic ulceration and gastric pathology. Lancet 338: 332-335.

Crabtree, J.E., Peichl, P., Wyatt, J.I., Stachl, U. and Lindley, I.J.D. 1993. Gastric IL-8 and IL-8 IgA autoantibodies in *Helicobacter pylori* infection. Scand. J. Immunol. 37: 65-70.

Crabtree, J.E., Wyatt, J.I., Trejdosiewicz, L.K., Peichl, P., Nichols, P.N., Ramsay, N., Primrose, J.N., and Lindley, I.J.D. 1994a. Interleukin-8 expression in *Helicobacter pylori*, normal and neoplastic gastroduodenal mucosa. J. Clin. Pathol. 47: 61 - 66.

Crabtree, J.E., Farmery, S.M., Lindley, I.J.D., Figura, N., Peichl, P., and Tompkins, D.S. 1994b. CagA/cytotoxic strains of *Helicobacter pylori* and interleukin-8 in gastric epithelial cells. J. Clin. Pathol. 47: 945 - 950

Crabtree, J.E., Covacci, A., Farmery, S.M., Xiang, Z., Tompkins, D.S., Perry, S., Lindley, I.J.D., and Rappuoli, R. 1995a. *Helicobacter pylori* induced interleukin-8 expression in gastric epithelial cells is associated with CagA positive phenotype. J. Clin. Pathol. 48: 41 - 45.

Crabtree, J.E, Xiang, Z., Lindley, I.J.D., Tompkins, D.S., Rappuoli, R., and Covacci, A. 1995b. Induction of interleukin-8 secretion from gastric epithelial cells by *cag*A negative isogenic mutant of *Helicobacter pylori*. J. Clin. Pathol. 48: 967 - 969.

Crabtree, J.E., Perry, S., and Lindley, I.J.D. 1996. IL-10 inhibits *H. pylori* induced neutrophil but not epithelial chemokine secretion. Gut 39 (suppl. 2): A42.

Crabtree, J.E. 1998a. The role of cytokines in *Helicobacter pylori* induced mucosal damage. Dig. Dis. Sci. 43: 46S-55S.

Crabtree, J.E., 1998b. The host inflammatory response to *Helicobacter pylori*. Eur. J. Gastroenterol. Hepatol. 10(suppl. 1): S9-S13.

Crabtree, J.E., Kersulyte, D., Li, S.D., Lindley, I.J.D., and Berg, D.E. 1999a. Modulation of *Helicobacter pylori* induced interleukin-8 synthesis in gastric epithelial cells mediated by *cag* PAI encoded VirD4 homologue. J.Clin. Pathol. 52: 653 -657.

Crabtree, J.E., Cox, J.M., Clayton, C.L., Tomita, T., Wallace, D.M., and Robinson PA. 1999b. *Helicobacter pylori cag* related changes in epithelial cell gene expression identified using high density cDNA array technology. Gastroenterology 116: A873.

Crabtree, J.E., Tomita, T., Yoshimura, T., Hayat, M., Cox, J.M., Clayton, C.L., and Robinson, P.A. 2000. Identification of novel early epithelial response gene hprg1 induced by *cag* positive *Helicobacter pylori*. Gut 46: A68.

Davies, G.R., Simmonds, N.J., Stevens, T.R.J., Sheaff, M.T., Banatvala, N., Laurenson, I.F., Blake, D.R., and Rampton, D.S. 1994. *Helicobacter pylori* stimulates antral mucosal reactive oxygen metabolite production *in vivo*. Gut 35: 179-185.

D' Elios, M.M., Manghetti, M., De Carli, M., Costa, F., Baldari, C.T., Burroni, D., Telford, J.L., Romagnani, S., and Del Prete, G. 1997. T helper 1 effector cells specific for *Helicobacter pylori* in gastric antrum of patients with peptic ulcer disease. J. Immunol. 158: 962-967.

Dinarello, C.A. 1999. IL-18: a TH1-inducing, proinflammatory cytokine and new member of the IL-1 family. J. Allergy Clin. Immunol. 103: 11-24.

Dixon, M.F., Genta, R.M., Yardley, J.H., and Correa, P. 1996. Classification and grading of gastritis. The updated Sydney System. Amer. J. Surg. Pathol. 20: 1161-1181.

Eaton, K.A., Ringler,S.R., and Danon, S.J. 1999. Murine splenocytes induce severe gastritis and delayed-type hypersensitivity and suppress bacterial colonization in *Helicobacter pylori*-infected SCID mice. Infect. Immun. 67: 4594-4602.

Falcone, M., and Sarvetnick, N. 1999. Cytokines that regulate autoimmune responses. Curr. Opin. Immunol. 11: 670-676.

Fan, X., Crowe, S.E., Behar, S., Gunasena, H., Ye, G., Haeberle, H., Houten, N.van, Gourley, W.K., Ernst, P.B., and Reyes, V.E.. 1998. The effect of class II major histocompatibility complex expression on adherence of *Helicobacter pylori* and induction of apoptosis in gastric epithelial cells: a mechanism of T helper cell type 1-mediated damage. J.Exp. Med. 187: 1659-1669.

Gewirtz, A.T., Siber, A.M., Madara, J.L., and McCormick, B.A. 1999. Orchestration of neutrophil movement by intestinal epithelial cells in response to *Salmonella typhimurium* can be uncoupled from bacterial internalization. Infect. Immun. 67: 608 - 617.

Ghiara, P., Rossi, M., Marchetti, M.,Di Tommaso, A, Vindigni, C., Ciampolini, F., Covacci, A., Telford, J.L., De Magistris, M.T., Pizza, M., Rappuoli, R., and Del Giudice, G. 1997. Therapeutic intragastric vaccination against *Helicobacter pylori* in mice eradicates an otherwise chronic infection and confers protection against reinfection. Infect. Immun. 65: 4997-5002.

Glocker, E., Lange, C., Covacci, A., Bereswill, S., Kist, M., and Pahl H.L. 1998. Proteins encoded by the *cag* pathogenicity island of *Helicobacter pylori* are required for NF-κB activation. Infect. Immun. 66: 2346 - 2348.

Groux H, and Powrie F. 1999. Regulatory T cells and inflammatory bowel disease. Immunol. Today 20: 442-445.

Harris, P.R., Mobley, H.L.T., Perez-Perez, G.I., Blaser, M.J., and Smith, P.D. 1996. *Helicobacter pylori* urease is a potent stimulus of mononuclear phagocyte activation and inflammatory cytokine production. Gastroenterology 111: 419-425.

Harris, P.R., Ernst, P.B., Kawabata, S., Kiyono, H., Graham, M.F., and Smith , P.D. 1998. Recombinant *Helicobacter pylori* urease activates primary mucosal macrophages. J. Infect. Dis. 178: 1516-1520.

Hida, N., Shimoyama, T. Jnr, Neville, P., Dixon, M.F., Axon, A.T.R., Shimoyama, T. Snr,, and Crabtree, J.E. 1999. Increased expression of interleukin 10 and IL-12(p40) mRNA in *Helicobacter pylori* infection in gastric mucosa: relationship to bacterial *cag* and peptic ulceration. J. Clin. Pathol. 52: 658-664.

Houghton, J., Korah, R.M., Condon, M.R., and Kim, K.H. 1999. Apoptosis in *Helicobacter pylori*-associated gastric and duodenal ulcer disease is mediated via the Fas antigen pathway. Dig. Dis. Sci. 44: 465-478.

Jiang, B., Jordan, M., Xing, Z., Smaill, F., Snider, D.P., Borojevic, R., Steele-Norwood, D., Hunt, R.H., and Croitoru, K. 1999. Replication-defective adenovirus infection reduces *Helicobacter felis* colonization in the mouse in a gamma interferon and interleukin-12 dependent manner. Infect. Immun. 67: 4539-4544.

Jones-Blackett, S., Hull, M.A., Davies, G.R., and Crabtree, J.E. 1999. Nonsteroidal anti-inflammatory drugs inhibit *Helicobacter pylori*-induced human neutrophil reactive oxygen metabolite production *in vitro*. Aliment. Pharmacol. Ther. 13: 1653-1661.

Kagnoff, M.F, and Eckmann, L. 1997 Epithelial cells as sensor of microbial infection. J. Clin. Invest. 100: 6-10.

Karita, M., Tummuru, M.K.R., Wirth, H.P., and Blaser, M.J. 1996. Effect of growth phase and acid shock on *Helicobacter pylori cag*A expression. Infect. Immun. 64: 4501-4507.

Karttunen, R., Kattunen, T., Ekre, H.P.T., and MacDonald, T.T.1995. Interferon gamma and interleukin 4 secreting cells in the gastric antrum in *Helicobacter pylori* positive and negative gastritis. Gut 36: 341-345.

Karttunen, R.A., Karttunen, T.J., Yousfi, M.M., El-Zimaity, H.M.T., Graham, D.Y., and El-Zaatari, F.A.K. 1997. Expression of mRNA for interferon gamma, interleukin-10 and interleukin-12 (p40) in normal gastric mucosa and in mucosa infected with *Helicobacter pylori*. Scand. J. Gastroenterol.32: 22-27.

Kasama, T., Strieter, R.M., Lukacs, N.W., Burdick, M.D., and Kunkel, S.L. 1994. Regulation of neutrophil-derived chemokine expression by IL-10. J. Immunol. 152: 3559 - 3569.

Keates, S., Keates, A.C., Mizoguchi, E., Bhan, A., and Kelly, C.P. 1997a. Enterocytes are the primary source of the chemokine ENA-78 in normal colon and ulcerative colitis. Am. J. Physiol. 273: G75-82.

Keates, S., Hitti, Y.S., Upton, M., and Kelly, C.P. 1997b. *Helicobacter pylori* infection activates NF-κB in gastric epithelial cells. Gastroenterology 113: 1099-1109.

Khulusi, S., Mendall, M.A., Patel, P., Levy, J., Badve, S., and Northfield, T.C. 1995. *Helicobacter pylori* infection density and gastric inflammation in duodenal ulcer and non-ulcer subjects. Gut 37: 319-324.

Kim, J.S., Jung, H.C., Kim, J.M., Song, I.S., Kim, C.Y. 1998. Interleukin-8 expression by human neutrophils activated by *Helicobacter pylori* soluble proteins. Scand. J. Gastroenterol. 33: 1249-1255.

Kuipers, E.J., Perez-Perez, G.I., Meuwissen, S.G., Meuwissen, S.G.M., and Blaser, M.J. 1995. *Helicobacter pylori* and atrophic gastritis: importance of the *cag*A status. J. Natl. Cancer Inst. 87: 1777-1780.

Kunstmann, E., Epplen, C., Elitok, E., Harder, M., Suerbaum, S., Peitz, U., Schmiegel, W., and Epplen, J.T. 1999. *Helicobacter pylori* infection and polymorphisms in the tumour necrosis factor region. Electrophoresis 20: 1756-1761.

Kurose, I., Granger, D.N., Evans, D.J., Evans, D.G., Graham, D.Y., Miyasaka, M., Anderson, D.C., Wolf, R.E., Cepinskas, G., and Kvietys, P.R. 1994. *Helicobacter pylori*-induced microvascular protein leakage in rats: role of neutrophils, mast cells, and platelets. Gastroenterology 107: 70-79.

Lehmann, F.S., Golodner, E.H., Wang, J., Chen, M.C.Y., Avedian, D., Calam, J., Walsh, J.H., Dubinett, S., and Soll, A.H. 1996. Mononuclear cells and cytokines stimulate gastrin release from canine antral cells in primary culture. Am. J. Physiol. 270: G783-788.

Li, S.D., Kersulyte, D., Lindley, I.J.D., Neelam, B.,Berg, D.E, and Crabtree, J.E. 1999. Multiple genes in the left half of the *cag* pathogenicity island genes of *Helicobacter pylori* are required for tyrosine kinase-dependent transcription of interleukin-8 in gastric epithelial cells. Infect. Immun. 67: 3893-3899.

Lord, P.C.W., Wilmoth, L.M.G., Mizel, S.B., and McCall, C.E. 1991. Expression of interleukin-1α and β genes by human blood polymorphonuclear leukocytes. J. Clin. Invest. 87: 837-839.

Madara, J.L., Stafford, J. 1989. Interferon gamma directly affects barrier function of cultured intestinal epithelial cells. J. Clin. Invest. 83: 724-727.

Maeda, S., Yoshida, H., Ikenoue, T., Ogura, K., Kanai, F., Kato, N., Shiratori, Y., and Omata, M. 1999. Structure of *cag* pathogenicity island in Japanese *Helicobacter pylori* isolates. Gut 44: 336-341.

Mai, U.E.H., Perez-Perez, G.I., Allen, J.B., Wahl, S.M., Blaser, M.J., and Smith, P.D. 1992. Surface proteins from *Helicobacter pylori* exhibit chemotactic activity for human leukocytes and are present in gastric mucosa. J. Exp. Med. 175: 517-525.

Marshall, B., Armstrong, J., McGechie, D., and Glancy, R. 1995. Attempt to fulfil Koch's postulate for pyloric *Campylobacter*. Med. J. Aust. 152: 436-439.

Masamune, A., Shimosegawa, T., Masamune, O., Mukaida, N., Koizmi, M., and Toyota,T. 1999. *Helicobacter pylori*-dependent ceramide production may mediate interleukin-8 expression in human gastric cancer cell lines. Gastroenterology 116: 1330-1341.

McCormick, B.A., Coglan, S.P., Delp-Archer, C., Miller, S.I., and Madara, J.L. 1993. *Salmonella typhimurium* attachment to human intestinal epithelial monolayers: transcellular signalling to subepithelial neutrophils. J. Cell Biol. 123: 895-907.

McCormick, B.A., Hofman, P.M., Kim, J., Carnes, D.K., Miller, S.I., and Madara, J.L. 1995. Surface attachment of *Salmonella typhimurium* to intestinal epithelia imprints the subepithelial matrix with gradients chemotactic for neutrophils. J. Cell Biol. 131: 1599 - 1608.

McCormick, B., Parkos, C.A., Colgan, S.P., Carnes D.K., and Madara, J.L. 1998. Apical secretion of a pathogen-elicited epithelial chemoattractant activity in response to surface colonization of intestinal epithelial by *Salmonella typhimurium*. J. Immunol. 160: 455-466.

Meyaard, L., Hovenkamp, E., Otto, S.A., and Miedema, F. 1996. IL-12-induced IL-10 production by human T cells as a negative feedback for IL-12 induced immune responses. J. Immunol. 156: 2776-2782.

Mohammadi, M., Czinn, S., Redline, R., and Nedrud, J. 1996. *Helicobacter*specific cell-mediated responses display a predominant Th1 phenotype and promote a delayed-type hypersensitivity response in the stomach of mice. J. Immunol. 156: 4729-4738.

Mohammadi, M., Nedrud, J., Redline, R., and Czinn, S.1997. Murine CD4 T-cell response to *Helicobacter* infection: TH1 cells enhance gastritis and TH2 cells reduce bacterial load. Gastroenterology 113: 1848-1857.

Monteleone, G., MacDonald, T.T., Wathen, N.C., Pallone, F., and Pender S.L.F. 1999. Enhancing lamina propria Th1 cell responses with interleukin 12 produces severe tissue injury. Gastroenterology 117; 1069-1077.

Moss, S.F., Legon,S., Davies, J., and Calam, J. 1994. Cytokine gene expression in *Helicobacter pylori* associated antral gastritis. Gut 35: 1567 -1570.

Muntzenmaier, A., Lange, C., Glocker, E., Covacci, A., Moran, A. Bereswill, S, Baeuerle, P.A., Kist, M., and Pahl, H.L. 1997. A secreted/shed product of *Helicobacter pylori* activates transcription factor nuclear factor-kB. J. Immunol. 159: 6140 -6147.

Naumann, M., Wessler, S., Bartsch, C.,Wieland, B., Covacci, A., Haas, R., and Meyer, T.F. 1999. Activation of activator protein 1 and stress response kinases in epithelial cells colonized by *Helicobacter pylori* encoding the *cag* pathogenicity island. J. Biol. Chem. 274: 31655-31662.

Noraard, A., Andersen, L.P., and Nielsen, H. 1995. Neutrophil degranulation by *Helicobacter pylori* proteins. Gut 36: 354-357.

Peek, R.M. Jr, Miller, G.G., Tham, K.T., Perez-Perez, G.I., Zhao, X., Atherton, J.C., and Blaser, M.J. 1995. Heightened inflammatory response and cytokine expression *in vivo* to CagA+ *Helicobacter pylori* strains. Lab. Invest. 73: 760-770.

Prinz, C., Neumayer, N., Mahr, S., Classen, M., and Schepp, W. 1997. Functional impairment of rat enterochromaffin-like cells by interleukin-1β. Gastroenterology 112: 364-375.

Rokutan, K., Teshima, S., Miyoshi, M., Nikawa, T., and Kishi, K. 1997. Oxidant-induced activation of nuclear factor-kappa B in cultured guinea pig gastric epithelial cells. Dig. Dis. Sci. 42: 1880-1889.

Roth, K.A., Kapadia, S.B., Martin, S.M., and Lorenz, R.G. 1999. Cellular immune responses are essential for the development of *Helicobacter felis*-associated gastric pathology. J. Immunol. 163: 1490-1497.

Rudi, J., Kuck, D., von Herbray, A., Mariana, S.M., Krammer, P.H., Galle, P.R., and Stremmel, W. 1998. Involvement of the CD95 (APO-1/Fas) receptor and ligand system in Helicobacter pylori induced gastric epithelial apoptosis. J. Clin. Invest. 102: 1506-1514.

Sansonetti, P.J., Arondel, J., Huerre, M., Harada, A., and Matsushima, K. 1999. Interleukin-8 controls bacterial transepithelial translocation at the cost of epithelial destruction in experimental shigellosis. Infect. Immun. 67: 1471-1480.

Segal, E.D., Falkow, S., and Tompkins, D.S. 1996. *Helicobacter pylori* attachment to gastric cells induces cytoskeletal rearrangements and tyrosine phosphorylation of host cell proteins. Proc. Natl. Acad. Sci. USA. 93: 1259-1264.

Segal, B.M., Dwyer, B.K., and Shevach, E.M. 1998. An interleukin (IL)-10/IL-12 immunoregulatory circuit controls susceptibility to autoimmune disease. J. Exp. Med. 187: 537-546.

Segal, E.D., Lange, C., Covacci, A., Tompkins, L.S. and Falkow, S. 1997. Induction of host signal transduction pathways by *Helicobacter pylori*. Proc. Natl. Acad. Sci. USA. 94: 7595-7599.

Segal, E.D., Cha, J., Falkow, S., and Tompkins, L.S. 1999. Altered states: involvement of phosphorylated CagA in the induction of host cellular growth changes by *Helicobacter pylori*. Proc. Natl. Acad. Sci. USA. 96: 14559-14564.

Serrano, M.T., Lanas, A.I., Lorente, S. and Sainz, R. 1997. Cytokine effects on pepsinogen secretion from human peptic cells. Gut 40: 42-48.

Sharma, S.A., Tummuru, M.K.R., Miller, G.G., and Blaser, M.J. 1995. Interleukin-8 response of gastric epithelial cell lines to *Helicobacter pylori* stimulation *in vitro*. Infect. Immun. 63: 1681 - 1687.

Sharma, S.A., Tummuru, M.K.R., Blaser, M.J., and Kerr, L.D. 1998. Activation of IL-8 gene expression by *Helicobacter pylori* is regulated by transcription factor nuclear factor-κB in gastric epithelial cells. J. Immunol. 160: 2401-2407.

Shibata, J., Goto, H., Arisawa, T. Niwa, Y., Hayakawa, T., Nakayama, A., and Mori, N. 1999. Regulation of tumour necrosis factor (TNF) induced apoptosis by soluble TNF receptors in *Helicobacter pylori* infection. Gut 45: 24-31.

Shimada T, Watanabe, N., Hiraishi, H., and Terano, A. 1999. Redox regulation of interleukin-8 expression in MKN28 cells. Dig. Dis. Sci. 44: 266-273.

Shimoyama, T., Everett, S.M., Dixon, M.F., Axon, A.T.R., and Crabtree, J.E. 1998. Chemokine mRNA expression in gastric mucosa is associated with *Helicobacter pylori cag*A positivity and severity of gastritis. J. Clin. Pathol. 51: 765-770.

Sobala, G.M., Crabtree, J.E., Dixon, M.F., Scorah, C.J., Taylor, J.D., Rathbone, B.J., Heatley, R.V., and Axon, A.T.R. 1991. Acute *Helicobacter pylori* infection. Clinical features, local and systemic immune responses, gastric mucosal histology and gastric juice ascorbic acid concentrations. Gut 32: 1415-1418.

Takahashi, S., Nakamura, E., and Okabe, S. 1998. Effects of cytokines, without and with *Helicobacter pylori* components, on mucus secretion by cultured gastric epithelial cells. Dig. Dis. Sci. 43: 2301-2308.

Takeda, K., Tsutsui, H., Yoshimoto, T., Adachi, O., Yoshida, N., Kishimoto, T., Okamura, H., Nakanishi, K., and Akira, S. 1998. Defective NK cell activity and Th1 response in IL-18 deficient mice. Immunity 8: 383-390.

Tarkkanen, J., Kosunen, T.U., and Saksela, E. 1993. Contact of lymphocytes with *Helicobacter pylori* augments natural killer cell activity and induces production of gamma interferon. Infect. Immun. 61: 3012-3016.

Telford, J.L., Ghiara, P., Dell'Orco, M., Comanducci, M., Burroni, M., Bugnoli, M., Teece, M.F., Censini, S., Covacci, A., Xiang, Z., Papini, E., Montecucco, C., Parente, L., and Rappuoli, R. 1994. Gene of the *Helicobacter pylori* cytotoxin and evidence of its key role in gastric disease. J. Exp. Med. 179: 1653-1658.

Teshima, S., Rokutan, K., Nikawa, T., and Kishi, K. 1998. Guinea pig gastric mucosal cells produce abundant superoxide anion through an NADPH oxidase-like system. Gastroenterology 115: 1186-1196.

Terres, A.M., Pajares, J.M., Hopkins, A.M., Murphy, A., Moran, A., Baird, A.W., and Kelleher, D. 1998. *Helicobacter pylori* disrupts epithelial barrier function in a process inhibited by protein kinase C activators. Infect. Immun. 66: 2943-2950.

Tomita, T., Hida, N., Dixon, M.F., Axon, A.T.R., Hyatt, M., and Crabtree, J.E. 1999. IL-18 mRNA expression is elevated in *Helicobacter pylori* infected gastric mucosa. Gastroenterology 116: A693.

Tomb, J.R., White, O., Kerlavage, A.R. *et al.* 1997. The complete genome sequence of the gastric pathogen *Helicobacter pylori*. Nature 388: 539-547.

Trinchieri, G. 1996. Interleukin-12: A proinflammatory cytokine with immunoregulatory functions that bridge innate resistance and antigenspecific adaptive immunity. Annu. Rev. Immunol. 13: 251-276.

Tummuru, M.K.R., Sharma, S.A., and Blaser, M.J. 1995. *Helicobacter pylori picB*, a homologue of the *Bordetella pertussis* toxin secretion protein, is required for induction of IL-8 in gastric epithelial cells. Mol. Microbiol. 18: 867-876.

Valnes, K., Huitfeldt, H.S., and Brandtzaeg, P. 1990. Relation between T cell number and epithelial HLA class II expression quantified by image analysis in normal and inflamed human gastric mucosa. Gut 31: 647-652.

Wagner, S., Beil, W., Westermann, J., Logan, RPH, Bock, C.T., Trautwein, C., Bleck, J.S., and Manns, M.P. 1997. Regulation of gastric epithelial cell growth by *Helicobacter pylori*: evidence for a major role of apoptosis. Gastroenterology 113: 1836-1847.

Wang, P., Wu, P., Siegel, M.I., Egan, R.W., and Billah, M.M. 1995. Interleukin (IL)-10 inhibits nuclear factor kB (NF-kB) activation in human monocytes. J. Biol. Chem. 270: 9558-9563.

Webb, L.M.C., Ehrengruber, M.U., Clark-Lewis, I., Baggiolini, M., and Rot, A. 1993. Binding to heparin sulfate or heparin enhances neutrophil responses to interleukin-8. Proc. Natl. Acad. Sci. USA. 90: 7158 - 7162.

Webb, P., Crabtree, J.E., and Forman D. 1999. Gastric cancer, cytotoxin associated gene A positive *Helicobacter pylori* and serum pepsinogens: an international study. Gastroenterology 116: 269-276.

Weigert, N., Schaffer, K., Schusdziarra, V., Classen, M., and Schepp, W. 1996. Gastrin secretion from primary cultures of rabbit antral G cells: stimulation by inflammatory cytokines. Gastroenterology 110: 147-154.

Yamaoka, Y., Kita, M., Kodama, T., Sawai, N., and Imanishi, J. 1996. *Helicobacter pylori cagA* gene and expression of cytokine messenger RNA in gastric mucosa. Gastroenterology 110: 1744 - 1752.

Yamaoka, Y., Kita, M., Kodama, T., Sawai, N., Tanahashi, T., Kashima, K., and Imanishi, J. 1998. Chemokines in the gastric mucosa in *Helicobacter* infection. Gut 42: 609-617.

Yang, S.K., Eckmann, L., Panja, A., and Kagnoff, M.F. 1997. Differential and regulated expression of C-X-C, C-C and C-chemokines by human colon epithelial cells. Gastroenterology 113: 1214-1223.

Yashimoto, K., Okamoto, S., Mukaida, N., Murakami, S., Mai, M., and Matsushima, K. 1992. Tumour necrosis factor-α and interferon-gamma induce interleukin-8 production in human gastric cancer cell line through acting concurrently on AP-1 and NF-kB-like binding sites of the IL-8 gene. J. Biol. Chem. 267: 22506-11.

Yasunaga, Y., Shinomura, Y., Kanayama, S., Higashimoto, Y., Yaba, M., Miyazaki, Y., Kondo, S., Murayama, Y., Nishibayashi, H., Kitamura, S., and Matsuzawa, Y. 1996. Increased production of interleukin 1ß and hepatocyte growth factor may contribute to foveolar hyperplasia in enlarged fold gastritis. Gut 39: 787-794.

Ye, G., Barrera, C., Fan, X., Gourley, W.K., Crowe, S.E., Ernst, P.B, and Reyes, Y.E. 1997. Expression of B7-1 and B7-2 costimulatory molecules by human gastric epithelial cells. J. Clin. Invest. 99: 1628-1636.

From: *Helicobacter pylori: Molecular and Cellular Biology*
ISBN 1-898486-25-5 © 2001 Horizon Scientific Press, Wymondham, UK.

6

Role of Th1/Th2 Cells in *H. pylori*–induced Gastric Disease

Mario M. D'Elios[*] and Gianfranco Del Prete

Abstract

H. pylori infection represents the major cause of gastroduodenal diseases such as chronic gastritis, peptic ulcer, gastric cancer or B-cell lymphoma of the mucosa-associated lymphoid tissue (MALT), but only a minority of infected patients ever develop these pathologies. The type of the host immune response against *H. pylori* seems to be crucial for the outcome of infection.

A model has been developed to investigate the gastric T-cell response to *H. pylori* occurring in infected patients. A predominant *H. pylori* -specific Th1 response, characterized by high IL-12, IFN-γ and TNF-α production was found in the gastric antrum of patients with peptic ulcer. In uncomplicated chronic gastritis most gastric *H. pylori* -specific T cells secrete both Th1- and Th2-type cytokines, showing a Th0 profile. Th0 was also the predominant cytokine profile of T cells derived from gastric MALToma, but these gastric T cells exhibit abnormal help for autologous B-cell proliferation and reduced perforin- and Fas-Fas Ligand-mediated killing of B cells, unlike T cells from chronic gastritis with or without ulcer.

These data suggest that the host gastric immune response to *H. pylori* can influence the clinical picture. Peptic ulcer may be an immunopathological consequence of a Th1-polarized response to some *H. pylori* antigens, whereas deregulated *H. pylori* -induced T cell-dependent B-cell activation may support the onset of low-grade B-cell lymphoma.

Introduction

Helicobacter pylori, a pathogen infecting the gastric antrum of half of the adult population worldwide, is thought to be the major cause of acute and chronic gastroduodenal pathologies, including gastric and duodenal ulcer, gastric cancer and gastric B-cell lymphoma of mucosa-associated lymphoid tissue (MALT) (Marshall *et al.*, 1994; Parsonnet *et al.*, 1991; Wotherspoon *et al.*, 1991). Despite a vigorous humoral response against *H. pylori* antigens, most infected subjects fail to eliminate the pathogen spontaneously. As in other infectious diseases, both the natural and the specific immune responses of the host are crucial for determining the outcome of the infection. The immune system has evolved various defence mechanisms against pathogens. The first defensive line is provided by 'natural immunity', including phagocytes, T cell receptor (TCR) γδ+ T cells, natural killer (NK) cells, mast cells, neutrophils and

[*]corresponding author, email: delios@cesit1.unifi.it

Table 1. Distinctive features of human Th1 and Th2 cells

	Th1	Th2
Cytokine production		
IFN-γ, TNF-β	+++	-
IL-2, TNF-α	+++	+
IL-3, GM-CSF	++	+++
IL-6, IL-13, IL-10	+	+++
IL-4, IL-5	-	+++
Cytolytic activity	+++	-
Interaction with monocytes for		
Procoagulant activity (PCA)	+++	-
Tissue Factor production	+++	-
Delayed Type Hypersensitivity (DTH)	+++	-
Inhibition of PCA and DTH	-	+++
Help to B cells for Ig synthesis		
IgM, IgG, IgA (low T/B ratio)	+++	++
(high T/B ratio)	-	+++
IgE	-	+++
Surface expression		
mIFN-γ, IL-12Rb2, IL-18R	+++	-/+
CD26, LAG-3, CCR5, CXCR3	+++	+/-
CD30, CCR8, CRTH2	-/+	+++
CD62L, CCR3, CCR4, CXCR4	+/-	+++

T cells can be classified as Th1 or Th2 on the basis of their cytokines profile and related effector functions. Th1 cells are responsible for the phagocyte-dependent host response macrophage activation and production of opsonizing antibodies. Th2 cells are responsible of phagocyte-independent host response, provide optimal help for both humoral and mucosal immunity and inhibit several macrophage functions. Th1 and Th2 differ for their cytolytic potential and helper functions for B-cell antibody synthesis. Upon antigen stimulation, Th2 clones, usually devoid of cytolytic activity, induce in a dose-dependent fashion immunoglobulin synthesis by autologous B cells. In contrast Th1 cells, most of which are cytolytic, provide B-cell help for immunoglobulin synthesis (except IgE) at low T-B cell ratios. When the T-B cell ratio is high, there is a decline in B-cell help, due to Th1-mediated lysis of autologous antigen-presenting B cells. Th1 and Th2 cells exhibit different ability to activate monocytes. Membrane interactions with activated Th1 cells and Th1 cytokines induce monocytes to express tissue factor and related procoagulant activity, as well as DTH. T cells expressing both Th1- and Th2-type cytokines are coded as Th0 cells, whose effector functions depend on the balance between the cytokines locally produced and the nature of responding cells.

eosinophils, as well as complement components and pro-inflammatory cytokines, such as interferons (IFNs), interleukin (IL)-1, IL-6, IL-12, IL-18 and tumor necrosis factor (TNF)-α. The more specialized TCR $\alpha\beta^+$ T lymphocytes provide the second level of defence. These cells account for the specific immunity, which results in specialized types of immune responses which allow vertebrates to recognize and clear (or at least control) infectious agents in different body compartments. Virus infected cells are killed by $CD8^+$ cytotoxic T lymphocytes (CTL). Most microbial components are processed and presented preferentially to $CD4^+$ T helper (Th) cells after endocytosis by antigen-presenting cells (APC). Th cells cooperate with B cells for the production of antibodies that opsonize extracellular microbes and neutralize their exotoxins. This branch of the specific Th cell-mediated immune response is known as humoral immunity. However, other microbes survive within macrophages in spite of the unfavorable microenvironment and can only be killed after activation of the macrophages to produce reactive metabolites and TNF-α by antigen-activated $CD4^+$ Th cells. This branch of the specific Th cell-mediated response is known as cell-mediated immunity (CMI).

Most successful immune responses involve both humoral and cell-mediated immunity, but in some conditions the two types of effector reactions tend to be mutually exclusive.

CD4$^+$ Th cells can develop different polarized patterns of cytokine production (type-1 or Th1 and type-2 or Th2) that account for prevalent CMI or antibody response, respectively (Mosmann *et al.*, 1986; Del Prete *et al.*, 1991).

The Th1/Th2 Network

Th1 cells produce IFN-γ, IL-2 and TNF-β, activate macrophage and elicit delayed-type hypersensitivity (DTH) reactions. Th2 cells produce IL-4, IL-5, IL-10 and IL-13, which act as growth/differentiation factors for B cells, eosinophils and mast cells and inhibit several macrophage functions (Table 1)(Del Prete, 1998). Similar heterogeneous cytokine profiles were also observed for CD8$^+$ cytotoxic T cells (Tc1, Tc2), γδ$^+$ T cells and NK cells (Mosmann *et al.*, 1996; Ferrick *et al.*, 1995). However, most T cells do not express a polarized type-1 or type-2 cytokine profile; such T cells (called Th0) represent a heterogenous population of partially differentiated effector cells consisting of multiple subsets which secrete different combinations of both Th1 and Th2 cytokines. The cytokine response of the host remains mixed unless polarizing signals from the microenvironment stimulate the differentiation into the Th1 or the Th2 pathway. Human Th1 and Th2 cells also differ in their responsiveness to cytokines.

Th1 and Th2 cells differ substantially in their cytolytic potential and the mode of help for B-cell antibody synthesis. Th2 clones, usually devoid of cytolytic activity, induce IgM, IgG, IgA, and IgE synthesis by autologous B cells in the presence of the specific antigen, with a response that is proportional to the number of Th2 cells present. In contrast, Th1 clones, most of which are cytolytic, provide B-cell help for IgM, IgG, IgA (but not IgE) synthesis at low T-cell/B-cell ratios. At high T-cell/B-cell ratios, B-cell help declines due to the Th1-mediated lytic activity against autologous antigen-presenting B-cells (Del Prete *et al.*, 1991b). Th1 and Th2 cells also differ in their abilities to activate cells of the monocyte-macrophage lineage. Th1, (but not Th2), cells help monocytes to express tissue factor (TF) production and procoagulant activity (PCA). In this type of Th cell-monocyte cooperation, both Th1 cytokines (namely IFN-γ) and cell-to-cell contact with activated T cells and are required for optimal TF and PCA synthesis, whereas Th2-derived IL-4, IL-10 and IL-13 are strongly inhibitory (Del Prete *et al.*, 1995a).

Polarized human Th1 and Th2 cells also exhibit preferential expression of some activation molecules or chemokine/cytokine receptors. CD30 is primarily expressed by IL-4-producing Th0/Th2 cells both *in vitro* (Del Prete *et al.*, 1995b) and *in vivo* (D'Elios *et al.*, 1997), whereas lymphocyte activation gene (LAG)-3 is preferentially expressed by IFN-γ-producing Th0 and Th1 cells (Annunziato *et al.*, 1996). LAG-3 expression and release is up-regulated by IFN-γ and down-regulated by IL-4, whereas CD30 expression/release, which is strictly dependent on IL-4, is lost by fully differentiated, Th1 cells that lack IL-4 receptors (Nakamura *et al.*, 1997). IFN-γ is transiently expressed on the cell surface during secretion and can serve as a marker of ongoing Th1 function (Assenmacher *et al.*, 1996). The chemokine receptors CCR3, CCR4, CCR8, CXCR4 and CRTH2 are preferentially associated with Th2 cells, whereas CXCR3 and CCR5 are preferentially expressed by Th1 cells (Baggiolini, 1998; Annunziato *et al.*, 1999). In addition, the β chain of the IL-12 receptor is selectively expressed by activated Th1 cells because this protein is induced during the differentiation of naive T cells in the Th1, but not the Th2, pathway (Rogge *et al.*, 1997). IL-18 receptor is also selectively and consistently expressed on polarized murine Th1, but not Th2, cells (Xu *et al.*, 1998).

The factors responsible for Th1 or Th2 polarization of Th cell profile have been investigated. Th1 and Th2 cells seem to develop from the same Th-cell precursor under the influence of mechanisms associated with antigen presentation (Kamagawa *et al.*, 1993). Both environmental and genetic factors influence Th1 or Th2 differentiation mainly by determining the 'leader cytokine' in the microenvironment of the responding Th-cell. IL-4 is the most powerful stimulus for Th2 differentiation, whereas IL-12, IL-18 and IFNs favor Th1 development (D'Elios *et al.*, 1999). The site of antigen presentation, the physical form of the immunogen, the type of adjuvant, and the dose of antigen all play roles in polarization (Constant *et al.*, 1997). Microbial products (particularly from intracellular bacteria) induce Th1-dominated responses because they stimulate IL-12 production. IFN-γ and IFN-α favor Th1 development by enhancing IL-12 secretion by macrophages and by maintaining the expression of functional IL-12 receptors on Th cells (Szabo *et al.*, 1995). IL-18 sustaines the expression of IL-12Rβ, indicating that IL-12 and IL-18 synergistically induce and maintain Th1 development (Xu *et al.*, 1998). The genetic mechanisms that interact with environmental factors in controlling the type of Th cell differentiation still remain elusive. Binding of IL-12 to its receptor on naive Th cells results in the triggering of intracellular signals (rapidly transduced into the nucleus), such as the Jak-signal transducer and activator of transcription (called STAT)-4 (Bacon *et al.*, 1995). Signaling by IL-4 occurs through activation of STAT-6 and disruption of the STAT-6 gene results in deficient Th2 responses (Rincon *et al.*, 1997).

Th1/Th2 Cells in Health and Disease

In most human infections, specific immunity is of crucial importance, but an inappropriate response may not only result in lack of protection, but can even contribute to the induction of immunopathology. In human leishmaniasis, the lack of IFN-γ and high levels of IL-4 are predictive for progression into fulminant visceral disease whereas individuals whose cells produce large amounts of IFN-γ, usually remain asymptomatic. Th1 cytokine mRNA signals were found in the skin of patients with localized and mucocutaneous leishmaniasis, whereas Th2 cytokine mRNA molecules were strongly expressed by patients with destructive forms of cutaneous or active visceral disease. Interestingly, IFN-γ in combination with pentavalent antimony was effective in treating severe or refractory visceral leishmaniasis (Reiner *et al.*, 1995).

Parasitic infections that usually elicit Th2 cytokines are characterized by eosinophilia and elevated IgE levels. The pathology resulting from *Schistosoma mansoni* infection is predominantly caused by the host Th2 response leading to granulomatous reaction and consequent damage of intestine and liver (Sher *et al.*, 1992; Contigli *et al.*, 1999).

In the immune response to bacterial infections, Th2 cells face toxin-producing bacteria because Th2 cytokines favor B-cell maturation and the production of neutralizing antibodies. Intracellular bacteria (e.g. *Leisteria monocytogenes, Mycobacteria, Salmonellae*) are more properly faced by Th1 cells, which produce cytokines able to activate macrophages and cytotoxic T cells. Mice with disrupted IFN-γ or IFN-γ receptor (IFN-γR) genes succumb to mycobacterial infections whereas mice resistant to *M. bovis* produce high IFN-γ and low IL-4 levels (Flynn *et al.*, 1993). Likewise, patients with IFN-γR or IL-12R deficiency are extremely sensitive to mycobacterial infections and develop severe and often fatal disease (Newport *et al.*, 1996; de Jong *et al.*, 1998).

Th0 cells, which secrete a combination of both Th2- and Th1-type cytokines, should be the best effector cells in the immune response to extracellular bacteria since both antibodies (which neutralize adhesion/invasion and opsonize bacteria) and phagocytosis are required. The predominance of the Th1 or Th2 responses in any infectious disease is probably modulated by the pathogen plus the genetic background and innate immunity of the host. Since bacteria possess several components which can trigger IL-12 production by macrophages, it is not surprising that most bacterial infections favor Th1 development. These "Th1 inducers" include viral polynucleotides, the lipoarabinomannan of *Mycobacteria,* teichoic acids of gram-positive bacteria and lipopolysaccharides of gram-negative bacteria. In genetically predisposed individuals, strong and persistent Th1 responses against bacteria may result in immunopathological reactions, such as Lyme disease following infection with *Borrelia burgdorferi,* or reactive arthritis following infection with *Yersinia enterocolitica* (Yssel *et al.*, 1992; Lahesma *et al.*, 1992).

Th1-type effector mechanisms, such as DTH and CTL activity are supposed to play a central role in acute allograft rejection. Proteins and transcripts for intragraft IL-2, IFN-γ and the CTL-associated granzyme B, were found in allografts undergoing acute rejection. T-cell clones derived from kidney grafts of patients during the acute phase of rejection showed a clear Th1 cytokine profile and effector functions (D'Elios *et al.*, 1997a). The production of Th2-type cytokines seems to be critical for the induction and maintenance of allograft tolerance. *In vivo* studies on the cytokine network during tolerance induction showed decreased IL-2/IFN-γ and increased IL-4/IL-10 transcription (Dallmann, 1995). Embryo, because of the expression of paternal histocompatibility antigens, resembles an allograft which is not rejected by the maternal immune system till the time of delivery. This apparent paradox has recently been ascribed to a Th2 switch occurring at the materno-fetal interface, allowing foetal survival through inhibition of Th1 response (Piccinni *et al.*, 1998).

Organ-specific autoimmune diseases (such as Hashimoto's thyroiditis, multiple sclerosis, type 1 diabetes mellitus) (Del Prete, 1998), and Crohn's disease are good examples of Th1 polarization. Increased expression of mRNA for IL-2 and IFN-γ was found in the gut mucosa of patients with active Crohn's disease, whereas IL-4 mRNA expression was frequently undetectable (Niessner *et al.*, 1995). T-cell clones generated from the mucosa of patients with Crohn's disease produced higher levels of IFN-γ, but significantly reduced IL-4 and IL-5, in comparison with T-cell clones derived from patients with noninflammatory bowel disorders. High numbers of both IFN-γ-producing CD4 T cells and IL-12-containing macrophages were found in inflammatory gut infiltrates from patients with Crohn's disease, suggesting that in situ IL-12 production plays a role in the development of Th1 cells at intestinal level (Parronchi *et al.*, 1997).

Th1/Th2 Cells in *H. pylori* Infection

Most *H. pylori*-infected patients are unable to clear the pathogen, suggesting that *H. pylori* might somehow hamper the host immune response. *H. pylori* may interfere with protective immunity by acting on professional APC through its vacuolating cytotoxin (VacA). VacA impairs antigen processing and the subsequent priming of an efficient immune response (Molinari *et al.*, 1998). The failure to clear *H. pylori* from the gastric environment almost invariably leads to chronic antral gastritis. Colonization of the stomach by *H. pylori* is consistently accompanied by inflammation of the gastric mucosa, whose intensity depends on the host immune reaction against the pathogen. Colonization activates the natural immunity cellular compartement, represented by neutrophils and macrophages (Figure 1). The gastric epithelium expresses IL-8 following contact with *H. pylori* (Crabtree *et al.*, 1993). IL-8 is important for the initial host response to the bacterium, because it is a strong chemotactic and activating factor for neutrophils, which in turn can initiate and expand the inflammatory cascade. Furthermore, certain *H. pylori* components, such as the lipopolysaccharide, VacA, the cytotoxin-associated protein CagA, urease and the heat shock proteins (hsp) can cross the damaged layer of gastric cells and activate macrophages. Activation of macrophages results in the release of several cytokines, including IL-12, IL-1, IL-6, TNF-α, IFN-α and chemokines, such as IL-8. Moreover, neutrophils are also able to produce IL-12 in response to bacterial

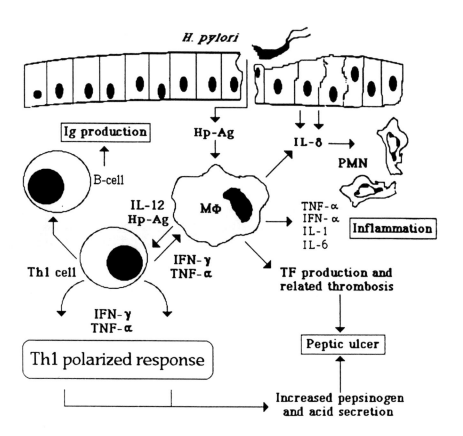

Figure 1. Schematic representation of the Th1 hypothesis in peptic ulcer. Upon contact with the gastric epithelium, *H. pylori* induces epithelial cells to synthesize and secrete several chemokines, as IL-8, attracting and activating polymorphonuclear (PMN) leukocytes. *H. pylori* and its antigens (Hp-Ag) activate macrophages (MF), and trigger IL-12 production. Processing and presentation of Hp-Ag to specific T cells in presence of IL-12 induces the preferential development of *H. pylori* -specific T-cell response into the Th1 pathway, which is further polarized by IFN-γ and TNF-α secreted by activated Th1 cells. Recognition of Hp-Ag induces *H. pylori* -specific Th1 cells to help local immunoglobulin (Ig) production by B cells. Th1 polarization may lead to peptic ulcer through gastric hyperacidity elicited, at least in part, by cytokines such as IFN-γ and TNF-α able to induce functional alterations of gastric cells. Another possible mechanism by which a polarized Th1 reponse may induce alterations of epithelial cell integrity is the induction of monocyte production of tissue factor and expression of procoagulant activity leading to microvascular thrombosis. According to "Th1 hypothesis", in non ulcer chronic gastritis, Th1-induced immunopathology would be quenched by the concomitant production of Th2 cytokines, which may exert a protective effect by inhibiting some of the detrimental effects exerted by Th1 cytokines on gastric chief cells as well as the tissue factor production by activated monocytes.

products (Trinchieri, 1995). The local cytokine "milieu", particularly the IL-12 produced by cells of the natural immunity is crucial in driving the subsequent specific T-cell response into a more or less polarized Th1 pattern.

The patterns of cytokines produced in the antral mucosa of *H. pylori* -infected patients with peptic ulcer were analyzed by RT-PCR. Antral biopsies from patients with ulcers expressed IL-12, IFN-γ, and TNF-α (but not IL-4) mRNAs, whereas no cytokine mRNA was found in the mucosa of *H. pylori* -negative controls (D'Elios *et al.*, 1997b). Immunohistochemistry of the same biopsies confirmed these observations (D'Elios *et al.*, 1997c).

Antigen Specificity and Cytokine Profiles of Gastric T Cells

Recent studies have examined the antigen specificity and the cytokines produced by the *H. pylori*-specific T-cell clones derived from the antral mucosa of 21 *H. pylori* -infected patients with chronic antral gastritis (CG). Six of the patients also had peptic ulcer (CG-PU), and five developed low grade gastric B-cell lymphoma (MALToma) (D'Elios *et al.*, 1997b, 1997d, 1999). Gastric biopsies were pre-cultured in IL-2-conditioned medium in order to preferentially expand T cells activated *in vivo,* and T-cell blasts were cloned according to an efficient technique that allows the growth of virtually every single T cell (D'Elios *et al.*, 1997b). A total number of 174 CD4 and 84 CD8 T-cell clones were obtained from CG-PU patients, 165 CD4 and 21 CD8 clones from MALToma patients and 289 CD4 and 58 CD8 clones were obtained from CG patients (D'Elios *et al.*, 1997b, 1997d, 1999). All clones were screened for their ability to proliferate in response to crude *H. pylori* lysate, urease, CagA, VacA and hsp in the presence of autologous APC. None of the CD8 clones obtained was responsive to the *H. pylori* antigens tested, whereas a proportion of CD4 clones did respond (CG-PU: 19%, CG: 29%, MALToma: 17%). As summarized in Figure 2, among the *H. pylori* -reactive CG-PU clones, 50% were specific for CagA, 9% for VacA, 3% for urease, 6% for hsp, and 32%

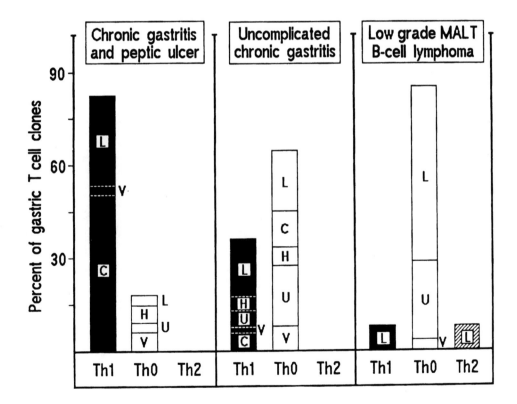

Figure 2. Antigen repertoire and cytokine profile of *H. pylori*–reactive T-cell clones derived from the gastric mucosa of *H. pylori*–infected patients with uncomplicated chronic gastritis, peptic ulcer or low grade B-cell lymphoma of the gastric MALT. *In vivo* activated T cells were recovered from biopsy specimens of gastric mucosa and cloned by limiting dilution. T-cell blasts from each clone were seeded in triplicate cultures with irradiated autologous peripheral blood mononuclear cells as APC in the presence of medium or optimal doses of crude *H. pylori* lysate (L), CagA (C), VacA (V), urease (U) or hsp (H). After appropriate culture periods, the proliferative response and the cytokine profile of each clone were assessed.

did not respond to any of the purified antigens tested, but proliferated in response to the bacterial lysate. Among the *H. pylori* -reactive clones from CG patients, 16% were specific for CagA, 10% for VacA, 25% for urease, 9% for hsp, and 39% responded only to the *H. pylori* lysate. Among the *H. pylori* -reactive clones from MALToma patients, 25% were specific for urease, 4% for VacA, and 71% responded only to *H. pylori* lysate. These data suggest that in both uncomplicated chronic gastritis and MALToma, urease is an important target for the gastric T-cell response, whereas CagA seems to be the immunodominant antigen recognized by local *H. pylori* -reactive T cells in patients with peptic ulcer (D'Elios *et al.*, 1997b). On the other hand, most T-cell clones from MALT lymphoma responded only to *H. pylori* lysate, suggesting that still undefined antigens of *H. pylori* are involved.

In peptic ulcer patients, *in vitro* stimulation with the appropriate *H. pylori* antigens induced a polarized Th1 profile in which the great majority (82%) of *H. pylori*-reactive Th clones produced IFN-γ but not IL-4 (D'Elios *et al.*, 1997b) (Figure 2). In contrast, most (64%) *H. pylori*-specific T-cell clones derived from uncomplicated chronic gastritis showed a Th0 phenotype, producing IL-4 and/or IL-5 together with IFN-γ, whereas only one third of *H. pylori*-specific gastric T cells were polarized Th1 effectors (D'Elios *et al.*, 1997d). Similarly, most of *H. pylori*-specific T cell clones derived from the gastric mucosa of MALToma patients were able to produce both Th1 and Th2 cytokines (D'Elios *et al.*, 1999).

Deficient Cytotoxic Control of *H. pylori*-induced B-cell Growth in MALT Lymphoma

Stimulation of T-cell clones derived from the gastric antrum of infected patients with the appropriate *H. pylori* antigen results in helper function for B-cell proliferation and Ig production (D'Elios *et al.*, 1997b). This can explain the intense B-cell activation in the lymphoid tissue associated with the antral mucosa during chronic *H. pylori* infection and the resulting high levels of specific antibodies found in the serum (Rathbone *et al.*, 1986; Crabtree *et al.*, 1995). In chronic gastritis patients either with or without ulcer, the helper function exerted by *H. pylori* antigen-stimulated gastric T-cell clones on B cells was negatively regulated by the concomitant cytolytic killing of B cells (D'Elios *et al.*, 1997b). In contrast, gastric T-cell clones from MALToma patients were unable to down-modulate their antigen-induced help for B cell proliferation (D'Elios *et al.*, 1999). Indeed, none of the gastric *H. pylori*-specific T-cell clones from MALToma was able to express perforin-mediated cytotoxicity against autologous B cells. Moreover, most (88%) Th clones from uncomplicated chronic gastritis induced Fas-Fas ligand-mediated apoptosis in target cells, whereas only a minority (21%) of *H. pylori* -specific gastric clones from MALToma patients were able to induce apoptosis in target cells, including autologous B cells (D'Elios *et al.*, 1999).

H. pylori-related Immunopathology

Although infection of the gastric antrum by *H. pylori* represents the primary cause of gastroduodenal diseases (Cover *et al.*, 1996), the reason why only 20-30% of patients with *H. pylori*-induced chronic gastritis present clinical consequences of the infection still remains unclear. There are a number of postulated mechanisms whereby identified products of *H. pylori* can induce mucosal injury (Telford *et al.*, 1994; Tomb *et al.*, 1997). Several independent studies agree that Th1 polarization of the immune response to *H. pylori* is associated with more severe disease (D'Elios *et al.*, 1997b; Hauer *et al.*, 1997; Bamford *et al.*, 1998; Sommer *et al.*, 1998). The preferential activation of Th1 cells and the subsequent production of their cytokines, namely IFN-γ and TNF-α, can potentiate gastrin secretion and pepsinogen release, as observed in animal models (Weigert *et al.*, 1996; Fiorucci *et al.*, 1996) (Figure 1). Data from our laboratory indicate that TNF-α and IFN-γ stimulate pepsinogen release from isolated human gastric chief cells in a dose-dependent manner. The simultaneous addition of IL-4 inhibits the IFN-γ-induced pepsinogen release by chief cells, rather than being synergistic, whereas it has no effect on the pepsinogen release induced by TNF-α (D'Elios *et al.*, 1998). Moreover, Th1 cells are able to induce both tissue factor production by monocytes (Del Prete *et al.*, 1995a) and the activation of the coagulation cascade, followed by microvascular thrombosis and the consequent alteration of epithelial cell integrity. Indirect support of the hypothesis that the Th1-type of gastric immune response against *H. pylori* contributes to the pathogenesis of peptic ulcer comes from the observation that in kidney graft recipients (undergoing strong immunosuppression) peptic ulcer and active inflammatory lesions were virtually absent, despite a higher prevalence of *H. pylori* colonization (Hruby *et al.*, 1997).

The results obtained so far demonstrate that the gastric T-cell response to *H. pylori* antigens characterized by a mixed Th1-Th2 cytokine profile is associated with a low rate of ulcer. The concept that Th2 cytokines, particularly IL-4 and IL-10, are important in balancing and quenching some immunopathological effects of polarized Th1 responses is supported by clinical and experimental observations. Clinicians have long known that dyspeptic symptoms are reduced during pregnancy in patients suffering from peptic ulcer and these patients tend to undergo remission during pregnancy (Cappel *et al.*, 1998). This might be an indirect effect of the preferential Th2 "switch" occurring in pregnancy, which enables the mother to "tolerate" her offspring by inhibiting Th1 responses, that would otherwise promote "graft" (fetus) rejection.

In mice, T-cell dependent immune responses are needed for protection against *H. pylori* whereas an antibody response is not essential for protective immunity (Ermak *et al.*, 1998). However, if an inappropriate T-cell response is induced against *H. pylori,* it may even result in damage to the host. SCID mice develop gastric ulcer after receiving T cells from *H. pylori* infected patients demonstrating that host immunity is involved in the development of peptic ulcers (Yokota *et al.*, 1999). In *H. felis*-infected mice, neutralization of IFN-γ significantly reduced the severity of gastritis, supporting the concept that preferential activation of a Th1 response contributes to the development and maintenance of gastric immunopathology (Mohammadi *et al.*, 1996). The magnitude of *H. felis*-induced inflammation in IL-4-deficient mice was higher than in their wild-type counterparts. Moreover, infection with *H. felis* induced only minimal inflammation in BALB/c mice, whose genetic background is prone to high IL-4 production in response to different antigens (Mohammadi *et al.*, 1996).

The results of these studies provide further evidence that a polarized Th1 response is associated with gastric inflammation and disease whereas a mixed Th1/Th2 response is able to reduce the unbalanced proinflammatory Th1 response. If the hypothesis that some local IL-4 production may result in protection from ulcer is correct, the so-called "African enigma" (i.e. the discrepancy between high rates of *H. pylori* infection and a low prevalence of peptic ulcer) (Holcombe *et al.*, 1992) may reflect that Africans live in endemic areas of helminth infections, which elicit strong and persistent Th2-dominated responses. A Th2-oriented host immunological background would hamper an efficient defence against mycobacteria, while limiting the inflammatory responses to pathogens like *H. pylori*. Thus, peptic ulcer may be regarded as the immunopathological outcome of a chronic inflammatory process induced by some *H. pylori* strains in subjects genetically and/or environmentally biased to develop strong Th1-polarized responses.

Imbalanced *H. pylori*-induced T-cell Help may Result in B-cell Lymphoma

Low-grade gastric MALT lymphoma is a very rare complication of *H. pylori* infection and represents a model to study the interplay between chronic infection, immune response and lymphomagenesis. Low grade MALT lymphoma is the first neoplasia that regresses after antibiotic therapy (Wotherspoon *et al.*, 1993). A prerequisite for lymphomagenesis is the development of secondary inflammatory MALT induced by chronic *H. pylori* challenge (Isaacson, 1994). The tumor cells of low-grade gastric MALT lymphoma are memory B cells that are still responsive to differentiation signals (such as CD40 costimulation and cytokines produced by antigen-stimulated T helper cells) and remain dependent for their growth on the stimulation by *H. pylori* -specific T cells (Hussel *et al.*, 1996; Greiner *et al.*, 1997). In early phases, this tumor is sensitive to the withdrawal of *H. pylori*-induced T-cell help, providing an explanation for both the tendency of this tumor to remain localized at the primary site and its regression after eradication of *H. pylori* (Wotherspoon *et al.*, 1993; Bayerdoffer *et al.*, 1995). The growth of neoplastic B cells may depend on their evading T cell-mediated cytotoxicity. Gastric T cells from MALT lymphoma showed both defective perforin-mediated cytotoxicity and poor ability to induce Fas-Fas ligand-mediated apoptosis, thus providing a possible explanation for their enhanced helper activity on B-cell proliferation. Both defects were restricted to MALT lymphoma-infiltrating T cells, because specific T helper cells from the peripheral blood of the same patients expressed the same degree of cytolytic potential and pro-apoptotic activity as did T cells from chronic gastritis patients (D'Elios *et al.*, 1999). The reason why gastric T cells of MALT lymphoma deliver full help to B cells, but are deficient in the concomitant control of B-cell growth, remains unclear. It has been shown that VacA toxin inhibits antigen processing in APC, but not the exocytosis of perforin-containing granules of NK cells (Molinari *et al.*, 1998). Possibly, in some *H. pylori*-infected individuals other bacterial components affect the development in gastric T cells of regulatory cytotoxic mechanisms for B-cell proliferation, allowing exhaustive and inbalanced B-cell help and lymphomagenesis to occur.

Conclusions

The concept of Th1/Th2 cytokine network provides an useful model for explaining both protection and pathogenesis for several immunopathological disorders. The development of polarized Th1 or Th2 responses depends on both the individual genetic background and environmental factors, especially cytokines of the natural immunity at the time of antigen presentation. Th1-dominated responses are potentially effective in eradicating infectious agents, particularly those hidden within the host cells. When Th1 response is poorly effective or exhaustively prolonged, it may result in host pathology. Polarized Th2 responses provide incomplete protection against the majority of infectious agents. They may be regarded as a part of down-regulatory mechanisms for exaggerated and inappropriate Th1 responses. Considering *H. pylori* infection, current evidence suggests that manipulating Th1/Th2 balance might be useful for future therapeutic strategies and that a balanced combination of Th1- and Th2-type specific responses might represent the goal for protective immunization.

Acknowledgments

Part of this work was supported by grants from the Associazione Italiana per la Ricerca sul Cancro and the University of Florence.

References

Annunziato, F., Manetti, R., Tomasevic, L., Giudizi, M.G., Biagiotti, R., Giannò, V., Germano, P., Mavilia, C., Maggi, E., and Romagnani, S. 1996. Expression and release of LAG-3-associated protein by human CD4$^+$ T cells are associated with IFN-γ. FASEB J. 10: 767-776.

Annunziato, F., Cosmi, L., Galli, G., Beltrame, C., Romagnani, P., Manetti, R., Romagnani, S., and Maggi, E. 1999. Assessment of chemokine receptor expression by human Th1 and Th2 cells *in vitro* and *in vivo*. J. Leukoc. Biol. 65: 691-699

Assenmacher, M., Scheffold, A., Schmitz, J., Segura Checa, J.A., Miltenyi, S., and Radbruch, A. 1996. Specific expression of surface interferon-γ on interferon-γ-producing T cells from mouse and man. Eur. J. Immunol. 26: 263-267.

Bacon, C.M., Petricoin, E.F., Ortaldo, J.E., Rees, R.C., Larner, A.C.,Johnston, J.A., and O'Shea, J.J. 1995. Interleukin 12 induces tyrosine phosphorylation and activation of STAT4 in human lymphocytes. Proc. Natl. Acad. Sci. USA. 92: 7307-7311

Baggiolini, M. 1998. Chemokines and leukocyte traffic. Nature. 392: 565-568

Bamford, K.B., Fan, X., Crowe, S.E., Leary, J.F., Gourley, W.K., Luthra, G.K., Brooks, E.G., Graham, D.Y., Reyes, V.E., and Ernst, P.B. 1998. Lymphocytes in the human gastric mucosa during Helicobacter pylori have a T helper cell 1 phenotype. Gastroenterology. 114: 482-492.

Bayerdorffer, E., Neubauer, A., Rudolph, B., Thiede, C., Lehn, N., Eidt, S., and Stolte, M. 1995. Regression of primary gastric lymphoma of mucosa-associated lymphoid tissue after cure of *H. pylori* infection. Lancet 345: 1591-1594.

Cappell, M.S., and Garcia, A. 1998. Gastric and duodenal ulcers during pregnancy. Gastroenterol. Clin. North America. 27: 169-183.

Constant, S.L., and Bottomly, K. 1997. Induction of Th1 and Th2 CD4$^+$ T cell responses: The alternative approaches. Annu. Rev. Immunol. 15: 297-322.

Contigli, C., Silva-Teixeira, D.N., Del Prete, G., D'Elios, M.M., De Carli, M.. Manghetti, M., Amedei, A., Almerigogna, F., Lambertucci, and J.R., Goes, A.M. 1999. Phenotype and cytokine profile of Schistosoma mansoni specific T cell lines and clones derived from schistosomiasis patients with distinct clinical forms. Clin. Immunol. 91: 338-344.

Cover, T.L., and Blaser, M.J. 1996. *Helicobacter pylori* infection, a paradigm for chronic mucosal inflammation: pathogenesis and implications for eradication and prevention. Adv. Intern. Med. 41: 85-117.

Crabtree, J.E., Eyre, D., Levy, L., Covacci, A., Rappuoli, R., and Morgan, A.G. 1995. Serological evaluation of *Helicobacter pylori* eradication using recombinant CagA protein. Gut. 36:A46.

Crabtree, J.E., Peichl, P., Wyatt, J.I., Stachl, U., and Lindley, I.J. 1993. Gastric interleukin-8 and IgA IL-8 autoantibodies in *Helicobacter pylori* infection. Scand. J. Immunol. 37: 65-70.

Dallman, M.J. 1995. Cytokines and transplantation: Th1/Th2 regulation of the immune response to solid organ transplants in the adult. Curr. Opin. Immunol. 7:632-638.

D'Elios, M.M., Josien, R., Manghetti, M., Amedei, A., De Carli, M., Cuturi, M.C., Blancho, G., Buzelin, F., Del Prete, G., and Soulillou, J.-P. 1997a. Predominant Th1 infiltration in acute rejection episodes of human kidney grafts. Kidney Int. 51: 1876-1884.

D'Elios, M.M., Manghetti, M., De Carli, M., Costa, F., Baldari, C.T., Burroni, D., Telford, J.L., Romagnani, S., and Del Prete, G. 1997b. Th1 effector cells specific for *Helicobacter pylori* in the gastric antrum of patients with peptic ulcer disease. J. Immunol. 158: 962-967.

D'Elios, M.M., Romagnani, P., Scaletti, C., Annunziato, F., Manghetti, M., Mavilia, C., Parronchi, P., Pupilli, C., Pizzolo, G., Maggi, E., Del Prete, G., and Romagnani S. 1997c. *In vivo* CD30 expression in human diseases with predominant activation of Th2-like T cells. J. Leukoc. Biol. 61: 539- 544.

D'Elios, M.M., Manghetti, M., Almerigogna, F., Amedei, A., Costa, F., Burroni, D., Baldari, C.T., Romagnani, S., Telford, J.L., and Del Prete G. 1997d. Different cytokine profile and antigen-specificity repertoire in *Helicobacter pylori*-specific T cell clones from the antrum of chronic gastritis patients with or without peptic ulcer. Eur. J. Immunol. 27: 1751-1755.

D'Elios, M.M., Andersen, L.P., and Del Prete G. 1998. Inflammation and host response. The year in *Helicobacter pylori*. Curr. Opin. Gastroenterol. 14 (S1): 15-19.

D'Elios, M.M., Amedei, A., Manghetti, M., Costa, F., Baldari, C.T., Quazi A.S., Telford, J.L., Romagnani, S., and Del Prete G. 1999. Impaired T-cell regulation of B-cell growth in Helicobacter pylori-related gastric low-grade MALT lymphoma. Gastroenterology. 117: 1105-1112.

de Jong, R., Altare, F., Elferink, D.G., and Ottenhoff, T.H. 1998. Severe mycobacterial and Salmonella infections in IL-12 receptor-deficient patients. Science. 280: 1435-1438.

Del Prete, G., De Carli, M., Mastromauro, C., Macchia, D., Biagiotti, R., Ricci, M., and Romagnani, S. 1991a. Purified protein derivative of *Mycobacterium tuberculosis* and excretory-secretory antigen(s) of *Toxocara canis* expand *in vitro* human T cells with stable and opposite (type 1 T helper or type 2 T helper) profile of cytokine production. J. Clin. Invest. 88: 346-351.

Del Prete, G.F., De Carli, M., Ricci, M., and Romagnani, S. 1991b. Helper activity for immunoglobulin synthesis of T helper type 1 (Th1) and Th2 human T cell clones: the help of Th1 clones is limited by their cytolytic capacity. J. Exp. Med. 174: 809-813.

Del Prete, G., De Carli, M., Lammel, R.M., D'Elios, M.M., Daniel, K.C., Giusti, B., Abbate, R., and Romagnani S. 1995a. Th1 and Th2 T-helper cells exert opposite regulatory effects on procoagulant activity and tissue factor production by human monocytes. Blood. 86: 250-257.

Del Prete, G., De Carli, M., Almerigogna, F., Daniel, K.C., D'Elios, M.M., Zancuoghi, G., Vinante, E., Pizzolo, G., and Romagnani S. 1995b. Preferential expression of CD30 by human CD4 T cells producing Th2-type cytokines. FASEB J. 9: 81-86.

Del Prete, G. 1998. The concept of Type-1 and Type-2 helper T cells and their cytokines in humans. Intern. Rev. Immunol. 16: 427-455.

Ermak, T.H., Giannasca, P.J., Nichols, R., Myers, G.A., Nedrud, J., Weltzin, R., Lee, C.K., Kleanthous, H., and Monath, T.P. 1998 Immunization of mice with urease vaccine affords protection against Helicobacter pylori infection in the absence of antibodies and is mediated by MHC class II-restricted responses. J. Exp. Med. 188: 2277-2288.

Ferrick, D.A., Schrenzel, M.D., and Mulvania T. 1995. Differential production of IFN-γ and IL-4 in response to Th1- and Th2-stimulating pathogens by gamma delta T cells *in vivo*. Nature. 373: 255-258.

Fiorucci, S., Santucci, L., Migliorati, G., Riccardi, C., Amorosi, A., Mancini, A., Roberti, R., and Morelli, A. 1996. Isolated guinea pig gastric chief cells express tumor necrosis factor receptors coupled with the sphingomyelin pathway. Gut. 38: 182-189.

Flynn, J.L., Chan, J., Trieblod, K.J., Dalton, D.K., Stewart, T.A., and Bloom, B.R. 1993. An essential role for IFN-γ in resistance to *Mycobacterium tuberculosis*. J. Exp. Med. 178: 2249-2254.

Greiner, A., Knorr, C., Qin, Y., Sebald, W., Schimpl, A., Banchereau, J., and Muller-Hermelink, H.K. 1997. Low-grade B cell lymphomas of mucosa-associated lymphoid tissue (MALT-type) require CD40-mediated signaling and Th2-type cytokines for *in vitro* growth and differentiation. Am. J. Pathol. 150: 1583-1593.

Hauer, A.C., Finn, T.M., MacDonald, T.T., Spencer, J., and Isaacson, P.G. 1997. Analysis of TH1 and TH2 cytokine production in low grade B cell gastric MALT-type lymphomas stimulated *in vitro* with *Helicobacter pylori*. J. Clin. Pathol. 50: 957-959.

Holcombe, C. 1992. *Helicobacter pylori*: the African enigma. Gut. 33: 429-431.

Hruby, Z., Myszka-Bijak, K., Gosciniak, G., Blaszczuk, J., Czyz, W., Kowalski, P., Falkiewicz, K., Szymanska, G., and Przondo-Mordarska, A. 1997. *Helicobacter pylori* in kidney allograft recipients: high prevalence of colonization and low incidence of active inflammatory lesions. Nephron. 75: 25-29.

Hussell, T., Isaacson, P.G., Crabtree, J.E., and Spencer, J. 1996. *Helicobacter pylori* –specific tumour-infiltrating T cells provide contact dependent help for the growth of malignant B cells in low-grade gastric lymphoma of mucosa-associated lymphoid tissue. J. Pathol. 178: 122-127.

Isaacson, P.G. 1994. Gastrointestinal lymphoma. Hum. Pathol. 25:1020-1029.

Kamogawa, Y., Minasi, L.E., Carding, S.R., Bottomly, K., and Flavell, R.A. 1993. The relationship of IL-4- and IFN-γ-producing T cells studied by lineage ablation of IL-4-producing cells. Cell. 75: 985-995

Lahesmaa, R., Yssel, H., Batsford, S., Luukkainen, R., Mootonen, T., Steinman, L., and Peltz, G. 1992. *Yersinia enterocolitica* activates a T helper type 1-like T cell subset in reactive arthritis. J. Immunol. 148: 3079-3085.

Marshall, B.J. 1994. *Helicobacter pylori*. Am. J. Gastroenterol. 89: 116-128.

Mohammadi, M., Czinn, S., Redline, R., and Nedrud, J. 1996. *Helicobacter*-specific cell-mediated immune response display a predominant Th1 phenotype and promote a delayed-type hypersensitivity response in the stomachs of mice. J. Immunol. 156: 4729-4738.

Molinari, M., Salio, M., Galli, C., Norais, N., Rappuoli, R., Lanzavecchia, A., and Montecucco, C. 1998. Selective inhibition of Ii-dependent antigen presentation by *Helicobacter pylori* toxin VacA. J. Exp. Med. 187: 135-140.

Mosmann, T.R., and Cherwinski, H., Bond, M.W., Giedlin, M.A., and Coffman R.L. 1986. two types of murine T cell clone. I. Definition according to profiles of lymphokine activities and secreted proteins. J. Immunol. 136: 2348-2357.

Mosmann, T.R., and Sad, S. 1996. The expanding universe of T-cell subsets: Th1, Th2 and more. Immunol. Today. 17: 138-146.

Nakamura, T., Lee, R.K., Nam, S.Y., Al-Ramadi, B.K., Koni, P.A., Bottomly, K., Podack, E.R., and Flavell, R.A. 1997. Reciprocal regulation of CD30 expression on CD4[+] T cells by IL-4 and IFN-γ. J. Immunol. 158: 2090-2098.

Newport, M.J., Huxley, C.M., Huston, S., and Levin, M. 1996. A mutation in IFN-γ receptor gene and sensitivity to mycobacterial infection. N. Engl. J. Med. 335: 1941-1949.

Niessner, M., and Volk, B.A. 1995. Altered Th1/Th2 cytokine profiles in the intestinal mucosa of patients with inflammatory bowel disease as assessed by quantitative transcribed polymerase chain reaction. Clin. exp. Immunol. 101: 428-435.

Parronchi, P., Romagnani, P., Annunziato, F., Sampognaro, S., Becchio, A., Giannarini, L., Maggi, E., Pupilli, C., Tonelli, F., and Romagnani, S. 1997. Type 1 T-helper cells predominance and interleukin-12 expression in the gut of patients with Crohn's disease. Am. J. Pathol. 150: 823-831.

Parsonnet, J., Friedman, G.D., Vandersteen, D.P., Chang, Y., Vogelman, J.H., Orentreich, N., and Sibley, R.K. 1991. *Helicobacter pylori* infection and the risk of gastric cancer. New Engl. J. Med. 325: 1127-1131.

Piccinni, M.P., Beloni, L., Livi, C., Maggi, E., Scarselli, G., and Romagnani, S. 1998. Role of type 2 T helper (Th2) cytokines and leukemia inhibitory factor (LIF) produced by decidual T cells in unexplained recurrent abortions. Nature Med. 4: 1020-1024.

Rathbone, B.J., Wyatt, J.I., Worsley, B.W., Shires, S.E., Trejdosiewicz, L.K., and Heatley, R.V. 1986. Systemic and local antibody responses to gastric *Campylobacter pyloridis* in non-ulcer dyspepsia. Gut 27: 642-647.

Reiner, S.L., and Locksley, R.M. 1995. The regulation of immunity to *Leishmania Major*. Annu. Rev. Immunol. 13: 151-177.

Rincon, M., and Flavell, R.A. 1997. T cell subsets: transcriptional control in the Th1/Th2 decision. Curr. Biol. 7: 729-732.

Rogge, L., Barberis-Maino, L., Biffi, M., Passini, N., Presky, D.H., Gubler, U., and Senigaglia, F. 1997. Selective expression of an interleukin-12 receptor component by human T helper 1 cells. J. Exp. Med. 185: 825-831

Sher, A., Gazzinelli, R.T., Oswald, I.P., Clerici, M., Kullberg, M., Pearce, E.J., Berzofsky, J.A., Mosmann, T.R., James, S.L., and Morse, H.C. 1992. Role of T cell derived cytokines in the downregulation of immune responses in parasitic and retroviral infection. Immunol. Rev. 127: 183-204

Sommer, F., Faller, G., Konturek, P., Kirchner, T., Hahn, E.G., Zeus, J., Röllinghoff, M., and Lohoff, M. 1998. Antrum- and corpus mucosa-infiltrating CD4(+) lymphocytes in *Helicobacter pylori* gastritis display a Th1 phenotype. Infect. Immun. 66: 5543-5546.

Szabo, S., Jacobson, N.G., Dighe, A.S., Gubler, U., and Murphy, K.M. 1995. Developmental committment to the Th2 lineage by extinction of IL-12 signaling. Immunity. 2: 665-675

Telford, J.L., Ghiara, P., Dell'Orco, M., Comanducci, M., Burroni, D., Bugnoli, M., Tecce, M.F., Covacci, A., Xiang, Z., Papini, E., Montecucco, C., Parente, L., and Rappuoli, R. 1994. Gene structure of the *Helicobacter pylori* cytotoxin and evidence of its key role in gastric disease. J. Exp. Med. 179: 1653-1670.

Tomb, J.F., White, O., Kerlavage, A.R., Clayton, R.A., Sutton, G.G., and Venter, C.J. et al. 1997. The complete genome sequence of the gastric pathogen *Helicobacter pylori* . Nature. 388: 539-547.

Trinchieri, G. 1995. Interleukin-12: a proinflammatory cytokine with immunoregulatory functions that bridge innate resistance and antigen-specific adaptive immunity. Annu. Rev. Immunol. 13: 251-276.

Weigert, N., Schaffer, K., Schusdziarra, V., Classen, M., Schepp, W. 1996. Gastrin secretion from primary cultures of rabbit antral G cells: stimulation by inflammatory cytokines. Gastroenterology. 110: 147-154.

Wotherspoon, A.C., Ortiz-Hidalgo, C., Falzon, M.F., Isaacson, P.G. 1991. *Helicobacter pylori* associated gastritis and primary B-cell lymphoma. Lancet 338: 1175-1176

Wotherspoon, A.C., Doglioni, C., Diss, T.C., Pan, L., Moschini, A., De Boni, M., Isaacson, P.G. 1993. Regression of primary low grade B cell gastric lymphoma of mucosa-associated lymphoid tissue type after eradication of *Helicobacter pylori*. Lancet. 342: 575-577.

Xu, D., Chan, W.L., Leung, B.P., Hunter, D., Schulz, K., Carter, R.W. McInnes, IB., Robinson, J., and Liew, F.Y. 1998. Selective expression and functions of interleukin 18 receptor on T helper (Th) type 1 but not Th2 cells. J. Exp. Med. 188: 1485-1492

Yokota, K., Kobayashi, K., Kahawara, Y., Hayashi, S., Hirai, Y., Mizuno, M., Okada, H., Akagi, T., Tsuji, T., and Oguma, K. 1999. Gastric ulcers in SCID mice induced by *Helicobacter pylori* infection after transplanting lymphocytes from patients with gastric lymphoma. Gastroenterology. 117: 893-899.

Yssel, H., Shanafelt, M.C., Soderberg, C., Schneider, P.V., Anzola, J., Peltz, G. 1992. *Borrelia burgdorferi* activates a T helper type 1-like T cell subset in Lyme arthritis. J. Exp. Med. 174: 593-601.

From: *Helicobacter pylori: Molecular and Cellular Biology*
ISBN 1-898486-25-5 © 2001 Horizon Scientific Press, Wymondham, UK.

7

Helicobacter pylori Induced Apoptosis

Siegfried Wagner[*] and Winfried Beil

Abstract

Helicobacter pylori gastritis is accompanied by a reversible induction of gastric epithelial cell apoptosis and proliferation which leads to an increased rate of gastric epithelial cell turnover. Enhanced apoptotic rates were found at the margin of peptic ulcers, while inadequately low apoptotic rates were documented in foci of intestinal metaplasia, suggesting a possible role of the apoptosis to proliferation ratio as a factor determining the clinical outcome of *H. pylori* infection. *In vitro* studies have demonstrated that *H. pylori* is capable of directly inducing apoptosis and a synergistic role of proinflammatory cytokines has also been shown. All clinical isolates of *H. pylori*, irrespective of their *cagA* and *vacA* status, induced DNA fragmentation in gastric epithelial cells, whereas *Campylobacter jejuni* failed to do so. *H. pylori* induced apoptosis was associated with an up-regulation of the cell death receptor CD95/APO-1/Fas and of the pro-apoptotic Bcl-2 homologues Bak and Bax in gastric epithelial cells. *H. pylori* induced DNA fragmentation was preceded by the sequential activation of caspase-9, caspase-8, and caspase-3. The molecular nature of the apoptosis inducing factor(s) and its interaction with the apoptotic signalling pathway are unknown and need to be addressed in future studies.

Introduction

Programmed cell death or apoptosis is a genetically controlled active process of the single cell where certain signals initiate a series of events leading to cell self elimination (Kerr *et al*, 1972; White, 1996, Evan and Littlewood, 1998). Apoptosis is defined by a highly characteristic sequence of morphological changes, resulting in the death of a cell without inflammatory sequelae. During apoptosis, the cell shrinks, microvilli (if present) disappear, and the cell detaches from neighbouring cells. Concurrently, chromatin condensation and DNA fragmentation occur. Finally, the cell fragments into intact, membrane-bound bodies, termed apoptotic bodies, which are rapidly phagocytosed by neighbouring cells and mononuclear cells. Thus, apoptosis is often inconspicuous and difficult to detect *in vivo*.

Apoptosis is essential in the development and homeostasis of multicellular organisms. The gastrointestinal tract is characterised by high epithelial cell turnover, with cell division exactly balancing cell death (Que and Gores, 1996). *H. pylori* infection causes alterations in the epithelial cell turnover of the gastric mucosa, which may have important implications for

[*]corresponding author, email: wagner.siegfried@mh-hannover.de

the pathogenesis of *H. pylori* associated diseases like peptic ulcer or gastric cancer (Peek *et al.*, 1997; Shirin and Moss, 1998; Kohda *et al.*, 1999; Scotiniotis *et al.*, 1999). Understanding the molecular mechanisms of *H. pylori* induced apoptosis has just begun and may provide the basis for new therapeutic strategies for gastroduodenal disorders linked to *H. pylori* infection.

Epithelial Cell Turnover and Homeostasis in the Stomach

Cell proliferation and apoptosis are the key factors for maintaining gastric tissue homeostasis. Under physiological conditions, both processes are spatially separated within the gastric gland. The proliferating zone is located in the neck and isthmus, whereas the apoptotic zone is located at the mucosal surface and towards the base of the gastric gland (Hall *et al.*, 1994). During maturation, the new gastric epithelial cells migrate towards the two extremities of the gland. Mucous cells move up into the foveolar zone while parietal and chief cells move down into the base of the gland. The cells differentiate during this bidirectional migration and gradually lose their replicative capacity. Finally the senescent cells die by apoptosis. Hall *et al.* (1994) have shown that apoptosis may account for the bulk of cell loss in the gut including the stomach, and that apoptosis is a central feature of the regulation of cell number in gastric tissue, unlike earlier views claiming that cells were shed passively into the gut lumen. Two forms of apoptosis may contribute to normal cell loss in the stomach (Shirin and Moss, 1998): (i) senescent apoptosis which occurs at the end of the cell's natural lifespan and represents the major route of physiological cell loss; (ii) altruistic apoptosis which occurs prematurely in response to severe DNA damage. Altruistic apoptosis serves to eliminate mutated cells before they can proliferate to form potentially neoplastic clones.

Disturbances in the delicate balance between cell proliferation and apoptosis should lead to disease states: acute excessive cell loss would disrupt mucosal integrity leading to gastric erosions or ulcer formation, while chronic unbalanced cell loss would result in atrophy. Conversely, tissue proliferation and neoplasia may be associated with the inhibition of apoptosis. Since *H. pylori* can cause very diverse clinical outcomes, including peptic ulcer, atrophy, and neoplasm, distinct regulatory effects of *H. pylori* on the balance between gastric epithelial cell apoptosis and proliferation could contribute to these spectrum of disease manifestations (Shirin and Moss, 1998).

Gastric Epithelial Cell Apoptosis in Patients with *H. pylori* Infection

It is well known that *H. pylori* causes reversible hyperproliferation of gastric epithelial cells (Fraser *et al.*, 1994; Cahill *et al.*, 1995; Lynch *et al.*, 1995; Fan *et al.*, 1996; Havard *et al.*, 1996; Correa and Miller, 1998). In contrast, strong evidence has recently accumulated also demonstrating the induction of apoptosis by *H. pylori*. Induction of apoptosis by *H. pylori* has been reported in patients with peptic ulcer (Moss *et al.*, 1996; Houghton *et al.*, 1999; Kohda *et al.*, 1999; Konturek *et al.*, 1999) and gastritis (Mannick *et al.*, 1996; Jones *et al.*, 1997; Rudi *et al.*, 1998; Steininger *et al.*, 1998, Zhu *et al.*, 1998). The majority of studies reported a three to five-fold increase in mean apoptotic rates in duodenal ulcer patients (Figure 1a), which dropped significantly after *H. pylori* eradication (Figure 1b) (Moss *et al.*, 1996; Kohda *et al.*, 1999; Konturek *et al.*, 1999). Only one small study by Anti *et al.* (1998) failed to show a significant increase of epithelial apoptosis in gastritis patients infected with *H. pylori*. Thus, it is now well accepted that *H. pylori* causes a reversible increase of epithe-

a

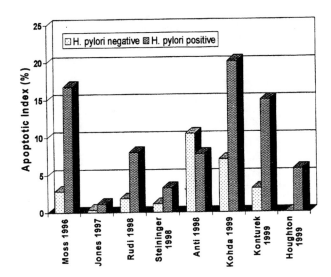

b

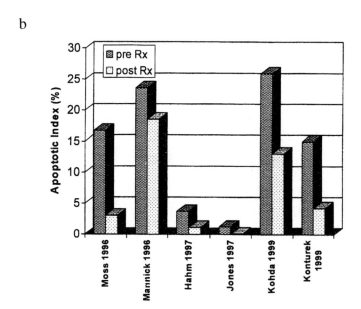

Figure 1. The influence of *H. pylori* infection on gastric epithelial apoptosis in patients with gastritis and duodenal ulcer. a) A compilation of apoptotic indices in subjects with and without *H. pylori* infection. Apoptotic rates are expressed as percentage of counted cells, except for the data of Houghton *et al.* (1999) which are expressed per 1000 counted cells. b) Apoptotic indices in *H. pylori* positive patients before (pre Rx) and after (post Rx) eradication therapy.

a

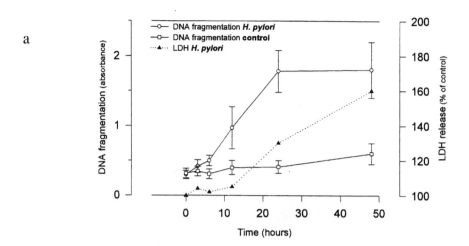

b

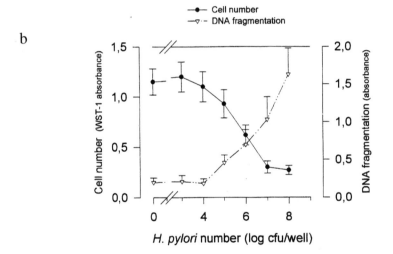

c

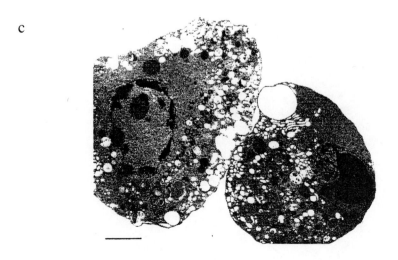

lial apoptosis in gastritis and duodenal ulcer patients. Interestingly, *H. pylori* not only induces gastric epithelial apoptosis, but also increases apoptosis of mucosal B-lymphocytes in patients with gastritis (Reinacher-Schick *et al.*, 1998).

It has been suggested that increased epithelial apoptosis may be restricted to those subjects colonised by *H. pylori* that lack the *cag* pathogenicity island (Peek *et al.*, 1997). This contrasts with data obtained from duodenal ulcer patients who do have increased epithelial apoptotic rates although their *H. pylori* virtually always carry the *cag* pathogenicity island (Moss *et al.*, 1996; Konturek *et al.*, 1999). Further *in vivo* studies are needed to define the role of the *cag* pathogenicity island in the induction of apoptosis by *H. pylori*. *In vitro*, it has been clearly demonstrated that *cagA* positive *H. pylori* isolates are able to induce gastric epithelial apoptosis (see below) (Wagner *et al.,* 1997; Rudi *et al.*, 1998; Peek *et al.*, 1999). Mechanisms and regulatory factors involved in *H. pylori* induced apoptosis are difficult to investigate *in vivo* since it is impossible to distinguish direct bacterial effects from those induced by the associated inflammatory response.

Mechanisms and Pathways of *H. pylori* Induced Apoptosis

H. pylori Directly Induces Apoptosis *In Vitro*

Cell culture studies have the advantage of examining the effects of *H. pylori in vitro* without the associated immune response. Cultivation of gastric cancer cells with *H. pylori* caused a rapid and concentration dependent induction of DNA fragmentation (Figure 2), indicating that *H. pylori* is capable of directly triggering apoptosis (Wagner *et al.*, 1997). Apoptosis could be induced with either intact bacteria or soluble extracts of *H. pylori* (Chen *et al.*, 1997; Rudi *et al.*, 1998; Peek *et al.*, 1999). Immortalised gastric tumour cell lines are less differentiated than freshly isolated non-malignant gastric epithelial cells, and the induction of apoptosis by *H. pylori* in these cell lines could be due to their malignant derivation. However, very similar results have been obtained with freshly isolated gastric epithelial cells in primary culture, confirming the data from studies with cancer cell lines.

All *H. pylori* Isolates Induce Apoptosis

To investigate whether the induction of apoptosis correlates with the expression of bacterial virulence factors, 60 clinical *H. pylori* isolates with different *cagA* (34 positive, 26 negative) and *vacA* (26 s1m1, 12 s1m2, 21 s2m2) phenotypes were co-cultured with the gastric cancer cell line AGS cells. All strains induced DNA fragmentation to varying extent, and there was no association with the *cagA* or *vacA* genotype (Wagner *et al.*, 1998). The induction of gastric epithelial apoptosis is specific for *H. pylori* because control experiments showed that *Campylobacter jejuni* failed to induce DNA fragmentation in gastric HM02 cells (Wagner *et al.*, 1997).

Figure 2. *H. pylori* induces apoptosis in the gastric cancer cell line HM02. a) Time course of DNA fragmentation (oligonucleosome-bound DNA in the 20,000 g supernatant of cell lysates) and LDH release. b) Effect of increasing numbers of *H. pylori* on cell number and DNA fragmentation. c) Transmission electron micrograph of two apoptotic cells. Their nuclei display electron dense chromatin along the inner surface of the nuclear membrane with deeply stained patches of chromatin condensation. The cell membrane is intact, whereas the cytoplasm contains numerous vacuoles (size bar = 2 μm). The figures are adapted from Wagner *et al.*, 1997.

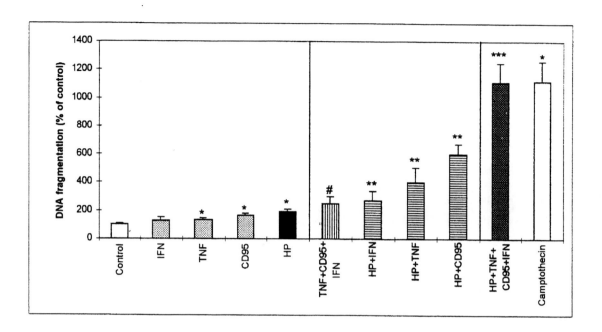

Figure 3. Regulation of *H. pylori* induced DNA fragmentation by the proinflammatory factors TNFα, CD95 ligand and interferon γ. DNA fragmentation was determined with a histone ELISA. HM02 cells (8x10⁴/well) were incubated for 24 h with TNF((20 ng/ml), antibody to CD95 (20 ng/ml), interferon γ (10 ng/ml), *H. pylori* (10⁶ cfu/well) or camptothecin (4μg/ml) in the combinations indicated. The data represent mean ± SD of 3 determinations. * and #, p<.05 versus control; **, p<.05 versus *H. pylori*; ***, p<.05 versus *H. pylori* or any of the cytokines. (adapted from Wagner *et al.*, 1997).

Synergistic Effect of Proinflammatory Cytokines

H. pylori infection results in increased concentrations in the mucosa of many proinflammatory and immunoregulatory cytokines (Crabtree, 1998). Among these are tumour necrosis factor (TNF)α and interferon (IFN)γ (Bamford *et al.*, 1998). CD95/APO-1/Fas ligand is known to be involved in the initiation of immune-mediated apoptosis (Rudi *et al.,* 1998). Therefore we studied the *in vitro* effects of these cytokines in *H. pylori* mediated gastric epithelial apoptosis (Wagner *et al.*, 1997). Treatment of gastric epithelial cells with TNFα or an agonistic CD95/APO-1/Fas antibody in the absence of *H. pylori* had a modest stimulatory effect on DNA fragmentation, which was less than DNA fragmentation induced by *H. pylori* (Figure 3). When *H. pylori* was combined with these factors, DNA fragmentation was stimulated dramatically, indicating that *H. pylori* acts synergistically with proinflammatory cytokines to induce gastric epithelial apoptosis. The extent of apoptosis is determined by both the bacterial load and the local concentration of proinflammatory cytokines.

Parallels Between the *In Vitro* Findings and the Situation *In Vivo*

Figure 4 presents a model of gastric epithelial cell homeostasis which reconciles the *in vitro* findings in co-culture systems with the clinical observations in *H. pylori* infected patients (Wagner *et al.*, 1997). In the healthy stomach, cell loss due to apoptosis is small and is matched by the number of newly formed cells in the proliferating zone. In *H. pylori* gastritis, en-

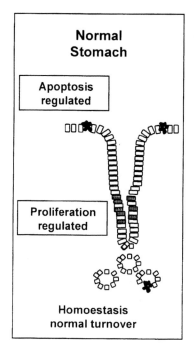
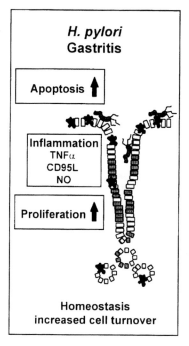
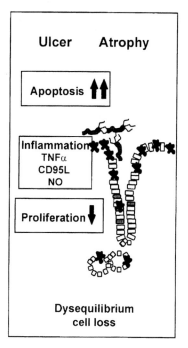

Figure 4. Model of gastric epithelial cell turnover and mucosal homeostasis in health and *H. pylori*-associated gastric disease. The figure illustrates the roles of apoptosis and epithelial proliferation in gastric glands in the normal stomach and during gastritis and ulcer/atrophy induced by *H. pylori*. (adapted from Wagner *et al.*, 1997).

hanced apoptosis is found primarily in the superficial compartment of the gastric mucosa where the bulk of the bacteria is found. Apoptosis is induced directly by *H. pylori* and is further upregulated by the associated inflammatory response. This enhanced apoptosis causes a compensatory hyperproliferation and an enlargement of the proliferating zone, resulting in epithelial homeostasis but at a higher rate of cell turnover. A higher density of *H. pylori* or an exaggerated immune response could disrupt this equilibrium due to excessive apoptosis and/ or an inhibition of epithelial proliferation, ultimately resulting in cell loss promoting the development of peptic ulcers (in the acute phase) or atrophy (in the chronic phase). Indeed, the density of *H. pylori* during infection has been recently identified as an important factor for the development of duodenal ulcer and a baseline infection density seems to be necessary for ulcer formation (Khulusi *et al.*, 1995). Moreover, an increased epithelial apoptotic rate has been documented at the ulcer margin both in peptic ulcer patients and in a rat model of gastric ulcers (Li *et al.*, 1997, 1998; Kohda *et al.*, 1999). In duodenal ulcers, the patients at the active stage exhibited a significantly higher level of apoptosis than those at the healing and scarring stage (Kohda *et al.,* 1999). In *H. pylori* infected patients with atrophic gastritis, an increased apoptotic rate was observed which correlated with the expression of anti-canalicular autoantibodies (Steininger *et al.*, 1998). In a small case-control study apoptosis and proliferation were increased 31 years before the development of gastric atrophy (Moss *et al.*, 1999). Conversely, a reduced apoptotic rate has been reported in foci of intestinal metaplasia in *H. pylori* infected patients and it has been suggested that this contributes to *H. pylori* associated gastric carcinogenesis (Scotiniotis *et al.*, 2000). Taken together, these findings suggest that in patients with *H. pylori* infection the apoptosis to proliferation ratio may contribute to the broad spectrum of disease manifestations. How *H. pylori* regulates the balance between apoptosis and proliferation is currently unknown.

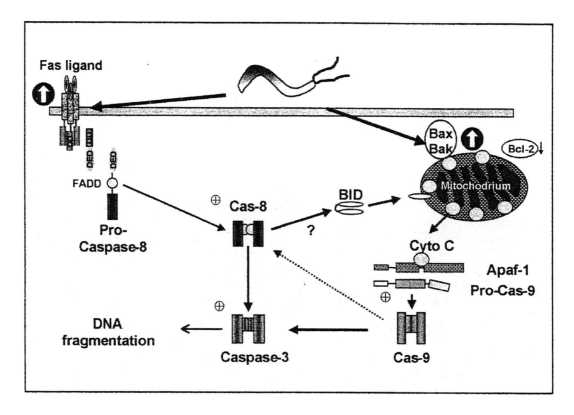

Figure 5. Model of interaction between *H. pylori* and the apoptotic signalling pathway. Two distinct apoptotic signalling pathways are depicted, via the Fas death receptor and via the mitochondrial route, as well as their downstream events.

Interaction Between *H. pylori* and the Apoptotic Pathway

A variety of triggers have been identified which can initiate apoptosis: growth factor withdrawal, cell cycle perturbations, DNA damage, alterations of the extracellular matrix, oxidative stress, nitric oxide, immunologic-mediated processes, activation of specific cell death receptors. These triggers have been grouped into two distinct signalling pathways for the induction of apoptosis, namely the activation of specific cell death receptors and the exposure to cellular stress (Nagata, 1997; Adams and Cory 1998; Ashkenazi and Dixit, 1998; Green 1998; Vaux and Korsmeyer, 1999; Mehmet, 2000). Preliminary evidence suggests that *H. pylori* may influence apoptosis in a number of subtle ways although the molecular nature of the pro-apoptotic factor is still unknown (Figure 5).

Cell Death Receptors

Activation of the cell death receptor CD95/APO-1/Fas augments apoptosis induced by *H. pylori* (Wagner *et al.*, 1997; Rudi *et al.*, 1998; Jones *et al.*, 1999). In addition, *H. pylori* up-regulates the expression of CD95 in infected patients and upon infection of AGS cells *in vitro* (Rudi *et al.*, 1998; Jones *et al.*, 1999). A ligand for the CD95 receptor is normally expressed by T lymphocytes and natural killer cells, which suggests that increased CD95 expression enhances the susceptibility of gastric epithelial cells to T cell killing. Moreover, in *H. pylori* gastritis, gastric epithelial cells are also capable of expressing the CD95 ligand on their sur-

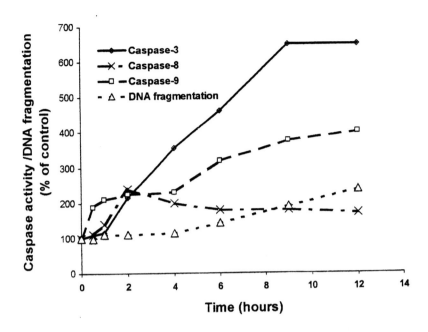

Figure 6. Time course of caspase activity and DNA fragmentation in gastric epithelial cells (AGS) incubated with *H. pylori*. Caspase activity was determined fluorometrically using specific AFC-labelled probes. DNA fragmentation was analysed with a histone ELISA. The data represent the mean of three independent experiments.

faces. This implies that in addition to lymphocyte mediated apoptosis, the gastric epithelial cells may also be capable of fratricide (Rudi *et al.*, 1998).

Bcl-2 Family

Apoptosis involves the mitochondrial signalling pathway, including the recruitment of pro-apoptotic bcl-2 homologues and the release of cytochrome C to help form apoptosomes (oligomeric complexes formed by Apaf-1, pro-caspase-9, and cytochrome C; Figure 5) (Vaux 1997; Adams and Cory 1998; Green and Reed 1998; Martinou 1999). The bcl-2 family can be subdivided in three subfamilies: (i) the bcl-2 subfamily with prosurvival activity, e.g. bcl-2, bcl-x$_L$, CED9, (ii) the bax subfamily with proapoptotic activity, e.g. bax, bak, bok and (iii) the BH3 subfamily with proapoptotic activity, e.g. bik, bid, bad, EGL1 (Adams and Cory, 1998). All bcl-2 subfamilies possess at least one of four conserved motifs known as the bcl-2 homology domains (BH1 to BH4). Those bcl-2 homologues most similar to bcl-2 promote cell survival by inhibiting adapters needed for the activation of downstream caspases. More distant relatives promote apoptosis, apparently through mechanisms that include displacing the adapters from the prosurvival proteins. Thus, for many but not all apoptotic signals, the balance between these competing activities determines cell fate. *H. pylori* induced apoptosis has been shown to be accompanied by the increased expression of the pro-apoptotic bax subfamily members bax and bak (Chen *et al.*, 1997; Konturek *et al.*, 1999). In duodenal ulcer patients, *H. pylori* upregulates the transcription of bax mRNA and the expression of bax in the mucosa of the antrum and body, whereas the expression of bcl-2 is downregulated (Konturek *et al.*, 1999). The eradication of *H. pylori* failed to normalise the levels of bax and bcl-2. Chen *et al.*, (1997) demonstrated an increased expression of bak in patients with *H.*

pylori gastritis and in *H. pylori* infected AGS cells. Thus, *H. pylori* seems to shift the balance between the prosurvival and proapoptotic members of the bcl-2 family towards apoptosis.

Caspases

The induction of apoptosis through the binding of cell death receptors leads to the activation of initiator caspases (caspase-8, -2, -10), which then activate downstream effector caspases (caspase-3, -6, -7) (Ashkenazi and Dixit 1998; Green 1998; Thornberry and Lazebnik 1998; Mehmet 2000). This mitochondrial signalling route is also accompanied by the activation of caspase-9 (Figure 5). Therefore, we investigated whether *H. pylori* induced apoptosis is associated with the activation of specific caspases (Wagner *et al.*, 1999). The incubation of gastric epithelial cells with *H. pylori* led to a time- and dose-dependent sequential activation of caspase-9, caspase-8, and caspase-3 which preceded the induction of DNA fragmentation (Figure 6). *H. pylori* did not induce activation of caspase-1 indicating that its activation of caspases is specific. The inhibition of caspase-3 and caspase-9 completely blocked *H. pylori* induced DNA fragmentation, whereas the inhibition of caspase-1 failed to block DNA fragmentation. These findings indicate that the activation of caspase-9 and caspase-3 are essential for *H. pylori* induced apoptosis. A central role of caspase-3 in *H. pylori* induced apoptosis has been independently reported recently (Kim *et al.*, 2000).

Other Priming Signals

Shirin *et al.* (1999) reported that the induction of apoptosis by *H. pylori* stops the cell cycle at G_1-S. Fan *et al.* (1998) suggested that the binding to class II MHC molecules may induce the apoptosis of gastric epithelial cells but did not supply details on the signalling pathway. High doses of purified *H. pylori* lipopolysaccharide can also induce apoptosis (Piotrowski *et al.*, 1996, Slomiany *et al.*, 1998). Other priming signals from *H. pylori* or the inflammatory response associated with infection that can potentially induce apoptosis include oxygen radicals, ammonia, urease, and ceramide (Lin *et al.*, 1998; Shirin and Moss, 1998; Fiers *et al.*, 1999). These potential priming signals have not yet been studied in detail.

Outlook

Alterations in the ratio of apoptosis to proliferation may be important for the different clinical disease manifestations of *H. pylori* infection. Understanding the molecular mechanisms involved in the control of apoptosis and proliferation might form the basis for new therapeutic strategies. The recent advent of improved animal models for *H. pylori* infection provides a promising tool to clarify these issues.

References

Adams, J.M. and Cory, S. 1998. The Bcl-2 protein family: arbiters of cell survival. Science 281: 1322-1326.

Anti, M., Armuzzi, A., Iascone, E., Valenti, A., Lippi, M.E., Covino, M., Vecchio, F.M., Pierconti, F., Buzzi, A., Pignataro, G., Bonvicini, F., and Gasbarrini, G. 1998. Epithelial-cell apoptosis and proliferation in *Helicobacter pylori*-related chronic gastritis. Ital. J. Gastroenterol. Hepatol. 30: 153-159.

Ashkenazi, A. and Dixit, V.M. 1998. Death receptors: signaling and modulation. Science 281: 1305-1308.

Bamford, K.B., Fan, X., Crowe, S.E., Leary, J.F., Gourley, W.K., Luthra, G.K., Brooks, E.G., Graham, D.Y., Reyes, V.E., and Ernst, P.B. 1998. Lymphocytes in the human gastric mucosa during *Helicobacter pylori* have a T helper cell 1 phenotype. Gastroenterology 114: 482-492.

Cahill, R.J., Xia, H., Kilgallen, C., Beattie, S., Hamilton, H., and O'Morain, C. 1995. Effect of eradication of *Helicobacter pylori* infection on gastric epithelial cell proliferation. Dig. Dis. Sci. 40: 1627-1631.

Chen, G., Sordillo, E.M., Ramey, W.G., Reidy, J., Holt, P.R., Krajewski, S., Reed, J.C., Blaser, M.J., and Moss, S.F. 1997. Apoptosis in gastric epithelial cells is induced by *Helicobacter pylori* and accompanied by increased expression of BAK. Biochem. Biophys. Res. Commun. 239: 626-632.

Correa, P. and Miller, M.J. 1998. Carcinogenesis, apoptosis and cell proliferation. Br. Med. Bull. 54: 151-162.

Crabtree, J.E. 1998. Role of cytokines in pathogenesis of *Helicobacter pylori*-induced mucosal damage. Dig. Dis. Sci. 43: 46S-55S.

Evan, G. and Littlewood, T. 1998. A matter of life and cell death. Science 281: 1317-1322.

Fan, X., Crowe, S.E., Behar, S., Gunasena, H., Ye, G., Haeberle, H., Van Houten, N., Gourley, W.K., Ernst, P.B., and Reyes, V.E. 1998. The effect of class II major histocompatibility complex expression on adherence of *Helicobacter pylori* and induction of apoptosis in gastric epithelial cells: A mechanism for T helper cell type 1-mediated damage. J. Exp. Med. 187: 1659-1669.

Fan, X.G., Kelleher, D., Fan, X.J., Xia, H.X., and Keeling, P.W. 1996. *Helicobacter pylori* increases proliferation of gastric epithelial cells. Gut 38: 19-22.

Fraser, A.G., Sim, R., Sankey, E.A., Dhillon, A.P., and Pounder, R.E. 1994. Effect of eradication of *Helicobacter pylori* on gastric epithelial cell proliferation. Aliment. Pharmacol. Ther. 8: 167-173.

Green, D.R. 1998. Apoptotic pathways: the roads to ruin. Cell 94: 695-698.

Green, D.R. and Reed, J.C. 1998. Mitochondria and apoptosis. Science 281: 1309-1312.

Fiers, W., Beyaert, R., Declercq, W., and Vandenabeele, P. 1999. More than one way to die: apoptosis, necrosis and reactive oxygen damage. Oncogene 18: 7719-7730.

Hahm, K.B., Lee, K.J., Choi, S.Y., Kim, J.H., Cho, S.W., Yim, H., Park, S.J., and Chung, M.H. 1997. Possibility of chemoprevention by the eradication of *Helicobacter pylori*: oxidative DNA damage and apoptosis in H. pylori infection. Am. J. Gastroenterol. 92: 1853-1857.

Hall, P.A., Coates, P.J., Ansari, A., and Hopwood, D. 1994. Regulation of cell number in the mammalian gastrointestinal tract: the importance of apoptosis. J. Cell Sci. 107: 3569-3577.

Havard, T.J., Sarsfield, P., Wotherspoon, A.C., and Steer, H.W. 1996. Increased gastric epithelial cell proliferation in *Helicobacter pylori* associated follicular gastritis. J. Clin. Pathol. 49: 68-71.

Houghton, J., Korah, R.M., Condon, M.R., and Kim, K.H. 1999. Apoptosis in *Helicobacter pylori*-associated gastric and duodenal ulcer disease is mediated via the Fas antigen pathway. Dig. Dis. Sci. 44: 465-478.

Jones, N.L., Shannon, P.T., Cutz, E., Yeger, H., and Sherman, P.M. 1997. Increase in proliferation and apoptosis of gastric epithelial cells early in the natural history of *Helicobacter pylori* infection. Am. J. Pathol. 151: 1695-1703.

Jones, N.L., Day, A.S., Jennings, H.A., and Sherman, P.M. 1999. *Helicobacter pylori* induces gastric epithelial cell apoptosis in association with increased Fas receptor expression. Infect. Immun. 67: 4237-4242.

Kerr, J.F., Wyllie, A.H., and Currie, A.R. 1972. Apoptosis: a basic biological phenomenon with wide-ranging implications in tissue kinetics. Br. J. Cancer 26: 239-257.

Khulusi, S., Mendall, M.A., Patel, P., Levy, J., Badve, S., and Northfield, T.C. 1995. *Helicobacter pylori* infection density and gastric inflammation in duodenal ulcer and non-ulcer subjects. Gut 37: 319-324.

Kim, J.M., Kim, J.S., Jung, H.C., Song, I.S., and Kim, C.Y. 2000. Apoptosis of human gastric epithelial cells via caspase-3 activation in response to *Helicobacter pylori* infection: possible involvement of neutrophils through tumor necrosis factor alpha and soluble Fas ligands. Scand. J. Gastroenterol. 35: 40-48.

Kohda, K., Tanaka, K., Aiba, Y., Yasuda, M., Miwa, T., and Koga, Y. 1999. Role of apoptosis induced by *Helicobacter pylori* infection in the development of duodenal ulcer. Gut 44: 456-462.

Konturek, P.C., Pierzchalski, P., Konturek, S.J., Meixner, H., Faller, G., Kirchner, T., and Hahn, E.G. 1999. *Helicobacter pylori* induces apoptosis in gastric mucosa through an upregulation of Bax expression in humans. Scand. J. Gastroenterol. 34: 375-383.

Li, H., Mellgard, B., and Helander, H.F. 1997. Inoculation of VacA- and CagA- *Helicobacter pylori* delays gastric ulcer healing in the rat. Scand. J. Gastroenterol. 32: 439-444.

Li, H., Kalies, I., Mellgard, B., and Helander, H.F. 1998. A rat model of chronic *Helicobacter pylori* infection. Studies of epithelial cell turnover and gastric ulcer healing. Scand. J. Gastroenterol. 33: 370-378.

Lin, Y.L., Liu, J.S., Chen, K.T., Chen, C.T., and Chan, E.C. 1998. Identification of neutral and acidic sphingomyelinases in *Helicobacter pylori*. FEBS Lett. 423: 249-253.

Lynch, D.A., Mapstone, N.P., Clarke, A.M., Sobala, G.M., Jackson, P., Morrison, L., Dixon, M.F., Quirke, P., and Axon, A.T. 1995. Cell proliferation in *Helicobacter pylori* associated gastritis and the effect of eradication therapy. Gut 36: 346-350.

Mannick, E.E., Bravo, L.E., Zarama, G., Realpe, J.L., Zhang, X.J., Ruiz, B., Fontham, E.T., Mera, R., Miller, M.J., and Correa, P. 1996. Inducible nitric oxide synthase, nitrotyrosine, and apoptosis in *Helicobacter pylori* gastritis: effect of antibiotics and antioxidants. Cancer Res. 56: 3238-3243.

Martin, S.J. and Green, D.R. 1995. Protease activation during apoptosis: death by a thousand cuts? Cell 82: 349-352.

Martinou, J.C. 1999. Apoptosis. Key to the mitochondrial gate. Nature 399: 411-412.

Mehmet, H. 2000. Caspases find a new place to hide. Nature 403: 29-30.

Moss, S.F., Calam, J., Agarwal, B., Wang, S., and Holt, P.R. 1996. Induction of gastric epithelial apoptosis by *Helicobacter pylori*. Gut 38: 498-501.

Moss, S.F., Valle, J., Abdalla, A.M., Wang, S., Siurala, M., and Sipponen, P. 1999. Gastric cellular turnover and the development of atrophy after 31 years of follow-up: a case-control study. Am. J. Gastroenterol. 94: 2109-2114.

Nagata, S. 1997. Apoptosis by death factor. Cell 88: 355-365.

Peek, R.M.J., Moss, S.F., Tham, K.T., Perez-Perez, G.I., Wang, S., Miller, G.G., Atherton, J.C., Holt, P.R., and Blaser, M.J. 1997. *Helicobacter pylori* cagA+ strains and dissociation of gastric epithelial cell proliferation from apoptosis. J. Natl. Cancer Inst. 89: 863-868.

Peek, R.M.J., Blaser, M.J., Mays, D.J., Forsyth, M.H., Cover, T.L., Song, S.Y., Krishna, U., and Pietenpol, J.A. 1999. *Helicobacter pylori* strain-specific genotypes and modulation of the gastric epithelial cell cycle. Cancer Res. 59: 6124-6131.

Piotrowski, J., Skrodzka, D., Slomiany, A., and Slomiany, B.L. 1996. *Helicobacter pylori* lipopolysaccharide induces gastric epithelial cells apoptosis. Biochem. Mol. Biol. Int. 40: 597-602.

Que, F.G. and Gores, G.J. 1996. Cell death by apoptosis: basic concepts and disease relevance for the gastroenterologist. Gastroenterology 110: 1238-1243.

Reinacher-Schick, A., Petrasch, S., Burger, A., Suerbaum, S., Kunstmann, E., and Schmiegel, W. 1998. *Helicobacter pylori* induces apoptosis in mucosal lymphocytes in patients with gastritis. Z. Gastroenterol. 36: 1021-1026.

Rudi, J., Kuck, D., Strand, S., von Herbay, A., Mariani, S.M., Krammer, P.H., Galle, P.R., and Stremmel, W. 1998. Involvement of the CD95 (APO-1/Fas) receptor and ligand system in *Helicobacter pylori*-induced gastric epithelial apoptosis. J. Clin. Invest. 102: 1506-1514.

Salvesen, G.S. and Dixit, V.M. 1997. Caspases: intracellular signaling by proteolysis. Cell 91: 443-446.

Scotiniotis, I.A., Rokkas, T., Furth, E.E., Rigas, B., and Shiff, S.J. 2000. Altered gastric epithelial cell kinetics in *Helicobacter pylori*-associated intestinal metaplasia: implications for gastric carcinogenesis. Int. J. Cancer 85: 192-200.

Shirin, H. and Moss, S.F. 1998. *Helicobacter pylori* induced apoptosis. Gut 43: 592-594.

Shirin, H., Sordillo, E.M., Oh, S.H., Yamamoto, H., Delohery, T., Weinstein, I.B., and Moss, S.F. 1999. *Helicobacter pylori* inhibits the G1 to S transition in AGS gastric epithelial cells. Cancer Res. 59: 2277-2281.

Slomiany, B.L., Piotrowski, J., and Slomiany, A. 1998. Induction of caspase-3 and nitric oxide synthase-2 during gastric mucosal inflammatory reaction to *Helicobacter pylori* lipopolysaccharide. Biochem. Mol. Biol. Int. 46: 1063-1070.

Steininger, H., Faller, G., Dewald, E., Brabletz, T., Jung, A., and Kirchner, T. 1998. Apoptosis in chronic gastritis and its correlation with antigastric autoantibodies. Virchows Arch. 433: 13-18.

Thompson, C.B. 1995. Apoptosis in the pathogenesis and treatment of disease. Science 267: 1456-1462.

Thornberry, N.A. and Lazebnik, Y. 1998. Caspases: enemies within. Science 281: 1312-1316.

Vaux, D.L. 1997. CED-4—the third horseman of apoptosis. Cell 90: 389-390.

Vaux, D.L. and Korsmeyer, S.J. 1999. Cell death in development. Cell 96: 245-254.

Wagner, S., Beil, W., Westermann, J., Logan, R.P.H., Bock, C.T., Trautwein, C., Bleck, J.S., and Manns, M.P. 1997. Regulation of gastric epithelial cell growth by *Helicobacter pylori*: evidence for a major role of apoptosis. Gastroenterology 113: 1836-1847.

Wagner, S., Wingbermühle, D., Beil, W., Obst, B., Sobek-Klocke, I., Schmidt, H., Bleck, J., and Manns, M.P. 1998. Significance of *H. pylori cagA* and *vacA* genotypes on apoptosis of gastric epithelial cells. Gastroenterology 114: A326.

Wagner, S., Sobek-Klocke, I., Obst, B., Schmidt, H., Bleck, J.S., Kirchner, G., Göke, M., Manns, M.P., and Beil, W. 1999. Activation of caspase-8 and -3 mediates apoptosis in gastric epithelial cells induced by *Helicobacter pylori*. Gastroenterology 116: A347.

White, E. 1996. Life, death, and the pursuit of apoptosis. Genes Dev. 10: 1-15.

Zhu, G.H., Yang, X.L., Lai, K.C., Ching, C.K., Wong, B.C., Yuen, S.T., Ho, J., and Lam, S.K. 1998. Nonsteroidal antiinflammatory drugs could reverse *Helicobacter pylori*-induced apoptosis and proliferation in gastric epithelial cells. Dig. Dis. Sci. 43: 1957-1963.

8

Animal Models for *Helicobacter pylori* Gastritis

Kathryn A. Eaton[*]

Abstract

Almost since the first successful culture of *Helicobacter pylori* and its association with human disease, animal models have been central in the progression of *Helicobacter* research. Studies in animals have illuminated numerous aspects of disease including the roles of bacterial virulence factors in colonization, the role of the host immune response in clearing infection as well as contributing to disease, and mechanisms of carcinogenesis. In addition, the development of effective treatment regimes, and the current progress in development of safe and effective vaccines are largely products of the use of animal models.

In recent years, as our understanding of *Helicobacters* progressed, animal models have become increasingly refined. The introduction of robust mouse models of disease due to *Helicobacter pylori* has led to an explosion of studies using these models in recent years, and progress has been rapid in the elucidation of all aspects of disease. This review describes the history and use of animal models of *H. pylori* gastritis, with emphasis on recently developed mouse and other rodent models. Features of animal models are described as are their uses in and contributions to the investigation of bacterial virulence factors, carcinogenesis, and host responses and immunology.

Philosophy of Animal Models

In *Helicobacter pylori* research, as in biomedical research in general, animal models have enjoyed a mixed status since their inception. On one hand, invasive and manipulative studies, critical to our understanding of host-pathogen interactions, can only be done in animals. On the other hand, animal model studies must always be tempered by the realization that there are species differences both between host species and humans and between the bacteria that normally infect animals and humans. These potential differences could have major effects on experimental outcomes.

Indeed, there is a curious bimodal perception in the biomedical community. Firstly, there is the perception that all mammals represent little furry humans and can be expected to respond in the same way. Secondly, no animal is believed to adequate represent the complexities of human disease. In fact, these perceptions are both true and false. It is true that the basic mechanisms of physiology and disease are remarkably similar between species. From the responses to pathogens through to the complexities of development, animal species are con-

[*] email: eaton.1@osu.edu

sistently similar in pattern, even where they differ in detail. However, species differences do exist and must be considered when designing animal experiments. These considerations are complicated by the fact that many subtle and possibly unknown differences exist that may have large effects on experimental results.

So what is a biomedical researcher to do? Fortunately, the answer lies in the same considerations applied to all biologic research. All models, be they *in vitro* systems, *in vivo* non-human systems, epidemiologic systems, or computer simulations, must be interpreted in light of the known characteristics of the models. Differences between host species and humans and between non-human *Helicobacter* and *H. pylori* may affect experimental findings, but such differences may also be helpful in dissecting mechanisms of human disease. For example, investigation of known species and strain differences in response to gastric helicobacters (Eaton *et al.*, 1993b; Mohammadi *et al.*, 1996b; Sakagami *et al.*, 1996, 1997; Dubois *et al.*, 1999; Sutton *et al.*, 1999) could lead to insights which will explain differences in responses between individual human patients. It is the goal of this review to provide an overview of the current state of the art in animal models of helicobacter gastritis while highlighting their role in understanding the pathogenesis of human disease.

Non-Rodent Models

Animal model studies used in *H. pylori* research between 1987 and 1997 have been extensively reviewed (Dubois, 1998; Eaton, 1999; Lee, 1999; Nedrud, 1999), and will be only briefly described here. Until quite recently, most animal models depended on hosts other than rodents and/or *Helicobacter* species other than *H. pylori* (for a review of the *Helicobacter* species used in animal models, see Eaton, 1999). The first animal models involved germ-free piglets and non-human primates (Krakowka *et al.*, 1987; Lambert *et al.*, 1987; Newell *et al.*, 1987). These species are susceptible to *H. pylori* of human origin, and develop persistent chronic gastritis in response to colonization (Krakowka *et al.*, 1987; Eaton *et al.*, 1989, 1990; Dubois *et al.*, 1994., 1996, 1999; Handt *et al.*, 1997; Stadtlander *et al.*, 1998; Reindel *et al.*, 1999). Their use is expensive and requires facilities and expertise that are not widely available. Nevertheless, these models have been extremely valuable in the study of bacterial virulence factors (Eaton *et al.*, 1989, 1991, 1992b, 1994, 1995a, 1996c, 1997, 1999; Akopyants *et al.*, 1995; Mankoski *et al.*, 1999), therapy (Krakowka *et al.*, 1998; Dubois *et al.*, 1999; Lee *et al.*, 1999a), and vaccination procedures (Stadtlander *et al.*, 1996; Dubois *et al.*, 1998; Lee *et al.*, 1999a, 1999b). "Post-vaccination gastritis", a phenomenon whereby vaccination actually exacerbates *H. pylori* gastritis, was first described in piglets (Eaton *et al.*, 1992a).

Germ-free dogs and specific pathogen-free cats can also be colonized by *H. pylori*, and a few studies have been done in these species (Radin *et al.*, 1990; Handt *et al.*, 1994, 1995; Fox *et al.*, 1995, 1996b; Perkins *et al.*, 1996). Lesions are mild, however, and practical considerations have limited the use of these models to a few laboratories. Recently, experimental *H. pylori* colonization of conventional laboratory beagles was described (Rossi *et al.*, 1999). Non-*pylori* animal *Helicobacter*s have also had significant influence on the study of *Helicobacter* gastritis. Cats and dogs are commonly colonized by endogenous helicobacters of the *heilmannii/felis* group (see below) (Geyer *et al.*, 1993; Eaton *et al.*, 1996b; Papasouliotis *et al.*, 1997; Honda *et al.*, 1998a; Jalava *et al.*, 1998; Neiger *et al.*, 1998; Peyrol *et al.*, 1998; Yamasaki *et al.*, 1998; Norris *et al.*, 1999; Simpson *et al.*, 1999). While naturally infected animals have been used to investigate pathogenesis, the general consensus among most researchers is that these organisms are commensal and do not cause disease in their natural hosts (Otto *et al.*, 1994; Eaton *et al.*, 1996b; Papasouliotis *et al.*, 1997; Happonen *et al.*, 1998; Neiger *et al.*, 1998; Peyrol *et al.*, 1998; Yamasaki *et al.*, 1998; Norris *et al.*, 1999; Simpson *et al.*, 1999). An interesting exception is the captive cheetah, which can be colonized by both *Helicobacter acinonychus,* previously *Helicobacter acinonyx* (Eaton *et al.*, 1993a,1993c; Truper and De Clari, 1997), and unculturable *Helicobacter* of the *heilmannii/felis* group. Infected cheetahs develop severe gastritis and wasting disease which is specifically associated with captivity. Wild cheetahs do not develop lesions of gastritis in response to gastric *Helicobacter* (Terio *et al.*, 1999), and an examination of the differences between captive and wild cheetahs might lead to insights into the role of the host response in *Helicobacter*-associated disease.

Ferrets and other mustelids are naturally colonized with an endogenous *Helicobacter, H. mustelae* (Fox *et al.*, 1988, 1990; Tompkins *et al.*, 1988; Gottfried *et al.*, 1990; Otto *et al.*, 1990). In some ferrets, *H. mustelae* induces gastritis similar to *H. pylori*-associated gastritis in humans. Interestingly, not all infected ferret colonies have gastritis (Tompkins *et al.*, 1988), but the mechanisms underlying these differences have not been investigated. It has been suggested that *H. mustelae* also predisposes ferrets to gastric carcinoma (Fox *et al.*, 1997) and MALT lymphoma (Erdman *et al.*, 1997), and infection with *H. mustelae* may increase the incidence of gastric carcinoma in ferrets treated with N'-methyl- N'-nitro-N-nitrosoguanidine (MNNG), a mucosal carcinogen (Fox *et al.*, 1993b). The relatively long lifespan of the ferret (6-8 years), and the difficulties associated with generating and maintaining *H. mustelae*-free ferrets (Otto *et al.*, 1990) has limited use of this model. However, *H. mustelae* is a natural infection that causes disease in the native host, colonization rates are high and persistent, and the bacterial species is well adapted to the host.

Non-rodent models have been and continue to be useful in many aspects of *H. pylori* research. Many of them closely replicate the range of *H. pylori*-associated human disease, and they are less contrived than rodent models. Some models use naturally occurring host-adapted *Helicobacter*s, and all use naturally outbred host populations, more similar to human infection than are rodent populations. That having been stated, recently developed rodent models offer new and exciting opportunities for the elucidation of molecular and immunologic mechanisms of *H. pylori* related disease. Rodents allow experimental control of host genetic background, bacterial strain and genetics. Also, germ-free and specific pathogen free rodents can be used to eliminate interactions with non-pylori helicobacters and other enteric flora

Thus, as indicated by the large number of recent mouse studies, murine models of disease are likely to be at the forefront of *H. pylori* research for some time to come.

Rodent Models: Mice

Probably the most influential non-*pylori* model is the conventional or germ-free mouse colonized with *Helicobacter felis* and related bacteria. These Helicobacters of animal origin colonize diverse host species and are useful for experimentation because of their wide host range compared to *H. pylori*. These bacteria are larger than *H. pylori* and colonize deeper in the gastric glands. Some, like *H. felis*, can be cultured *in vitro* while others, like *Helicobacter heilmannii* (Solnick *et al.*, 1993) (also called *Gastrospirillum*) are as yet unculturable. Mice are highly susceptible to bacteria of the *felis/heilmannii* group and host responses to all these organisms appear similar (Lee *et al.*, 1990, 1993; Eaton *et al.*, 1993b, 1995b; Fox *et al.*, 1993a; Moura *et al.*, 1993). Gastric lesions vary somewhat depending on mouse strain and bacterial species (Mohammadi *et al.*, 1996b; Sakagami *et al.*, 1996), but in general bacteria of the *felis/heilmannii* group cause moderate to severe chronic gastritis (lymphocytic and plasmacytic) or chronic active (lymphocytic, plasmacytic, and neutrophilic) gastritis by 4-8 weeks after inoculation. Long-term colonization (1 year or more) is accompanied by more severe gastritis and marked proliferation of gastric epithelium with loss of normal peptic gland structure and replacement by undifferentiated mucus-type cells. Grossly, the mucosa appears either diffusely thickened or contains 1-2 mm raised nodules with or without gastric ulcers (Eaton *et al.*, 1995b; Fox *et al.*, 1996a). Histologically, epithelial proliferation is either polypoid or sessile, and is associated with the loss of parietal and other specialized cell types coupled with their replacement by less differentiated mucus-type cells (Lee *et al.*, 1990, 1993, 1994; Eaton *et al.*, 1993b, 1995b; Fox *et al.*, 1996a). In the most severely affected regions, parietal and chief cells are largely absent and are replaced by mucus-type glands. This change is variously referred to as "metaplasia", "mucus metaplasia", "atrophy" (referring to loss of parietal cells), and "hyperplasia" (because it is accompanied by increased cellular proliferation and increased thickness of the gastric mucosa). It should not be confused with true atrophy in which glands disappear and are replaced by fibrous connective tissue stroma and the gastric mucosa decreases in thickness, or intestinal metaplasia which is characterized by replacement of gastritis-type epithelium by intestinal type epithelium containing goblet cells. True intestinal metaplasia is an important precursor of gastric carcinogenesis in humans and is presumed to be due to *H. pylori*, but is not consistently present in mice.

Mice are not susceptible to most human *H. pylori* strains. Early studies in mice resulted in failure of colonization, weak colonization, and/or the lack of persistence (Ehlers *et al.*, 1988; Cantorna *et al.*, 1990; Karita *et al.*, 1991; Marchetti *et al.*, 1995), leading researchers to pursue other models. The introduction of mouse-adapted *H. pylori* was something of a breakthrough in animal model research. In 1997, Adrian Lee and colleagues described a mouse-adapted bacterial strain which they called SS1 (Sydney Strain 1) (Lee *et al.*, 1997). This strain, isolated from a patient with gastritis and a history of peptic ulcer disease, consistently colonizes mice well (Lee *et al.*, 1997; Ferrero *et al.*, 1998; Chen *et al.*, 1999; Eaton *et al.*, 1999d). Other human *H. pylori* isolates have also been shown to colonize mice weakly, but strain SS1 appears to be the best colonizer described so far. The density of colonization is 10^2 - 10^4 fold higher than that achieved with other strains (Marchetti *et al.*, 1995; Chen *et al.*, 1999; Sawai *et al.*, 1999), and, in contrast to previously described strains (Telford *et al.*, 1994; Ghiara *et al.*, 1995; Marchetti *et al.*, 1995), SS1 consistently produces gastritis in some strains of mice. Therefore, SS1 has become the most widely used bacterial strain in recent murine studies and it is likely that introduction of this strain and its wide dissemination has led to the marked increase in animal model studies seen in recent years.

Most strains of mice are susceptible to infection with *H. pylori* strain SS1. C57BL/6 mice support between 10^6 and 10^7 cfu/gram of gastric mucosa (Lee *et al.*, 1997; Chen *et al.*, 1999; Eaton *et al.*, 1999d), or about 10^5 - 10^6 cfu/stomach. This level of colonization results in the presence of bacteria within gastric pits, and often large numbers of bacteria can be seen packed within glands. Other strains of mice support lower and more variable colonization, about 10^3 -10^5 cfu/g (Lee *et al.*, 1997; Ferrero *et al.*, 1998). In these mice, bacteria are rarely visible in histologic sections, and if present are confined to the antrum-body border and the cardia adjacent to the squamous border. In addition to density of colonization, the severity of gastritis varies with the mouse strain. "Reactor" mouse strains, such as C57BL/6 and C3H, develop moderate, multifocal to focally severe gastritis by about 16 weeks after inoculation, and gastritis continues to worsen over the course of infection. Lesions are similar in character to those induced by *H. felis* but less severe. Foci of more intense inflammation may be accompanied by neutrophilic gland abscesses, destruction of glands, and superficial mucosal erosions, although true ulceration is not seen. In areas of more intense inflammation, loss of the normal fundic gland morphology ("mucus metaplasia") as described above for *H. felis*, is present. In some mouse strains, such as BALB/c and DBA, little to no gastritis develops (Sakagami *et al.*, 1996).

Uses of Mouse Models

Most recent animal model studies have used mice colonized with *H. felis* or *H. pylori*. The use of *H. felis* is somewhat limited by genetic and physiologic differences from *H. pylori*, and an examination of the role of specific bacterial virulence factors requires the use of *H. pylori* itself. Animal models using *H. felis* and *H. pylori* have yielded several consistent findings: the essential nature of flagellae for colonization (Eaton *et al.*, 1992b, 1996c; Josenhans *et al.*, 1999), host T and B cells responses, cytokine profiles and immunoglobulin responses (Mohammadi *et al.*, 1996b, 1997; Ferrero *et al.*, 1997, 1998; Saldinger *et al.*, 1998; Eaton *et al.*, 1999c, 1999d; Roth *et al.*, 1999), suppression but not clearance of colonization by conventional vaccination (Lee *et al.*, 1994, 1995; Blanchard *et al.*, 1995; Ferrero *et al.*, 1995; Pappo *et al.*, 1995; Lee, 1996; Guy *et al.*, 1998, 1999), post-immunization gastritis (Ermak *et al.*, 1997; Goto *et al.*, 1999), and differences in the host response depending on host species and strain (Mohammadi *et al.*, 1996b; Sakagami *et al.*, 1996). The consistency of these results indicate that studies in mice colonized by *H. felis* remain relevant and can reveal general mechanisms of disease which are applicable to humans.

Bacterial Virulence Factors

The virulence factors of *H. pylori* thus far examined in mice include the *cag* proteins that are potentially involved in the production of disease(Ghiara *et al.*, 1995; Eaton *et al.*, 1999a). In addition, virulence factors involved in colonization have also been investigated, such as urease (Ghiara *et al.*, 1995; Karita *et al.*, 1995; Skouloubris *et al.*, 1998), flagellin and flagellar control proteins (Foynes *et al.*, 1999; Kim *et al.*, 1999), gamma glutamyl transpeptidase (Blanchard *et al.*, 1999b; Chevalier *et al.*, 1999), and other factors (Janvier *et al.*, 1999; Raudonikiene *et al.*, 1999). Development of new molecular techniques such as shuttle vectors (Heuermann *et al.*, 1998) and methods of conditional expression of targeted genes (Bijlsma *et al.*, 1999) are likely to greatly facilitate virulence factor studies in mice.

Bacterial virulence factor studies in mice are limited by the fact that strain SS1 is somewhat less transformable than other bacterial strains (unpublished observations). In addition, experimental *in vivo* results using *H. pylori* mutants must be interpreted in light of the high spontaneous mutation rate of this organism. The genetic heterogeneity of *H. pylori* is well known, and functional studies have revealed rapid phenotypic variation in motility (Eaton *et al.*, 1992b), LPS expression (Appelmelk *et al.*, 1999), and antibiotic resistance (Jenks *et al.*, 1999). The ability to rapidly alter its phenotype may represent an adaptive feature for survival in different environments, but recent data suggest that it may also adversely affect colonization experiments. For example, conflicting data on the importance for colonization of gamma-glutamyl transpeptidase (Blanchard *et al.*, 1999b; Chevalier *et al.*, 1999) and *cag*-region genes (Bjorkholm *et al.*, 1999; Eaton *et al.*, 1999a) may be explained by random loss of colonization factors during laboratory passage, rather than loss of colonization associated with a specific targeted mutation. Thus, colonization by several independent transformants of a genetic construct should be examined before the lack of colonization can confidently be attributed to loss of the targeted gene.

Carcinogenesis

As described above, long-term colonization for 12 months or more of mice by bacteria of the *felis/heilmannii* group leads to marked polypoid or sessile adenomatous proliferation, sometimes associated with gastric ulceration. These lesions, while never proven to be neoplastic, show some atypical features and have been used to investigate carcinogenesis and mechanisms of *Helicobacter*-associated epithelial proliferation.. Co-carcinogenesis studies in mice have shown increased proliferation of gastric epithelium when *H. felis* infection was combined with nitrosamines such as MNNG and methyl-nitroso urea (MNU) (Danon *et al.*, 1998a; Shimizu *et al.*, 1998; Barreto-Zuniga *et al.*, 1999) or with a high salt diet (Shimizu *et al.*, 1998; Fox *et al.*, 1999b). Whether these proliferative lesions are actually pre-neoplastic is still not clear, but examination of their pathogenesis has revealed some intriguing findings. For example, the proliferation and development of a pre-neoplastic cell type in mice colonized with *H. felis* is correlated with the presence of a murine gene encoding secreted phospholipase A2 (sPLA2). This genetic locus, *mom1*, is associated with colonic polyp formation in mice with multiple intestinal neoplasia (Wang *et al.*, 1998) and may be important in gastric carcinogenesis as well. In another study, mice with a single, interrupted p53 allele responded to *H. felis* with more pronounced proliferation than normal mice, suggesting that loss of tumor suppressor genes may contribute to pre-neoplastic proliferation (Fox *et al.*, 1996a). However, no increase in expression of p53 has been described in *Helicobacter*-related gastric polyps in mice (Danon *et al.*, 1998a), and the role of this tumor suppressor gene remains uncertain. Finally, it has been shown that (at least in BALB/c mice) mucosal polypoid proliferation is dependent on normal T cell function (Eaton *et al.*, 1996a), and that the diffuse proliferation associated with *H. pylori* in mice is dependent on an intact host adaptive immune response (Eaton *et al.*, 1999d). Thus, if epithelial proliferation does represent a preneoplastic lesion, it is likely to be due at least in part to a host immune responses to bacterial products. These results lay an important foundation for experimental examination of the roles in preneoplastic gastric epithelial proliferation of bacterial products and host immune responses and inflammation.

Most cancer studies in mice have focused on the pathogenesis of epithelial neoplasms, but a few have investigated the role of gastric *Helicobacter* in lymphoid proliferation. In virtually all species, gastric Helicobacters induce the development of submucosal lymphoid follicles. These can become widespread in the stomach, and in chronically infected mice, they become confluent and invade not only the lamina propria but also the muscularis mucosae. This marked proliferation of mucosal-associated lymphoid tissue (MALT) in chronically infected mice has been used to model MALT lymphoma in humans (Enno *et al.*, 1995, 1998).

Role of the Host in Gastritic Disease

Mice colonized with *H. felis* have proven to be a valuable tool for investigating many aspects of host immune response to both infection and immunization. Genetic studies have demonstrated that host resistance to *H. felis*-associated gastritis is not MHC linked (Sakagami *et al.*, 1996), and is inherited in a dominant fashion (Sutton *et al.*, 1999). The strong dominant inheritance suggests that resistance may be attributable to one or a few genes. Host strain differences have also been revealing in other ways. The finding, for example, that C3H/HeJ LPS non-responder mice fail to develop gastritis in response to *H. felis* suggests that LPS may contribute to disease (Sakagami *et al.*, 1997) These findings are consistent with studies in humans suggesting that LPS antigens may correlate with increased severity of disease (Moran *et al.*, 1996). Further investigation of the genetics of murine host response to gastric Helicobacters could reveal important host markers for susceptibility to disease.

Perhaps the most exciting aspect of the mouse-*H. pylori* model of disease is the opportunity it affords for studying host immune responses. Mice are widely used for immunologic studies, and most of the significant discoveries in immunology over the latter half of this century have depended on murine models of disease. Many different inbred strains of mice are available, allowing comparison between strains and interbreeding experiments to isolate genetic factors which contribute to disease, as mentioned above. In addition, reagents and methods for evaluating mouse immune responses are well defined and easily available. Thus, both *in situ* and *in vitro* evaluation of humoral and cellular responses to specific antigens and epitopes is achievable, as is the isolation of single or small groups of relevant immune cells. Most immunology studies in *H. pylori* research have focused on responses to vaccination. However, many of those analyses are also relevant for understanding responses to infection. The first results with mice infected with *H. felis* were pivotal in demonstrating that gastritis due to *H. pylori* is largely Th-1 polarized as indicated by high levels of interferon gamma and interleukin-1, and low levels of interleukin-4 (Mohammadi *et al.*, 1996a, 1997) (see also Chapter 6). Similar responses of naturally infected humans (D'Elios *et al.*, 1997; Lindholm *et al.*, 1998) indicates that murine responses represent a good model for examination of the role of host immune response in pathogenesis. The Th-1 polarization of the host *H. pylori* immune response led to the suggestion that the stimulation of a more Th-2 polarized component by immunization may lead to more protection with less inflammation (Mohammadi *et al.*, 1997; Saldinger *et al.*, 1998). Other data, however, appear to suggest that both the Th-1 and Th-2 components may be at least in part protective (Guy *et al.*, 1998, Zevering *et al.*, 1999). Clearly, further study will be necessary to examine the roles of the various cytokines in protection and gastritis.

The use of genetically targeted mutant mice (cytokine "knockout" mice), described more fully in Chapter 9, has contributed supporting evidence regarding host immune responses, and will be briefly mentioned here. These studies have confirmed that IFNγ is essential for gastritis due to *H. pylori* (Sawai *et al.*, 1999) and that IL-10 is essential for dampening gastritis due to *H. felis* (Berg *et al.*, 1998), confirming the role of Th-1 polarized responses in contributing to disease. Results with IL-4 deficient and transgenic mice have not revealed marked differences, leaving the role of Th-2 cytokines unclear (Chen *et al.*, 1999). Examination of the response of cytokine knockout mice to vaccination has revealed that protective response to vaccination is MHCII- restricted. In those studies, MHCI mutant mice had normal responses to immunization, but MHCII mutant mice were not protected (Ermak *et al.*, 1998; Pappo *et al.*, 1999). Finally, studies with knockout mice have indicated that cellular rather than humoral immune responses are responsible for both gastritis and protection. μMT knockout mice, which fail to produce antibody but which have normal cellular immune responses, were indistinguishable from wild-type mice both in terms of gastritis, *H. pylori* colonization, and their response to vaccination (Ermak *et al.*, 1998; Blanchard *et al.*, 1999a). These studies confirm adoptive transfer studies, described below, which indicate that CD4 T cells are both necessary and sufficient for the production of gastritis due to *H. pylori*.

Our laboratory has recently developed an adoptive transfer model for examination of host immune responses to *H. pylori* (Eaton *et al.*, 1999d). In this model, splenocytes from C57BL/6 mice are adoptively transferred into infected C57BL/6 *scid/scid* (SCID, severe combined immunodeficient) mice. SCID mice fail to recombine either immunoglobulins or T cell receptors and thus do not have mature T or B lymphocytes. They completely lack an adaptive immune system, although their innate immunity is intact. Use of this model has revealed some intriguing results. First, in the absence of transferred cells, SCID mice do not develop gastritis in response to *H. pylori* infection. Bacterial colonization in these mice is high (approximately 10-fold-higher than colonization in normal mice), but no inflammation develops. More surprising is that adoptive transfer of normal splenocytes from uninfected or infected donors results in widespread, severe, and rapidly developing gastritis in infected (but not uninfected) recipient mice. In contrast to the mild, multifocal gastritis present in C57BL/6 mice 16 weeks after infection, gastritis in recipient mice is detectable by 2 weeks after transfer, and by 4 weeks after transfer it is severe and may involve up to 100% of the gastric mucosa. Finally, recipient mice develop a delayed-type hypersensitivity footpad response to bacterial antigen; a response which is not present in uninfected mice or in infected C57BL/6 mice. Thus it is clear that regulatory factors which suppress gastritis in normal mice are absent in recipient SCID mice. Recent findings have indicated that CD4+ T cells are responsible for the severe gastritis in recipient mice infected with either *H. pylori* (Eaton *et al.*, 1999c) or *H. felis* (Roth *et al.*, 1999), underscoring the important role of cellular immunity in *Helicobacter*-related disease. This adoptive transfer mouse model offers promise not only for investigation of host factors which contribute to or protect from gastritis, but will also be useful in identifying *H. pylori*-specific bacterial factors responsible for the induction of severe gastritis and host cellular immune response.

Morphologic Assessment of Murine Models

One factor which renders a comparison of murine models difficult are the widely different schemes used for morphologic assessment of gastritis. Morphologic assessment is an important part of the evaluation of host immune and inflammatory responses, but morphology can be difficult to quantify. In addition, different investigators use different systems of quantification, thus further hindering interpretation. In general, morphologic evaluation includes three factors: severity of mononuclear infiltration, severity of neutrophilic infiltration, and the extent of inflammation (that is, the amount of gastric tissue involved). Some investigators include epithelial changes as well as inflammation. Grading is usually semi-quantitative, using a 3 or 6-point scale based on subjective assessment, and a total score is usually calculated by combining several individual scores in various ways. Thus, systems vary not only in tissue preparation methods (something which is difficult to control in an asymmetric structure such as the stomach), but also in scoring methods, lesions scored, and the method of interpretation. Clearly a standardized system would benefit all researchers in this field.

We have developed a method of scoring of mouse gastric tissues based on the extent of gastritis (Eaton *et al.*, 1999d) that is still being standardized (Eaton *et al.*, 1999b). Individual lesions (gastritis, neutrophil influx, adenitis, and metaplasia) are scored according to specific definitions (Eaton *et al.*, 1999d). Each 20x microscopic field is examined, and the presence or absence of each lesion is ascertained. The number of affected fields divided by the total number of fields is multiplied by 100 to yield the percentage of affected mucosa. Verification of the system has indicated that it is reproducible among readers of varying experience and educational background, and that it accurately identifies mice according to treatment group. Standardization of scoring in this way is likely to greatly facilitate comparison of results between laboratories and further advance progress in the use of murine models of *H. pylori* gastritis.

Non-murine Rodent Models

Rats are larger than mice and yet retain some of the same advantages (inbred strains and readily available immunologic reagents), rendering them an attractive host species for animal model development. Gastritis due to *H. felis* in rats has been examined in a few studies (Fox *et al.*, 1991; Danon *et al.*, 1998b), and rats can also be colonized by murine-adapted *H. pylori* (Li *et al.*, 1998, 1999).However, rats appear to support only low levels of colonization by *H. pylori*, and respond with minimal to absent gastritis. It is likely, however, that like mice, different rat strains will respond differently, and it may only be a matter of time before useable rat models are developed.

Recently, guinea pigs were shown to be susceptible to mouse-adapted *H. pylori* (Shomer *et al.*, 1998). These animals too supported only low levels of colonization, but they developed mild gastritis. A second study described severe gastritis in guinea pigs (Sturegard *et al.*, 1998), and it is likely that like mice, different strains of guinea pigs respond differently to *H. pylori*. The large size of guinea pigs compared to mice and their dietary requirement for vitamin C (not present in rats and mice) may make these animals suitable for investigation of the role of vitamin C in protection from neoplastic transformation in association with gastric *Helicobacter* infection.

Mongolian gerbils, *Meriones unguiculatus*, have achieved some notoriety in *H. pylori* research because of the assertion (Watanabe *et al.*, 1998) that infection of these animals by *H. pylori* causes gastric cancer. The first report (Yokota *et al.*, 1991) of *H. pylori* in gerbils described minimal gastritis and no other lesions for up to 2 months after inoculation. Neither gastritis nor colonization were quantified in this study, however. Subsequent studies have reported different results regarding both colonization and the number of lesions induced by *H. pylori* in gerbils. Some studies report frequent, rapid development of severe gastritis, ulcers, intestinal metaplasia, and neoplastic transformation (Honda *et al.*, 1998a, 1998b; Watanabe *et al.*, 1998; Ikeno *et al.*, 1999) while others report few ulcers and no metaplasia or neoplasia (Wirth *et al.*, 1998). In all cases in which ulcers were depicted, they were morphologically similar to stress ulcers (Matsumoto *et al.*, 1997; Takahashi *et al.*, 1998; Wirth *et al.*, 1998; Ikeno *et al.*, 1999), like the ulcers associated with H. *heilmannii* infection in mice (Eaton *et al.*, 1995b). The reported densities of bacterial colonization and virulence of *H. pylori* strains for gerbils also vary. In spite of these differences, most studies agree that, in contrast to the first report, colonization is accompanied by gastritis and that gastric stress-like ulcers occur.

While standardization is clearly necessary, the gerbil model promises to offer another opportunity for investigation of *H. pylori* related disease. Gerbils are larger than mice, allowing for more manipulations, and if claims of enhanced susceptibility and carcinogenesis prove reproducible, this model may be useful in evaluating bacterial factors which contribute to such lesions. Gerbils have already been used in some studies of colonization factors (Wirth *et al.*, 1998), co-carcinogenesis (Sugiyama *et al.*, 1998; Tatematsu *et al.*, 1998; Shimizu *et al.*, 1999; Tokieda *et al.*, 1999), and treatment (Kusuhara *et al.*, 1998; Suzuki *et al.*, 1998; Keto *et al.*, 1999), and the overall results concur with those found in mice. An excellent concise review of the use of Mongolian gerbils in *Helicobacter* research has recently been published (Wirth *et al.*, 1998).

Conclusions and Future Directions

Recent rapid progress in the development and use of animal models has been striking. The development of a good mouse model has put animal studies within the reach of most laboratories, and has greatly facilitated immunologic and genetic studies. A few additional, unusual but intriguing new models will also bear watching. Recently, human embryonic stomachs were implanted in nude mice and shown to mature and support colonization by *H. pylori* (Lozniewski *et al.*, 1999b). Use of this model offers an unusual opportunity to examine the role of *H. pylori* in the human stomach *ex vivo*, and could be useful for studies of treatment (Lozniewski *et al.*, 1999a) or colonization. Co-infection studies with *Helicobacter* and viruses (Jiang *et al.*, 1999) or intestinal nematodes (Fox *et al.*, 1999a, 1999c) promise to further illuminate the roles of host T cell responses in native and vaccine-induced immunity to gastric helicobacters. Thus, animal models continue to lend themselves to novel approaches to host-pathogen interactions.

As animal model studies progress and lead to new insights, they also lead to new questions. For example, the existence and mechanism of post vaccination gastritis is an important subject of study. Understanding this phenomenon will be vital to development of appropriate vaccination protocols for human use. The role of T helper cytokines in both gastritis and protection is also under intense investigation, and determination of the roles of Th-1 and Th-2 polarized responses may lead to novel therapeutic approaches. The field is rapidly changing, and within the next few years is likely to markedly alter, if not revolutionize, our understanding of *H. pylori* related disease.

References

Akopyants, N.S., Eaton, K.A. and Berg, D.E. 1995. Adaptive mutation and cocolonization during *Helicobacter pylori* infection of gnotobiotic piglets. Infect. Immun. 63: 116-121.

Appelmelk, B.J., Martin, S.L., Monteiro, M.A., Clayton, C.A., McColm, A.A., Zheng, P., Verboom, T., Maaskant, J.J., van den Eijnden, D.H., Hokke, C.H., Perry, M.B., Vandenbroucke-Grauls, C.M. and Kusters, J.G. 1999. Phase variation in *Helicobacter pylori* lipopolysaccharide due to changes in the lengths of poly(C) tracts in alpha3-fucosyltransferase genes. Infect. Immun. 67: 5361-5366.

Barreto-Zuniga, R., Kato, Y., Yamamoto, N., Nomura, K., Marotta, F., Giussepe, S., Shiotani, J., Shiozaw, K., Okuyama, M., Kitagawa, T. and Masakazu, M. 1999. *Helicobacter pylori* increases development of neoplastic lesion of glandular stomach in mice treated with N-methyl-N-nitrosourea. Gastroenterology 118: A375.

Berg, D.J., Lynch, N.A., Lynch, R.G. and Lauricella, D.M. 1998. Rapid development of severe hyperplastic gastritis with gastric epithelial dedifferentiation in *Helicobacter felis*-infected IL-10(-/-) mice. Amer. J. Pathol. 152: 1377-1386.

Bijlsma, J.J.E., Lie-A-Ling, M., Vandenbroucke-Grauls, C.M.J.E. and Kusters, J.G. 1999. Identification of genes essential for the growth of *Helicobacter pylori* at low ph. Gastroenterology 118: A123.

Bjorkholm, B., Engstrand, L., Rubio, C. and Falk, P. 1999. Cag pathogenicity island is required for colonization of *Helicobacter pylori* in a transgenic mouse model. Proceedings of the American Society for Microbiology, Chicago, ILL.

Blanchard, T., Czinn, S., Redline, R., Sigmund, N., Harriman, G. and Nedrud, J. 1999a. Antibody-independent protective mucosal immunity to gastric helicobacter infection in mice. Cellular Immunology 191: 74-80.

Blanchard, T.G., Czinn, S.J., Maurer, R., Thomas, W.D., Soman, G. and Nedrud, J.G. 1995. Urease-specific monoclonal antibodies prevent *Helicobacter felis* infection in mice. Infect. Immun. 63: 1394-1399.

Blanchard, T.G., McGovern, K.J., Youngman, P., Gutierrez, J., G, N.J. and S J Czinn, S.J. 1999b. Gamma-glutamyl transpeptidase is not essential for infection of the gastric mucosa by *H. pylori*. Gastroenterology 18: A143.

Cantorna, M.T. and Balish, E. 1990. Inability of human clinical strains of *Helicobacter pylori* to colonize the alimentary tract of germfree rodents. Can. J. Microbiol. 36: 237-241.

Chen, W., Shu, D. and Chadwick, V.A. 1999. *Helicobacter pylori* infection in interleukin-4-deficient and transgenic mice. Scand. J. Gastroenterol. 34: 987-992.

Chevalier, C., Thiberge, J., Ferrero, R. and Labigne, A. 1999. Essential role of *Helicobacter pylori* gamma-glutamyltranspeptidase for the colonization of the gastric mucosa of mice. Molec. Microbiol. 31: 1359-1372.

Danon, S. and Eaton, K. 1998a. The role of gastric Helicobacter and N-methyl-N'-nitro-N-nitrosoguanidine in carcinogenesis of mice. Helicobacter 3: 260-268.

Danon, S., Moss, N., Larsson, H., Arvidsson, S., Ottosson, S., Dixon, M. and Lee, A. 1998b. Gastrin release and gastric acid secretion in the rat infected with either *Helicobacter felis* or *Helicobacter heilmannii*. J. Gastroenterol. Hepatol. 13: 95-103.

D' Elios, M.M., Manghetti, M., Almerigogna, F., Amedei, A., Costa, F., Burroni, D., Baldari, C.T., Romagnani, S., Telford, J.L. and Delprete, G. 1997. Different cytokine profile and antigen-specificity repertoire in *Helicobacter pylori*-specific T cell clones from the antrum of chronic gastritis patients with or without peptic ulcer. Eur. J. Microbiol. 27: 1751-1755.

Dubois, A. 1998. Animal models of Helicobacter infection. Lab. Anim. Sci. 48: 596-603.

Dubois, A., Berg, D.E., Incecik, E.T., Fiala, N., Hemanackah, L.M., Delvalle, J., Yang, M.Q., Wirth, H.P., Perezperez, G.I. and Blaser, M.J. 1999. Host specificity of *Helicobacter pylori* strains and host responses in experimentally challenged nonhuman primates. Gastroenterology 116: 90-96.

Dubois, A., Berg, D.E., Incecik, E.T., Fiala, N., Hemanackah, L.M., Perezperez, G.I. and Blaser, M.J. 1996. Transient and persistent experimental infection of nonhuman primates with *Helicobacter pylori*: Implications for human disease. Infect. Immun. 64: 2885-2891.

Dubois, A., Fiala, N., Heman, A.L., Drazek, E.S., Tarnawski, A., Fishbein, W.N., Perez, P.G. and Blaser, M.J. 1994. Natural gastric infection with *Helicobacter pylori* in monkeys: a model for spiral bacteria infection in humans. Gastroenterology 106: 1405-1417.

Dubois, A., Lee, C.K., Fiala, N., Kleanthous, H., Mehlman, P.T. and Monath, T. 1998. Immunization against natural *Helicobacter pylori* infection in nonhuman primates. Infect. Immun. 66: 4340-4346.

Eaton, K.A. 1999. Animal models of *Helicobacter* gastritis. In: Gastroduodenal Disease and *Helicobacter pylori*. T. U. Westblom, S. J. Czinn and J. G. Nedrud. Eds. Heidelberger Platz 3, D-14197 Berlin, Germany, Springer-Verlag Berlin. p.123-154.

Eaton, K.A., Brooks, C.L., Morgan, D.R. and Krakowka, S. 1991. Essential role of urease in pathogenesis of gastritis induced by *Helicobacter pylori* in gnotobiotic piglets. Infect. Immun. 59: 2470-2475.

Eaton, K.A., Catrenich, C.E., Makin, K.M. and Krakowka, S. 1995a. Virulence of coccoid and bacillary forms of *Helicobacter pylori* in gnotobiotic piglets. J. Infect. Dis. 171: 459-462.

Eaton, K.A., Cover, T.L., Tummuru, M.K.R., Blaser, M.J. and Krakowka, S. 1997. Role of vacuolating cytotoxin in gastritis due to *Helicobacter pylori* in gnotobiotic piglets. Infect. Immun. 65: 3462-3464.

Eaton, K.A., Danon, S.J., Hoepf, T. and Ringler, S. 1996a. Gastric epithelial hyperplasia and hypertrophy caused by helicobacter gastritis in mice. Vet. Pathol. 33: 615.

Eaton, K.A., Danon, S.J., Kersulyte, D. and Berg, D.E. 1999a. *In vivo* relevance of *Helicobacter pylori cag* genes in two animal models. Gastroenterology 118: A152.

Eaton, K.A., Danon, S.J., Weisbrode, S. and Krakowka, S. 1999b. A new grading system for reproducible inter-laboratory histologic quanification of gastritis in mice infected with *Helicobacter pylori*. Proceedings of the Tenth International Workshop on Campylobacter, Helicobacter, and Related Organisms, Baltimore, Maryland.

Eaton, K.A., Dewhirst, F.E., Paster, B.J., Tzellas, N., Coleman, B.E., Paola, J. and Sherding, R. 1996b. Prevalence and varieties of Helicobacter species in dogs from random sources and pet dogs: Animal and public health implications. J. Clin. Microbiol. 34: 3165-3170.

Eaton, K.A., Dewhirst, F.E., Radin, M.J., Fox, J.G., Paster, B.J., Krakowka, S. and Morgan, D.R. 1993a. *Helicobacter acinonyx* sp. nov., isolated from cheetahs with gastritis. Int. J. Syst. Bacteriol. 43: 99-106.

Eaton, K.A. and Krakowka, S. 1992a. Chronic active gastritis due to *Helicobacter pylori* in immunized gnotobiotic piglets. Gastroenterology 103: 1580-1586.

Eaton, K.A. and Krakowka, S. 1994. Effect of gastric pH on urease-dependent colonization of gnotobiotic piglets by *Helicobacter pylori*. Infect. Immun. 62: 3604-3607.

Eaton, K.A., Mefford, M. and Ringler, S.S. 1999c. CD4+ T cells are necessary and sufficeint for gastritis in *Helicobacter pylori*-infected SCID mice. Vet. Pathol. 36(5): 510.

Eaton, K.A., Morgan, D.R. and Krakowka, S. 1989. *Campylobacter pylori* virulence factors in gnotobiotic piglets. Infect. Immun. 57: 1119-1125.

Eaton, K.A., Morgan, D.R. and Krakowka, S. 1990. Persistence of *Helicobacter pylori* in conventionalized piglets. J. Infect. Dis. 161: 1299-1301.

Eaton, K.A., Morgan, D.R. and Krakowka, S. 1992b. Motility as a factor in the colonisation of gnotobiotic piglets by *Helicobacter pylori*. J. Med.Microbiol. 37: 123-127.

Eaton, K.A., Radin, M.J. and Krakowka, S. 1993b. Animal models of bacterial gastritis: the role of host, bacterial species and duration of infection on severity of gastritis. Zlbt. Bakteriol. 280: 28-37.

Eaton, K.A., Radin, M.J. and Krakowka, S. 1995b. An animal model of gastric ulcer due to bacterial gastritis in mice. Vet. Pathol. 32: 489-497.

Eaton, K.A., Radin, M.J., Kramer, L., Wack, R., Sherding, R., Krakowka, S., Fox, J.G. and Morgan, D.R. 1993c. Epizootic gastritis associated with gastric spiral bacilli in cheetahs (*Acinonyx jubatus*). Vet. Pathol. 30: 55-63.

Eaton, K.A., Ringler, S.R. and Danon, S.J. 1999d. Murine splenocytes induce severe gastritis and delayed-type hypersensitivity and suppress bacterial colonization in *Helicobacter pylori*-infected SCID mice. Infect. Immun. 67: 4594-4602.

Eaton, K.A., Suerbaum, S., Josenhans, C. and Krakowka, S. 1996c. Colonization of gnotobiotic piglets by *Helicobacter pylori* deficient in two flagellin genes. Infect. Immun. 64: 2445-2448.

Ehlers, S., Warrelmann, M. and Hahn, H. 1988. In search of an animal model for experimental *Campylobacter pylori* infection: administration of *Campylobacter pylori* to rodents. Zblt Bakteriol 268: 341-346.

Enno, A., Orourke, J., Braye, S., Howlett, R. and Lee, A. 1998. Antigen-dependent progression of mucosa-associated lymphoid tissue (MALT)-type lymphoma in the stomach: Effects of antimicrobial therapy on gastric MALT lymphoma in mice. Amer. J. Pathol. 152: 1625-1632.

Enno, A., Orourke, J.L., Howlett, C.R., Jack, A., Dixon, M.F. and Lee, A. 1995. MALTomalike lesions in the murine gastric mucosa after long-term infection with *Helicobacter felis:* A mouse model of *Helicobacter pylori*-induced gastric lymphoma. Amer. J. Pathol. 147: 217-222.

Erdman, S.E., Correa, P., Coleman, L.A., Schrenzel, M.D., Li, X. and Fox, J.G. 1997. Helicobacter mustelae-associated gastric MALT lymphoma in ferrets. Amer. J. Pathol. 151: 273-280.

Ermak, T., Giannasca, P., Nichols, R., Myers, G., Nedrud, J., Weltzin, R., Lee, C., Kleanthous, H. and Monath, T. 1998. Immunization of mice with urease vaccine affords protection against *Helicobacter pylori* infection in the absence of antibodies and is mediated by MHC class II-restricted responses. J. Exper. Med. 188: 2277-2288.

Ermak, T.H., Ding, R., Ekstein, B., Hill, J., Myers, G.A., Lee, C.K., Pappo, J., Kleanthous, H.K. and Monath, T.P. 1997. Gastritis in urease-immunized mice after *Helicobacter felis* challenge may be due to residual bacteria. Gastroenterology 113: 1118-1128.

Ferrero, R.L., Thiberge, J.M., Huerre, M. and Labigne, A. 1998. Immune responses of specific-pathogen-free mice to chronic *Helicobacter pylori* (strain SS1) infection. Infect. Immun. 66: 1349-1355.

Ferrero, R.L., Thiberge, J.M., Kansau, I., Wuscher, N., Huerre, M. and Labigne, A. 1995. The GroES homolog of *Helicobacter pylori* confers protective immunity against mucosal infection in mice. Proc. Nat. Acad. Sci. 92: 6499-6503.

Ferrero, R.L., Thiberge, J.M. and Labigne, A. 1997. Local immunoglobulin G antibodies in the stomach may contribute to immunity against *Helicobacter* infection in mice. Gastroenterology 113: 185-194.

Fox, J., Li, X., Cahill, R., Andrutis, K., Rustgi, A., Odze, R. and Wang, T. 1996a. Hypertrophic gastropathy in *Helicobacter felis*-infected wild-type C57BL/6 mice and p53 hemizygous transgenic mice. Gastroenterology 110: 155-166.

Fox, J.G., Batchelder, M., Marini, R., Yan, L., Handt, L., Li, X., Shames, B., Hayward, A., Campbell, J. and Murphy, J.C. 1995. *Helicobacter pylori*-induced gastritis in the domestic cat. Infect. Immun. 63: 2674-2681.

Fox, J.G., Beck, P., Dangler, C.A., Wang, T., Whary, M.T., Shi, H.N. and Nagler-Anderson, C. 1999a. Modulation of the Th1/Th2 Response with *Heligomosomoides pylogyrus* and *Helicobacter felis* Coinfection in Mice: An Animal Model to Study the "African Enigma". Proceedings of the Tenth International Workshop on Campylobacter, Helicobacter, and Related Organisms, Baltimore, MD.

Fox, J.G., Blanco, M., Murphy, J.C., Taylor, N.S., Lee, A., Kabok, Z. and Pappo, J. 1993a. Local and systemic immune responses in murine *Helicobacter felis* active chronic gastritis. Infect. Immun. 61: 2309-2315.

Fox, J.G., Cabot, E.B., Taylor, N.S. and Laraway, R. 1988. Gastric colonization by *Campylobacter pylori* subsp. *mustelae* in ferrets. Infect. Immun. 56: 2994-2996.

Fox, J.G., Correa, P., Taylor, N.S., Lee, A., Otto, G., Murphy, J.C. and Rose, R. 1990. *Helicobacter mustelae*-associated gastritis in ferrets. An animal model of *Helicobacter pylori* gastritis in humans. Gastroenterology 99: 352-361.

Fox, J.G., Dangler, C.A., Sager, W., Borkowski, R. and Gliatto, J.M. 1997. Helicobacter mustelae-associated gastric adenocarcinoma in ferrets (Mustela putorius furo). Vet. Pathol. 34: 225-229.

Fox, J.G., Dangler, C.A., Taylor, N.S., King, A., Koh, T.J. and Wang, T.C. 1999b. High-salt diet induces gastric epithelial hyperplasia and parietal cell loss, and enhances *Helicobacter pylori* colonization in C57BL/6 mice. Cancer Res. 59: 4823-4828.

Fox, J.G., Dangler, C.A., Wang, T., Beck, P., Shi, H.N. and Nagler-Anderson, C. 1999c. Solving the African Enigma: Coinfection with an Intestinal Helminth Modulates Inflammation and Reduces Gastric Atrophy in a Mouse Model of Helicobacter Infection. Proceedings of the Tenth International Workshop on Campylobacter, Helicobacter, and Related Organisms.

Fox, J.G., Lee, A., Otto, G., Taylor, N.S. and Murphy, J.C. 1991. *Helicobacter felis* gastritis in gnotobiotic rats: an animal model of *Helicobacter pylori* gastritis. Infect. Immun. 59: 785-791.

Fox, J.G., Perkins, S., Yan, L., Shen, Z., Attardo, L. and Pappo, J. 1996b. Local immune response in *Helicobacter pylori*-infected cats and identification of *H-pylori* in saliva, gastric fluid and faeces. Immunology 88: 400-406.

Fox, J.G., Wishnok, J.S., Murphy, J.C., Tannenbaum, S.R. and Correa, P. 1993b. MNNG-induced gastric carcinoma in ferrets infected with *Helicobacter mustelae*. Carcinogenesis 14: 1957-1961.

Foynes, S., Dorrell, N., Ward, S.J., Zhang, Z.W., Mccolm, A.A., Farthing, M.J.G. and Wren, B.W. 1999. Functional analysis of the roles of FliQ and FlhB in flagellar expression in *Helicobacter pylori*. Fems Microbiol. Lett. 174: 33-39.

Geyer, C., Colbatzky, F., Lechner, J. and Hermanns, W. 1993. Occurrence of spiral-shaped bacteria in gastric biopsies of dogs and cats. Vet. Rec. 133: 18-19.

Ghiara, P., Marchetti, M., Blaser, M.J., Tummuru, M.K.R., Cover, T.L., Segal, E.D., Tompkins, L.S. and Rappuoli, R. 1995. Role of the *Helicobacter pylori* virulence factors vacuolating cytotoxin, CagA, and urease in a mouse model of disease. Infect. Immun. 63: 4154-4160.

Goto, T., Nishizono, A., Fujioka, T., Ikewaki, J., Mifune, K. and Nasu, M. 1999. Local secretory immunoglobulin A and postimmunization gastritis correlate with protection against *Helicobacter pylori* infection after oral vaccination of mice. Infect. Immun. 67: 2531-2539.

Gottfried, M.R., Washington, K. and Harrell, L.J. 1990. *Helicobacter pylori*-like microorganisms and chronic active gastritis in ferrets. Amer. J. Gastroenterol. 85: 813-818.

Guy, B., Hessler, C., Fourage, S., Haensler, J., Vialonlafay, E., Rokbi, B. and Millet, M.J.Q. 1998. Systemic immunization with urease protects mice against *Helicobacter pylori* infection. Vaccine 16: 850-856.

Guy, B., Hessler, C., Fourage, S., Rokbi, B. and Millet, M. 1999. Comparison between targeted and untargeted systemic immunizations with adjuvanted urease to cure *Helicobacter pylori* infection in mice. Vaccine 17: 1130-1135.

Handt, L.K., Fox, J.G., Dewhirst, F.E., Fraser, G.J., Paster, B.J., Yan, L.L., Rozmiarek, H., Rufo, R. and Stalis, I.H. 1994. *Helicobacter pylori* isolated from the domestic cat: Public health implications. Infect. Immun. 62: 2367-2374.

Handt, L.K., Fox, J.G., Stalis, I.H., Rufo, R., Lee, G., Linn, J., Li, X.T. and Kleanthous, H. 1995. Characterization of feline *Helicobacter pylori* strains and associated gastritis in a colony of domestic cats. J. Clin. Microbiol. 33: 2280-2289.

Handt, L.K., Fox, J.G., Yan, L.L., Shen, Z., Pouch, W.J., Ngai, D., Motzel, S.L., Nolan, T.E. and Klein, H.J. 1997. Diagnosis of *Helicobacter pylori* infection in a colony of rhesus monkeys (Macaca mulatta). J. Clin. Microbiol. 35: 165-168.

Happonen, I., Linden, J., Saari, S., Karjalainen, M., Hanninen, M., Jalava, K. and Westermarck, E. 1998. Detection and effects of helicobacters in healthy dogs and dogs with signs of gastritis. J. Amer. Vet. Med. Assn. 213: 1767-1774.

Heuermann, D. and Haas, R. 1998. A stable shuttle vector system for efficient genetic complementation of *Helicobacter pylori* strains by transformation and conjugation. Molec. Gen. Genet. 257: 519-528.

Honda, S., Fujioka, T., Tokieda, M., Gotoh, T., Nishizono, A. and Nasu, M. 1998a. Gastric ulcer, atrophic gastritis, and intestinal metaplasia caused by *Helicobacter pylori* infection in Mongolian gerbils. Scand. J. Gastroenterol. 33: 454-460.

Honda, S., Fujioka, T., Tokieda, M., Satoh, R., Nishizono, A. and Nasu, M. 1998b. Development of *Helicobacter pylori*-induced gastric carcinoma in Mongolian gerbils. Cancer Res. 58: 4255-4259.

Ikeno, T., Ota, H., Sugiyama, A., Ishida, K., Katsuyama, T., Genta, R.M. and Kawasaki, S. 1999. *Helicobacter pylori*-induced chronic active gastritis, intestinal metaplasia, and gastric ulcer in Mongolian gerbils. Amer. J. Pathol. 154: 951-960.

Jalava, K., On, S.L.W., Vandamme, P.A.R., Happonen, I., Sukura, A. and Hanninen, M.L. 1998. Isolation and identification of Helicobacter spp, from canine and feline gastric mucosa. Appl. Env. Microbiol. 64: 3998-4006.

Janvier, B., Grignon, B., Audibert, C., Pezennec, L. and Fauchere, J.L. 1999. Phenotypic changes of *Helicobacter pylori* components during an experimental infection in mice. FEMS Immunol. Med. Microbiol. 24: 27-33.

Jenks, P., Labigne, A. and Ferrero, R. 1999. Exposure to metronidazole *in vivo* readily induces resistance in *Helicobacter pylori* and reduces the efficacy of eradication therapy in mice. Antimicrob. Agents. Chemother. 43: 777-781.

Jiang, B., Jordana, M., Xing, Z., Smaill, F., Snider, D.P., Borojevic, R., Steelenorwood, D., Hunt, R.H. and Croitoru, K. 1999. Replication-defective adenovirus infection reduces *Helicobacter felis* colonization in the mouse in a gamma interferon- and interleukin-12-dependent manner. Infect. Immun. 67: 4539-4544.

Josenhans, C., Ferrero, R.L., Labigne, A. and Suerbaum, S. 1999. Cloning and allelic exchange mutagenesis of two flagellin genes of *Helicobacter felis*. Molec. Microbiol. 33: 350-362.

Karita, M., Kouchiyama, T., Okita, K. and Nakazawa, T. 1991. New small animal model for human gastric *Helicobacter pylori* infection: success in both nude and euthymic mice. Amer. J. Gastroenterol. 86: 1596-1603.

Karita, M., Tsuda, M. and Nakazawa, T. 1995. Essential role of urease *in vitro* and *in vivo* *Helicobacter pylori* colonization study using a wild-type and isogenic urease mutant strain. J. Clin. Gastroenterol. 21 Suppl. 1: S160-163.

Keto, Y., Takahashi, S. and Okabe, S. 1999. Healing of *Helicobacter pylori*-induced gastric ulcers in Mongolian gerbils combined treatment with omeprazole and clarithromycin.Dig. Dis. Sci. 44: 257-265.

Kim, J.S., Chang, J.H., Chung, S. and Yum, J.S. 1999. Molecular cloning and characterization of the *Helicobacter pylori fliD* gene, an essential factor in flagellar structure and motility. J. Bacteriol. 181: 6969-6976.

Krakowka, S., Eaton, K.A. and Leunk, R.D. 1998. Antimicrobial therapies for *Helicobacter pylori* infection in gnotobiotic piglets. Antimicrob. Agents. Chemother. 42: 1549-1554.

Krakowka, S., Morgan, D.R., Kraft, W.G. and Leunk, R.D. 1987. Establishment of gastric *Campylobacter pylori* infection in the neonatal gnotobiotic piglet. Infect. Immun. 55: 2789-2796.

Kusuhara, H., Hirayama, F., Matsuyuki, H., Hisadome, M. and Ikeda, Y. 1998. Evaluation of combined antibiotic-omeprazole therapies in *Helicobacter pylori*-infected Mongolian gerbils. J. Gastroenterol. 33: 14-17.

Lambert, J., Borromeo, M., Pinkard, K., Turner, H., Chapman, C. and Smith, M. 1987. Colonization of gnotobiotic piglets with *Campylobacter pyloridis*—an animal model? [letter]. J. Infect. Dis. 155: 1344.

Lee, A. 1994. The use of a mouse model in the study of *Helicobacter* sp.-associated gastric cancer. Eur. J. Gastroenterol. Hepatol. 6 Suppl 1: S67-71.

Lee, A. 1996. Therapeutic immunization against Helicobacter infection. Gastroenterology 110: 2003-2006.

Lee, A. 1999. Animal models of Helicobacter infection. Molec. Med. Today 5: 500-501.

Lee, A., Chen, M., Coltro, N., O'Rourke, J., Hazell, S., Hu, P. and Li, Y. 1993. Long term infection of the gastric mucosa with Helicobacter species does induce atrophic gastritis in an animal model of *Helicobacter pylori* infection.Zbl. Bakt. 280: 38-50.

Lee, A. and Chen, M.H. 1994. Successful immunization against gastric infection with Helicobacter species: Use of a cholera toxin B-subunit-whole-cell vaccine. Infect. Immun. 62: 3594-3597.

Lee, A., Fox, J.G., Otto, G. and Murphy, J. 1990. A small animal model of human *Helicobacter pylori* active chronic gastritis. Gastroenterology 99: 1315-1323.

Lee, A., Orourke, J., Deungria, M.C., Robertson, B., Daskalopoulos, G. and Dixon, M.F. 1997. A standardized mouse model of *Helicobacter pylori* infection: Introducing the Sydney strain. Gastroenterology 112: 1386-1397.

Lee, C., Soike, K., Hill, J., Georgakopoulos, K., Tibbitts, T., Ingrassia, J., Gray, H., Boden, J., Kleanthous, H., Giannasca, P., Ermak, T., Weltzin, R., Blanchard, J. and Monath, T. 1999a. Immunization with recombinant *Helicobacter pylori* urease decreases colonization levels following experimental infection of rhesus monkeys. Vaccine 17 1493-1505.

Lee, C.K., Soike, K., Giannasca, P., Hill, J., Weltzin, R., Kleanthous, H., Blanchard, J. and Monath, T.P. 1999b. Immunization of rhesus monkeys with a mucosal prime, parenteral boost strategy protects against infection with *Helicobacter pylori*. Vaccine 17: 3072-3082.

Lee, C.K., Weltzin, R., Thomas, W.D., Kleanthous, H., Ermak, T.H., Soman, G., Hill, J.E., Ackerman, S.K. and Monath, T.P. 1995. Oral immunization with recombinant *Helicobacter pylori* urease induces secretory IgA antibodies and protects mice from challenge with *Helicobacter felis*. J. Infect. Dis. 172: 161-172.

Li, H., Andersson, E.M. and Helander, H.F. 1999. Reactions from rat gastric mucosa during one year of *Helicobacter pylori* infection.Dig. Dis. Sci. 44: 116-124.

Li, H., Kalies, I., Mellgard, B. and Helander, H.F. 1998. A rat model of chronic *Helicobacter pylori* infection - Studies of epithelial cell turnover and gastric ulcer healing. Scand. J. Gastroenterol. 33: 370-378.

Lindholm, C., Quidingjarbrink, M., Lonroth, H., Hamlet, A. and Svennerholm, A.M. 1998. Local cytokine response in *Helicobacter pylori*-infected subjects. Infect. Immun. 66: 5964-5971.

Lozniewski, A., Duprez, A., Renault, C., Muhale, F., Conroy, M., Weber, M., Le, F.A. and Jehl, F. 1999a. Gastric penetration of amoxicillin in a human *Helicobacter pylori*-infected xenograft model. Antimicrob. Agents. Chemother. 43: 1909-1913.

Lozniewski, A., Muhale, F., Hatier, R., Marais, A., Conroy, M.C., Edert, D., Lefaou, A., Weber, M. and Duprez, A. 1999b. Human embryonic gastric xenografts in nude mice: a new model of *Helicobacter pylori* infection. Infect. Immun. 67: 1798-1805.

Mankoski, R., Hoepf, T., Krakowka, S. and Eaton, K.A. 1999. flaA mRNA transcription level correlates with *Helicobacter pylori* colonisation efficiency in gnotobiotic piglets. J. Med. Microbiol. 48: 395-399.

Marchetti, M., Arico, B., Burroni, D., Figura, N., Rappuoli, R. and Ghiara, P. 1995. Development of a mouse model of *Helicobacter pylori* infection that mimics human disease. Science 267: 1655-1658.

Matsumoto, S., Washizuka, Y., Matsumoto, Y., Tawara, S., Ikeda, F., Yokota, Y. and Karita, M. 1997. Induction of ulceration and severe gastritis in Mongolian gerbil by *Helicobacter pylori* infection. J. Med. Microbiol. 46: 391-397.

Mohammadi, M., Czinn, S., Redline, R. and Nedrud, J. 1996a. Helicobacter-specific cell-mediated immune responses display a predominant Th1 phenotype and promote a delayed-type hypersensitivity response in the stomachs of mice. J. Immunol. 156: 4729-4738.

Mohammadi, M., Nedrud, J., Redline, R., Lycke, N. and Czinn, S.J. 1997. Murine CD4 T-cell response to Helicobacter infection: TH1 cells enhance gastritis and TH2 cells reduce bacterial load. Gastroenterology 113: 1848-1857.

Mohammadi, M., Redline, R., Nedrud, J. and Czinn, S. 1996b. Role of the host in pathogenesis of Helicobacter-associated gastritis: *H. felis* infection of inbred and congenic mouse strains. Infect. Immun. 64: 238-245.

Moran, A.P., Prendergast, M.M. and Appelmelk, B.J. 1996. Molecular mimicry of host structures by bacterial lipopolysaccharides and its contribution to disease. FEMS Immunol. Med. Microbiol. 16: 105-115.

Moura, S.B., Queiroz, D.M., Mendes, E.N., Nogueira, A.M. and Rocha, G.A. 1993. The inflammatory response of the gastric mucosa of mice experimentally infected with "*Gastrospirillum suis*". J. Med. Microbiol. 39: 64-68.

Nedrud, J.G. 1999. Animal models for gastric Helicobacter immunology and vaccine studies.Fems Immunol. Med. Microbiol. 24: 243-250.

Neiger, R., Dieterich, C., Burnens, A., Waldvogel, A., Corthesytheulaz, I., Halter, F., Lauterburg, B. and Schmassmann, A. 1998. Detection and prevalence of Helicobacter infection in pet cats. J. Clin. Microbiol. 36: 634-637.

Newell, D.G., Hudson, M.J. and Baskerville, A. 1987. Naturally occurring gastritis associated with *Campylobacter pylori* infection in the rhesus monkey. Lancet 2: 1338.

Norris, C.R., Marks, S.L., Eaton, K.A., Torabian, S.Z., Munn, R.J. and Solnick, J.V. 1999. Healthy cats are commonly colonized with "*Helicobacter heilmannii*" that is associated with minimal gastritis. J. Clin. Microbiol. 37: 189-194.

Otto, G., Fox, J.G., Wu, P.Y. and Taylor, N.S. 1990. Eradication of *Helicobacter mustelae* from the ferret stomach: an animal model of *Helicobacter (Campylobacter) pylori* chemotherapy. Antimicrob. Agents. Chemother. 34: 1232-1236.

Otto, G., Hazell, S.H., Fox, J.G., Howlett, C.R., Murphy, J.C., O'Rourke, J.L. and Lee, A. 1994. Animal and public health implications of gastric colonization of cats by Helicobacter-like organisms. J. Clin. Microbiol. 32: 1043-1049.

Papasouliotis, K., Gruffydd-Jones, T.J., Werrett, G., Brown, P.J. and Pearson, G.R. 1997. Occurrence of "gastric Helicobacter-like organisms" in cats. Vet. Rec. 140: 369-370.

Pappo, J., Thomas, W.D., Kabok, Z., Taylor, N.S., Murphy, N.S. and Fox, J.G. 1995. Effect of oral immunization with recombinant urease on murine *Helicobacter felis* gastritis. Infect. Immun. 63: 1246-1252.

Pappo, J., Torrey, D., Castriotta, L., Savinainen, A., Kabok, Z. and Ibraghimov, A. 1999. *Helicobacter pylori* infection in immunized mice lacking major histocompatibility complex class I and class II functions. Infect. Immun. 67: 337-341.

Perkins, S.E., Yan, L.L., Shen, Z., Hayward, A., Murphy, J.C. and Fox, J.G. 1996. Use of PCR and culture to detect *Helicobacter pylori* in naturally infected cats following triple antimicrobial therapy. Antimicrob. Agents. Chemother. 40: 1486-1490.

Peyrol, S., Lecoindre, P., Berger, I., Deleforge, J. and Chevallier, M. 1998. Differential pathogenic effect of two Helicobacter-like organisms in dog gastric mucosa. J. Submicrosc. Cytol. Pathol. 30: 425-433.

Radin, M.J., Eaton, K.A., Krakowka, S., Morgan, D.R., Lee, A., Otto, G. and Fox, J. 1990. *Helicobacter pylori* gastric infection in gnotobiotic beagle dogs. Infect. Immun. 58: 2606-2612.

Raudonikiene, A., Zakharova, N., Su, W., Jeong, J., Bryden, L., Hoffman, P., Berg, D. and Severinov, K. 1999. *Helicobacter pylori* with separate beta- and beta'-subunits of RNA polymerase is viable and can colonize conventional mice. Molec. Microbiol. 32: 131-138.

Reindel, J.F., Fitzgerald, A.L., Breider, M.A., Gough, A.W., Yan, C., Mysore, J.V. and Dubois, A. 1999. An epizootic of lymphoplasmacytic gastritis attributed to *Helicobacter pylori* infection in cynomolgus monkeys (*Macaca fascicularis*). Vet. Pathol. 36: 1-13.

Rossi, G., Rossi, M., Vitali, C.G., Fortuna, D., Burroni, D., Pancotto, L., Capecchi, S., Sozzi, S., Renzoni, G., Braca, G., Del Giudice, G., Rappuoli, R., Ghiara, P. and Taccini, E. 1999. A conventional beagle dog model for acute and chronic infection with *Helicobacter pylori*. Infect. Immun. 67: 3112-3120.

Roth, K., Kapadia, S., Martin, S. and Lorenz, R. 1999. Cellular immune responses are essential for the development of *Helicobacter felis*-associated gastric pathology. J. Immunol. 163: 1490-1497.

Sakagami, T., Dixon, M., J, O.R., Howlett, R., Alderuccio, F., Vella, J., Shimoyama, T. and Lee, A. 1996. Atrophic gastric changes in both *Helicobacter felis* and *Helicobacter pylori* infected mice are host dependent and separate from antral gastritis. Gut 39: 639-648.

Sakagami, T., Vella, J., Dixon, M.F., O'Rourke, J., Radcliff, F., Sutton, P., Shimoyama, T., Beagley, K. and Lee, A. 1997. The endotoxin of *Helicobacter pylori* is a modulator of host-dependent gastritis. Infect. Immun. 65: 3310-3316.

Saldinger, P.F., Porta, N., Launois, P., Louis, J.A., Waanders, G.A., Bouzourene, H., Michetti, P., Blum, A.L. and Corthesytheulaz, I.E. 1998. Immunization of BALB/c mice with Helicobacter urease B induces a T helper 2 response absent in Helicobacter infection. Gastroenterology 115: 891-897.

Sawai, N., Kita, M., Kodama, T., Tanahashi, T., Yamaoka, Y., Tagawa, Y.I., Iwakura, Y. and Imanishi, J. 1999. Role of gamma interferon in *Helicobacter pylori*-induced gastric inflammatory responses in a mouse model. Infect. Immun. 67: 279-285.

Shimizu, N., Inada, K., Nakanishi, H., Tsukamoto, T., Ikehara, Y., Kaminishi, M., Kuramoto, S., Sugiyama, A., Katsuyama, T. and Tatematsu, M. 1999. *Helicobacter pylori* infection enhances glandular stomach carcinogenesis in Mongolian gerbils treated with chemical carcinogens. Carcinogenesis 20: 669-676.

Shimizu, N., Kaminishi, M., Tatematsu, M., Tsuji, E., Yoshikawa, A., Yamaguchi, H., Aoki, F. and Oohara, T. 1998. *Helicobacter pylori* promotes development of pepsinogen-altered pyloric glands, a preneoplastic lesion of glandular stomach of BALB/c mice pretreated with N-methyl-N-nitrosourea. Cancer Lett. 123 63-69.

Shomer, N., Dangler, C., Whary, M. and Fox, J. 1998. Experimental *Helicobacter pylori* infection induces antral gastritis and gastric mucosa-associated lymphoid tissue in guinea pigs. Infect. Immun. 66: 2614-2618.

Simpson, K., McDonough, P., D., S.-A., Chang, Y., Harpending, P. and Valentine, B. 1999. *Helicobacter felis* infection in dogs: effect on gastric structure and function. Vet. Pathol. 36: 237-248.

Skouloubris, S., Thiberge, J., Labigne, A. and De, R.H. 1998. The *Helicobacter pylori* UreI protein is not involved in urease activity but is essential for bacterial survival *in vivo*. Infect. Immun. 66: 4517-4521.

Solnick, J.V., O'Rourke, J., Lee, A., Paster, B.J., Dewhirst, F.E. and Tompkins, L.S. 1993. An uncultured gastric spiral organism is a newly identified Helicobacter in humans. J. Infect. Dis. 168: 379-385.

Stadtlander, C.T.K.H., Gangemi, J.D., Khanolkar, S.S., Kitsos, C.M., Farris, H.E., Fulton, L.K., Hill, J.E., Huntington, F.K., Lee, C.K. and Monath, T.P. 1996. Immunogenicity and safety of recombinant *Helicobacter pylori* urease in a nonhuman primate.Dig. Dis. Sci. 41: 1853-1862.

Stadtlander, C.T.K.H., Gangemi, J.D., Stutzenberger, F.J., Lawson, J.W., Lawson, B.R., Khanolkar, S.S., Elliottraynor, K.E., Farris, H.E., Fulton, L.K., Hill, J.E., Huntington, F.K., Lee, C.K. and Monath, T.P. 1998. Experimentally induced infection with *Helicobacter pylori* in squirrel monkeys (*Saimiri* spp.): Clinical, microbiological, and histopathologic findings. Lab. Anim. Sci. 48: 303-309.

Sturegard, E., Sjunnesson, H., Ho, B., Willen, R., Aleljung, P., Ng, H.C. and Wadstrom, T. 1998. Severe gastritis in guinea-pigs infected with *Helicobacter pylori*. J. Med. Microbiol. 47: 1123-1129.

Sugiyama, A., Maruta, F., Ikeno, T., Ishida, K., Kawasaki, S., Katsuyama, T., Shimizu, N. and Tatematsu, M.1998. *Helicobacter pylori* infection enhances N-methyl-N- nitrosourea-induced stomach carcinogenesis in the Mongolian gerbil. Cancer Res 58: 2067-2069.

Sutton, P., Wilson, J., Genta, R., Torrey, D., Savinainen, A., Pappo, J. and Lee, A. 1999. A genetic basis for atrophy: dominant non-responsiveness and helicobacter induced gastritis in F-1 hybrid mice. Gut 45: 335-340.

Suzuki, H., Mori, M., Kai, A., Suzuki, M., Suematsu, M., Miura, S. and Ishii, H. 1998. Effect of rebamipide on *H-pylori*-associated gastric mucosal injury in Mongolian gerbils. Dig. Dis. Sci. 43 (9 Suppl.): S181-S187.

Takahashi, S., Keto, Y., Fujita, H., Muramatsu, H., Nishino, T. and Okabe, S. 1998. Pathological changes in the formation of *Helicobacter pylori*-induced gastric lesions in Mongolian gerbils.Dig. Dis. Sci. 43: 754-765.

Tatematsu, M., Yamamoto, M., Shimizu, N., Yoshikawa, A., Fukami, H., Kaminishi, M., Oohara, T., Sugiyama, A. and Ikeno, T. 1998. Induction of glandular stomach cancers in *Helicobacter* pylori-sensitive Mongolian gerbils treated with N-methyl-N-nitrosourea and N-methyl-N'-nitro-N-nitrosoguanidine in drinking water. Japan. J. Cancer Res. 89: 97-104.

Telford, J., Ghiara, P., M., D.O., Comanducci, M., Burroni, D., Bugnoli, M., Tecce, M., Censini, S., Covacci, A., Xiang, Z. and et, a. 1994. Gene structure of the *Helicobacter pylori* cytotoxin and evidence of its key role in gastric disease. J. Exper. Med. 179: 1653-1658.

Terio, K., Munson, L. and Solnick, J. 1999. Chronic Helicobacter-associated gastritis in cheetahs. Vet. Pathol. 36: 510.

Tokieda, M., Honda, S., Fujioka, T. and Nasu, M. 1999. Effect of *Helicobacter pylori* infection on the N-methyl-N'-nitro-N-nitrosoguanidine-induced gastric carcinogenesis in mongolian gerbils. Carcinogenesis 20: 1261-1266.

Tompkins, D.S., Wyatt, J.I., Rathbone, B.J. and West, A.P. 1988. The characterization and pathological significance of gastric Campylobacter-like organisms in the ferret: a model for chronic gastritis? Epidemiol. Infect. 101(2): 269-278.

Truper, H.G.., and De Clari, L. 1997. Taxonomic note: Necessary correction of specific epithets formed as substantives (nouns) "in apposition". Int. J. Syst. Bacteriol. 47: 908-909.

Wang, T.C., Goldenring, J.R., Dangler, C., Ito, S., Mueller, A., Jeon, W.K., Koh, T.J. and Fox, J.G. 1998. Mice lacking secretory phospholipase A(2) show altered apoptosis and differentiation with *Helicobacter felis* infection. Gastroenterology 114: 675-689.

Watanabe, T., Tada, M., Nagai, H., Sasaki, S. and Nakao, M. 1998. *Helicobacter pylori* infection induces gastric cancer in Mongolian gerbils. Gastroenterology 115(3): 642-648.

Wirth, H.P., Beins, M.H., Yang, M.Q., Tham, K.T. and Blaser, M.J. 1998. Experimental infection of Mongolian gerbils with wild-type and mutant *Helicobacter pylori* strains. Infect. Immun. 66: 4856-4866.

Yamasaki, K., Suematsu, H. and Takahashi, T. 1998. Comparison of gastric lesions in dogs and cats with and without gastric spiral organisms. J. Amer. Vet. Med. Assn. 212: 529-533.

Yokota, K., Kurebayashi, Y., Takayama, Y., Hayashi, S., Isogai, H., Isogai, E., Imai, K., Yabana, T., Yachi, A. and Oguma, K. 1991. Colonization of *Helicobacter pylori* in the gastric mucosa of Mongolian gerbils. Microbiol. Immunol. 35: 475-480.

Zevering, Y., Jacob, L., and Meyer, T.F. 1999. Naturally acquired human immune responses against *Helicobacter pylori* and implications for vaccine development. Gut 45: 465-474.

From: *Helicobacter pylori: Molecular and Cellular Biology*
ISBN 1-898486-25-5 © 2001 Horizon Scientific Press, Wymondham, UK.

9

Transgenic Mouse Models for Studying the Relationship between *Helicobacter pylori* Infection and Gastric Cancer

Per G. Falk, Andrew J. Syder, Janaki L.Guruge,
Lars G. Engstrand, and Jeffrey I. Gordon[*]

Abstract

Sero-epidemiologic studies suggest that *H. pylori* infection may increase the risk of developing gastric adenocarcinoma. Nonetheless, the host and microbial factors that may produce this outcome are poorly understood. Transgenic mouse models offer one approach for developing and conducting direct tests of hypotheses about the mechanisms that underlie *H. pylori*-associated tumorigenesis.

Sero-epidemiologic Evidence for the Association between *H. pylori* Infection and Gastric Adenocarcinoma

In 1994, the International Agency for Research on Cancer designated *H. pylori* as a human carcinogen (IARC Monograph, 1994). Three meta-analyses have concluded that there is an increased risk for gastric adenocarcinoma among individuals who have serologic evidence of *H. pylori* infection. In one meta-analysis of 19 epidemiological studies, the pooled odd ratios (OR) for gastric adenocarcinoma among *H. pylori* infected versus non-infected individuals was 1.92, with a lower 95% confidence interval (CI) of 1.32 (Huang, *et al.,* 1998). In another meta-analysis, based on 34 retrospective studies of 3300 gastric cancer cases, the pooled OR was 2.5 (95% CI = 1.9-3.4; Danesh, 1999). A third analysis, based on 42 independent studies, also found a positive association (pooled OR = 2.04; 95% CI 1.69-2.45; Eslick *et al.,* 1999).

Cohort studies typically are nested (patients with a disease are identified within the cohort and controls are picked randomly but matched for age and gender), not restricted to a defined geographic locale, and involve an ethnically diverse population. Nyrén (1998) has reviewed eight nested case-control cohort studies and noted that five provided compelling evidence of a significant risk for gastric cancer (OR = 2.8-6), while 2 showed a tendency (OR = 1.5-1.6), and one was negative (OR = 0.9).

Given the genetic diversity of humans and *H. pylori* (Wang *et al.,* 1999; Doig *et al.,* 1999; Marais *et al.,* 1999), arguably the ideal epidemiologic study of the association between

[*]corresponding author, email: jgordon@molecool.wustl.edu

infection and tumorigenesis would be a prospective case-control study, restricted to a defined catchment area that is representative of a relatively homogenous indigenous population. Additional mechanistic insights could be obtained if bacteria were isolated from individual cases and controls over a period of time. In this way, individuals would serve as their own controls and the evolution of genetic changes in the bacterium could be correlated with the evolution of histopathologic changes in the gastric ecosystem of the host.

To date, no reported study has fulfilled these idealized design critieria. Nyren examined five serology-based case-control studies, some of which were limited to hospital-based patients, and others that were population-based. Two studies had well-defined population strata, a discrete sampling interval, and controls representative of the study base from which cases were derived. These studies revealed statistically insignificant OR values (1.3 and 1.4). In contrast, a prospective, hospital-based study of a Swedish population with a 2-year sampling period found that *H. pylori* is an independent risk factor for developing gastric cancer (OR = 2.6; 95% CI = 1.4-5.0; Hansson *et al.*, 1993).

Microbial and Host Co-factors

The divergent results obtained from various epidemiologic studies suggest that the association between *H. pylori* and gastric adenocarcinoma is influenced by environmental, microbial, and host co-factors. Unfortunately, our knowledge of these co-factors is limited. We do know that, depending upon geographic locale, 30-90% of individuals with recently diagnosed gastric adenocarcinomas will harbor *H. pylori* (Webb and Forman, 1995). Childhood appears to be a time when the gastric cancer risk pattern is established (Staszewski and Haenszel, 1965; King and Locke, 1980a, 1980b). Populations at high risk for development of gastric cancer also have a high colonization rate among children (Mitchell *et al.*, 1992). Moreover, risk factors for early acquisition of a colonizing strain, such as low socio-economic status, multiple siblings, and crowded living conditions (Sitas *et al.*, 1991; Mendall *et al.*, 1992; Galpin *et al.*, 1992; Webb *et al.*, 1994), are also risk factors for gastric tumorigenesis (Barker *et al.*, 1990; Blaser *et al.*, 1995).

Most epidemiological data about microbial co-factors are based on serologic studies. These serologic surveys are not accompanied by systematic genotypic characterization of colonizing strain(s), especially as a function of the histopathologic changes that occur in their host's stomachs. Although seropositivity is indicative of infection, it does not provide definitive information about the current status of infection. For example, a recent Swedish study of cases and controls presenting to a hospital-based endoscopy clinic disclosed that 29 of 72 patients with gastric cancer (cases) were culture-positive (40%), while 66% had a positive IgG ELISA test of serum for bacterial cell-associated proteins (OR 2.5 for the serologic test; CI = 1.4-4.4; Enroth *et al.*, 2000). 126 of 324 (39%) of controls in this study were also culture-positive. Immunoblot analysis of sera from cases and controls disclosed that expression of CagA, but not VacA, correlated with cancer (OR = 2.3 and 1.3 respectively; reference = infected controls without cancer). When bacterial isolates from cases and controls were subdivided into type I, intermediate, and type II strains, based on their levels of CagA and VacA expression, the OR for cancer in the corpus or antrum of the stomach was 2.1 in cases harboring type I strains, 2.2 in cases containing intermediate strains, and 0.3 in cases with type II strains. The distributions of each of the three types of strains were similar for both the intestinal and diffuse histologic types of adenocarcinoma (Enroth, *et al.*, 2000). Remarkably, the OR for cases with adenocarcinoma in the proximal portion of stomach (cardia) was ≤1 for each of the three designated strain types. These results raise the question of what microbial and host factors influence tumor location in the gastric ecosystem.

Atrophic Gastritis

A population-based cohort study of hospitalized Swedish patients with unoperated peptic ulcer disease disclosed a statistically significant 80% increase in the risk of gastric cancer among gastric ulcer patients relative to the general population. There was a significant 40% *decrease* in risk among duodenal ulcer patients (Hansson *et al.,* 1996). Since virtually every patient with duodenal ulcer has *H. pylori*-induced antral gastritis (typically with Cag⁻ isolates, even when duodenal isolates are Cag⁺, Hamlet *et al.,* 1999), it is unlikely that inflammation *per se* causes cancer. Multifocal atrophic gastritis correlates with gastric ulcer disease and gastric cancer, especially the intestinal type (Schrager *et al.,* 1967; Sipponen 1992; Lechago and Correa 1993). In contrast there is an inverse association between atrophic gastritis and duodenal ulcer disease (Sipponen, 1992). While *H. pylori* infection is related to atrophic gastritis in unselected patients (Kuipers *et al.,* 1995), the effect appears more pronounced in gastric ulcer patients and is less evident in patients with duodenal ulcer disease (e.g. Eidt and Stolte 1994; Lin *et al.,* 2000).

Atrophic gastritis is associated with loss of two of the principal epithelial cell lineages present in the corpus region of the stomach: acid-producing parietal cells and pepsinogen-producing zymogenic (chief) cells. Although development of atrophic gastritis may be an important event in *H. pylori*-associated carcinogenesis (reviewed in Genta, 1997), we lack direct information about why this should be the case. A number of hypotheses have been advanced. For example, host and microbial factors may cooperate to facilitate penetration of luminal carcinogens into the gastric mucosa: i.e., (i) atrophic gastritis is characterized by reduced acid production and attendant changes in the gastric microflora, both of which may promote formation of N-nitroso compounds implicated in development of gastric adenocarcinoma (Sjöstedt *et al.,* 1988; Bartsch *et al.,* 1988); (ii) atrophic gastritis correlates with altered mucus production and increased gastric absorption of small hydrophilic compounds (Molinari *et al.,* 1984; Söderholm *et al.,* 1994). (iii) *H. pylori* reduces mucus production and alters its physical-chemical properties (Dixon, 1994; Micots *et al.,* 1993; Sørbye and Svanes, 1995); and (iv) the bacterium may increase DNA damage by free radicals and reactive nitrogen species (Correa and Miller, 1998).

The Need for Genetically Defined Animal Models

The frequency and persistence of *H. pylori* infection in humans makes it an attractive model for examining the co-evolution and co-adaptation of a bacterium and its host over a significant fraction of an individual's lifespan. However, variations encountered among humans and colonizing strains of *H. pylori*, make it very difficult to develop hypotheses about the contributions of environmental, host and microbial factors to the evolution of gastric cancer. In the best of circumstances, a long-term prospective study would have to be conducted in a specified geographical area, with as much genetic homogeneity in the population as possible, where infected individuals with superficial gastritis and/or atrophic gastritis would be followed over time. Serial endoscopies would be needed so that cultures could be obtained and histopathologic surveys of the stomach performed. Whole genome genotyping of colonizing strains could then be performed as a function of the histopathologic status of the host. Broad-based gene expression profiling, using RNA prepared from gastric biopsy specimens, could yield a detailed molecular signature of these histopathologic states. Finally, since the evolution of gastric cancer appears to involve the same gene families as those found in colon cancer (see Chan *et al.,* 1999), a comparison of allelic variation among these genes between cases and controls in the catchment area may be informative.

It is very likely that even if such detailed molecular analyses of host- and microbial

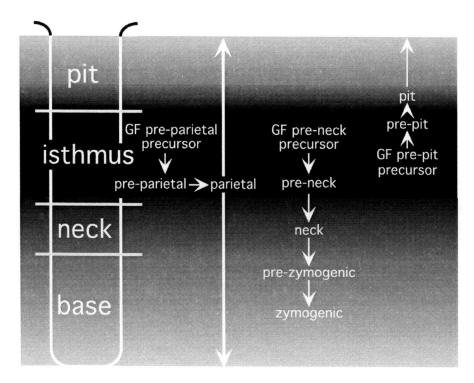

Figure 1. Summary of epithelial renewal in the gastric units of adult mice. See text for details.

responses to infections could be conducted, the results would reveal a very complex pattern of evolution of an infection within an individual, as well as variations between individuals within a geographically or ethnically constrained cohort. All of this complexity serves to confound the problem of defining pathogenetic mechanisms that link this bacterium to cancer. Genetically well defined and manipulatable animal models, employing extensively genotyped clinical isolates of *H. pylori*, provide an attractive *in vivo* setting for describing the molecular details of an evolving infection, from both the host and microbial perspective. These animal models also allow formal testing of the roles of specified microbial, host and/ or environmental factors in determining whether the outcome of infection is cancer. The relevance of any results obtained from these systems can then be examined in the context of human epidemiological studies.

Transgenic Mouse Models for Studying *H. pylori* Pathogenesis

Rationale for Using Mice

Chapter 8 describes the current status of nontransgenic animal models of *H. pylori* infection. Some of these models can recapitulate the progression of histopathologic changes postulated to occur in *H. pylori*-infected humans that develop cancer. For example, Honda *et al.*, (1998a, 1998b) reported that Mongolian gerbils infected with a CagA+, VacA+ strain develop atrophic gastritis and focal intestinal metaplasia within 6 months, severe intestinal metaplasia after 12 months, and well differentiated adenocarcinoma (2 of 5 animals) after 18 months. Watanabe *et al.*, (1998) found that 37% of gerbils infected with *H. pylori* exhibited gastric adenocarcinoma (intestinal type) after 62 weeks and that tumorigenesis was related to development of intestinal metaplasia in areas of chronically inflamed mucosa. Another group used Mongolian gerbils to demonstrate that infection with *H. pylori* and continuous exposure to *N*-methyl-*N*-nitrosurea in their drinking water induces gastric cancer (both intestinal and diffuse types) after 40 weeks (Sugiyama *et al.*, 1998). In this study, cancer did not develop in animals exposed to *H. pylori*, or carcinogen, alone for 40 weeks.

The Mongolian gerbil provides important evidence for a sequence of *H. pylori*-associated histopathologic changes, progressing from superficial gastritis to atrophic gastritis to intestinal metaplasia and ultimately adenocarcinoma. However, it has limitations, notably an inability to use this rodent species to conduct direct genetic tests of the role of host factors postulated to regulate and/or mediate the course of infection. The mouse offers a number of advantages over other animal models: i.e. extensive genetic definition, an ongoing aggressive effort to sequence its entire genome, and the ability to be readily manipulated genetically for both gain-of-function and loss-of-function experiments. In addition, there is more information about gastric epithelial renewal in the mouse than in any other mammal.

Epithelial Renewal in the Gastric Units of Adult Mice

The proximal third of the adult mouse stomach (forestomach) is lined with a squamous epithelium, while its distal two thirds is lined with a glandular epithelium containing thousands of tubular mucosal invaginations known as gastric units (Figure 1). Epithelial renewal occurs within these units. In the central, or corpus, region of the adult mouse stomach (also known as the zymogenic zone), each gastric unit contains an average steady state census of ~200 epithelial cells, representing three predominant lineages (pit, parietal, and zymogenic cell), and two minor lineages (enteroendocrine and caveolated cell) (Karam and Leblond, 1992).

Karam and Leblond used *in vivo* pulse labeling with tritiated thymidine, followed by electron microscopic autoradiography, to obtain a detailed morphologic description of the presumptive multipotent gastric stem cell, its immediate daughters, as well as the pathways their descendants follow during terminal differentiation to mature pit, parietal and zymogenic cells (Karam, 1993; Karam and Leblond, 1993a-d). The results of their studies are summarized in Figure 1.

The stem cell niche is located in a zone of proliferation positioned within the central portion (isthmus) of the gastric unit. The presumptive multipotent isthmal stem cell gives rise to a granule-free (GF) pre-pit cell precursor, a granule-free pre-neck cell precursor, and a granule-free pre-parietal cell precursor (see Figure 1 and Karam and Leblond, 1993a,1993d). Mucus-producing pit cells differentiate during a 3 d *upward* migration from the isthmus, through the upper portion of the gastric unit (pit region) to the surface epithelium. During the course of this rapid unidirectional migration, they undergo a sequence of morphologic changes (pre-pit to pit to surface epithelial cells; Figure 1). Upon arrival at the surface epithelium, they undergo an apoptotic or necrotic death (Karam and Leblond, 1993b). Members of the zymogenic cell lineage differentiate during a *downward* migration from the isthmus through the neck and base regions of the gastric unit, during which time they are transformed from pre-neck cells to mucus- and pepsinogen-producing neck cells, to pre-zymogenic cells, and finally to mature zymogenic cells (Figure 1). Zymogenic cells have a lifespan of ~190 d (Karam and Leblond, 1993c).

Unlike pit and zymogenic cells, differentiated parietal cells arise within the isthmus. Pre-parietal cells are converted to mature acid-producing parietal cells in 1 d (Karam, 1993). Mature parietal cells then undergo a bipolar migration. ~50% of parietal cells migrate from the isthmus up through the pit region where they are eliminated by necrosis, exfoliation, or phagocytosis. An equivalent fraction of parietal cells migrate from the isthmus down through the neck to the base were they are removed by apoptosis or phagocytosis. The lifespan of parietal cells averages 54 d (Karam, 1993). The distribution of parietal cells throughout the gastric unit places them in a strategic position to influence the proliferative activity of isthmal progenitors as well as the terminal differentiation programs of other lineages (see below).

Not all gastric units in the adult mouse stomach contain these three epithelial cell types. Units located in antrum of the mouse stomach (the pure mucus zone) are populated by pit cells and neck cells: they do not contain the differentiated zymogenic descendants of neck cells, nor do they contain parietal cells. A region interposed between the zymogenic and pure mucus zone has been named the mucoparietal zone (Karam and Leblond, 1992) because its component gastric units are populated by parietal, pit, *and* neck cells.

Gastric Epithelial Lineage Progenitors Predominate During Early Phases of Gastric Unit Morphogenesis

Morphogenesis of gastric units is not completed until the fourth postnatal week in the mouse. This process has been divided into four stages (Karam *et al.*, 1997). The first stage occurs during late fetal life when the endoderm undergoes cytodifferentiation to an epithelial monolayer with numerous short solid infoldings (termed 'primordial buds'). Each bud contains ~20 cells per longitudinal section. An average of 8% of these cells have the morphologic features of differentiated pit, neck, parietal, and enteroendocrine cells found in adult gastric units. The remainder resemble granule-free lineage progenitors (plus a few pre-pit, pre-neck, and pre-parietal cells). Postnatal day 1 (P1) to P7 encompass the second stage of gastric unit morphogenesis. During this period, the total number of cells per bud does not change but the fractional representation of differentiated pit, neck, and enteroendocrine cells increases 10-fold to ~80% of the total. The third stage occurs between P7 and P15, and is associated with an increase in the number of all cell types (precursors and their differentiated descendants), elongation of buds, and formation of nascent units. Granule-free precursors, pre-pit, pre-neck, and pre-parietal cells remain scattered along the developing gastric unit until P15. The last stage of gastric unit development occurs between P15 and P28, and is characterized by the formation of distinct pit, isthmus, neck, and base compartments.

Responses of Mice to Other Helicobacter Species

H. heilmannii plus N-methyl-N'-nitro-N-nitrosoguanidine produces nodular hyperplasia with parietal cell loss in mice sacrificed at 12 to 18 months of age (Danon and Eaton, 1998). *H. felis* infection in the C57Bl/6 strain of mice is associated with development of hypertrophic gastropathy, but not cancer, over the course of a 1 year infection (Fox *et al.*, 1996). Wang and coworkers (2000) used a transgenic mouse model (INS-GAS) to examine the impact of hypergastrinemia on the outcome of *H. felis* infection. In this model, forced expression of gastrin in pancreatic β-cells was used to mimic the hypergastrinemia that develops early in the course of *H. pylori* infections in humans, as well as in the later stages of infection when significant parietal cell loss has occurred. In the absence of *H. felis* infection, hypergastrinemic INS-GAS mice slowly lose parietal cells: significant depletion is evident by 10 months of age and is progressive through at least 20 months. Parietal cell loss is associated with an expansion of the zone of proliferation in gastric units, expression of two growth factors (heparin-binding EGF and TGFα), metaplasia, dysplasia, and development of carcinoma *in situ* followed by invasion into blood vessels. Dissemination into regional lymph nodes was not reported (Wang *et al.*, 2000). *H. felis* infection of INS-GAS mice accelerated the evolution of these histopathologic changes but did not promote progression to metastatic disease. These findings led this group to propose that hypergastrinemia is a host co-factor that contributes to parietal cell loss and may promote development of cancer in *Helicobacter*-infected individuals.

Unfortunately, there have been remarkably few studies that have used mice to explore how *H. pylori* infection may promote tumorigenesis. Shimizu *et al.,* (1998) reported that *H. pylori* enhances the development of pre-neoplastic lesions in the antrum of Balb/C mice exposed to *N*-methyl-*N*-nitrosourea.

A Transgenic Mouse Model that Examines the Impact of Bacterial Attachment to a Terminally Differentiated Gastric Epithelial Cell on the Outcome of Infection

Histologic studies indicate that the population of *H. pylori* in the epithelium is approximately two orders of magnitude lower than the population in the mucus (Nowak and Forouzandeh, 1997; Atherton *et al.*, 1996). The rate of mucus turnover is faster than the rate of epithelial shedding (e.g. Lipkin *et al.*, 1963; Rubinstein and Tirosh, 1994). Mathematical modeling studies (Kirschner and Blaser, 1995; Blaser and Kirschner, 1999) have led to the proposal that the bacterial population that adheres to the epithelium functions to sustain colonization. The ability of *H. pylori* to attach with high affinity to the gastric epithelium of its host involves both host- and microbial factors: (i) the ability of a colonizing strain to produce adhesins recognized by various gastric epithelial receptors; and (ii) the ability of the host to express these receptors in its various gastric epithelial lineages and/or their progenitors either prior to, or during the course of infection.

Attachment can also affect host-microbial cross talk and thereby influence host responses. For example, Segal *et al.* (1999) have shown that attachment of strains containing the *cag* PAI to a human gastric adenocarcinoma epithelial cell line induces formation of cup/pedestal structures at the site of attachment, phosphorylation of CagA, insertion of CagA into host cells where it forms a cylindrical structure in close proximity to areas of actin recruitment, and cellular spreading with generation of lamellipodia and filapodia. Introduction of CagA into host cells (Segal *et al.*, 1999; Stein *et al.*, 2000; Odenbreit *et al.*, 2000) via type IV secretion and subsequent effects on phosphorylation of host cell proteins (Odenbreit *et al.*, 2000) are reminiscent of strategies employed by invasive organisms, such as *Yersinia* (Cornelis and Wolf-Watz, 1997): i.e. they perturb host biology by delivery of specific bacterial proteins (Covacci and Rappuoli, 2000). Of course, attachment may affect the host through other mechanisms: for example, through changes in immune responses (see below), or by increasing bacterial population density at particular locales, thereby effecting microbial gene expression via quorum sensing (Costerton *et al.*, 1999).

In vitro studies indicate that *H. pylori* produces a variety of adhesins that recognize a number of different receptors (Dunn *et al.*, 1997). HpaA is a 183-residue protein reported to bind $\alpha 2,3$-sialyl glycans (Evans *et al.*, 1993), although data supporting this conclusion are controversial (O'Toole *et al.*, 1995; Jones *et al.*, 1997). Lingwood *et al.* (1993) have identified a 63 kDa exoenzyme-S-like polypeptide that interacts with phosphatidylethanolamine-, gangliotetraosylceramide-, and globotetraosylceramide-containing lipid 'receptors' *in vitro*. However, there are conflicting results about whether this protein functions as an adhesin *in vivo* (Odenbreit *et al.*, 1996). Hsp70 is a 616 amino acid bacteria protein whose expression is regulated by acid. It binds to sulfated oligosaccharides (Huesca *et al.*, 1998).

As noted by Gehard *et al.*, elsewhere in this volume, *babA2* encodes a 721 amino acid protein that binds with high affinity to the human histo-blood group antigen Lewis b (Fuc$\alpha 1,2$Gal$\beta 1,3$[Fuc$\alpha 1,4$]GlcNAcβ) (Ilver *et al.*, 1998). ~70% of all humans produce Lewis b-containing glycans: expression of these glycans in the gastric epithelium is restricted to the

Table I. *In situ* binding studies to the gastric epithelium of transgenic mice provides evidence that a clinical isolate of *H. pylori* (Hp1) expresses two distinct classes of adhesins: one specific for Lewis b- and the other for NeuAcα2,3Galβ1,4-containing glycans

Carbohydrate epitope		Inhibits Hp1 binding to epithelium	
		α1,3/4 FT	*tox*176
Lewis[a]	Galβ1-3GlcNAcβ 4 | Fucα1	no	no
Lewis[b]	Fucα1-2Galβ1-3GlcNAcβ 4 | Fucα1	yes	no
Lewis[x]	Galβ1-4GlcNAcβ 3 | Fucα1	no	no
Lewis[y]	Fucα1-2Galβ1-4GlcNAcβ 3 | Fucα1	no	no
3'-Sialyllactose	Neu5Acα2-3Galβ1-4Glc	no	yes
6'-Sialyllactose	Neu5Acα2-6Galβ1-4Glc	no	no
Sialyl-Lewis[a]	Neu5Acα2-3Galβ1-3GlcNAcβ 4 | Fucα1	no	no
Sialyl-Lewis[x]	Neu5Acα2-3Galβ1-4GlcNAcβ 3 | Fucα1	no	yes

pit cell lineage (Borén *et al.*, 1993; Ilver *et al.*, 1998). Surveys of strains recovered from individuals with diverse Lewis phenotypes indicated that a majority of clinical *H. pylori* isolates express Lewis b binding adhesins (Ilver *et al.*, 1998). Moreover, Gerhard *et al.* (1999) found that the presence of *babA2* is significantly associated with duodenal ulcer and gastric adenocarcinoma.

A transgenic mouse model has provided direct evidence that *H. pylori* attachment to Lewis b-expressing pit cells affects host immune responses and can lead to loss of parietal cells. Forced expression of the human α1,3/4 fucosyltransferase (α1,3/4 FT) in the pit cell lineage of neonatal and adult transgenic mice (genetic background = FVB/N) allows production of Lewis b-containing glycans (Falk *et al.*, 1995). A clinical isolate (Hp1), recovered from a Lewis b-positive patient with chronic gastritis, was able to colonize the stomachs of conventionally raised α1,3/4 FT transgenic animals and their normal age-matched non-transgenic, Lewis b-negative littermates with equivalent efficiency (~80% after a single gavage of 10^7 colony forming units; CFU), and to equivalent microbial density (~10^5-10^7 CFU/stomach after an 8 week infection) (Guruge *et al.*, 1998; Syder *et al.*, 1999).

Whole genome genotyping, using DNA microarrays containing amplified ORFs from the fully sequenced *H. pylori* 26695 and J99 genomes, indicates that Hp1 has lost *cag*8-26. The isolate also appears to have significant sequence divergence involving (i) 42 genes present in the 26695 (TIGR strain) but not the J99 strain, (ii) 35 genes present in the J99 but not the 26695 strain, and (iii) 25 genes present in both of these two fully sequenced *H. pylori* strains (Nina Salma *et al.*, Personal Commun.).

Hp1 expresses binding determinants *ex vivo* that allow it to interact with Lewis b-positive pit cells present in sections of α1,3/4 FT mouse stomach (Guruge *et al.*, 1998). Binding can be inhibited by pre-treatment of bacteria with purified Lewis b-human serum albumin (HSA) neoglycoconjugates, but not by pre-treatment with Lewis x-HSA, Lewis y-HSA, Lewis a-HSA, or HSA alone (see Table I for carbohydrate structures). Hp1 does not exhibit detectable binding to the gastric epithelium of normal FVB/N mice (Guruge, *et al.*, 1998). (Note that the lack of detectable binding to sections of gastric epithelium prepared from normal FVB/N mice is not a unique feature of Hp1: it is also true for the NCTC 11637 and 11638 strains, for 23 isolates recovered from Peruvian patients, and for 6 isolates from cases in the hospital endoscopy-based Swedish case-control study of Enroth and coworkers described above.)

After an 8-16 week infection of transgenic α1,3/4 FT mice, Hp1 can be readily observed bound to Lewis b-positive pit cells. It is also affiliated with the luminal mucus. In contrast, bacteria appear to be limited to luminal mucus gel of infected, Lewis b-negative nontransgenic mice (Guruge *et al.*, 1998). Studies of age-matched normal and transgenic mice, inoculated with Hp1 at 4-52 weeks of age and sacrificed 8-16 weeks later, revealed that *in vivo* binding of Hp1 to pit-cell Lewis b glycans leads to more severe gastritis, and enhanced production of antibodies that react with bacterial- and parietal cell Lewis x glycans (Guruge *et al.*, 1998; Syder *et al.*, 1999).

Why is Lewis x the target of this host humoral immune response? The majority of *H. pylori* isolates produce Lewis x (Galβ1,4[Fucα1,3]GlcNAcβ) and Lewis y (Fucα1,2Galβ1,4[Fucα1,3]GlcNAcβ) epitopes in their LPS (Wirth *et al.*, 1997; Appelmelk *et al.*, 1998b). In uninfected normal and α1,3/4 FT mice belonging to the FVB/N inbred strain, Lewis x immunodeterminants are produced by a subset of parietal cells (Guruge *et al.*, 1998). (Comparisons of normal and transgenic mice homozygous for wild type or null alleles of the *Rag-1* gene established that parietal cell Lewis x production does not require signals from mature mucosal T- or B-cells). After an 8 week infection, α1,3/4 FT transgenic mice have a significantly greater incidence (80% versus 15%), and higher titer of circulating Lewis x antibodies compared to infected age- and gender-matched normal littermates (Guruge *et al.*, 1998; Syder *et al.*, 1999). The extent of parietal cell loss correlates with antibody titer, suggesting the following pathogenetic scheme:

(1) Hosts that express gastric epithelial receptors that are recognized by adhesins produced by their colonizing strains of *H. pylori* have a greater risk of developing more severe gastritis.

(2) Bacterial attachment (e.g., to pit cells) can promote the development of cross reactive immunopathology if microbial immunodeterminants (e.g., LPS-associated Lewis x) resemble immunodeterminants produced by one of the host's gastric epithelial lineages . If these immunodeterminants are expressed in parietal cells, the destiny of the infection may be skewed towards greater parietal cell loss and the development of atrophic gastritis (Guruge *et al.*, 1998; Appelmelk *et al.*, 1998a).

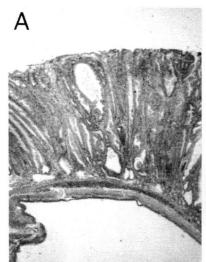 

Figure 2. Hyperplasia and cystic changes observed in the gastric epithelium of conventionally raised *tox*176 mice. (A) Hematoxylin and eosin-stained section from the corpus of an 8 month old transgenic mouse stomach showing thickened epithelium with some cystic changes. (B) Section prepared from a 10 month old mouse with more advanced cystic changes and a greater degree of inflammation. The arrow points to a cystic lesion that is not truly invasive, based on an analysis of serial sections.

Unfortunately, little information is available about the structural and functional properties of the gastric mucosal immune system. *H. pylori* infection of α1,3/4 FT mice provides a model for addressing the general question of the role of microbial attachment in development of immunopathology, and the specific issue of identifying microbial genes and host factors that contribute to enhanced immune responsiveness and responses when *H. pylori* binds to the gastric epithelium. These host factors may include cytokines such as IL-1β, IL-6, IL-8 and TNF-α or iNOS - all of which are increased in *H. pylori* positive humans (Wilson *et al.*, 1998; Fu *et al.*, 1999).

A Transgenic Mouse Model Demonstrating that an Engineered Ablation of Parietal Cells Results in an Amplification of Gastric Epithelial Lineage Progenitors

Another FVB/N-based transgenic mouse model has addressed the question of how parietal cell ablation may contribute to the development of tumorigenesis. This second model was an outgrowth of functional mapping studies of transcriptional regulatory elements present in the gene encoding the non-catalytic β subunit of mouse H+/K+ ATPase. These studies, performed in mice containing a transgene consisting of nucleotides -1035 to +24 of the β-subunit gene linked to nucleotides +3 to +2150 of a human growth hormone (hGH) reporter, showed that β-subunit $^{-1035 \text{ to } +24}$ restricts hGH expression to pre-parietal cells and their differentiated descendants (Lorenz and Gordon, 1993; Li *et al.*, 1996). Expression was not observed in granule-free isthmal precursors of the zymogenic or pit cell lineages, or in their differentiating and differentiated descendants. Expression begins coincident with the first appearance of parietal cell precursors in late fetal life and continues in parietal cells through at least the first two years of life.

H+/K+ ATPase β subunit$^{-1035 \text{ to } +24}$ was used to direct expression of an attenuated diphtheria toxin A fragment (*tox*176) to pre-parietal and parietal cells so that they would be destroyed and the effects on epithelial homeostasis defined. Parietal cells were completely eliminated from all gastric units in two pedigrees of *tox*176 mice (Syder *et al.*, 1999). Loss of parietal cells was accompanied by an inhibition of the terminal differentiation of neck to zymogenic (chief) cells, suggesting that parietal cells are the source of factors necessary for the terminal differentiation of this lineage. Importantly, EM analysis of the stomachs of these animals revealed that parietal cell ablation was also accompanied by enhanced proliferation of the presumptive multipotent stem cell and its granule-free committed daughters.

Long term EM studies of conventionally raised *tox*176 mice belonging to these two pedigrees revealed a similar pattern of progressive expansion of lineage progenitors. At 6-7 weeks of age, the mice have an average 4-fold increase in the number of granule-free progenitors, compared to nontransgenic littermates and cagemates (Li *et al.*, 1996). By 6 months, the progenitor population has increased to the point where it represents ~50% of all epithelial cells, compared to <2% in normal littermates. The pit cell population is not diminished. In contrast neck cells are depleted and zymogenic cells are virtually absent. By 8 months of age, the epithelium is markedly thickened, and contains large cyst-like structures (Figure 2). Surveys of twenty 8-12 month old *tox*176 mice revealed glandular-appearing structures insinuating themselves into the lamina propria in 45% of the animals. These structures are composed of epithelial monolayers that include cells which resemble granule-free lineage progenitors. Epithelial invasion through the gastric wall or into blood vessels was not detectable, and no metastasis to regional lymph nodes or visceral organs were noted in 15 conventionally raised *tox*176 mice at 12 and 18 months of age (Syder *et al.*, unpublished data.).

There has been some debate about whether development of intramucosal cysts represents a precancerous state in humans (Hizawa *et al.,* 1997). Rubio and Owen (1998) has found that the occurrence of cysts is significantly higher in patients with the intestinal type of gastric adenocarcinomas of their corpus and antrum: this association was evident in diverse geographic locations. Niv and Turani (1991) concluded that cystic changes are secondary to inflammation and do not represent a preneoplastic state.

*tox*176 mice are achlorhydric (gastric pH ≥ 7) (Li *et al.*, 1996; Syder *et al.*, 1999). The absence of acid production is accompanied by a marked expansion in the gastric microflora. Comparisons of 2-3 month old transgenic mice (belonging to the two pedigrees with complete parietal cell loss) with their age-matched nontransgenic cagemates established that parietal cell ablation was associated with an average 2-log increase in the number of bacteria recovered when gastric contents were cultured under aerobic conditions. Average increases of 3- and 7-logs were observed when bacteria were cultured under microaerophilic and anaerobic conditions, respectively. The gastric microflora of specified pathogen-free FVB/N mice in our mouse barrier facility is predominated by *Lactobacillus spp.* The microflora of 2-3 month old *tox*176 mice contains *Lactobacillus sp.*, coagulase-negative *Staphylococcus sp.*, *E. coli*, and *Proteus mirabilis* (Syder *et al.*, 1999). Given this pronounced expansion of the gastric microflora, it is not surprising that *tox176* mice develop a chronic diffuse gastritis by the time that they are 6-8 months old. FACS analysis revealed 47% TCRαβ+ and 22% TCRγδ+ T cells, 14% B-cells, 5% NK1.1 natural killer cells, and 2.5% macrophages. Gastritis is not evident in age-matched conventionally raised non-transgenic cagemates.

Conventionally raised *tox*176 transgenic mice have been re-derived as germ-free to examine the role of the microflora and the gastritis in promoting expansion of epithelial lineage progenitors (Syder *et al.*, 1999). As in conventionally raised *tox*176 mice, there is a progressive increase in the fractional representation of lineage progenitors and an accompanying inhibition of the terminal differentiation of zymogenic cells. During the first 2-4 months of life, the magnitude of the progenitor amplification is similar to the expansion noted in age-matched conventionally raised *tox*176 animals. However, compared to their conventionally raised counterparts, 8-10 month old germ-free *tox*176 mice have a milder gastritis and ~2-3 fold lower level of progenitor amplification (Syder *et al.*, unpublished data.).

Studies of conventionally raised 3-4 month old *tox*176 mice homozygous for the wild type or a null allele of *Rag1* indicated that the progenitor amplification associated with parietal cell ablation can occur in the absence of mature B- and T-cells (Syder *et al.*, 1999). However, the comparisons of older conventionally raised and gnotobiotic mice suggest that

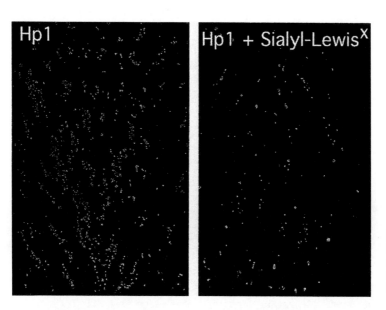

Figure 3. Binding of *H. pylori* to the gastric epithelium of a *tox*176 mouse is inhibited by pretreatment of bacteria with purified sialyl-Lewis x.

development of chronic gastritis in the former contributes in part to the continued expansion of lineage progenitors as animals age. This point will have to be explored further by crossing germ-free normal C57Bl/6 (B6) mice and B6 *Rag1* knockout mice to germ-free FVB/N *tox*176 animals to generate *tox*176 mice homozygous for the wild type or the null allele of *Rag1* (and with similar B6 and FVB/N contributions to their genetic background).

Another observation has provided insights about the regulation of progenitor amplification in *tox*176 mice. Several pedigrees of *tox*176 mice exhibited a heritable mosaic pattern of transgene expression that resulted in complete loss of parietal cells from some, but not all, of their gastric units. Remarkably, increases in progenitor cell proliferation were confined to those gastric units that lacked parietal cells (Syder *et al.*, 1999). The gastric unit autonomous nature of these changes indicates that parietal cells exert local environmental control over lineage progenitors. Pedigrees of *tox*176 transgenic mice with a mosaic pattern of parietal cell ablation have 2-3 fold elevations in serum gastrin levels. The gastric unit autonomous nature of progenitor proliferation suggests that it cannot be ascribed simply to the effects of gastrin (Syder *et al.*, 1999).

As noted previously, the parietal cell lineage is the only principal gastric epithelial lineage that completes its differentiation in the isthmus. As such, it is in an excellent position to deliver products that could regulate the proliferative activity of epithelial progenitors, thereby functioning as a part of a homeostatic feedback system that 'informs' progenitors about the status of epithelial renewal/differentiation in a gastric unit. It is tempting to speculate that parietal cells located near granule-free lineage progenitors normally secrete regulators of growth factor availability/activity (e.g. binding proteins; proteases): ablation of parietal cells could lead to increased delivery of active growth factors to isthmal progenitors and their subsequent expansion via increased proliferation.

The Amplified Lineage Progenitors in *tox176* Express Oncofetal Sialylated Carbohydrate Epitopes that Function as Receptors for *H. pylori* Adhesins

*tox*176 mice manifest several changes in their stomachs that are encountered in patients with chronic atrophic gastritis: i.e. loss of parietal and zymogenic cells, increased epithelial proliferation, and expression of NeuAcα2,3Galβ1,4 glycans (Farinati *et al.*, 1988). These NeuAcα2,3Galβ1,4 glycans, detectable with *Maackia amurensis* agglutinin (Syder *et al.*,

1999), are expressed by the expanded lineage progenitors. They are also normally produced by lineage progenitors that predominate in the nascent gastric units of embryonic day 18 (E18) normal mice. Moreover, *in situ* binding studies using sections prepared from the stomachs of uninfected E18 FVB/N mice, adult FVB/N α1,3/4 FT transgenic mice, and adult FVB/N *tox*176 animals revealed that the Hp1 clinical isolate constitutively expresses at least two functional classes of adhesins - one which interacts with pit cell-associated Lewis b and another which interacts with progenitor cell-associated NeuAcα2,3Galβ1,4-glycans (Syder *et al.*, 1999). Hp1 binding to sections of adult *tox*176 stomach is inhibited by pre-treatment of the bacteria with a purified sialyl-Lewis x-HSA (Figure 3). The Hp1 adhesin is specific for terminal sialic acid in an α2,3 linkage to galactose since neither 6'-sialyllactose-HSA nor purified neoglycoconjugates containing fucose linked to galactose (Lewis b-HSA; Lewis y-HSA) inhibit binding (Table I). The adhesin also discriminates between galactose linked β1,3 to N-acetylglucosamine and galactose linked β1,4 since binding is blocked by pre-incubation with sialyl-Lewis x but not with sialyl-Lewis a (Table I).

When 6 week old conventionally raised *tox*176 mice are infected with Hp1 for 8 weeks, there is a 15-fold greater number of gastric mucosal CD45+ lymphocytes (principally CD4+-TCRαβ+ T-cells) compared to age-, gender- and microbial density-matched normal littermates. (CD45+ cells are below the limits of quantitation by FACS in age-matched uninfected, conventionally raised *tox*176 mice and their normal littermates; Syder *et al.*, 1999). *tox*176 mice also exhibit an increased incidence and titer of Lewis x antibodies even though they lack Lewis x-containing parietal cells, indicating that this attachment-facilitated response must be elicited by bacterial LPS-affiliated, rather than host parietal cell-associated Lewis x immunodeterminants.

Germ-free *tox*176 mice have an expanded population of NeuAcα2,3Galβ1,4 glycan-positive progenitors, indicating that the normal gastric microflora is not a source of signals that are required to produce these sialylated glycans (Syder *et al.*, 1999). Hp1 also binds to the gastric epithelium of germ-free *tox*176 mice but not to the gastric epithelium of germ-free non-transgenic littermates. Binding to the germ-free *tox*176 stomach is inhibited by pre-treatment of Hp1 with the same neoglycoconjugates that inhibit bacterial binding to the gastric epithelium of conventionally raised *tox*176 mice (Syder *et al.*, Unpublished data.).

Gnotobiotic tox176 mice offer a number of advantages over conventionally raised mice. *First*, the efficiency of colonization of germ-free tox176 mice is greater (98-100% versus ~50%; Syder *et al.*, manuscript in preparation). This improved efficiency is likely due, at least in part, to the absence of a competing endogenous flora. *Second*, there is less animal-to-animal variability in the density of colonization (i.e. 10^6-10^8 CFU of Hp1/stomach 8 weeks after a single gavage of 10^7 bacteria). *Third*, it is possible to accurately define the cellular patterns of bacterial tropism in germ-free animals. EM studies of bacterial attachment, established that Hp1 binds *in vivo* to the apical membranes of granule-free lineage progenitors in 14 week old *tox*176 mice (Syder *et al.*, Unpublished data.).

Like conventionally raised animals, histochemical and immunohistochemical studies of stomach sections prepared from gnotobiotic 14 week old *tox*176, α1,3/4 FT, and normal mice 8 weeks after inoculation with Hp1, have revealed an increased cellular immune response associated with bacterial attachment. There is also an enhanced humoral response to bacterial LPS-associated Lewis x epitopes: ELISA assays have revealed significantly higher titers of these antibodies in the sera of *tox*176 mice compared to their normal littermates (Syder *et al.*, Unpublished data.). A similar enhanced humoral immune response was observed after Hp1 colonization of gnotobiotic α1,3/4FT transgenics. We are currently in the process of using gnotobiotic mice to compare bacterial Lewis x production as a function of residency in the gastric ecosystems of normal, α1,3/4FT and *tox*176 mice.

Chronic Atrophic Gastritis and Gastric Cancer in *H. pylori* Infection: Pathogenetic Considerations Arising out of the Analysis of *tox176* mice

Results obtained from the transgenic mouse models suggest the following three component pathogenetic mechanism that puts a host at risk for the development of cancer:

(1) A matching of bacterial adhesin production and host receptor production, either as a pre-existing condition or as a result of the evolution of an infection with induction of microbial and/or host gene expression, puts the host at risk for development of a more severe inflammatory response.

(2) An infection that results in loss of parietal cells, either through (i) cross-reactive immunopathology involving structurally similar epitopes produced by bacterial and parietal epitopes, (ii) hypergastrinemia, and/or (iii) other as yet unspecified microbial and host determinants, produces a coincident increase in the proliferative activity of isthmal lineage progenitors.

(3) If there is a matching of microbial adhesin production with expression of receptors (e.g. NeuAcα2,3Galβ1,4 glycans) by expanded lineage progenitors, bacterial attachment ensues resulting in the perturbation of the properties of these progenitors, such as increased proliferation, alterations in survival, or changes in genome stability. Microbial gene expression and the ability to deliver bacterial gene products to host cells is also likely to be affected as a result of attachment. Attachment may promote internalization of bacteria into long-lived isthmal progenitors, thereby providing a secure niche for these organisms and a chance to evolve a variety of genotypic changes (for evidence that *H. pylori* is internalized into cells see Evans *et al.*, 1992; Hultén *et al.*, 1996; Löfman *et al.*, 1997; Engstrand *et al.*, 1997; Wilkinson *et al.*, 1998; Ko *et al.*, 1999; Su *et al.*, 1999; Björkhoml et al., 2000).

In this envisioned scheme, the coincidence of host and microbial factors skews the destiny of *H. pylori* infection towards neoplasia. This proposed mechanism, and the findings obtained from *tox*176 mice, integrate several observations previously made by others in humans. *First*, *H. pylori* isolates have been reported to bind to NeuAcα2,3Galβ1,4 glycoconjugates expressed by human gastrointestinal adenocarcinoma-derived cell lines (Hirmo *et al.*, 1996; Simon *et al.*, 1997*). Second*, in the normal human stomach, rare isthmal cells have been noted to produce NeuAcα2,3Galβ1,4 glycans. *Third*, parietal cell loss in patients with chronic atrophic gastritis has been associated with increased accumulation of NeuAcα2,3Galβ1,4 glycoconjugates (Farinati *et al.*, 1988). *Finally*, NeuAcα2,3Galβ1,4 glycans are expressed on the epithelium of patients with intestinal metaplasia and with adenocarcinoma (note that carbohydrate epitopes terminating with the NeuAcα2,3Galβ1,4 disaccharide, such as sialyl-Lewis[x], and sialyl-T antigen, are also encountered in a variety of adenocarcinomas; [Sakamoto *et al.*, 1989; Amado *et al.*, 1998; Werther *et al.*, 1996; Hakomori, 1996]).

Prospects

Gnotobiotic-transgenic mouse models offer an opportunity to develop and test hypotheses about the microbial, host and environmental factors that may determine the outcome of *H. pylori* infection in humans. By using gnotobiotic animals, the microbial composition of the stomach can be precisely defined at the onset of infection. Genotypic changes that a strain undergoes during the ensuing course of infection can be monitored. Issues such as whether *H. pylori* strains readily exchange genetic material, or how antibiotic resistance affects fitness in the gastric ecosystem (Björkman *et al.*, 2000) can be addressed in the 'simplified' gastric ecosystems of these mice.

The question of the role of microbial attachment, or the impact of atrophic gastritis on the destiny of infection can be ascertained by introducing an isolate such as Hp1 into normal mice, α1,3/4FT mice ('attachment control' where parietal cells are present and attachment is limited to mature Lewis b-containing pit cells), and *tox*176 mice (no parietal cells; bacterial attachment is restricted to amplified lineage progenitors). The responses of these gnotobiotic hosts can be defined broadly in molecular terms using cDNA- or oligo-based microarrays to profile changes in mRNA accumulation. DNA microarrays can be used to generate comprehensive descriptions of genotypic changes in a colonizing strain as a function of the pre-existing and evolving features of its genetically engineered gastric ecosystem. Convenient methods for determining changes in microbial gene expression will, hopefully, be available in the near future so that there can be concommittent monitoring of microbial and host responses (Chiang *et al.*, 1999).

While analyses of the co-evolution of microbial and host responses in genetically defined and simplified gastric ecosystems will be very important for establishing a conceptual foundation for understanding pathogenesis, it is critical that such analyses be performed using strains with 'relevant' genotypes. This will require a commitment to human epidemiologic studies that survey microbial genotypes, at a whole genome level, initially among individuals enrolled in carefully designed case-control studies, and ultimately within individuals over the course of their infection, and as a function of histopathologic status. The goal of such molecular epidemiologic studies will be to identify genotypes characteristic of isolates associated with histopathologic states that precede the development of gastric cancer (e.g. chronic atrophic gastritis), and with gastric adenocarcinoma itself. It is likely that the former, rather than the latter strains will reveal microbial genes that are candidates for direct involvement in disease progression.

Together, information gathered from (gnotobiotic) transgenic mouse models and human epidemiologic studies should provide new insights about the molecular mechanisms underlying *H. pylori* pathogenesis, a set of microbial and host markers that may help identify patients at risk for development of severe pathology, and new potential therapeutic strategies.

Acknowledgements

We thank Olof Nyrén, Sherif Karam, Lisa Roberts, Helena Enroth, Nina Salma, Karen Guillemin, Tim McDaniel, Lucy Tompkins, and Stanley Falkow for helpful discussions and for providing us with experimental results prior to their publication.

References

Amado, M., Carneiro, F., Seixas, M., Clausen, H., and Sobrinho, M. 1998. Dimeric sialyl-Le(x) expression in gastric carcinoma correlates with venous invasion and poor outcome. Gastroenterology 114: 462-470.

Appelmelk B.J., Faller, G., Claeys, D., Kirchner, T. and Vandenbroucke-Grauls, C.M. 1998a. Bugs on trial: the case of *Helicobacter pylori* and autoimmunity. Immunol. Today 19: 296-299.

Appelmelk B.J., Shiberu, B., Trinks, C., Tapsi, N., Zheng, P.Y., Verboom, T., Maaskant, J., Hokke, C.H., Schiphorst, W.E., Blanchard, D., Simoons-Smit, I.M., van den Eijnden,

D.H., and Vandenbroucke-Grauls, C.M. 1998b. Phase variation in *Helicobacter pylori* lipopolysaccharide. Infect. Immun. 66: 70-76.

Atherton, J.C., Tham, K.T., Peek, R.M.Jr., Cover, T.L., and Blaser, M.J. 1996. Density of *Helicobacter pylori* infection *in vivo* as assessed by quantiative culture and histology. J. Infect. Dis. 174:552-556.

Barker, D.J.P., Coggon, D., Osmond, C., and Wickham, C. 1990. Poor housing in childhood and high rates of stomach cancer in England and Wales. Br. J. Cancer 61: 575-578.

Bartsch, H., Ohshima, H., and Pignatelli, B. 1988. Inhibitors of endogenous nitrosation. Mechanisms and implications in human cancer prevention. Mutat. Res. 202: 307-324.

Björkholm, B., Zhukhovitsky, V., Lofman, C., Hulten, K., Enroth, H., Block, M., Rigo, R., Falk, P., and Engstrand, L. 2000. *Helicobacter pylori* entry into human gastric epithelial cells: A potential deterimant of virulence, persistence, and treatment failures. Helicobacter. 5: 148-154.

Björkman, J., Nagaev, I., Berg, O.G., Hughes, D., and Andersson, D.I. 2000. Effects of environment on compensatory mutations to ameliorate costs of antibiotic resistance. Science 287: 1479-1482.

Blaser, M.J., and Kirschner, D. 1999. Dynamics of *Helicobacter pylori* colonization in relation to the host response. Proc. Natl. Acad. Sci. USA 96: 8359-8364.

Blaser, M.J., Chyou, P.H., and Nomura, A. 1995. Age at establishment of *Helicobacter pylori* infection and gastric carcinoma, gastric ulcer, and duodenal ulcer risk. Cancer Res. 55: 562-565.

Borén, T., Falk, P., Roth, K.A., Larsson, G., and Normark, S. 1993. Attachment of *Helicobacter pylori* to human gastric epithelium mediated by blood group antigens. Science. 262: 1892-1895.

Chan, A.O.-O., McLuk, J., Hui, W-M., and Lam, S-K. 1999. Molecular biology of gastric carcinoma: From laboratory to bedside. J. Gastroenterol. Hepatol. 14: 1150-1160.

Chiang, S.L., Mekalanos, J.J., and Holden, D.W. 1999. *In vivo* genetic analysis of bacterial virulence. Annu. Rev. Microbiol. 53: 129-154.

Cornelis, G.R., and Wolf-Watz, H. 1997. The Yersinia Yop virulon: a bacterial system for subverting eukaryotic cells. Mol. Microbiol. 23: 861-867.

Costerton, J.W., Stewart, P.S., and Greenberg, E.P. 1999. Bacterial biofilms: A common cause of persistent infections. Science 284: 1318-1322.

Covacci, A., and Rappuoli, R. 2000. Tyrosine-phosphorylated bacterial proteins: Trojan horses for the host cell. J. Exp. Med. 191: 587-592.

Correa, P., and Miller, M.J.S. 1998. Carcinogenesis, apoptosis and cell proliferation. Br. Med. Bull. 54: 151-162.

Danesh, J. 1999. *Helicobacter pylori* infection and gastric cancer: systematic review of the epidemiological studies. Aliment Pharmacol. Ther. 13: 851-856.

Danon, S.J., and Eaton, K.A. 1998. The role of gastric Helicobacter and N-methyl-N'-nitro-N-nitrosoguanidine in carcinogenesis of mice. Helicobacter 3: 260-268.

Dixon, M.F. 1994. Pathophysiology of *Helicobacter pylori* infection. Scand. J. Gastroenterol. 29 (suppl 201):7-10.

Doig, P., de Jonge, B.L., Alm, R.A., Brown, E.D., Uria-Nickelsen, M., Noonan, B., Mills, S.D., Tummino, P., Carmel, G., Guild, B.C., Moir, D.T., Vovis, G.F., and Trust, T.J. 1999. *Helicobacter pylori* physiology predicted from genomic comparison of two strains. Microbiol. Mol. Biol. Rev. 63: 675-707.

Dunn, B.E., Cohen, H., and Blaser, M.J. 1997. *Helicobacter pylori*. Clin. Microbiol. Rev. 10: 720-741.

Eidt, S., and Stolte, M. 1994. Antral intestinal metaplasia in *Helicobacter pylori* gastritis. Digestion 55: 13-18.

Engstrand, L., Graham, D., Scheynius, A., Genta, R.M., and El-Zaatari, F. 1997. Is the sanctuary where *Helicobacter pylori* avoids antibacterial treatment intracellular? Am. J. Clin Pathol. 108: 504-509.

Enroth, H., Kraaz, W., Engstrand, L., Nyren, O., and Rohan, T. 2000. *Helicobacter pylori* strain types and risk of gastric cancer: a case-control study. Cancer Epidemiol. Biomarkers Prev. 9: 981-985.

Eslick, G.D., Lim, L. L.-Y., Byles, J.E., Xia, H. H.-X., and Talley, N.J. 1999. Association of *Helicobacter pylori* infection with gastric carcinoma: a meta-analysis. Am. J. Gastroenterol. 94: 2373-2379.

Evans, D.G., Evans, D.J. Jr., and Graham, D.Y. 1992. Adherence and internalization of *Helicobacter pylori* by HEP-2 cells. Gastroenterology 102: 1557-1567.

Evans, D.G., Karjalainen, T.K., Evans, D.J. Jr., Graham, D.Y., and Lee, C.-H. 1993. Cloning, nucleotide sequence, and expression of gene encoding an adhesin subunit protein of *Helicobacter pylori.* J. Bacteriol., 175: 674-683.

Falk, P.G., Bry, L., Holgersson, J., and Gordon, J.I. 1995. Expression of a human α-1,3/4 fucosyltransferase in the pit cell lineage of FVB/N mouse stomach results in production of Lewisb-containing glycoconjugates: A potential transgenic mouse model for studying *Helicobacter pylori* infection. Proc. Natl. Acad. Sci. USA 92: 1515-1519.

Farinati, F., Nitti, D., Cardin, F., Di Mario, F., Costa, F., Rossi, C., Marchett, A., Lise, M., and Naccarato, R. 1988. CA 19-9 determination in gastric juice: role in identifying gastric cancer and high risk patients. Eur. J. Cancer Clin. Oncol. 24: 923-927.

Fox, J.G., Li, X., Cahill, R.J., Andrutis, K., Rustgi, A.K., Odze, R., and Wang, T.C. 1996. Hypertrophic gastropathy in Helicobacter felis-infected wild-type C57BL/6 mice and p53 hemizygous transgenic mice. Gastroenterology 110: 155-166.

Fu, S., Ramanujam, K.S., Wong, A., Fantry, G.T., Drachenberg, C.T., James, S.P., Meltzer, S.J., and Wilson, K.T. 1999. Increased expression and cellular localization of inducible nitric oxide synthase and cyclooxygenase 2 in *Helicobacter pylori* gastritis. Gastroenterology 116: 1319-1329.

Galpin, O.P., Whitaker, C.J., and Dubiel, A.J. 1992. *Helicobacter pylori* infection and overcrowding in childhood. Lancet 339: 619.

Genta, R.M. 1997. *Helicobacter pylori*, inflammation, mucosal damage, and apoptosis: pathogenesis and definition of gastric atrophy. Gastroenterology 113: S51-S55.

Gerhard, M., Lehn, N., Neumayer, N., Borén, T., Rad, R., Schepp, W., Miehlke, S., Classen, M., and Prinz, C. 1999. Clinical relevance of the *Helicobacter pylori* gene for blood-group antigen-binding adhesin. Proc. Natl. Acad. Sci. USA 96: 12778-12783.

Guruge, J.L., Falk, P.G., Lorenz, R.G., Dans, M., Wirth, H.P., Blaser, M.J., Berg, D.E., and Gordon, J.I. 1998. Epithelial attachment alters the outcome of *Helicobacter pylori* infection. Proc. Natl. Acad. Sci. USA 95: 3925-3930.

Hakomori, S. 1996. Tumor malignancy defined by aberrant glycosylation and Sphingo (glyco) lipid metabolism. Cancer Res. 56: 5309-5318.

Hamlet, A., Thoreson, A.C., Nilsson, O., Svennerholm, A.M., and Olbe, L. 1999. Duodenal *Helicobacter pylori* infection differs in cagA genotype between asymptomatic subjects and patients with duodenal ulcers. Gastroenterology. 6: 259-268.

Hansson, L.-E., Nyrén, O., Hsing, A.W., Bergström, R., Josefsson, S., Chow, W.-H., Fraumeni, J.F., Adami, H.-O. 1996. The risk of stomach cancer in patients with gastric or duodenal ulcer disease. New Eng. J. Med. 4: 242-249.

Hansson, L.-E., Engstrand, L., Nyrén, O., Evans, D.J., Jr., Lindgren, A., Bergstrom, R., Andersson, B., Athlin, L., Bendtsen, O., Tracz, P. 1993. *Helicobacter pylori* infection: independent risk indicator of gastric adenocarcinoma. Gastroenterology 105:1098-1103.

Hirmo, S., Kelm, S., Schauer, R., Nilsson, B., and Wadstrom, T. 1996. Adhesion *Helicobacter pylori strains* to α2,3-linked sialic acids. Glycoconj. J. 13: 1005-1011.

Hizawa, K., Suekane, H., Kawasaki, M., Yao, T., Aoyagi, K., and Fujishima, M. 1997. Diffuse cystic malformation and neoplasia associated cystic formation in the stomach. Endosonographic Features and diagnosis of tumor depth. J. Clin. Gastroenterol. 25: 634-639.

Honda, S., Fujioka, T., Tokieda, M., Gotoh, T., Nishizono, A., and Nasu, M. 1998a. Gastric ulcer, atrophic gastritis, and intestinal metalplasia caused by *Helicobacter pylori* infection in mongolian gerbils. Scand. J. Gastroenterol. 33: 454-460.

Honda, S., Fujioka,T., Tokieda, M., Satoh, R., Nishizono, A., and Nasu, M. 1998b. Development of *Helicobacter pylori*-induced gastric carcinoma in mongolian gerbils. Cancer Res. 58: 4255-4259.

Huang, J.-Q., Sridhar, S., Chen, Y., and Hunt, R. 1998. Meta-analysis of the relationship between *Helicobacter pylori* seropositivity and gastric cancer. Gastroenterology 114: 1169-1179.

Huesca, M., Goodwin, A., Bhagwansingh, A., Hoffman, P., and Lingwood, C.A. 1998. Characterization of an acidic-pH-inducible stress protein (hsp70), a putative sulfatide binding adhesin, from *Helicobacter pylori*. Infect. Immun. 66: 4061-4067.

Hultén, K., Rigo, R., Gustafsson, I., and Engstrand, L. 1996. New pharmacokinetic *in vitro* model for studies of antibiotic activity against intracellular microorganisms. Antimicrob. Agents Chemother. 40: 2727-2731.

Ilver, D., Arnqvist, A., Ogren, J., Frick, I.M., Kersulyte, D., Incecik, E.T., Berg, D.E., Covacci, A., Engstrand, L., and Borén, T. 1998. *Helicobacter pylori* adhesin binding fucosylated histo-blood group antigens revealed by retagging. Science. 279: 373-377.

IARC monographs on the evolution of carcinogenic risks to humans. 1994. Vol. 61 Schistosomes, Liver Flukes and *Helicobacter pylori*: Lyon, France IARC. p. 177-240.

Jones, A.C., Logan, R.P., Foynes, S., Cockayne, A., Wren, B.W., and Penn, C.W. 1997. A flagellar sheath protein of *Helicobacter pylori* is identical to HpaA, a putative N-acetylneuraminyllactose-binding hemagglutinin, but is not an adhesin for AGS cells. J. Bacteriol. 179: 5643-5647.

Karam, S.M. 1993. Dynamics of epithelial cells in the corpus of the mouse stomach. IV. Bidirectional migration of parietal cells ending in their gradual degeneration and loss. Anat. Rec. 236: 314-332.

Karam, S.M., and Leblond, C.P. 1992. Identifying and counting epithelial cell types in the corpus of the mouse stomach. Anat. Rec. 232: 231-246.

Karam, S.M., and Leblond, C.P. 1993a. Dynamics of epithelial cells in the corpus of the mouse stomach. I. Identification of proliferative cell types and pinpointing of the stem cell. Anat. Rec. 236: 259-279.

Karam, S.M., and Leblond, C.P. 1993b. Dynamics of epithelial cells in the corpus of the mouse stomach.II. Outward migration of pit cells. Anat. Rec. 236: 280-296.

Karam, S.M., and Leblond, C.P. 1993c. Dynamics of epithelial cells in the corpus of the mouse stomach. III. Inward migration of neck cells followed by progressive transformation into zymogenic cells. Anat. Rec. 236: 297-313.

Karam, S.M., and Leblond, C.P. 1993d. Dynamics of epithelial cells in the corpus of the mouse stomach. V. Behavior of entero-endocrine and caveolated cells: general conclusion on cell kinetics in the oxyntic epithelium. Anat. Rec. 236:333-340.

Karam, S.M., Li, Q., and Gordon, J.I. 1997. Gastric epithelial morphogenesis in normal and trangenic mice. Am. J. Physiol. 272:G1209-1220.

King, H., and Locke, F.B. 1980a. Cancer mortality among Chinese in the United States. J. Natl. Cancer Inst. 65: 1141-1148.

King, H., and Locke, F.B. 1980b. Cancer mortality among Japanese in the United States. J. Natl. Cancer Inst. 65: 1149-1156.

Kirschner, D., and Blaser, M.J. 1995. The dynamics of *H. pylori* infection of the human stomach. J. Theor. Biol. 176:281-290.

Ko, G.H., Kang, S.M., Kim, Y.K., Lee, J.H., Park, C.K., Youn, H.S., Baik, S.C., Cho, M.J., Lee, W.K., and Rhee, K.H. 1999 Invasiveness of *Helicobacter pylori* into human gastric mucosa. Helicobacter 4: 77-81.

Kuipers, E.J., Uyterlinde, A.M., Peña, A.S., Roosendaal, R., Pals, G., Nelis, G.F., Festen, H.P., and Meuwissen., S.G. 1995. Long-term sequelae of *Helicobacter pylori* gastritis. Lancet 345: 1525-1528.

Lechago, J., and Correa, P. 1993. Prolonged achlorhydria and gastric neoplasia: Is there a causal relationship? Gastroenterology 104: 1554-1557.

Li, Q., Karam, S.M., and Gordon, J.I. 1996. Diphtheria toxin-mediated ablation of parietal cells in the stomach of transgenic mice. J. Biol. Chem. 271: 3671-3676.

Lin, C.W., Wu, S.C., Lee, S.C., and Cheng, K.S. 2000. Genetic analysis and clinical evaluation of vacuolating cytotoxin gene A and cytotoxin-associated gene A in Taiwanese *Helicobacter pylori* isolates from peptic ulcer patients. Scand. J. Infect. Dis. 32: 51-57.

Lingwood, C.A., Wasfy, G., Han, H., and Huesca, M. 1993. Receptor affinity purification of a lipid-binding adhesin from *Helicobacter pylori*. Infect. Immun. 61: 2474-2478.

Lipkin, M., Sherlock, B., and Bell, B. 1963. Cell proliferation kinetics in the gastrointestinal tract of man. Gastroenterology, 46:721-735.

Löfman, C., Rigo, R., Block, M., Hulten, K., Enroth, H., and Engstrand, L. 1997. Bacterium-host interactions monitored by time-lapse photography. Nat. Med. 3:8 930-931.

Lorenz, R.G. and Gordon, J. I. 1993. Use of transgenic mice to study regulation of gene expression in the parietal cell lineage of gastric units. J. Biol. Chem 168: 26559-26570.

Marais, A., Mendz, G.L., Hazell, S.L., and Megraud, F. 1999. Metabolism and genetics of *Helicobacter pylori*: the Genome era. Micro. Mol. Biol. Rev. 63: 642-674.

Mendall, M.A., Goggin, P.M., Molineaux, N., Levy, J., Toosy, T., Strachan, D., and Northfield, T.C. 1992. Childhood living conditions and *Helicobacter pylori* seropositivity in adult life. Lancet 339: 896-897.

Micots, I., Augeron, C., Laboisse, C.L., Muzeau, F., Megraud, F. 1993. Mucin exocytosis: a major target for *Helicobacter pylori*. J. Clin. Pathol. 46:241-245.

Mitchell, H.M., Li, Y.Y., Hu, P.J., Liu, Q., Chen, M., Du, G.G., Wang, Z.J., Lee, A., and Hazell, S.L. 1992. Epidemiology of *Helicobacter pylori* in southern China: identification of early childhood as the critical period for acquisition. J. Infect. Dis. 166:149-153.

Molinari, F., Parodi, M.C., DeAngelis, P., and Cheli, R. 1984. Gastric mucus secretion in chronic gastritis. Scand. J. Gastroenterol. 19 (suppl. 92): 167-171.

Niv, Y., and Turani, H. 1991. Cystic changes in gastric glands after gastric surgery and in the intact stomach. J. Clin. Gastroenterol. 13: 465-469.

Nowak, J.A, Forouzandeh, B., and Nowak, J.A. 1997. Estimates of *H. pylori* densities in the gastric mucus layer by PCR, histological examination and CLO test. Am. J. Clin. Pathol. 108: 284-288.

Nyrén, O. 1998. Is *Helicobacter pylori* really the cause of gastric cancer? Sem. Cancer. Biol. 8: 275-283.

Odenbreit, S., Wieland, B., and Haas, R. 1996. Cloning and genetic characterization of *Helicobacter pylori* catalase and construction of a catalase-deficient mutant strain. J. Bacteriol. 178: 6960-6967.

Odenbreit, S., Püls, J., Sedlmaier, B., Gerland, E., Fischer, W., and Haas, R. 2000. Translocation of *Helicobacter pylori* CagA into gastric epithelial cells by type IV secretion. Science 287: 1497-1500.

O'Toole, P.W., Janzon, L., Doig, P., Huang, J., Kostrzynska, M., and Trust, T.J. 1995. The putative neuraminyllactose-binding hemagglutinin HpaA of *Helicobacter pylori* CCUG 17874 is a lipoprotein. J. Bacteriol. 177: 6049-6057.

Rubinstein, A., and Tirosh, B. 1994. Mucus gel thickness and turnover in the gastrointestinal tract of the rat: response to cholinergic stimulus and implication for mucoadhesion. Pharm. Res. 11: 794-799.

Rubio, C.A., and Owen, D.A. 1998. A comparative study between the gastric mucosa of wester Canadians and other dwellers of the Pacific Basin. Anticancer Rese. 18: 2463-2470.

Sakamoto, S., Watanabe, T., Tokumaru, T., Takagi, H., Nakazato, H., and Lloyd, K.O. 1989. Expression of Lewisa, Lewisb, Lewisx, Lewisy, siayl-Lewisa, and sialyl-Lewisx blood group antigens in human gastric carcinoma and in normal gastric tissue. Cancer Res. 49: 745-752.

Schrager, J., Spink, R., and Mitra, S. 1967. The antrum in patients with duodenal and gastric ulcers. Gut 8: 497-508.

Segal, E.D., Cha, J., Lo, J., Falkow, S., and Tompkins, L.S. 1999. Altered states: Involvement of phosphorylated CagA in the induction of host cellular growth changes by *Helicobacter pylori*. Proc. Natl. Acad. Sci. USA 96: 14559-14564.

Shimizu, N., Kaminishi, M., Tatematsu, M., Tsuji, E., Yoshikawa, A., Yamaguchi, H., Aoki, F., and Oohara, T. 1998. *Helicobacter pylori* promotes development of pepsinogen-altered pyloric glands, a preneoplastic lesion of glandular stomach of BALB/c mice pretreated with *N*-methyl-*N*-nitrosourea. Cancer Letts. 123: 63-69.

Simon, P.M., Goode, P.L., Mobasseri, A., and Zopf, D. 1997. Inhibition of *Helicobacter pylori* binding to gastrointestinal epithelial cells by sialic acid-containing oligosaccharides. Infect. Immun. 65: 750-757.

Sipponen, P. 1992. Long-term consequences of gastroduodenal inflammation. Eur. J. Gastroenterol. Hepatol. 4: S25-29.

Sitas F., Forman, D., Yarnell, J.W.G., Burr, M.I., Elwood, P.C., Pedley, S., and Marks, K.J. 1991. *Helicobacter pylori* infection rates in relation to age and social class in a population of Welsh men. Gut 32:25-28.

Sjöstedt, S., Kager, L., Heimdahl, A., and Nord, C.E. 1988. Microbial colonization of tumors in relation to the upper gastrointestinal tract in patients with gastric carcinoma. Ann. Surg. 207: 341-346.

Söderholm, J.D., Borch, K., Olaison, G., and Franzén, L. 1994. Gastric leakiness in patients with atrophic gastritis. Gut 35 Suppl4: A8.

Sørbye, H., and Svanes, K. 1995. Gastric mucosal protection against penetration of carcinogens into the mucosa. Scand. J. Gastroenterol. 30: 929-934.

Staszewski, J., and Haenszel, W. 1965. Cancer mortality among the Polish-born in the United States. J. Natl. Cancer Inst. 35: 291-297.

Stein, M., Rappuoli, R., and Covacci, A. 2000. Tyrosine phosphorylation of the *Helicobacter pylori* CagA antigen after *cag* –driven host cell translocation. Proc. Natl. Acad. Sci. USA 97: 1263-1268.

Su, B., Johansson, S., Fällman, M., Patarroyo, M., Granstrom, M., and Normark, S. 1999. Signal transduction-mediated adherence and entry of *Helicobacter pylori* into cultured cells. Gastroenterology 117: 595-604.

Sugiyama, A., Maruta, F., Ikeno, T., Ishida, K., Kawasaki, S., Katsuyama, T., Shimizu, N., and Tatematsu, M. 1998. *Helicobacter pylori* infection enhances N-Methyl-N-nitrosourea-induced stomach carcinogenesis in the mongolian gerbil. Cancer Res. 58: 2067-2069.

Syder, A.J., Guruge, J.L., Li, Q., Hu, Y., Oleksiewicz, C.M., Lorenz, R.G., Karam, S.M., Falk, P.G., and Gordon, J.I. 1999. *Helicobacter pylori* attached to NeuAcα2,Galβ1,4) Glycoconjugates produced in the stomach of transgenic mice lacking parietal cells. Molec. Cell 3: 263-274.

Wang, G., Humanyun, M.Z., and Taylor, D. E. 1999. Mutation as an origin of genetic variability in *Helicobacter pylori*. Trends Microbiol. 7: 488-492.

Wang, T.C., Dangler, C.A., Chen, D., Goldenring, J.R., Koh, T., Raychowdhury, R., Coffey, R.J., Ito, S., Varro, A., Dockray, G.J., and Fox, J.G. 2000. Synergistic interaction between hypergastrinemia and *helicobacter* infection in a mouse model of gastric cancer. Gastroenterology 118: 36-47.

Watanabe, T., Tada, M., Nagai, H., Sasaki, S., and Nakao, M. 1998. *Helicobacter pylori* infection induces gastric cancer in mongolian gerbils. Gastroenterology 115: 642-648.

Webb, P.M., Knight, T., Greaves, S., Wilson, A., Newell, D.G., Elder, J., and Forman, D. 1994. Relation between infection with *Helicobacter pylori* and living conditions in childhood: evidence for person to person transmission in early life. Brit. Med. J. 308: 750-753.

Webb, P.M., and Forman, D. 1995. *Helicobacter pylori* as a risk factor for cancer. Baillière's Clin. Gastroenterol. 9: 563-582.

Werther, J.L., Tatematsu, M., Klein, R., Kurihara, M., Kumaga, K., Llorens, P., Guidugli Neto, J., Bodian, C., Pertsemeidis, D., Yamachika, T., Kitou, T., and Itzkowitz, S. 1996. Sialosyl-Tn antigen as a marker of gastric cancer progression: an international study. Int. J. Cancer 69: 193-199.

Wilkinson, S.M., Uhl, J.R., Kline, B.C., and Cockerill, F.R. III 1998. Assessment of invasion frequencies of cultured HEp-2 cells by clinical isolates of *Helicobacter pylori* using an acridine orange assay. J. Clin. Pathol. 51: 127-133.

Wilson, M., Seymour, R., and Henderson, B. 1998. Bacterial perturbation of cytokine networks. Infect. Immun. 66: 2401-2409.

Wirth, H.P., Yang, M., Peek, R.M. Jr, Tham, K.T., and Blaser, M.J. 1997. *Helicobacter pylori* Lewis expression is related to the host Lewis phenotype. Gastroenterology, 113: 1091-1098.

From: *Helicobacter pylori: Molecular and Cellular Biology*
ISBN 1-898486-25-5 © 2001 Horizon Scientific Press, Wymondham, UK.

10

Helicobacter pylori Urease

Harry L.T. Mobley[*]

Abstract

Urease, a protein expressed in high quantity by *Helicobacter pylori*, catalyzes the hydrolysis of urea to yield ammonia and carbamate. The K_m ranges from 0.17-0.48 mM urea and the specific activity ranges from 1100 to 1700 µmol urea/min/mg protein. The apoenzyme is comprised of six copies of each of two structural subunits, UreA and UreB and the metalloenzyme contains two nickel ions in each of the six active sites. Seven contiguous genes, *ureABIEFGH*, are transcribed in the same direction from a 6.13 kb sequence and all of them except *ureI* are necessary for synthesis of a catalytically active enzyme. Active urease is required for the colonization of the gastric mucosa.

Introduction

Urease is produced by numerous taxonomically varied bacterial species. The enzyme is produced by normal commensal flora and non-pathogens but is also a potent virulence factor for some species including *Proteus mirabilis* (Jones *et al.*, 1990), *Staphylococcus saprophyticus* (Gatermann and Marre, 1989) and *Helicobacter pylori* (Eaton *et al.*, 1991).

Why is urease one of the most intensely studied proteins of *H. pylori*? Urease is used for taxonomic identification. Urease is necessary for colonization of the gastric mucosa, is a potent immunogen to which a vigorous immune response develops and is a vaccine candidate. Urease is also used for diagnosis and follow up after treatment, and represents an interesting model for metalloenzyme study. Thus urease is of interest to the microbiologist, biochemist, gastroenterologist, immunologist, and vaccinologist. Prior to the discovery of *H. pylori*, humans were thought to produce "gastric urease." Instead, it is now known that this bacterium that colonizes the gastric mucosa of humans is the source of this notable protein.

Enzymology

Enzymatic Reaction

Urease (urea amidohydrolase: EC3.5.1.5) catalyzes the hydrolysis of urea to yield ammonia and carbamate. The latter compound spontaneously decomposes to yield another molecule of ammonia and carbonic acid:

[*]email: hmobley@umaryland.edu

$$\overset{\displaystyle O}{\overset{\displaystyle \|}{H_2N - C - OH}} + 2H_2O \rightarrow NH_3 + H_2CO_3$$

In solution, the released carbonic acid and the two molecules of ammonia are in equilibrium with their deprotonated and protonated forms, respectively. The net effect of these reactions is an increase in pH.

$$H_2CO_3 \leftrightarrow H^+ + HCO_3^-$$

$$2NH_3 + 2H_2O \leftrightarrow 2NH_3^+ + 2OH^-$$

Kinetic Constants and Optima

The affinity constants of purified *H. pylori* urease have been calculated using rates of hydrolysis measured over a range of urea concentrations from 0-5 mM at 23°C. Reciprocal plots of $1/V$ vs $1/S$ yielded a K_m ranging from 0.17 to 0.48 mM (Dunn *et al.*, 1990; Hu and Mobley, 1990; Evans *et al.*, 1991). Using purified enzyme under saturating conditions, specific activities ranged from 1100 to 1700 µmol urea/min/mg protein (Dunn *et al.*, 1990; Hu and Mobley, 1990). Although *H. pylori* has a K_m that reflects a higher affinity for substrate than the ureases of other species, this appears to be appropriate to the niche of the bacterium. If one assumes that physiological concentrations of urea to which *H. pylori* is exposed are the same as in serum (1.7-3.4 mM), then the urease would be saturated and working at the Vmax. Optimal enzyme activity was observed at 43°C for urease in a cell lysate (Mobley *et al.*, 1988). Enzymatic activity was not inhibited by 0.02% azide (Mobley *et al.*, 1988).

Active Site and Catalysis

The active site of the enzyme is found in the UreB subunit and includes amino acid residues scattered throughout the primary sequence (Jabri *et al.*, 1995) that are brought into proximity by the tertiary structure. The identification of these residues in the active site has been achieved by site directed mutagenesis (Martin and Hausinger, 1992; Sriwanthana and Mobley, 1993; Park and Hausinger, 1993), studies of apoprotein activation (Park and Hausinger, 1995), and structural determinations by X-ray crystallography (Jabri *et al.*; 1992; Jabri *et al.*, 1995). Residues His-136, His-138, Lys-219, His-248, His-274, and Asp-362 come in direct contact with the two nickel ions, urea, or a water molecule within the active site (using the numerical scheme for *H. pylori* UreB of Labigne *et al.*, 1991). In addition, His-322 is near the active site and acts as a general base in the catalysis.

Urea is hydrolyzed according to the scheme first described by Zerner's group (Dixon *et al.*, 1980) for the jack bean urease. Urea binds in *O*-coordination to one nickel ion (aided by His-221). The active base His-322 then activates a water molecule bound to the other nickel ion. Attack by the metal-coordinated hydroxide on the substrate carbon atom results in a tetrahedral intermediate that bridges the two nickel sites. A proton is transferred to the intermediate with accompanying ammonia release and water then displaces the carbamate to complete the cycle.

Protein Structure

Urease is a high molecular weight multi-subunit metalloenzyme. It appears that all ureases are closely related and that the mechanisms of catalysis are similar for all of these proteins. It is curious that such a complex protein is required to carry out a simple hydrolysis.

Purification

H. pylori cultures contain extraordinarily large amounts of urease activity. The purified enzyme is not significantly more active than purified ureases from other species but rather simply represents a larger proportion of the total cell protein in this species. The two urease subunits (apparent molecular size 66 and 29.5 kDa) are extremely prominent after Coomassie Blue staining of SDS PAGE gels of crude lysates (Hu and Mobley, 1990). In contrast, the concentration of urease subunits in crude lysates of other urease-positive bacterial species is too low to detect such bands with this method.

The native urease protein has been purified from *H. pylori* by isolation from the cytosol (Hu and Mobley, 1990) or by elution from the cell surface with low ionic strength solvents (Dunn *et al.*, 1990; Evans *et al.*, 1991; Turbett *et al.*, 1992). For purification from the cytosol, French press cell lysates were chromatographed on DEAE-Sepharose, phenyl-Sepharose, Mono-Q, and Sepharose 6 resins. In this purification scheme, purified urease represented 6% of the soluble protein in the crude extracts and was estimated to have a native molecular size of 550 kDa. On the basis of subunit size, a 1: 1 subunit ratio as estimated from scanning densitometry of Coomassie Blue stained SDS-polyacrylamide gels, and the native molecular weight of 550 kDa, the data are indicate that the native enzyme contains six copies of each subunit. *H. pylori* urease has an isoelectric point of 5.90 to 5.99 (Dunn *et al.*, 1990; Evans *et al.*, 1991; Turbett *et al.*, 1992).

Appearance by Electron Microscopy

By transmission electron microscopy, purified native urease appears as a round, doughnut-shaped, hexagonal particle with a darkly staining core (Austin *et al.*, 1991; Turbett *et al.*, 1992; Dunn *et al.*, 1992). The enzyme is 13 nm in diameter and displays three-fold rotational symmetry. When it was eluted from the bacterial cell surface, urease copurified with a 60 kDa chaperonin heat-shock protein whose native molecular weight and macromolecular structure is very similar to that of urease (Austin *et al.*, 1991; Dunn *et al.*, 1992; Evans *et al.*, 1992).

X-Ray Diffraction Studies

The crystal structure of the related urease from *Klebsiella aerogenes* has been solved on the basis of X-ray diffraction studies (Jabri *et al.*, 1992; Jabri *et al.*, 1995). Because of the high degree of homology between all bacterial ureases, we can infer that the *H. pylori* urease shares a similar structure. The *K. aerogenes* enzyme is only about half the molecular weight of the *H. pylori* enzyme and it contains only three copies of each subunit. Therefore, although the precise arrangement for the *H. pylori* holoenzyme is not known, it may be comprised of two of the quaternary structures that were resolved for the *Klebsiella* enzyme (Figure 1).

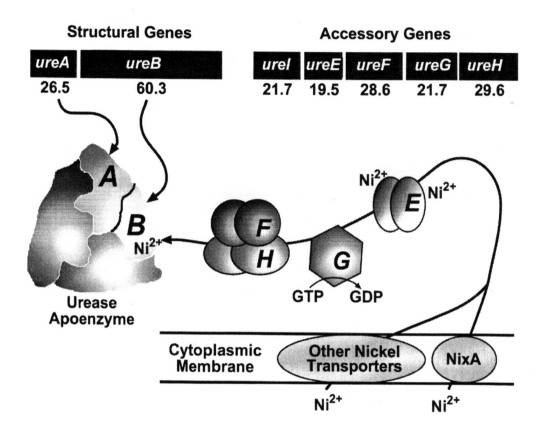

Figure 1. Model for synthesis of a catalytically active urease in *H. pylori*. The urease gene cluster, comprised of seven chromosomally encoded genes, is present as a single copy on the chromosome. *ureA* and *ureB* encode the 26.5 kDa and 60.3 kDa subunits, respectively. Six copies of each subunit spontaneously self assemble to form the catalytically inactive apoenzyme. The urease protein depicted shows three copies of each subunit and is adapted from the crystal structure of the *Klebsiella aerogenes* urease (Jabri *et al.*, 1995). The known molecular size of *H. pylori* urease (550 kDa) would require two of the depicted protein structures to be associated in some manner (Hu and Mobley, 1990). This arrangement has not yet been solved. Accessory genes *ureE*, *ureF*, *ureG* and *ureH* encode accessory proteins UreE, UreF, UreG and UreH which, by analogy to homologs from other species, serve to insert nickel ions (Ni^{2+}) into the apoenzyme in an energy requiring reaction (Mobley *et al.*, 1995). UreE is a nickel-binding dimer. UreG carries a GTP-binding site. The function of UreI is currently under investigation. Two nickels ions are coordinated into the active site of each UreB subunit. Thus, each *H. pylori* urease contains 12 nickel ions when fully activated. Nickel ions are transported into the cell by NixA, a high affinity membrane transport protein (Mobley *et al.*, 1995). Additional backup nickel transport proteins are probably also present. The net result of the interaction of these genes and proteins is a catalytically active urease.

Nickel

As for other ureases, *H. pylori* contains nickel ions in the active site. The nickel content of *H. pylori* urease has been measured using atomic adsorption spectroscopy (Hawtin *et al.*, 1991). All active ureases that have been subjected to analysis, whether from plants, fungi, or bacteria, contain nickel ions (Hausinger, 1987). Based on the sum of many studies, two nickel ions (Ni^{2+}) appear to be present in each active site. Because *H. pylori* has six copies of each of the two distinct subunits (UreA and UreB), there are therefore six active sites and 12 nickel ions per fully loaded enzyme molecule.

Effect of pH on Urease

Exposure of whole cells of *H. pylori* to buffer below pH 4 results in the loss of intracellular urease activity. In soluble enzyme preparations from lysed cells and supernatants, urease activity is not detectable after incubation at pH <5 for 30 minutes. Exposure of whole cells of *H. pylori* to pH 5 or below inhibits overall synthesis of total protein including nascent urease. At pH 6 or 7, urease represents 10% of total cell protein (Bauerfeind *et al.*, 1997).

Genetics

Genetic Organization

The genes encoding *H. pylori* urease are located as a single 6.13 kb gene cluster on the bacterial chromosome (Clayton *et al.*, 1990; Labigne *et al.*, 1991; Cussac *et al.*, 1993) (Figure 1). Seven contiguous genes, all transcribed in the same direction, are necessary for synthesis of an active enzyme (Cussac *et al.*, 1992; Hu *et al.*, 1992; Hu and Mobley, 1993). The genes have been designated *ureABIEFGH*. All of the genes except for *ureI* share homology with urease genes of other species including (but not limited to) *Bacillus sp. TB-90* (Maeda *et al.*, 1994) , *K. aerogenes* (Mulrooney *et al.*, 1990; Lee *et al*, 1992), *P. mirabilis* (Jones and Mobley, 1989; Nicholson *et al.*, 1993; Sriwanthana *et al.*, 1993), *Ureaplasma ureolyticum* (Blanchard, 1990; Neyrolles *et al.*, 1996), and *Yersinia enterocolitica* (Skurnik *et al.*, 1993; deKoning-Ward *et al.*, 1994).

Structural Genes

The two structural subunits are encoded by *ureA* and *ureB*, the first two genes of the gene cluster (Clayton *et al.*, 1990; Labigne *et al.*, 1991). UreA is predicted to be 26.5 kDa and UreB is predicted to be between 60.3 and 61.0 kDa (the sequences differed for two strains). The UreA subunit of *Helicobacter* species is somewhat unusual in that the primary amino acid sequence encoded by the *ureA* gene is always encoded by two separate genes in all other bacterial species, as if the two smaller urease subunit genes of other species may have fused to form *H. pylori ureA*. Expression of *ureA* and *ureB* is sufficient to produce an assembled apoenzyme (Hu *et al.*, 1992). Under these conditions, no nickel ions are inserted into the active site of the enzyme and the enzyme lacks catalytic activity.

Accessory Genes

For synthesis of a catalytically active urease, the accessory genes, *ureI, ureE, ureF, ureG,* and *ureH* must also be expressed (Cussac *et al.*, 1992) (Figure 1). While the *ureI* gene is unique to

H. pylori, the remaining accessory genes encode proteins that share homology with gene products of the urease gene clusters of other bacterial species. It is believed that these accessory proteins interact with the apoenzyme and deliver nickel ions to the active site in an energy dependent process (for review see Mobley *et al.*, 1995).

Homology to Other Ureases

Based on the nucleotide sequence of the urease genes of *H. pylori* and other bacterial species, it is certain that all ureases share a common ancestral gene. Despite the fact that ureases are composed of multiple copies of one (jack bean), two (all Helicobacters), or three (all other bacterial species) distinct subunits, the primary amino acid sequences are well conserved. For example, UreA of *H. pylori* shares 48% and 42% amino acid sequence identity with the corresponding N-terminal sequences of the jack bean urease subunit and the combined sequences of UreA and UreB of *P. mirabilis*, respectively (Jones and Mobley, 1989; Labigne *et al.*, 1991; Riddles *et al.*, 1991). When the entire amino acid sequence of the structural subunits is compared, *H. pylori* shares 58% identity with the urease of *K. aerogenes* (Mulrooney *et al.*, 1990; Cussac *et al.*, 1992). Residues that are conserved in all ureases and that are involved in catalysis have been discussed above in the section on Active site and catalysis (see above).

Expression of Catalytically Active Recombinant Urease

To express catalytically active *H. pylori* urease in *Escherichia coli* up to the level produced by wild type *H. pylori* strains, a number of requirements were identified (Cussac *et al.*, 1992; Hu and Mobley, 1993). Expressing only UreA and UreB was sufficient to produce a normally assembled apoenzyme with no catalytic activity. If accessory genes were coexpressed, a weakly active urease could be produced, but only when the bacteria were cultured on minimal salts medium containing no histidine or cysteine, which chelate free nickel ions. When the structural genes, *ureA* and *ureB*, were overexpressed *in trans* to the entire gene cluster, full catalytic activity and urease protein levels could be produced in *E. coli*, but again only on minimal salts medium lacking histidine and cysteine. It was necessary to also express NixA (a high affinity Ni^{2+} transporter) before full catalytic activity, comparable to that of wild type *H. pylori*, could be obtained in rich bacteriological media such as Luria broth.

Construction of Mutants

Urease mutants have been constructed by allelic exchange mutagenesis. Antibiotic resistance cassettes have been inserted into cloned *ureA* (McGee and Mobley, unpublished), *ureB* (Ferrero *et al.*, 1992), and *ureG* (Ferrero *et al.*, 1992) and these constructs were electroporated into wild type *H. pylori* strains. Antibiotic-resistant *H. pylori* were evaluated for double crossover mutations in which the wild type allele was exchanged for the insertionally inactivated allele. Such mutants that lacked detectable urease activity were readily selected and had no apparent alteration of growth rates, demonstrating that urease activity is not necessary for viability *in vitro*. These mutants are, however, uniformly avirulent in animal models of infection when tested in pathogenesis studies (see below).

Physiology

Substrate Availability

Urea is synthesized in the liver and is found in the serum, saliva and gastric juice at concentrations below 10 mM. Urea is excreted in urine at high concentrations ranging from 400-500 mM (Griffith *et al.*, 1976).

Enzyme Localization

Unlike the ureases of most bacterial species, the *H. pylori* enzyme is not strictly cytoplasmic. In aging cultures, urease can be found adherent to the cell surface or shed into the medium (Phadnis *et al.*, 1996); this appears to be due to lysis of a subset of the population and readsorption of the protein onto the cell surface of still-viable bacteria. The importance of the external urease has been debated. Two studies have shown contrasting roles of cytoplasmic versus surface-exposed urease. In the first study, *H. pylori* was cultured such that the bacteria possessed either cytoplasmic and surface urease or cytoplasmic urease alone. Assays were conducted in the presence and absence of flurofamide, a poorly diffusible urease inhibitor (Krishnamurthy *et al.*, 1998). Bacteria with only cytoplasmic urease were more susceptible to acid; likewise, *E. coli* expressing *H. pylori* urease cytoplasmically was also susceptible to acid. This suggests that surface exposed urease contributes to resistance to transient acid exposure. In contrast, Scott *et al.* (1998) argued that, because external urease is inactive below pH 5 and internal urease has its maximal activity below pH 5.5, internal urease is most likely to be responsible for acid tolerance.

Use of Ammonia Generated by Hydrolysis of Urea

Ammonia, the product of urea hydrolysis, is a preferred nitrogen source for bacteria. Ammonia is assimilated into protein and other nitrogenous compounds in bacteria by a single pathway (Reitzer and Magasanik, 1987). Glutamine synthetase (EC 6.3.1.2) catalyzes the reaction: NH_3 + glutamate +ATP $\rightarrow$ glutamine + ADP + P_i. Glutamine, in turn, serves as a nitrogen donor for other nitrogenous compounds including alanine, glycine, serine, histidine, tryptophan, CTP, AMP, carbamoyl-phosphate, and glucosamine 6-phosphate.

The 1443-bp glutamine synthetase gene, *glnA*, encodes a polypeptide of predicted 481 amino acid residues with a molecular weight of 54,317 (Garner *et al.*, 1998). Enzyme activity is regulated post-translationally in most genera by adenylation of the protein. The adenylation site found in most bacterial homologs (consensus sequence of NLYDLP) is replaced in *H. pylori* by NLFKLT (residues 405-410). Because the Tyr (Y) residue is the target of adenylation and the *H. pylori* glutamine synthetase lacks that residue in four strains examined, no adenylation should occur within this motif and the *H. pylori* enzyme is probably not regulated in this manner.It was not possible to isolate a glutamine synthetase-deficient mutant by allelic exchange (Garner *et al.*, 1998) and glutamine synthetase is probably essential for viability and critical for nitrogen assimilation in *H. pylori*. The enzyme appears to be active under all physiologic conditions consistent with the singular niche that *H. pylori* occupies.

NixA and Nickel Transport

All ureases contain two nickel ions in each of their active sites; *H. pylori* urease appears to have six active sites and twelve nickel ions. To overcome the nickel limitation that probably occurs in the host, *H. pylori* has developed a high affinity system to acquire nickel ions. It appears that nickel ions can be transported into *H. pylori* by at least two mechanisms. Firstly, NixA is a protein bound to the cytoplasmic membrane- of 36,991 molecular weight which transports nickel ions with a K_T of 11.3 nM (Mobley *et al.*, 1995) (Figure 1). *E. coli* expressing NixA transported nickel ions with a V_{max} of 1750 pmol/min/10^8 bacteria. Topology studies using *phoA* and *lacZ* translational fusions revealed the protein to possess eight transmembrane domains plus a large periplasmic loop (Fulkerson and Mobley, submitted for publication). Both the N-terminus and C-terminus reside in the cytosol. Negatively charged residues (Asp and Glu) and His residues that reside in the trans-membrane domains were shown to be critical for the active transport of nickel ions. Two domains, $GX_2HAXDADH$ in helix II and $GX_2FX_2GHSSVV$ in helix III, have been identified in NixA and the small family of high affinity nickel transport proteins that are critical for transport (Fulkerson *et al.*, 1998). Insertional inactivation of *nixA* in *H. pylori* ATCC 43504 resulted in a 69% decrease in the rate of nickel transport and a 42% reduction in urease activity as compared to the parent strain (Bauerfeind *et al.*, 1996). These rates varied among strains, but are generally reduced in the *nixA* mutants. The fact that nickel transport or urease activity were not totally abolished supports the existence of a second mechanism of nickel transport which remains to be elucidated.

Metal Binding Proteins that may Affect Urease Activity

Nickel ions are required for catalytic activity of urease, yet a *nixA* mutant of *H. pylori* still retains some urease activity (Bauerfeind *et al.*, 1996) (see above). It was concluded from this observation that additional proteins may play a role in nickel binding or transport. Three *H. pylori* proteins have been identified that are predicted to bind nickel or other metals.

Hpn is a small protein (7.1 kDa) of only 60 amino acid residues (Gilbert *et al.*, 1995). Nearly half (47%) of the amino acids in Hpn are histidine, indicating that this protein binds nickel and zinc ions strongly. However, although Hpn should serve as a sink for nickel and should activate urease, a mutant in which *hpn* was insertionally inactivated showed no reduction of urease activity following culture *in vitro* (Gilbert *et al.*, 1995). The effects of this mutation have not yet been examined *in vivo*.

A P-type ATPase has recently been identified in *H. pylori* (Melchers *et al.* 1996). A 686 amino acid long protein was predicted from the sequence which contains consensus sites for phosphorylation, ATP-binding, and, at the N terminus, sequences rich in histidine, methionine, glutamate, and aspartate residues which may act as a nickel binding domain. This protein is likely to be a heavy metal ion exporter that uses ATP as an energy source for transport. Mutation of the gene encoding the ATPase appears to diminish but not abolish (as originally reported) urease activity (K. Melchers, personal communication).

HspA (heat shock protein A) of *H. pylori* is a 118 amino acid residue long homolog of GroES-like heat shock proteins. It is encoded by the first gene, *hspA,* of a bicistronic cluster in *H. pylori* (Suerbaum *et al.*, 1994). While the predicted protein is highly conserved with respect to its homologs, the C-terminus contains a 25 amino acid metal binding motif that is rich in cysteine (4 residues) and histidine (6 residues) , unlike any of the other GroES homologs. When *hspA* is expressed in *E. coli in trans* to the *H. pylori* urease gene cluster, urease activity is slightly elevated, suggesting that this protein may affect urease activity in the wild type organism. Mutation of *hspA* appears to be lethal for *H. pylori* and thus the role of this heat shock protein in urease activity cannot be tested directly in the native organism.

Pathogenesis

Requirement for Colonization

While urease is not required for *in vitro* viability of *H. pylori*, it is clear that the enzyme is necessary for colonization of the gastric mucosa and represents a critical virulence determinant. A urease-negative mutant of an *H. pylori* strain, isolated after mutagenesis with nitrosoguanidine, was used to inoculate gnotobiotic piglets. The mutant, which retained only 0.4% of the urease activity of the parent strain, could not colonize any of ten orally challenged piglets as assessed at 3 and 21 days after challenge (Eaton *et al.*, 1991). Furthermore, no pathological symptoms were observed in these piglets. The parent strain successfully colonized all seven piglets tested and caused gastritis. This work was continued by comparing an allelic exchange mutant in which *ureG* had been interrupted with a kanamycin resistance cassette with its otherwise isogenic parent. In these studies, the piglets were treated with omeprazole, a proton pump inhibitor, which abolished acid secretion and yielded a neutral pH in the stomach of the piglets. The parent strain colonized normally in numbers ranging from a mean $\log_{10}$ CFU of 4.4-6.9. The urease-negative mutant was unable to colonize the gastric mucosa at normal physiological pH and was recovered only in low numbers (mean $\log_{10}$ CFU <2) from omeprazole-treated, achlorhydric piglets. These results confirmed that urease enzymatic activity was essential for colonization. Importantly, it also implied that the role of urease extended beyond that of acid protection. Other studies of *H. pylori* urease mutants demonstrated an inability to colonize the gastric mucosa of nude mice (Tsuda *et al.*, 1994) and *Cynomolgus* monkeys (Takahashi *et al.* 1993).

Detection of *H. pylori* Using Urease

Urease Biopsy Test

The high level of expression of urease by *H. pylori* allows ready detection of the bacterium in gastric biopsies. Biopsies obtained by endoscopy can be placed in a gel containing urea plus phenol red (a pH indicator). If *H. pylori* are present, preformed urease will hydrolyze the urea, raise the pH, and change the color of the phenol red from yellow to red. This concept, first developed by McNulty and Wise (1985) using Christensen's urea broth, was used for development of the CLO-test, the first commercially available test for the presence of *H. pylori* (Marshall *et al.*, 1987). The utility of this test has been repeatedly validated in the literature.

Urease-positive Colonies after Culture

H. pylori yields pinpoint colonies after 3 to 5 days of primary culture on serum-based medium from endoscopic biopsied that are virtually always strongly urease-positive A urease reaction, in the presence of positive oxidase and catalase reactions, is diagnostic for this species.

Urea Breath Test

Although the urease biopsy test reaction is simple, it does require biopsy by endoscopy, an invasive procedure. A noninvasive procedure, the urea breath test, has been developed which serves as a sensitive and specific, qualitative indicator of infection,. The patient is given an oral dose of labeled urea, either [^{13}C]urea (Graham *et al.*, 1987) or [^{14}C]urea (Bell *et al.*, 1987). If colonized by *H. pylori*, urea will be hydrolyzed and $^{13}CO_2$ or $^{14}CO_2$ will be liberated, enter the bloodstream, exchange in the lungs, and be exhaled. Exhaled CO_2 is trapped and quantitated using a mass spectrometer for $^{13}CO_2$ or a scintillation counter for $^{14}CO_2$. While a number of members of the normal anaerobic gut flora are urease-positive and could potentially interfere with this test, data collected thus far indicate that false-positive reactions are rare.

PCR Identification

PCR amplification of *H. pylori* urease genes has been used as a proxy for the presence of viable (or non-viable) *H. pylori*. Amplification of a 411-bp *ureA* fragment from boiled culture supernatants has been used in identification of *H. pylori* (Clayton *et al.*, 1992, 1993; van Zwet *et al.*, 1993). The primers were sensitive and specific for *H. pylori* and other *Helicobacter spp.* did not react. PCR amplification of a 365 bp PCR product, from an adjacent area of the 5' end of the *ureA* gene sequence, has been reported using a pair of PCR primers, of which one was degenerate (Westblom *et al.*, 1993). Consistent amplification from gastric juice was reported using this system. PCR amplification from regions of the *ureB* gene have also been used to detect *H. pylori* in paraffin-embedded biopsy samples (Westblom *et al.*, 1993).

Typing Systems Based on Urease Genes

There is a great deal of strain to strain DNA sequence variation among strains of *H. pylori* (see Chapter 19), resulting in heterogeneity of restriction sites and sequences. This has allowed the development of various typing methods based on urease genes. The first system that was developed was based on PCR amplification of urease genes (Foxall *et al.*, 1992, 1993). PCR-amplification of the urease structural subunit genes, *ureA* and *ureB*, from clinical isolates yielded 2.4 kb PCR products, which when subjected to *Hae*III digestion, produced distinct patterns on agarose gels. These patterns allow easy differentiation between strains. Another method amplified the *ureC* gene (no longer considered a urease gene) followed by direct DNA sequencing of the PCR product to distinguish between strains of *H. pylori* (Courcoux *et al.*, 1990). Numerous base pair changes were detected, and strains were easily differentiated. Seven subsequent reports (Bamford *et al.*, 1993; Clayton *et al.*, 1991; Desai *et al.*, 1994; Lopez *et al.*, 1993; Moore *et al.*, 1993; Hurtado and Owen, 1994; Owen *et al.*, 1994) confirmed that urease genes (*ureA*, *ureB*, and the former *ureC*) can be PCR amplified, subjected to restriction endonuclease digestion, and used to differentiate strains on the basis of patterns on agarose gels. These tests can identify the presence of multiple strains in a single biopsy specimen, check for reinfection with the same strain following eradication therapy, and identify similar strains among family members. This general strategy has been successfully applied to other *H. pylori* genes. (See Chapter 18 for an in-depth discussion of typing systems).

Urease in Vaccines

Because infected humans mount a significant immunoglobulin response to urease, it was reasoned that immunization with urease might protect against colonization but it was unclear which form of the protein should be used. The possible candidates included isolated subunits (UreA or UreB), apoenzyme, and holoenzyme. Two groups used similar strategies and tested the efficiency of immunization with a mouse model and an *H. felis* challenge as a surrogate for *H. pylori*. Michetti *et al.* (1994) found moderate but nevertheless significant protection when purified urease was used for immunization. However, when urease structural subunits UreA and UreB, purified separately as histidine-tagged fusion proteins on a Ni^{2+} nitrilotriacetic acid column, were used for immunization, mice were strongly protected from colonization of the gastic mucosa by *H. felis*. UreB, which harbors the enzyme active site, gave earlier and more complete protection from the challenge strain. In later studies (Corthesy-Theulaz *et al.*, 1995), oral immunization with UreB was used therapeutically for eradication. Likewise, in independent efforts, Ferrero *et al.* (1994) used *H. pylori* and *H. felis* UreA and UreB, purified as maltose-binding protein translational fusions, to immunize mice with cholera toxin as the adjuvant. UreB again provided more significant protection than UreA, and the *H. felis*-derived subunits conferred better protection than the *H. pylori* urease subunits.

Using the holoenzyme would have the advantage of retaining conformational epitopes, but ureolytic activity would be undesirable because of the ubiquitous presence of urea in humans. Catalytically inactive but fully assembled *H. pylori* urease was viewed as more desirable and was adopted as a vaccine candidate based on the observation of Hu *et al.* (1992). They observed that the *H. pylori* urease apoenzyme, lacking nickel ions in the active site and thus lacking enzymatic activity, could be isolated from *E. coli* expressing only the *ureA* and *ureB* structural subunit genes. These clones lacked the accessory genes necessary for nickel ion incorporation. The apoenzyme was purified according to the original scheme used for the isolation of the native enzyme (Hu *et al.*, 1992). Furthermore, the apoenzyme eluted from the four resins used for purification (DEAE-Sepharose, Phenyl-Sepharose, MonoQ, and Superose 6) at the identical fractions observed for native enzyme. This suggested that the apoenzyme possessed the identical charge, hydrophobicity, shape, and size of the native enzyme and thus was fully assembled except that it lacks the 12 nickel ions from the active site. Pappo *et al.* (1995) and Lee *et al.* (1995) used the recombinant apourease encoded by clones carrying only *ureAB* for immunization and also showed significant protection of mice against *H. felis* challenge. (For a more in-depth discussion of vaccines, see Chapter 16).

Acknowledgments

This work was supported in part by Public Health Service Grant AI25567 from the National Institutes of Health.

References

Austin, J.W., Doig, P., Stewart, M., and Trust T.J. 1991. Macromolecular structure and aggregation states of *Helicobacter pylori* urease. J. Bacteriol. 173: 5663-5667.

Bamford, K.B., Bickley,J., Collins, J.S., Johnston, B.T., Potts, S., Boston, V., Owen. R.J., and Sloan, J.M. 1993. *Helicobacter pylori*: comparison of DNA fingerprints provides evidence for intrafamilial infection. Gut 34: 1348-1350.

Bauerfeind P., Garner R., Dunn B.E., and Mobley H.L.T. 1997. Synthesis and activity of *Helicobacter pylori* urease and catalase at low pH. Gut 40: 25-30.

Bauerfeind P., Garner R.M., Mobley H.L.T. 1996. Allelic exchange mutagenesis of *nixA* in *Helicobacter pylori* results in reduced nickel transport and urease activity. Infect. Immun. 64: 2877-2880.

Bell, G.D., Weil, J., Harrison, G., Morden, A., Jones, P.H., Gant, P.N., Trowell, J.E., Yoong, A.K., Daneshmend, T.K., and Logan, R.F. 1987. ^{14}C-urea breath analysis, a non-invasive test for *Campylobacter pylori* in the stomach. Lancet 1: 1367-1368.

Blanchard, A. 1990. *Ureaplasma urealyticum* urease genes; use of a UGA tryptophan codon. Mol. Microbiol. 4: 669-678.

Clayton, C.L., Kleanthous, H., Coates, P.J., Morgan, D.D., and Tabaqchali, S. 1992. Sensitive detection of *Helicobacter pylori* by using polymerase chain reaction. J. Clin. Microbiol. 30: 192-200.

Clayton, C.L., Kleanthous, H., Morgan, D.D., Puckey, L., and Tabaqchali S. 1993. Rapid fingerprinting of *Helicobacter pylori* by polymerase chain reaction and restriction fragment length polymorphism analysis. J. Clin. Microbiol. 31: 1420-1425.

Clayton, C., Kleanthous, H., and Tabaqchali, S. 1991. Detection and identification of *Helicobacter pylori* by the polymerase chain reaction. J. Clin. Pathol. 44: 515-516.

Clayton, C.L., Pallen, M.J., Kleanthous, H., Wren, B.W., and Tabaqchali S. 1990. Nucleotide sequence of two genes from *Helicobacter pylori* encoding for urease subunits. Nucl. Acids Res. 18: 362.

Corthesy-Theulaz, I., Porta, N., Glauser, M., Saraga, E., Vaney, A.-C., Haas, R., Kraehenbuhl, J.-P., Blum, A.L., and Michetti, P. 1995. Oral immunization with *Helicobacter pylori* urease B as a treatment against *Helicobacter* infection. Gastroenterol. 109: 115-121.

Courcoux, P., Freuland, C., Piemont, Y., Fauchere, J.L., and Labigne, A. 1990. Polymerase chain reaction and direct DNA sequencing as a method for distinguishing between different strains of *Helicobacter pylori*. Rev. Esp. Enf. Digest. 78 (Suppl. 1): 29-30.

Cussac, V., Ferrero, R.L., and Labigne, A. 1992. Expression of *Helicobacter pylori* urease genes in *Escherichia coli* grown under nitrogen-limiting conditions. J. Bacteriol. 174: 2466-2473.

Dixon, N.E., Riddles, P.W., Gazzola, C., Blakeley, R.L., and Zerner, B. 1980. Jack bean urease (EC 3.5.1.5). V. On the mechanism of action of urease on urea, formamide, acetamide, *N*-methylurea, and related compounds. Can. J. Biochem. 58: 1335-1344.

Dunn, B.E., Campbell, G.P., Perez-Perez, G.I., Blaser, M.J. 1990. Purification and characterization of urease from *Helicobacter pylori*. J. Biol. Chem. 265: 9464-9469.

Dunn, B.E., Roop, R.M. 2nd, Sung, C.-C., Sharma, S.A., Perez-Perez, G.I., and Blaser, M.J. 1992. Identification and purification of a cpn60 heat shock protein homolog from *Helicobacter pylori*. Infect. Immun. 60: 1946-1951.

de Koning-Ward, T.F., Ward, A.C., and Robins-Browne, R.M. 1994. Characterization of the urease-encoding gene complex of *Yersinia enterocolitica*. Gene 145: 25-32.

Desai, M., Linton, D., Owen, R.J., and Stanley, J. 1994. Molecular typing of *Helicobacter pylori* isolates from asymptomatic, ulcer and gastritis patients by urease gene polymorphism. Epidemiol. Infect. 112: 151-160.

Eaton, K.A., Brooks, C.L., Morgan, D.R., Krakowka, S. 1991. Essential role of urease in pathogenesis of gastritis induced by *Helicobacter pylori* in gnotobiotic piglets. Infect. Immun. 59: 2470-2475.

Eaton, K.A. and Krakowka, S. 1994. Effect of gastric pH on urease-dependent colonization of gnotobiotic piglets by *Helicobacter pylori*. Infect. Immun. 62: 3604-3607.

Evans, Jr, D.J., Evans, D.G., Engstrand, L., and Graham, D.Y. 1992. Urease-associated heat shock protein of *Helicobacter pylori*. Infect. Immun. 60: 2125-2127.

Evans Jr., D.J., Evans, D.G., Kirkpatrick, S.S., and Graham, D.S. 1991. Characterization of the *Helicobacter pylori* urease and purification of its subunits. Microbial Path. 10: 15-26.

Ferrero, R.L., Cussac V., Courcoux P., and Labigne A. 1992. Construction of isogenic urease-negative mutants of *Helicobacter pylori* by allelic exchange. J. Bacteriol. 174: 4212-

4217.

Ferrero, R.L., Thiberge, J.-M., Huerre, M., and Labigne, A. 1994. Recombinant antigens prepared from the urease subunits of *Helicobacter spp.*: evidence of protection in a mouse model of gastric infection. Infect. Immun. 62: 4981-4989.

Foxall, P.A., Hu, L.-T., and Mobley, H.L.T. 1990. Amplification of the complete urease structural genes from *Helicobacteri pylori* clinical isolates and cosmid gene bank clones. Rev. Esp. Enf. Digest. 78 (Suppl. 1): 128-129.

Foxall, P.A., L.-T. Hu, and H.L.T. Mobley. 1992. Use of polymerase chain reaction-amplified *Helicobacter pylori* urease structural genes for differentiation of isolates. J.Clin. Microbiol. 30: 739-741.

Fulkerson, Jr., J.F., Garner, R.M., and Mobley, H.L.T. 1998. Conserved residues and motifs in the NixA protein of *Helicobacter pylori* are critical for the high affinity transport of nickel ions. J. Biol. Chem. 273: 235-241.

Garner, R.G., Fulkerson Jr, J.F., and Mobley, H.L.T. 1998. *Helicobacter pylori* glutamine synthetase lacks features associated with transcriptional and posttranslational regulation. Infect. Immun. 66: 1839-1847.

Gatermann, S. and Marre, R. 1989. Cloning and expression of *Staphylococcus saprophyticus* urease gene sequences in *Staphylococcus carnosus* and contribution of the enzyme to virulence. Infect. Immun. 57: 2998-3002.

Gilbert, J.V., Ramakrishna J., Sunderman Jr F.W., Wright A., and Plaut A.G. 1995. Protein Hpn: cloning and characterization of a histidine-rich metal-binding polypeptide in *Helicobacter pylori* and *Helicobacter mustelae*. Infect. Immun. 63: 2682-2688.

Graham, D.Y., Klein, P.D., Evans, D.J., Alpert, L.C., Opekun, A.R., and Boutton, T.W. 1987. *Campylobacter pyloridis* detected by the ^{13}C-urea test. Lancet 1: 1174-1177.

Griffith, D.P., Musher, D.M., and Itin, C. 1976. Urease: the primary cause of infection-induced urinary stones. Invest. Urol. 13: 346-350.

Hausinger, RP. 1987. Nickel utilization by microorganisms. Microbiol. Rev. 51: 22-24.

Hawtin, P.R., Delves, H.T., and Newell, D.G. 1991. The demonstration of nickel in the urease of *Helicobacter pylori* by atomic absorption spectroscopy. FEMS Microbiol. Lett. 77: 51-54.

Hu, L.-T., Foxall, P.A., Russell, R., and Mobley, HLT. 1992. Purification of recombinant *Helicobacter pylori* urease apoenzyme encoded by *ureA* and *ureB*. Infect. Immun. 60: 2657-2666.

Hu, L.-T. and Mobley, H.L.T. 1990. Purification and N-terminal analysis of urease from *Helicobacter pylori*. Infect. Immun. 58: 992-998.

Hu, L.-T. and Mobley H.L.T. 1993. Expression of catalytically active recombinant *Helicobacter pylori* urease at wild type levels in *Escherichia coli*. Infect. Immun. 61: 2563-2569.

Hurtado, A., and Owen. R.J. 1994. Identification of mixed genotypes in *Helicobacter pylori* from gastric biopsy tissue by analysis of urease gene polymorphisms. FEMS Immun. Med. Microbiol. 8: 307-313.

Jabri, E., Carr, M.B., Hausinger R.P., and Karplus P.A. 1995. The crystal structure of urease from *Klebsiella aerogenes* at 2 Å resolution. Science 268: 998-1004.

Jabri, E., Lee, M.H., Hausinger, R.P., and Karplus P.A. 1992. Preliminary crystallographic studies of ureases from jack bean and from *Klebsiella aerogenes*. J. Molec. Biol. 227: 934-937.

Jones, B.D., Lockatell, C.V., Johnson, D.E., Warren, J.W., and Mobley, H.L.T. 1990. Construction of a urease-negative mutant of *Proteus mirabilis*: analysis of virulence in a mouse model of ascending urinary tract infection. Infect. Immun. 58: 1120-1123.

Jones, BD and Mobley HLT. 1988. *Proteus mirabilis* urease: genetic organization, regulation, and expression of structural genes. J. Bacteriol. 170: 3342-3349.

Jones, B.D. and Mobley, H.L.T. 1989. *Proteus mirabilis* urease: nucleotide sequence determination and comparison with jack bean urease. J. Bacteriol. 171: 6414-6422.

Krishnamurthy, P., Parlow, M., Zitzer, J.B., Vakil, N.B., Mobley, H.L.T., Levy, M., Phadnis, S.H., and Dunn, B.E. 1998. *Helicobacter pylori* containing only cytoplasmic urease is susceptible to acid. Infect. Immun. 66: 5060-5066.

Labigne A., Cussac V., and Courcoux P. 1991. Shuttle cloning and nucleotide sequences of *Helicobacter pylori* genes responsible for urease activity. J. Bacteriol. 173: 1920-1931.

Lee, M.H., Mulrooney, S.B., Renner, M.J., Markowicz, Y., and Hausinger R.P. 1992. *Klebsiella aerogenes* urease gene cluster: sequence of *ureD* and demonstration that four accessory genes (*ureD, ureE, ureF, ureG*) are involved in nickel metallocenter biosynthesis. J. Bacteriol . 174: 4324-4330.

Lee, C.K., Weltzin, R., Thomas, Jr W.D., Kleanthous, H., Ermak, T.H., Soman, G., Hill, J.E., Ackerman, S.K., and Monath, T.P. 1995. Oral immunization with recombinant *Helicobacter pylori* urease induces secretory IgA antibodies and protects mice from challenge with *Helicobacter felis*. J. Infect. Dis. 172: 161-172.

Lopez, C.R., Owen, R.J., and Desai, M. 1993. Differentiation between isolates of *Helicobacteri pylori* by PCR-RFLP analysis of urease A and B genes and comparison with ribosomal RNA gene patterns. FEMS Microbiol. Lett. 110: 37-44.

Maeda, M., Hidaka, M., Nakamura, A., and Masaki, H., Uozumi T. 1994. Cloning, sequencing, and expression of thermophilic *Bacillus* sp. strain TB-90 urease gene complex in *Escherichia coli*. J. Bacteriol. 176: 432-442.

Marshall, B.J., Warren, J.R., Francis, G.J., Langton, S.R., Goodwin, C.S., and Blincow E.D. 1987. Rapid urease test in the management of *Campylobacter pyloridis*-associated gastritis. Am. J. Gastroenterol., 82: 200-210.

Martin, P.R. and Hausinger R.P. 1992. Site-directed mutagenesis of the active site cysteine in *Klebsiella aerogenes* urease. J. Biol. Chem. 267: 20024-20027.

McNulty, C.A.M. and Wise, R. 1985. Rapid diagnosis of Campylobacter-associated gastritis. Lancet. i: 1443.

Melchers, K., Weitzenegger, T., Buhmann, A., Steinhilber, W., Sachs, G., and Schafer, K. 1996. Cloning and membrane topology of a P type ATPase from *Helicobacter pylori*. J. Biol. Chem. 271: 446-457.

Michetti, P., Corthesy-Theulaz, I., Davin, C., Haas, R., Vaney, A.-C., Heitz, M., Bille, J., Kraehenbuhl, J.-P., Saraga, E., and Blum, A.L. 1994. Immunization of BALB/c mice against *Helicobacter felis* infection with *Helicobacter pylori* urease. Gastroenterol. 107: 1002-1011.

Mobley, H.L.T., Cortesia, M.J., Rosenthal, L.E., and Jones B.D. 1988. Characterization of urease from *Campylobacter pylori*. J. Clin. Microbiol. 26: 831-836.

Mobley, H.L.T., Garner, R.E., and Bauerfeind, P. 1995. *Helicobacter pylori* nickel transport gene *nixA*: synthesis of catalytically active urease in *E. coli* independent of growth conditions. Mol. Microbiol. 16: 97-109.

Mobley, H.L.T., Island, M.D., and Hausinger, R.P. 1995. Molecular biology of microbial ureases. Microbiol. Rev. 59: 451-480.

Moore, R.A., Kureishi, A., Wong, S., and Bryan, L.E. 1993. Categorization of clinical isolates of *Helicobacter pylori* on the basis of restriction digest analysis of PCR-amplified *ureC* genes. J. Clin. Microbiol. 31: 1334-1335.

Mulrooney, S.B. and Hausinger, R.P. 1990. Sequence of the *Klebsiella aerogenes* urease genes and evidence for accessory proteins facilitating nickel incorporation. J. Bacteriol.

172: 5837-5843.

Neyrolles, O., Ferris, S., Behbahani, N., Montagnier, L., and Blanchard, A. 1996. Organization of *Ureaplasma urealyticum* urease gene cluster and expression in a suppressor strain of *Escherichia coli*. J. Bacteriol. 178: 647-655.

Nicholson, E.B., Concaugh, E.A., Foxall, P.A., Island, M.D., and Mobley, H.L.T. 1993. *Proteus mirabilis* urease: transcriptional regulation by *ureR*. J. Bacteriol. 175: 465-473.

Owen, R.J., Bickley, J., Hurtado, A., Fraser, A., and Pounder, R.E. 1994. Comparison of PCR-based restriction length polymorphism analysis of urease genes with rRNA gene profiling for monitoring *Helicobacter pylori* infections in patients on triple therapy. J. Clin. Microbiol. 32: 1203-1210.

Pappo, J., Thomas Jr, W.D., Kabok, Z., Taylor, N.S., Murphy, J.C., and Fox, J.G. 1995. Effect of oral immunization with recombinant urease on murine *Helicobacter felis* gastritis. Infect. Immun. 63: 1246-1252.

Park, I.-S. and Hausinger, R.P. 1993. Site directed mutagenesis of *Klebsiella aerogenes* urease: Identification of histidine residues that appear to function in nickel ligation, substrate binding, and catalysis. Protein Sci. 2: 1034-1041.

Park, I.-S. and Hausinger R.P. 1995. Requirement of CO_2 for in vitro assembly of the urease nickel metallocenter. Science 267: 1156-1158.

Phadnis, S.H., Parlow, M.H., Levy, M., Ilver, D., Caulkins, C.M., Connors, J.B., and Dunn, B.E. 1996. Surface localization of *Helicobacter pylori* urease and a heat shock protein homolog requires bacterial autolysis. Infect. Immun. 64: 905-912.

Reitzer, L.J. and Magasanik, B. 1987. Ammonia assimilation and the biosynthesis of glutamine, glutamate, aspartate, asparagine, L-alanine, and D-alanine, p. 302-320. In: *Escherichia coli* and *Salmonella*. Cellular and Molecular Biology. Neidhardt, F.C, ed. First Edition. American Society for Microbiology, Washington, D.C. p. 305-311.

Riddles, P.W., Whan, V., Blakeley, R.L., and Zerner, B. 1991. Cloning and sequencing of a jack bean urease-encoding cDNA. Gene 108: 265-267.

Scott, D.R., Weeks, D., Hong, C., Postius, S., Melchers, K., and Sachs, G. 1998. The role of internal urease in acid resistance and *Helicobacter pylori*. Gastroenterology 114: 58-70.

Skurnik, M, Batsford, S, Mertz, A, Schiltz, E, and Toivanen, P. 1993. The putative arthritogenic cationic 19-kilodalton antigen of *Yersinia enterocolitica* is a urease β-subunit. Infect. Immun. 61: 2498-2504.

Sriwanthana, B., Island, M.D., and Mobley, H.L.T. 1993. Sequence of the *Proteus mirabilis* urease accessory gene *ureG*. Gene 129: 103-106.

Sriwanthana, B., Mobley, H.L.T. 1993. *Proteus mirabilis* urease: Histidine 320 of UreC is essential for urea hydrolysis and nickel ion binding within the native enzyme. Infect. Immun. 61: 2570-2577.

Suerbaum, S., Thiberge, J.-M., Kansau, I., Ferrero, R.L., Labigne, A. 1994. *Helicobacter pylori hspA-hspB* heat-shock gene cluster: nucleotide sequence, expression, putative function and imunogenicity. Molec. Microbiol. 14: 959-974.

Takahashi, S., Igarashi, H., Nakamura, K., Masubuchi, N., Saitos, S., Aoyagi, T., Itoh, T., and Hirata, I. 1993. *Helicobacter pylori* urease activity - Comparative study between urease positive and urease negative strain. Jap. J. Clin. Med 51: 3149-3153.

Tsuda, M., Karita, M., Morshed, M.G., Okita, K., and Nakasaki, T. 1994. A urease-negative mutant of *Helicobacter pylori* constructed by allelic exchange mutagenesis lacks the ability to colonize the nude mouse stomach. Infect. Immun. 62: 3586-3589.

Turbett, G.R. Hoj, P.B., Horne, R., and Mee, B.J. 1992. Purification and characterization of the urease enzymes of *Helicobacter species* from humans and animals. Infect Immun.

60: 5259-5266.

van Zwet, A.A., Thijs, J.C., Kooistra-Smid, A.M.D., Schirm, J., and Snijder J.A.M. 1993. Sensitivity of culture compared with that of polymerase chain reaction for detection of *Helicobacter pylori* from antral biopsy specimens. J. Clin. Microbiol. 31: 1918-1920.

Westblom, T.U., Phadnis, S., Yang, P., and Czinn, S.J. 1993. Diagnosis of *Helicobacter pylori* infection by means of a polymerase chain reaction assay for gastric juice aspirates. Clin. Infect. Dis. 16: 367-371.

From: *Helicobacter pylori: Molecular and Cellular Biology*
ISBN 1-898486-25-5 © 2001 Horizon Scientific Press, Wymondham, UK.

11

Helicobacter Motility and Chemotaxis

Christine Josenhans and Sebastian Suerbaum[*]

Abstract

All gastric *Helicobacter* species are highly motile. Their characteristic sheathed flagellar filaments are composed of two copolymerized flagellins, FlaA and FlaB. Experiments in three different animal models with *Helicobacter pylori*, *Helicobacter mustelae* and *Helicobacter felis* showed that flagellar motility is essential for *Helicobacter* species to colonize the gastric mucus. *H. pylori* has homologs of all flagellar structural proteins known from *Salmonella* but the regulatory network of *H. pylori* motility genes lacks a typical *flhCD* master operon and a *flgM* sigma factor antagonist gene. Although *H. pylori* manifests chemotaxis towards urea and bicarbonate, it is still unknown which chemical gradients are used by *H. pylori in vivo* to maintain an optimal position in the gastric mucus layer. The chemotaxis system of *H. pylori* is genetically similar to the *Salmonella* system as based on genomic analysis but extensive functional analyses have not yet been performed.

Introduction

A prominent characteristic of all gastric *Helicobacter* species is flagella-driven motility. Motility is essential for colonization by the three gastric *Helicobacter* species, *Helicobacter pylori*, *Helicobacter mustelae*, and *Helicobacter felis* that have been tested (Eaton *et al*., 1996; Andrutis *et al*., 1997; Josenhans *et al*., 1999). Motility is probably required for the initial stages of infection, when *H. pylori* must move from the acidic gastric lumen into the only slightly acidic layer of the gastric mucus that is colonized during chronic infection. Motility is also considered essential for maintaining a constant reservoir of bacteria in the gastric mucus which is subject to constant and rapid turnover (Blaser, 1994; Suerbaum, 1995).

Motility is common among both pathogenic and non-pathogenic bacteria. Motile bacteria are propelled by one or multiple flagella. Flagella consist of three main structural parts: the flagellar motor plus other components of the basal body, the hook and the filament. The basal body spans the cell envelope, anchors the flagellum in the membranes, and is linked to the flagellar rotary motor at the base. The basal body is connected to the filament *via* the curved hook substructure. The filament is a long appendage that consists mainly of polymeric flagellin subunits. When the motor turns in the correct direction (e.g. counterclockwise for a left-handed helix), one or several filaments form a spiral and propel the bacterium. In all known motile bacteria, motility is linked to a chemotactic signal transduction system that

[*]corresponding author, email: ssuerbaum@hygiene.uni-wuerzburg.de

enables bacteria to move towards attractants or away from repellents. Three known mechanisms for implementing this directed movement are alternation between periods of "runs" and "tumbles", backward movement after a change of rotary direction, and start-stop movements. The flagellar apparatus and the chemotaxis system have been studied thoroughly in several different organisms. The organism for which the most is known is *Salmonella typhimurium*, but extensive information is also available for the motility systems of *Bacillus subtilis* and *Caulobacter crescentus* (Shapiro, 1995; Macnab, 1996; Aizawa, 1996).

The flagellar apparatus of *H. pylori* differs substantially from that of other motile bacteria. The investigation of *H. pylori* motility might provide clues to structural and regulatory adaptations of this common bacterial organelle to the particular requirements of the gastric niche. The chemotactic machinery that *H. pylori* uses to orient itself in the gastric mucus is similar to that of other chemotactic eubacteria. It consists of small sensory systems, which can sense the presence of attractants and repellents in the outside milieu and transduce signals to the bacterial flagellar basal body via several steps of a molecular chain reaction (for reviews see Parkinson, 1993; Stock and Surette, 1994).

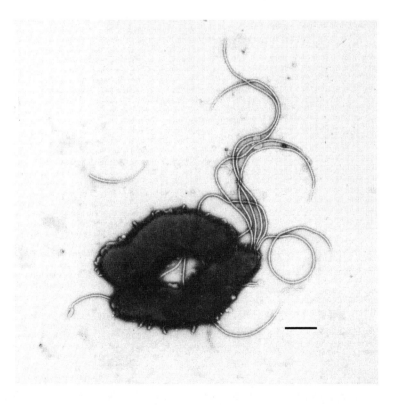

Figure 1. Transmission electron micrograph of *H. pylori* LSU1040-1. Negative staining was with 1% phosphotungstate (pH 7.0). Note the two curved bacteria with unipolar sheathed flagella.

Motility and Flagella - Molecular Biology and Structure

Gastric *Helicobacter* species possess multiple sheathed flagella arranged in different patterns in various species: unipolar polytrichous (*H. pylori*), bipolar polytrichous (*H. felis*) and peritrichous (*H. mustelae*). Together with slight differences in cell body morphology, these different patterns are thought to offer optimum motility in the stomach mucus of different hosts. Each *Helicobacter* flagellar filament is enveloped by a membraneous sheath. This sheath is continuous with the outer membrane and has a similar composition of lipopolysaccharides, phospholipids and proteins (Geis *et al.*, 1993). The sheath-associated lipoprotein, HpaA, is exclusively located in sheath membranes according to immunogold electron microscopy (Jones *et al.*, 1997). Other sheath-specific components have not yet been identified. The sheath is thought to protect the acid-labile flagellar filaments against low pH in the stomach because sheath vesicles are more resistant to dissociation under acidic conditions than are naked filaments (Geis *et al.*, 1993; Geis, Forsthoff and Suerbaum, unpublished).

All genes known to code for components of the flagellar substructures, i.e. the extended basal body (including the C-ring), the hook and the filament, are present in the *H. pylori* genome sequences (Table 1). Several genes encoding putative flagellar components that are not present in Enterobacteria were identified by homology to motility genes of other bacterial species (Table 1). However, the list of *H. pylori* motility genes does lack several that are known in *Salmonella*.

The flagellar filaments of all tested *Helicobacter* species are composed of two distinct flagellin subunits, FlaA and FlaB (Suerbaum *et al.*, 1993; Josenhans *et al.*, 1995; Josenhans *et al.*, 1999). These two flagellins share only 50% amino acid identity but probably evolved from a common ancestral flagellin molecule. Their molecular weights (53-54 kDa) are similar to those of enterobacterial flagellins. It has been suggested that *Helicobacter* FlaB is an exclusively hook proximal flagellin subunit that is solely expressed early during flagellar assembly (Kostrzynska *et al.*, 1991) but the possibility that both flagellin subunits are present throughout the length of the filaments has not been excluded. The ratio of FlaA to FlaB changes in a growth phase dependent manner, and probably also in response to environmental conditions (Josenhans *et al.*, 1995b). The two flagellins of *H. felis* undergo posttranslational glycosylation, but the function of this modification is still unknown (Josenhans *et al.*, 1999).

Mutants lacking the hook protein FlgE (O'Toole *et al.*, 1994), the flagellar cap protein FliD (Kim *et al.*, 1999), the export-associated proteins FlhA, FlhB, FliQ (Schmitz *et al.*, 1997; Foynes *et al.*, 1999) and the energy-providing ATPase of the flagellar type III system FliI (Jenks *et al.*, 1997; Porwollik *et al.*, 1999) lacked flagella. Thus, all these proteins are also needed in *H. pylori* for the synthesis of flagella.

The energy for flagellar rotation in *H. pylori* is derived from the proton motive force, which is thought to be composed of i) a membrane potential at the bacterial inner membrane and ii) a pH gradient between the environment and the bacterial cytoplasm (Nakamura *et al.*, 1998).

The genomic sequences of *H. pylori* strains 26695 and J99 (Tomb *et al.*, 1997; Alm *et al.*, 1999) have resulted in the tentative identification of many putative motility-associated genes. These assignments should be treated with caution until they are verified experimentally. An updated list of the putative motility-associated genes in the genome sequences is provided in Table 1, together with a comparison of the sets of motility genes in *Helicobacter* and *Salmonella*.

Including the tentative assignments, 40 genes in the *H. pylori* genomes are related to motility and the structural assembly of flagella and 15 genes are involved in chemotaxis and

Table 1. Motility- and flagella-associated genes in the complete genome sequences of *H. pylori* 26695 and J99

Gene no. in TIGR genome	Gene no in J99 genome (JHP)	Gene designation	Protein designation and function/homolog in *S. typhimurium* -yes/no (gene designation in *S. typhimurium*, if different from HP nomenclature)	Highest homology values in databases	Promotor if known	Source of homologous gene in databases
I: Genes Encoding Structural Proteins of the Flagellar Apparatus						
HP0115	107	*flaB*	Flagellin B /no	99.0	sigma54	*Helicobacter pylori*
HP0173	159	*fliR*	Regulatory protein of flagellar biogenesis flagellar basal body protein /yes	60.3		*Bacillus subtilis*
HP0232	217	-	Secreted protein involved in motility /no	99.2		*Helicobacter pylori*
HP0246	231	*flgI*	P ring forming protein of basal body /yes	62.9	sigma70 (?)	*Salmonella choleraesuis*
HP0295	280	*flgL*	HAP-3 homolog /yes	53.7	sigma54 (?)	*Bacillus subtilis*
HP0325	308	*flgH*	L ring forming protein of basal body /yes	55.1	sigma70 (?)	*Salmonella choleraesuis*
HP0327	310	*flaG1(flmH)*	Flagella associated protein /no	54.1	sigma 70 (?)	*Caulobacter crescentus*
HP0351	325	*fliF*	M ring-forming protein of basal body /yes	56.8	sigma70(?)	*Salmonella choleraesuis*
HP0352	326	*fliG*	Flagellar motor switch protein, putative C ring component /yes	63.6	sigma70 (?)	*Borrelia bugdorferi*
HP0353	327	*fliH*	Protein involved in flagellar export apparatus /yes	54.2	sigma70 (?)	*Borrelia bugdorferi*
HP0410	971	*hpa2*	Paralog of HP hemagglutinin protease - sheath associated protein /no	54.4		*Helicobacter pylori*
HP0584	531	*fliN*	Flagellar motor switch protein, putative C ring component /yes	68.5	sigma70 (?)	*Borrelia bugdorferi*
HP0601	548	*flaA*	Major flagellin A /yes (*fliC*)	99.8	sigma28	*Helicobacter pylori*
HP0684	625	*fliP'*	N-Terminal domain of basal body protein FliP involved in flagellar export apparatus /yes	80.3		*Bacillus subtilis*
HP0685	625	*fliP"*	Second half of basal body protein FliP[a] /yes	77.5		*Bacillus subtilis*
HP0751	688	*flaG2*	Gene of unknown function involved in biogenesis of polar flagellum /no	47.6	sigma70 (?)	*Vibrio parahaemolyticus*
HP0752	689	*fliD*	Hook-associated protein-2 (HAP-2), filament cap protein FliD /yes	49.2	sigma70 (?)	*Salmonella choleraesuis*
HP0753	690	*fliS*	Flagellar chaperone protein FliS /yes	55.6	sigma70 (?)	*Bacillus subtilis*
HP0770	707	*flhB1*	Flagellar basal body protein involved in export /yes	65.3	sigma70 (?)	*Bacillus subtilis*
HP0797	733	*hpaA*	Flagellar sheath "adhesin" protein -neuraminyllactose binding protein /no	100		*Helicobacter pylori*
HP0815	751	*motA*	Flagellar motor protein A /yes	63.0	sigma70 (?)	*Bacillus subtilis*
HP0816	752	*motB*	Flagellar motor protein B/yes	53.0	sigma70 (?)	*Bacillus subtilis*
HP0840	778	*flaA1*	Flagella-associated protein of unknown function /no	74.6		*Caulobacter crescentus*
HP0870	804	*flgE1*	Flagellar hook protein /yes	99.6	sigma54	*Helicobacter pylori*
HP0907	843	*flgD*	Flagellar hook scaffolding protein /yes	50.5	sigma54	*Salmonella choleraesuis*
HP0908	844	*flgE2(flgF)*	Putative flagellar hook basal body protein homolog /no	51.8	sigma54	*Helicobacter pylori*
HP1030	394	*fliY*	FliY protein /no	55.7		*Bacillus subtilis*
HP1031	393	*fliM*	Flagellar motor switch protein putative C ring component /yes	64.4		*Borrelia bugdorferi*
HP1035	389	*flhF*	Flagellar biosynthesis protein of unknown function /no	58.1		*Bacillus subtilis*

Table 1. *Continued*

HP1041	383	*flhA*	Flagellar basal body protein involved in export /yes	67.1	no consensus, sigma70 without the -35 region (?)	*Salmonella choleraesuis*
HP1092	333	*flgG*1	Homologous to distal rod protein of flagellar basal body, putative functional FlgF substitute /no-yes or rod-hook junction protein	51.6	sigma54	*Salmonella choleraesuis*
HP1119	1047	*flgK*	Hook-associated protein 1 (HAP 1) /yes	49.0	sigma54	*Salmonella choleraesuis*
HP1192	1117	-	Secreted protein involved in motility /no	89.0		*Helicobacter pylori*
HP1419	1314	*fliQ*	Flagellar basal body protein involved in export /yes	70.9		*Escherichia coli*
HP1420	1315	*fliI*	Flagella-associated ATPase, energizer of flagellar export /yes	66.0		*Bacillus subtilis*
HP1462	1355	-	Secreted protein involved in motility /no	97.1		*Helicobacter pylori*
HP1557	1465	*fliE*	Flagellar basal body protein /yes	54.0	sigma70 (?)	*Bacillus subtilis*
HP1558	1466	*flgC*	Proximal rod protein of basal body /yes	69.1	sigma54	*Borrelia bugdorferi*
HP1559	1467	*flgB*	Proximal rod protein of basal body /yes	56.6	sigma54	*Salmonella choleraesuis*
HP1575	1483	*flhB2*	Homolog of FlhB protein basal body /no	57.0		*Yersinia enterocolitica*
HP1585	1492	*flgG2*	Distal rod protein of flagellar basal body /yes	66.9	sigma70-like/sigma54	*Salmonella choleraesuis*

II: Genes Involved in Chemotaxis

HP0019	17	*cheV1*	CheW/CheY hybrid chemotaxis protein /no	54.6		*Bacillus subtilis*
HP0082	75	*tlpC*	Receptor protein of chemotaxis signalling system /specificity unknown	51.6		*Bacillus subtilis*
HP0099	91	*tlpA*	Chemotaxis receptor /specificity unknown	55.4		*Bacillus subtilis*
HP0103	95	*tlpB*	Chemotaxis receptor /specificity unknown	49.5		*Bacillus subtilis*
HP0391	990	*cheW*	Chemotaxis adaptor protein /yes	56.9		*Escherichia coli*
HP0392	989	*cheA*	Receptor coupled histidine kinase /yes, but fused to CheY2 domain in HP	68.0		*Borrelia bugdorferi*
HP0393	988	*cheV3*	Hybrid CheW-CheY chemotaxis protein /no	59.1		*Bacillus subtilis*
HP0599	546	*tlp*	Soluble chemotaxis receptor /specificity unknown			
HP0616	559	*cheV2*	Hybrid cheW-CheY chemotaxis protein	52.2		*Bacillus subtilis*
HP1067	358	*cheY1*	Chemotaxis effector protein - link to basal body motor /yes	99.2	sigma70	*Helicobacter pylori*

III: Regulatory Functions

HP0703	643	*flgR*	NtrC Homolog transcriptional activator of sigma54-regulated motility genes	69.1	sigma70	*Pseudomonas aeruginosa*
HP0714	652	*rpoN*	Sigma54 subunit of DNA-dependent RNA polymerase	58.6		*Bacillus subtilis*
HP1032	392	*fliA*	Sigma28 subunit of DNA-dependent RNA polymerase /"flagellar" sigma factor /yes	59.7		*Escherichia coli*

[a]HP0684 and HP0685 are one continuous ORF in strains J99. See text for details.

the regulation of motility. Table 1 includes several genes that are absent or poorly characterised in the *Salmonella* or *B. subtilis* motility systems: i) homologs of two *Caulobacter crescentus* flagella-associated genes whose precise functions are unknown, *flaG* (designated *flaG1* in Table 1) and *flaA1*, ii) the gene encoding the putative flagellar sheath protein HpaA is present in two copies and is unique to *H. pylori* and iii) a gene that is weakly homologous to *flaG* (*flaG2* in Table 1) in *Vibrio parahaemolyticus* where it is needed for the biogenesis of polar flagella. Genes which play an important role for flagellar assembly in *Salmonella* (see Figure 2) but are missing from the *H. pylori* genomes, include *fliK*, involved in hook length control, as well as *flgN, fliJ, flgA* and *fliT*, encoding chaperones for the hook protein FlgE, for hook/rod type proteins, P-ring assembly, and the flagellar cap protein FliD, respectively (Fraser *et al.*, 1999; Minamino and Macnab, 2000; Nambu and Kutsukake, 2000). Also missing from the genomes are genes encoding FlgJ, a flagella-specific muraminidase with cell-wall lytic activity (Nambu *et al.*, 1999), the master flagellar regulators FlhC and FlhD and the sigma28 antagonist FlgM. The gene encoding the hook protein FlgE, the export-linked gene *flhB*, which is involved in switching of export type specificities (Minamino and Macnab, 2000), the distal rod-associated gene *flgG* and the *hpa*A genes are each present in two copies.

Regulation of Motility Genes

The assembly of flagella requires the coordinated synthesis and sequential assembly of more than 40 proteins. In *Salmonella typhimurium*, this is achieved by a hierarchical control system that ensures that the components of the filament are only synthesized if an intact basal body and hook can be assembled (Aizawa, 2000). Major components of the Salmonella motility regulon are i) the master regulatory operon *flhCD,* which encodes transcriptional regulators that link motility to crucial cell functions like cell division (Kutsukake *et al.*, 1990), ii) the alternative sigma factor sigma 28 (sigmaD), which controls the expression of most late flagellar genes (e.g. flagellin genes) and iii) the sigma28 antagonist FlgM (Hughes *et al.*, 1993), a stop valve to ensure that late flagellar genes are only transcribed after intact basal body-hook structures have been assembled.

Although the regulation of *Helicobacter* motility genes is far from being solved, some distinctive features have emerged. Most of the motility associated genes of *H. pylori* are not organized in operons, which makes the search for expression networks and coregulated genes more complicated. A *flhCD* master operon and an anti sigma factor FlgM are absent in *H. pylori*. Consistent with that, *H. pylori* flagellar protein synthesis is not dependent on successful assembly of hook-basal bodies because an *H. pylori* mutant lacking the hook protein FlgE still expressed flagellin proteins (O'Toole *et al.*, 1994). Unlike Salmonella, the *H. pylori* late flagellar genes are not exclusively under the control of sigma28. Most of the late flagellar genes characterised (seven so far) are under the control of the sigma54 RNA polymerase cofactor. And many of the early flagellar genes are controlled by a sigma70-like promoter, which seems to be the most common flagella-associated promoter in *H. pylori* (see below, Table 1 and Figure 3). The sigma70 homolog of the *H. pylori* DNA-dependent RNA polymerase has been purified and found to be an 80 kDa protein (hence sigma80) (Beier *et al.*, 1998), which seems to have a slightly different RNA substrate specificity than does the *Escherichia coli* sigma70 subunit. A consensus sigma80 promoter sequence was recently defined by computational analysis of the genomic sequence data (Vanet *et al.*, 2000). Some motility-associated genes in *H. pylori* are coregulated with housekeeping genes, unlike in other bacteria (Beier *et al.*, 1997; Spohn and Scarlato, 1999; Porwollik *et al.*, 1999). The transcriptional

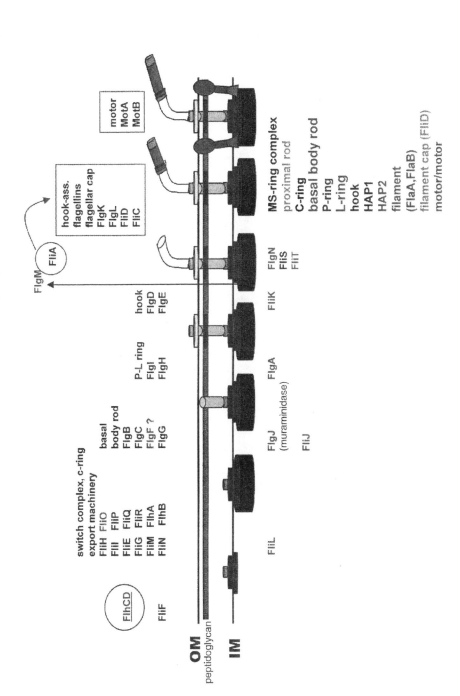

Figure 2. The pathway of flagellar assembly in *S. typhimurium*. Proteins lacking in the *H. pylori* system are highlighted in red. The position of the bacterial membranes (OM = outer membrane; IM = inner membrane) and peptidoglycan layer are indicated. Gene products that form part of the structure are shown above the membranes, gene products with putative assistant functions during flagellar assembly (e.g. chaperones) are noted below the membranes. The figure has been modified from the model of Aizawa (2000). A color coded list of the main structural parts of the flagellum is given in the lower right corner. A color version of this figure is printed on page 324 of this volume.

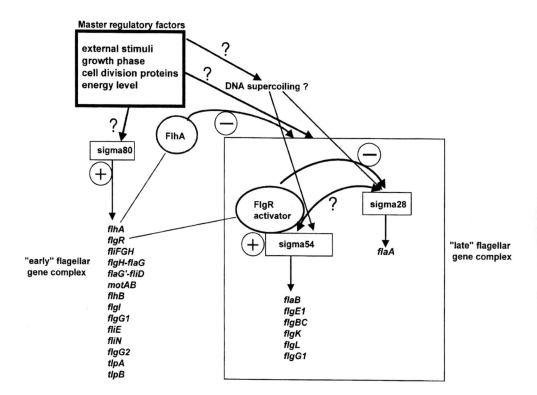

Figure 3. A model of the motility regulatory network in *H. pylori*. The model integrates the three different sigma factors involved in regulation of motility associated genes and the genes they control into a tentative scheme together with factors which are known to play a role in expression of flagellar components (Schmitz *et al*., 1997; Beier *et al*., 1997; Spohn *et al*., 1999, Vanet *et al*.; 2000, Josenhans *et al*., 1995b; Suerbaum *et al*., 1998). *flhA* was previously named *flbA* (Schmitz *et al*., 1997). The system is different from the hierarchical regulation present in *Salmonella*. Gene names and their functions are listed in Table 1. Arrows marked with "+" indicate positive regulation, arrows marked with "-" indicate negative regulation. Question marks indicate hypothetical connections. A color version of this figure is printed on page 324 of this volume.

activator FlgR, a homologue of the nitrogen-dependent regulator NtrC in *E. coli,* interlinks the expression of a number of sigma54 dependent motility genes. FlaA, which is governed by a typical flagellar sigma28 promoter, is slightly repressed by this transcription factor (Spohn and Scarlato, 1999). By alignment to consensus promoter sequences, Spohn and Scarlato also identified about 15 additional genes in nine transcriptional units which seem to be controlled by the sigma80 promoter (Table 1, Figure 3). One operon involved in motility has been shown to be expressed *in vivo* by screening for specific mRNA in biopsies of infected animal hosts and human patients (Porwollik *et al*., 1999).

DNA supercoiling may also be involved in the regulation of motility genes because the expression of both flagellin genes is influenced by (local or global) changes of supercoiling (Suerbaum *et al*., 1998; Spohn and Scarlato, 1999). *flaB* maps next to the gene encoding topoisomerase I, an enzyme, which resolves negative supercoiling of DNA and the gene encoding the sigma54 activating protein FlgR is in the same operon with the *gyrA* gene (encoding gyrase, which introduces negative DNA supercoils). These genetic linkages provide further support for a role of supercoiling in regulation of flagella synthesis. However, the mechanisms underlying a possible regulation of motility by DNA topology remain unknown.

Expression of *H. pylori* flagellin genes is differentially regulated in a growth phase dependent way (Josenhans *et al.*, 1995b). Consistent with that, *H. pylori* is only motile during the exponential growth phase, not in lag or stationary phase (Worku *et al.*, 1999). Evidence for differential regulation of *flaA* and *flaB* gene expression by environmental conditions is still lacking but the transcription of an operon containing chemotaxis genes is influenced by the presence of copper in the medium and by temperature shifts from 37° to 45°C (Beier *et al.*, 1997).

All clinical isolates of *H. pylori* are motile but non-motile variants accumulate rapidly during repeated *in vitro* passage. Non-motility can result from the slipped-strand mispairing mutagenesis of a poly-cytidine repeat within the coding region of *fliP*, a gene that encodes a structural component of the export machinery in the basal body (Josenhans *et al.*, 2000). In the genome sequence of *H. pylori* strain 26695, the *fliP* repeat is C9, which leads to a premature stop codon. 26695_{C9} bacteria are non-motile and deficient in flagella production. Motile revertants contain a C8 repeat that restores the reading frame. Switching from the non-motile to motile repeat length occurs with a frequency of approximately 10^{-4} to 10^{-5} per generation (Josenhans *et al.*, 2000). ON-OFF switching regulation of the motility system warrants further investigation because it might be important for *H. pylori* persistence and microecology in the host.

Chemotaxis

The chemotaxis system of *Helicobacter* species is still only poorly characterised although it has long been presumed that *H. pylori* must react to attractants in order to find its optimal niche within the stomach mucus.

A novel technology permits taking minute samples of mucus from defined positions within the mucus layer in order to study the distribution of bacteria within the mucus of anesthetized *H. felis*-infected mice. The bacteria accumulated in a distinct layer of the stomach mucus, about 20 μm above the stomach epithelium (Schreiber *et al.*, 1999). The simplest explanation for such a specific location is the existence of two or more convergent chemical gradients that are detected by the bacteria.

Genes in the complete *H. pylori* genomes have high homologies to chemotaxis genes from the model bacteria *S. typhimurium* and *C. crescentus*. Homologs of all major components required for chemotaxis have been identified, including the receptor coupled autokinase, CheA, the linker protein, CheW, and the chemotaxis effector, CheY (Beier *et al.*, 1997; Tomb *et al.*, 1997; Alm *et al.*, 1999; Foynes *et al.*, 2000). However, chemotaxis in *H. pylori* does differ somewhat from other chemotaxis systems. Only four genes with homology to chemotaxis receptors and of unknown specificity have been identified. One of these (HP599) seems to be a soluble cytoplasmic orthologue lacking the membrane anchor domain. This paucity of chemotaxis receptors suggests that *H. pylori* uses few chemical gradients for orientation. A second major difference from the *Salmonella* system is the lack of the CheB methylase and CheR methyltransferase. In *Salmonella*, these proteins regulate the chemotactic response by methylation of the membrane-bound receptors that are called methyl-accepting chemotaxis proteins (MCPs). This indicates that the *H. pylori* chemotaxis cascade works without methylation of the receptor proteins, similar to *Sinorhizobium meliloti* (Sourjik and Schmitt, 1998) and *Rhodobacter sphaeroides* (Shah *et al.*, 1999). Instead, *H. pylori* possesses not only CheW and CheY (CheY1) homologs but also three *cheV* genes. In *B. subtilis, cheV* corresponds to a hybrid of CheW and CheY, implying that *H. pylori* has multiple CheW plus CheY equivalents. Furthermore, *H. pylori* expresses still another CheY (CheY2) ortholog which consists of a CheY domain fused to CheA (Foynes *et al.*, 2000). Some or all of the five CheY domains might serve as high affinity "phosphate sinks" and act as terminators of the chemotaxis system. These would exert their function by dephosphorylating the specific CheY homologue (which remains to be characterized) which interacts with the flagellar motor, thereby terminating the tumbling response and inducing the return to a straight run movement. Such a mechanism has been described for *S. meliloti* (Sourjik and Schmitt, 1998) and *R. sphaeroides* (Shah *et al.*, 2000) and could compensate for the lack of an *H. pylori* homolog of the response terminator gene *cheZ*.

In vitro, *H. pylori* displays positive chemotaxis to moderate concentrations of urea, bicarbonate and a structural analogue of urea, the urease inhibitor flurofamide (Mizote *et al.*, 1997). This positive chemotaxis behavior depends on a functional urease enzyme, especially in viscous media (Nakamura *et al.*, 1998). Intrabacterial but not extracellular urease activity was crucial for this function. A positive chemotactic movement towards hog gastric mucin has also been observed (Foynes *et al.*, 2000). The *Helicobacter* motility pattern has been studied in liquid culture and in motility agar plates by phase contrast microscopy and computerized motility tracking with a Hobson BacTracker (Hazell *et al.*, 1986; Eaton *et al.*, 1992; Karim *et al.*, 1998). *Helicobacter* chemotactic behaviour does not consist of alternating phases of "runs" and "tumbling" as in *E. coli* or *Salmonella*, but rather of a combination of backward and forward movements, random darting movements with frequent changes in direction, and short runs.

Role of Motility in Colonization and Pathogenesis

Motility is not a general requirement for successful gastrointestinal pathogens. For example, all Shigella are non-motile, as are some of the most widely spread serovars of enterohemorrhagic *E. coli* (O157: H-, O26: H-, *etc.*). Despite the lack of motility, the infectious doses of these bacteria are extremely low. In contrast, three different animal models have proven that motility is essential for *H. pylori* infection. Early studies by Eaton and coworkers had shown that a non-motile laboratory strain (Tx30a) was unable to colonize gnotobiotic piglets. Animal passage selects for motile and highly flagellated variants of *H. pylori* (Eaton *et al.*, 1992), accompanied by a concomitant increase of flagellin mRNA (Mankoski *et al.*, 1999).

Non-motile/aflagellate *H. pylori flaA/flaB* double mutants were unable to colonize gnotobiotic piglets for more than two days (Eaton *et al.*, 1996). *flaA/flaB* double mutants of *H. mustelae* were completely unable to colonize specific-pathogen free (SPF) ferrets at any time point (Andrutis *et al.*, 1997).

The effects of the inactivation of single flagellin genes differed between *H. pylori* and *H. mustelae*. Both FlaA and FlaB flagellins are necessary for infectivity of *H. pylori* in a gnotobiotic piglet model. Single isogenic *flaA* and *flaB* mutants of the *H. pylori* strain N6 were recovered from gnotobiotic piglets 2 days and 4 days after inoculation, but not after 10 days. In contrast, single flagellin mutants of *H. mustelae* could establish persistent colonization in SPF ferrets. The density of colonization was initially much lower than that of the wild-type strain, but increased with time. For *H. felis*, inactivation of the major flagellin gene, *flaA*, was sufficient to prevent colonization of mice (Josenhans *et al.*, 1999).

The reasons for the differences between the three models are unknown. Since *H. felis* does not adhere to murine epithelial cells, it makes sense that any reduction of motility should have pronounced effects on its ability to colonize. The differences between the *H. pylori/*

piglet and the *H. mustelae*/ferret model may simply reflect the fact that ferrets are the natural host of *H. mustelae,* while the piglet is an artificial host for *H. pylori.* Hence, one of the differences between the models is that all physiologic adhesin receptors are present for the *H. mustelae*/ferret model but not necessarily for *H. pylori*/piglets. This interpretation is in agreement with mathematical models of *H. pylori* colonization. Although the vast majority of the bacterial population is within the mucus, these models predict that adherent bacteria do make an important contribution to maintaining colonization (Blaser and Kirschner, 1999). The stages of *H. pylori* infection at which motility is needed are unknown, and it remains unknown if there is also a phase of infection where motility is not required.

Flagellar filaments do not seem to play a role in the adhesion of *Helicobacter*, because flagellin mutants of *H. pylori* and *H. mustelae* were unaffected in their ability to adhere to primary gastric epithelial cells (Clyne *et al.*, 2000).

Perspectives

Numerous questions regarding motility and chemotaxis remain open. What substances does *H. pylori* detect for orientation in the mucus, and which chemotaxis sensor proteins are involved in sensing these gradients? Identification of these gradients and signal transduction pathways may provide an interesting new avenue for the treatment of *H. pylori* infection. In the near future, microarray hybridization technology might facilitate the complete description of the regulatory interconnection of flagellar genes, and improve our understanding of how the *H. pylori* flagellar network is coordinated without a master regulon and the FlgM anti-sigma factor. Finally, we do not understand the selective advantage in certain environments of a complex filament composed of two or more flagellins, and the different *Helicobacter* species provide good model systems to investigate this question.

Acknowledgments

We thank Chi Aizawa and Mark Achtman for critical reading of the manuscript and helpful suggestions. Our work on *H. pylori* motility was supported by a Gerhard Hess award and other grants from the Deutsche Forschungsgemeinschaft to S.S. and by a Benningsen-Foerder award from the Ministry of Science and Education of North-Rhine Westphalia to C.J.

References

Aizawa, S.-I. 2000. Flagella. In: J. Lederberg. (ed.), Encyclopedia of Microbiology. Academic Press, New York.

Aizawa, S.I. 1996. Flagellar assembly in *Salmonella typhimurium*. Mol. Microbiol. 19: 1-5.

Alm, R.A., Ling L.-S.L., Moir D.T., King, B.L., Brown, E.D., Doig, P.C., Smith, D.R., Noonan B., Guild, B.C., deJonge, B.L., *et al.* 1999. Genomic-sequence comparison of two unrelated isolates of the human gastric pathogen *Helicobacter pylori*. Nature 397: 176-180.

Andrutis, K.A., Fox, J.G., Schauer, D.B., Marini, R.P., Li, X., Yan, L., Josenhans, C., and Suerbaum, S. 1997. Infection of the ferret stomach by isogenic flagellar mutant strains of *Helicobacter mustelae*. Infect. Immun. 65: 1962-1966.

Beier, D., Spohn, G., Rappuoli, R., and Scarlato, V. 1997. Identification and characterization of an operon of *Helicobacter pylori* that is involved in motility and stress adaptation. J. Bacteriol. 179: 4676-4683.

Beier, D., Spohn, G., Rappuoli, R., and Scarlato, V. 1998. Functional analysis of the *Helicobacter pylori* principal sigma subunit of RNA polymerase reveals that the spacer region is important for efficient transcription. Mol. Microbiol. 30: 121-134.

Blaser, M.J. 1994. *Helicobacter pylori*: microbiology of a "slow" bacterial infection. Trends Microbiol. 1: 255-260.

Blaser, M.J. and Kirschner, D. 1999. Dynamics of *Helicobacter pylori* colonization in relation to the host response. Proc. Natl. Acad. Sci. USA. 96: 8359-8364.

Clyne, M., Ocroinin, T., Suerbaum, S., Josenhans, C., and Drumm, B. 2000. Adherence of isogenic flagellum-negative mutants of *Helicobacter pylori* and *Helicobacter mustelae* to human and ferret gastric epithelial cells. Infect. Immun. 68: 4335-4339.

Eaton, K.A., Morgan, D.R., and Krakowka, S. 1992. Motility as a factor in the colonisation of gnotobiotic piglets by *Helicobacter pylori*. J. Med. Microbiol. 37: 123-127.

Eaton, K.A., Suerbaum, S., Josenhans, C., and Krakowka, S. 1996. Colonization of gnotobiotic piglets by *Helicobacter pylori* deficient in two flagellin genes. Infect. Immun. 64: 2445-2448.

Foynes, S., Dorrell, N., Ward, S.J., Stabler, R.A., McColm, A.A., Rycroft, A.N., and Wren, B.W. 2000. *Helicobacter pylori* possesses two CheY response regulators and a histidine kinase sensor, CheA, which are essential for chemotaxis and colonization of the gastric mucosa. Infect. Immun. 68: 2016-2023.

Foynes, S., Dorrell, N., Ward, S.J., Zhang, Z.W., McColm, A.A., Farthing, M.J., and Wren, B.W. 1999. Functional analysis of the roles of FliQ and FlhB in flagellar expression in *Helicobacter pylori*. FEMS Microbiol. Lett. 174: 33-39.

Fraser, G.M., Bennett, J.C., and Hughes, C. 1999. Substrate-specific binding of hook-associated proteins by FlgN and FliT, putative chaperones for flagellum assembly. Mol. Microbiol. 32: 569-580.

Geis, G., Suerbaum, S., Forsthoff, B., Leying, H., and Opferkuch, W. 1993. Ultrastructure and biochemical studies of the flagellar sheath of *Helicobacter pylori*. J. Med. Microbiol. 38: 371-377.

Hazell, S.L., Lee, A., Brady, L., and Hennessy, W. 1986. *Campylobacter pyloridis* and gastritis: association with intercellular spaces and adaptation to an environment of mucus as important factors in colonization of the gastric epithelium. J. Infect. Dis. 153: 658-663.

Hughes, K.T., Gillen, K.L., Semon, M.J., and Karlinsey, J.E. 1993. Sensing structural intermediates in bacterial flagellar assembly by export of a negative regulator. Science 262: 1277-1280.

Jenks, P.J., Foynes, S., Ward, S.J., Constantinidou, C., Penn, C.W., and Wren, B.W. 1997. A flagellar-specific ATPase (FliI) is necessary for flagellar export in *Helicobacter pylori*. FEMS. Microbiol. Lett. 152: 205-211.

Jones, A.C., Logan, R.P.H., Foynes, S., Cockayne, A., Wren, B.W., and Penn, C.W. 1997. A flagellar sheath protein of *Helicobacter pylori* is identical to HpaA, a putative *N*-acetylneuraminyllactose-binding hemagglutinin, but is not an adhesin for AGS Cells. J. Bacteriol. 179: 5643-5647.

Josenhans, C., Eaton, K.A., Thevenot, T., and Suerbaum, S. 2000. Switching of flagellar motility in *Helicobacter pylori* by reversible length variation of a short homopolymeric sequence repeat in *fliP*, a gene encoding a basal body protein. Infect. Immun. 68: 4598-4603.

Josenhans, C., Ferrero, R.L., Labigne, A., and Suerbaum, S. 1999. Cloning and allelic exchange mutagenesis of two flagellin genes from *Helicobacter felis*. Mol. Microbiol. 33: 350-362.

Josenhans, C., Labigne, A., and Suerbaum, S. 1995. Comparative ultrastructural and functional studies of *Helicobacter pylori* and *Helicobacter mustelae* flagellin mutants: Both flagellin subunits, FlaA and FlaB, are necessary for full motility in *Helicobacter* species. J. Bacteriol. 177: 3010-3020.

Josenhans, C., A. Labigne, and S. Suerbaum. 1995b. Reporter gene analyses show that expression of both *H. pylori* flagellins is dependent on the growth phase. Gut 37 (Suppl. 1): A62

Karim, Q.N., Logan, R.P., Puels, J., Karnholz, A., and Worku, M.L. 1998. Measurement of motility of *Helicobacter pylori, Campylobacter jejuni*, and *Escherichia coli* by real time computer tracking using the Hobson BacTracker. J. Clin. Pathol. 51: 623-628.

Kim, J.S., Chang, J.H., Chung, S.I., and Yum, J.S. 1999. Molecular cloning and characterization of the *Helicobacter pylori fliD* gene, an essential factor in flagellar structure and motility. J. Bacteriol. 181: 6969-6976.

Kostrzynska, M., Betts, J.D., Austin, J.W., and Trust, T.J. 1991. Identification, characterization, and spatial localization of two flagellin species in *Helicobacter pylori* flagella. J. Bacteriol. 173: 937-946.

Kutsukake, K., Ohya, Y., and Iino, T. 1990. Transcriptional analysis of the flagellar regulon of *Salmonella typhimurium*. J. Bacteriol. 172: 741-747.

Macnab, R.M. 1996. Flagella and Motility. In: *Escherichia coli* and *Salmonella:* Cellular and Molecular Biology. F.C. Neidhardt, ed.. ASM Press, Washington D. C.

Mankoski, R., Hoepf, T., Krakowka, S., and Eaton, K.A. 1999. *flaA* mRNA transcription level correlates with *Helicobacter pylori* colonisation efficiency in gnotobiotic piglets. J. Med. Microbiol. 48: 395-399.

Minamino, T. and Macnab, R.M. 2000. Interactions among components of the *Salmonella* flagellar export apparatus and its substrates. Mol. Microbiol. 35: 1052-1064.

Mizote, T., Yoshiyama, H., and Nakazawa, T. 1997. Urease-independent chemotactic responses of *Helicobacter pylori* to urea, urease inhibitors, and sodium bicarbonate. Infect. Immun. 65: 1519-1521.

Nakamura, H., Yoshiyama, H., Takeuchi, H., Mizote, T., Okita, K., and Nakazawa, T. 1998. Urease plays an important role in the chemotactic motility of *Helicobacter pylori* in a viscous environment. Infect. Immun. 66: 4832-4837.

Nambu, T. and Kutsukake, K. 2000. The *Salmonella* FlgA protein, a putative periplasmic chaperone essential for flagellar P ring formation. Microbiology 146: 1171-1178.

Nambu, T., Minamino, T., Macnab, R.M., and Kutsukake, K. 1999. Peptidoglycan-hydrolyzing activity of the FlgJ protein, essential for flagellar rod formation in *Salmonella typhimurium*. J. Bacteriol. 181: 1555-1561.

O'Toole, P.W., Kostrzynska, M., and Trust, T.J. 1994. Non-motile mutants of *Helicobacter pylori* and *Helicobacter mustelae* defective in flagellar hook production. Mol. Microbiol. 14: 691-703.

Parkinson, J.S. 1993. Signal transduction schemes of bacteria. Cell 73: 857-871.

Porwollik, S., Noonan, B., and O'Toole, P.W. 1999. Molecular characterization of a flagellar export locus of *Helicobacter pylori*. Infect. Immun. 67: 2060-2070.

Schmitz, A., Josenhans, C., and Suerbaum, S. 1997. Cloning and characterization of the *Helicobacter pylori flbA* gene, which codes for a membrane protein involved in coordinated expression of flagellar genes. J. Bacteriol. 179: 987-997.

Schreiber, S., Stüben, M., Josenhans, C., Scheid, P., and Suerbaum, S. 1999. *In vivo* distribution of *Helicobacter felis* in the gastric mucus of the mouse: Experimental method and results. Infect. Immun. 67: 5151-5156.

Shah, D.S., Porter, S.L., Harris, D.C., Wadhams, G.H., Hamblin, P.A., and Armitage, J.P. 2000. Identification of a fourth *cheY* gene in *Rhodobacter sphaeroides* and interspecies interaction within the bacterial chemotaxis signal transduction pathway. Mol. Microbiol. 35: 101-112.

Shah, S., Qaqish, R., Patel, V., and Amiji, M. 1999. Evaluation of the factors influencing stomach-specific delivery of antibacterial agents for *Helicobacter pylori* infection. J. Pharm. Pharmacol 51: 667-672.

Shapiro, L. 1995. The bacterial flagellum: from genetic network to complex architecture. Cell 80: 525-527.

Sourjik, V. and Schmitt, R. 1998. Phosphotransfer between CheA, CheY1, and CheY2 in the chemotaxis signal transduction chain of *Rhizobium meliloti*. Biochemistry 37: 2327-2335.

Spohn, G. and Scarlato, V. 1999. Motility of *Helicobacter pylori* is coordinately regulated by the transcriptional activator FlgR, an NtrC homolog. J. Bacteriol. 181: 593-599.

Stock, J. and Surette, M. 1994. Bacterial chemotaxis. The motor connection. Curr. Biol. 4: 143-144.

Suerbaum, S. 1995. The complex flagella of gastric *Helicobacter* species. Trends Microbiol. 3: 168-170.

Suerbaum, S., Brauer-Steppkes, T., Labigne, A., Cameron, B., and Drlica, K. 1998. Topoisomerase I of *Helicobacter pylori*: juxtaposition with a flagellin gene (*flaB*) and functional requirement of a fourth zinc finger motif. Gene 210: 151-161.

Suerbaum, S., Josenhans, C., and Labigne, A. 1993. Cloning and genetic characterization of the *Helicobacter pylori* and *Helicobacter mustelae flaB* flagellin genes and construction of *H. pylori flaA*- and *flaB*-negative mutants by electroporation-mediated allelic exchange. J. Bacteriol. 175: 3278-3288.

Tomb, J.-F., White, O., Kerlavage, A.R., Clayton, R.A., Sutton, G.G., Fleischmann, R.D., Ketchum, K.A., Klenk, H.P., Gill, S., Dougherty, B.A., *et al.* 1997. The complete genome sequence of the gastric pathogen *Helicobacter pylori*. Nature 388: 539-547.

Vanet, A., Marsan, L., Labigne, A., and Sagot, M.F. 2000. Inferring regulatory elements from a whole genome. An analysis of *Helicobacter pylori* sigma(80) family of promoter signals. J. Mol. Biol. 297: 335-353.

Worku, M.L., Sidebotham, R.L., Walker, M.M., Keshavarz, T., and Karim, Q.N. 1999. The relationship between *Helicobacter pylori* motility, morphology and phase of growth: implications for gastric colonization and pathology. Microbiology 145: 2803-2811.

From: *Helicobacter pylori: Molecular and Cellular Biology*
ISBN 1-898486-25-5 © 2001 Horizon Scientific Press, Wymondham, UK.

12

Helicobacter pylori, an Adherent Pain in the Stomach

Markus Gerhard, Siiri Hirmo, Torkel Wadström,
Halina Miller-Podraza, Susann Teneberg, Karl-Anders Karlsson,
Ben Appelmelk, Stefan Odenbreit, Rainer Haas,
Anna Arnqvist and Thomas Borén[*]

Abstract

Modern strains of *Helicobacter pylori* are the result of selection for life in the human gastric mucosa. This is a most demanding environment, with high acidity, peristalsis, gastric empty-ing and shedding of cells and mucus. In order to ensure efficient long-term survival and colonization, these microbes have developed arrays of adherence properties for homing to the optimal set of host cells. Survival demands an adequate balance between movement due to flagellar motility and attachment by adherence. This balance allows the microbes to take advantage of continuous supplies of nutrients without cutting off escape from the mucus layer to evade the cellular immune responses. *H. pylori* possesses a multitude of genetic mechanisms for the flexible shifting of adherence specificities to allow adaptation to the host and the local environment. The detailed characterization of the mechanisms that support and maintain bacterial adherence will identify key-elements for the persistence of infection that can be targeted for anti-microbial drug strategies.

Introduction

The gastrointestinal tract is colonized by a multitude of non-pathogenic bacterial species, i.e. commensal bacteria, adapted to a diverse set of micro-environments in the mucus layers and on the epithelial surfaces. Bacterial adherence and corresponding binding sites in the host tissue allow the bacteria to colonize the gastrointestinal tract. After the initial contact be-tween the microbes and host tissue, the bacteria face a multitude of host responses and anti microbial defense systems. Bacteria express surface molecules, adhesins, that recognize a defined selection of binding sites, receptors, such as cell surface proteins, glycoproteins and glycolipids. The microbial affinity for specific receptor structures and the tissue-specific dis-tribution of receptors limits the number of hosts, tissues and cell lineages that can be colo-nized (Karlsson, 1989). Such preferential tissue targeting by bacteria is referred to as host and tissue tropism (see Falk *et al.*, 1994a). One of the best defined systems for bacterial tissue tropism is that of P-fimbriated uropathogenic *Escherichia coli*. The G-adhesin, located

[*]corresponding author, email: Thomas.Boren@odont.umu.se

on the tip of the P-fimbriae, binds to the Galα(1-4)Gal-containing glycolipid receptors expressed in kidney tissue. Presence of the G-tip adhesin is needed for pyelonephritis to occur (Roberts *et al.*, 1994). In contrast, the G-adhesins provide no competitive advantages for *E. coli* in the gastrointestinal tract, because it lacks the appropriate glycolipid receptors (Winberg *et al.*, 1995). These results also demonstrate the consequence of the microbial adherence process for the development of disease.

Tropism of *H. pylori*

H. pylori exhibits both distinct host tropism and detailed tissue tropism. *H. pylori* is a primate-host-specific pathogen *i.e.* natural infections are limited to humans and primates (Dubois *et al.*, 1991). *H. pylori* has adapted to the acidic environment of the human stomach where competition by other microbial species is at a minimum. *H. pylori* is normally found in the lower part of the stomach, the antrum, where the bacteria exhibit tropism for the gastric surface mucous cells. In this region, the pH is less acidic compared to the central part of the stomach, the corpus, where acid generating cells are abundant.

The slimy mucus layer lining the mucosa of the stomach is rich in urea and bicarbonate, which can efficiently buffer the acid pH from the lumen. This mucus barrier provides a protected niche in which *H. pylori* can colonize the gastric mucosa for long terms. An established infection can persist for the lifetime of the host (Blaser, 1993). However, continuous shedding of the mucus layer and the rapid turnover of epithelial cells also present problems for *H. pylori*.

H. pylori has developed efficient features such as flagella and spiral morphology for high motility and chemosensing of host chemotaxins, to direct the bacterium towards the epithelial lining. *In vivo*, the majority of *H. pylori* cells are found in the mucus layer but part of the population adheres intimately to the epithelial lining. The microbes probably adhere to gain nutrients and to modulate immune responses (Hessey *et al.*, 1990). Mathematical modeling suggests the importance of the microbial adherence process for the persistence of *H. pylori* infection (Kirschner *et al.*, 1995). In the mucus layer, *H. pylori* interacts with the highly glycosylated high-molecular-weight mucin molecules. In particular, sialylated-, fucosylated- and sulfated- carbohydrate epitopes are available as attachment sites. In order to establish chronic infection and inflammation processes, *H. pylori* penetrates the mucus layer and adheres directly to binding sites in the epithelial lining. Further interactions of *H. pylori* with components of the connective tissues have also been demonstrated (Wadström *et al.*, 1996). In contrast, the closely related *Helicobacter felis,* that is specific for feline and canine hosts, demonstrates a non-adherent type of colonization pattern in the mucus layer (Schreiber *et al.*, 1999), suggesting that different microbes use different patterns of colonization depending on the gastric environment.

Adherence of *H. pylori* and Promotion of Disease

Binding of *H. pylori* to gastric epithelial cells is considered to be an important virulence factor for the colonization of the human stomach. Adherence of the virulent *H. pylori* type 1 strains that contain the *cag* pathogenicity island (*cag*PAI) stimulates bacterial transmembrane signaling and cell activity. The membrane density increases and the Type IV-secretion apparatus is activated. The CagA protein is secreted into the host cells where it becomes tyrosine-phosphorylated (Segal *et al.*, 1999; Stein *et al.*, 2000; Odenbreit *et al.*, 2000). This is followed by localized actin polymerization in the mucosal cells and pedestal formation in the cells beneath the attached bacteria (Smoot *et al.*, 1993; Segal *et al.*, 1996; Stein *et al.*, 2000;

Covacci *et al.*, 2000). *H. pylori* also activates signal cascades in epithelial cells inducing cellular inflammatory responses, such as the induction of IL-8 secretion (Segal *et al.*, 1997). Attachment and effacement, very similar to those described for enteropathogenic *E. coli* (EPEC), are also caused by genes encoded within the *cag*PAI. Thus, adherence of *H. pylori* not only supports efficient and permanent colonization of the gastric mucosa, but might also be necessary for the development of disease (Guruge *et al.*, 1998).

One challenge for investigators studying adherence of *H. pylori* is the complexity of glycoconjugate-binding specificities. *H. pylori* is unique among known adherent bacteria in that it possesses at least ten binding specificities for different carbohydrates (Karlsson, 1998). Another important parameter would be the bacterial interaction with non-epithelial cells. *H. pylori* cells induce inflammatory responses and phagocytosis upon contact with polymor-phonuclear leukocytes (PMNs) (see Rautelin *et al.*, 1993). PMNs are rich in various sialylated glycoconjugates and strongly bind the bacteria. In contrast, normal human gastric epithelial cells are essentially devoid of such glycoconjugates (Filipe, 1979; Madrid *et al.*, 1990) and express sulfatide and Lewis b antigens, which are absent from PMNs. Therefore, specificity for sialic acids may be essential for interaction with PMNs. Bacterial cells are not necessarily killed by PMNs; they may escape damage, and can even use degradation products of the inflammatory responses as a nutritional source (Hessey *et al.*, 1990; Blaser, 1996).

So far only the Lewis b antigen binding BabA adhesin has been identified (Ilver *et al.*, 1998), and correlated to a special group of disease promoting strains, the triple positive *H. pylori* strains. This group of strains are characterized by the presence of the Vacuolating Cytotoxin, VacA, the Cytotoxin Associated antigen, CagA, and the Blood group Antigen Binding Adhesin, BabA. The triple positive strains are found significantly more frequently in patients with ulcer disease or gastric adenocarcinoma, respectively, than in patients with gastritis only ($p<0,05$). The results suggest that tight adherence to the epithelium mediated by Lewis b antigen binding might be an important factor for the induction of pathogenicity when additional factors as *vacA* and *cagA* are present (Gerhard *et al.*, 1999). This hypothesis has been followed further by investigating the genotypes of these virulence factors in gastric biopsies from patients with *H. pylori* associated gastritis (n=132) (Gerhard *et al.*, unpublished observations). In these patients, the expressed *babA2* gene was detected in 32% of *H. pylori*[+] biopsies, and was more frequent during severe antral and corpus gastritis (G0; 11%, G1; 30%, G2; 35%, and G3; 63%). In biopsies of patients with atrophy or intestinal metaplasia, *babA2* was found in 70% and 56%, respectively. The triple positive strains discriminated significantly better between low grade gastritis or severe gastritis (G3), atrophy and intesti-nal metaplasia, respectively, than the current type 1 classification. These data support the view that presence of *babA2*, particularly in combination with *vacAs1* and *cagA,* could be useful as a parameter to detect patients who are at higher risk for developing severe histologi-cal changes or cancer precursor lesions such as intestinal metaplasia. Prospective follow-up studies are needed to determine whether patients infected with triple positive strains will benefit from eradication in early stages of gastritis. With the rapidly developing techniques in proteomics, genomics, and transgenics, in combination with development of tools for ad-vanced *H. pylori* genetics, the existence of additional adhesins and corresponding host receptors may soon be revealed. Such a powerful combination of techniques will lay the foundation for the rational basis of how to find out the relationships of defined binding specificities and their consequences for development of disease.

In the following review, we will first describe the many different binding sites or receptors in the gastric mucosa, together with the corresponding bacterial adhesins, involved in adher-ence and colonization of *H. pylori* and how adherence to specific receptors may relate to disease. Then, in the second part of this review we will describe two strategies for protein purification and genetic manipulations that allowed the identification of specific adhesins.

H. pylori Interactions with Sialylated Binding Sites

Red cells from humans and animal sources differ in their cell surface glycoconjugates and can be used to reveal subtle differences in receptor specificity. Bacterial agglutination of erythrocytes (hemagglutination) has been used as an initial step to study the detailed microbial binding specificity and identify the bacterial adhesin proteins. This is well illustrated by the different classes of P-fimbriated uropathogenic *E. coli* which can bind to the Galα(1-4)Gal-series of glycolipids in kidney tissue, the carbohydrate epitope that constitutes the P-blood group antigen. The specificities for hemagglutination mirror the fine differences in receptor epitope recognition that distinguish these bacteria (Strömberg *et al.*, 1991).

A number of microbe-host interactions depend on the recognition of sialylated glycoconjugates on host cells (Kelm *et al.*, 1997). The initial description of the binding of *H. pylori* to sialylated carbohydrates (Evans *et al.*, 1988) was followed by a number of papers on sialic acid-dependent binding of *H. pylori*.

On the basis of hemagglutination and hemagglutination inhibition data with a variety of sialyl-glycoconjugates, *H. pylori* strains have been grouped into sialic acid dependent and independent classes (Lelwala-Guruge *et al.*, 1992; Hirmo *et al.*, 1996). The binding of most strains of *H. pylori* is highly specific for certain sialic acid modifications and glycosidic linkages (Kelm *et al.*, 1997). The use of re-sialylated human erythrocytes showed that *H. pylori* strains preferentially bind to terminal α2,3-linked sialic acids on both *N*- and *O*-glycans, rather than α 2,6-linkages (Hirmo *et al.*, 1996).

Hemagglutination by *H. pylori* and binding of *H. pylori* to various tissues can be blocked by a range of sialylated oligosaccharides (e.g. sialyl-lactose or sialyl-*N*-acetyllactosamine) or various sialo-glycoproteins (Lelwala-Guruge *et al.*, 1992; Evans *et al.*, 1995; Hirmo *et al.*, 1996; Simon *et al.*, 1997). Data on the adherence of radiolabeled bacteria to glycoconjugates immobilized on artificial surfaces indicate that the sialylated epitopes recognized by *H. pylori* are also associated with natural glycolipids, such as the ganglioside GM3 (Slomiany *et al.*, 1989), 3'sialylparagloboside, NeuAcα3Galβ4GlcNAcβ3Galβ4GlcCer and polyglycosylceramides, the PGCs (Miller-Podraza *et al.* 1999).

Sialylated Polyglycosylceramides

PGCs are highly glycosylated glycosphingolipids containing branched poly-N-acetyllactosamine chains with repeated binding epitopes (Miller-Podraza *et al.* 1996 and references therein). Human granulocytes, the inflammatory cells engaged in *H. pylori*-related infections, contain PGCs as well as other sialylated glycolipids and glycoproteins (Miller-Podraza *et al.*, 1999). Contact with bacterial cells results in changes in the surface molecules (Enders *et al.*, 1995) and subsequent activation of PMNs (Rautelin *et al.*, 1993). Adherence of *H. pylori* to sialylated structures could be of importance for bacterial survival during phagocytosis (Chmiela *et al.*, 1994), and one of the current ideas on the dynamics of *H. pylori*-infections is that *H. pylori* exacerbates the mucosal inflammation to thrive on the inflammatory products (Blaser, 1992, 1996). The structure of the sialylated epitope on PGCs that is recognized by *H. pylori* is not yet known, but it has been shown that the PGCs on human erythrocytes to which *H. pylori* binds are heterogeneous in regard to molecular weight and their degrees of fucosylation and sialylation. The number of monosaccharides per chain can vary from 15 to 40, according to MALDI-TOF mass spectrometry (Karlsson *et al.*, 1999). The interaction of *H. pylori* with human PGCs is strictly dependent on the presence of sialic acid and its glycerol side chain because mild periodate oxidation completely eliminates binding (Miller-Podraza *et al.*, 1998). Results from molecular modeling indicate that the epitope may include NeuAc linked close to a branch in the primary chain (Karlsson, 1998). In addition, those fractions to which *H. pylori* can bind contain NeuAc linked to Gal by α2,3 glycosidic linkages (Johansson *et al.*, 1999). The ability of *H. pylori* cells to recognize sialylated sequences is strongly influenced by growth conditions. Bacteria grown on agar plates bind to a range of sialylated glycoconjugates while *H. pylori* grown in liquid media (broth) preferentially bind to PGCs (Miller-Podraza *et al.*, 1996, 1997). These results suggest that *H. pylori* expresses distinctly different sialic acid receptors in different environments, possibly to ensure colonization under a variety of conditions.

Sialylated Epitopes in Response to Inflammation and Disease

The binding of *H. pylori* to sialylated glycoconjugates of different complexities and topographies suggests that these might represent receptors that are important for colonization of the gastric mucosa. Although, the level of sialylation in the epithelial cell lining in the human gastric mucosa is normally low (Filipe, 1979; Madrid *et al.*, 1990), neuraminidase treatment does reduce the binding of *H. pylori* strains to human gastric mucin (Tzouvelekis *et al.*, 1991; Hirmo *et al.*, 1998). These results indicate that direct adherence to the gastric mucus layer is dependent on binding to sialic acid residues.

Sialic acid-dependent hemagglutination is a feature of one third of clinical *H. pylori* isolates (Lelwala-Guruge *et al.*, 1992) but no correlation has been reported between the hemagglutination titers and the cell adhesion properties of *H. pylori* (Clyne *et al.*, 1993; Kobayashi *et al.*, 1993). Furthermore, no correlation was found between sialic acid-dependent binding properties of *H. pylori* and the degree of disease (Taylor *et al.*, 1992). These observations do not negate a role of binding to sialic acid residues for colonization and disease because, as described above, chronic inflammation of the mucosa might modulate the glycosylation patterns of epithelial cells. Increased levels of the sialyl-Lewis a antigen were shown to be stimulated as a response to *H. pylori* infection (Ota *et al.*, 1998). This is a dynamic process and after antibiotic treatment and *H. pylori* eradication, the sialyl-Lewis a antigen was again expressed by fewer epithelial cells. Similarly, the sialyl-Lewis x antigen is over-expressed in bronchial mucins from patients with cystic fibrosis or chronic bronchitis who are infected by *Pseudomonas aeruginosa* (Davril *et al.*, 1999). Thus, up-regulation of sialyl-Lewis antigens in epithelial cells might be a general mechanism of inflammatory response to infectious agents, similar to the up-regulation of the sialyl-Lewis a/x antigens by the endothelial cell lining that is triggered by inflammation (Varki *et al.*, 1994). The levels of *O*-acetylated sialic acids are also increased during the development of gastric metaplasia that is correlated with *H. pylori* infection (Mullen *et al.*, 1995) and sialylated oligosaccharides efficiently inhibit the binding of fresh clinical *H. pylori* strains to gastrointestinal epithelial cells (Simon *et al.*, 1997). Sialyl-glycoprotein conjugates were more potent as adherence inhibitors than free sialylated oligosaccharides, indicating a preference for multivalent binding. *In vivo*, sialylated mucin receptors provide such multivalent receptor presentations. Encouraging results on the suppression of infection using anti-adhesive therapy with 3´-sialyllactose have been obtained with *H. pylori*-infected Rhesus monkeys (Mysore *et al.*, 1999). Furthermore, a genetically engineered ablation of parietal cells in mice results in the increased production of sialylated/oncofetal carbohydrate epitopes which function as binding sites for *H. pylori* in the deeper parts of the gastric mucosa, i.e. in the cell proliferation region (Syder *et al.*, 1999). High-salt diets are associated with multifocal reduction of parietal cell numbers and enhanced *H. pylori* colonization (Fox *et al.*, 1999). Taken together, these data suggest that the sialic acid-dependent adhesion properties are relevant to disease promoting processes, and might direct the infection to certain ecological niches such as the mucus layer, or to specific sets of cell lineages with up-regulated sialylation patterns.

H. pylori Adhesin Proteins with Specificity for Sialylated Glycoconjugates

Several approaches have been used to identify sialic acid-dependent hemagglutinins. The *hpaA* gene encoding the HpaA *N*-acetylneuraminyllactose-binding fimbrial haemagglutinin (Evans *et al* 1988) was cloned and sequenced (Evans *et al,* 1993). However, the HpaA protein is probably localized within the bacterial cells and is thereby unlikely to be involved in adhesion (O'Toole *et al.*, 1995; Jones *et al.*, 1997). Another non-fimbrial surface hemagglutinin of 19.6 kDa that recognises sialic acid residues on fetuin was also isolated (Lelwala-Guruge *et al.*, 1993). In addition, an *H. pylori* protein of 25 kDa that binds to sialic acid has been characterized and affinity purified by adsorption to erythrocytes (Huang *et al.*, 1992). Binding of *H. pylori* to the extracellular matrix protein laminin can be initiated through interaction with lipopolysaccharide (LPS) by a protein of similar size with sialic acid binding specificity (Valkonen *et al.*, 1997). The variety of adhesin protein candidates for sialylated binding sites might reflect non-specific interactions that allow purification by affinity chromatography. Immobilized sialylated and sulfated oligosaccharide compounds such as fetuin and heparan sulfate are highly charged and can result in ionic interactions in addition to more specific ligand-receptor binding. Affinity chromatography using immobilized fetuin have resulted in the purification of iron-binding proteins, such as the *H. pylori* ferritin protein Pfr (Frazier *et al.*, 1993). It is therefore very important to use genetic techniques in order to confirm the importance for adherence of the proteins that have been purified.

The Neutrophil-activating Protein of *H. pylori*

The HP-NAP protein (see also Chapter 15) is a neutrophil-activating protein first isolated from water extracts of bacterial cells by Evans *et al.*, and the gene for this protein was found in all 21 strains analyzed by PCR (Evans *et al.*, 1995). HP-NAP induces the production of reactive oxygen radicals by human neutrophil granulocytes, and increases the expression of CD11b/CD18 molecules on the neutrophil surface, resulting in the increased adhesion of neutrophils to human endothelial cells.

Because the activation of neutrophils/PMNs might depend on specific lectin-carbohydrate interactions, potential carbohydrate receptors for HP-NAP were investigated by binding HP-NAP to immobilized glycosphingolipids (Teneberg *et al.*, 1997). The data demonstrated selective interaction of the protein with two minor gangliosides of human neutrophil granulocytes. The binding of HP-NAP to gangliosides requires a terminal α2.3-linked sialic acid plus a repetitive *N*-acetyllactosamine element (NeuAcα2.3Galβ1.4GlcNAcβ1.3Galβ1.4GlcNAcβ-) such as are found in extended sialyllactosamine variants. The protein also binds to sulfated ceramides (sulfatide), but binding is merely dependent on the sulfate group, since no binding of HP-NAP to the corresponding non-sulfated glycosphingolipid (galactosylceramide) was obtained. Namavar *et al.* (1998) have suggested that HP-NAP is an adhesin that confers binding of *H. pylori* to mucins, such as salivary mucin molecules.

Similar to the ferritin protein, HP-NAP also exhibits iron-binding properties (Tonello *et al.*, 1999). These results suggest that HP-NAP is important for the persistence of chronic infection and inflammatory processes by providing *H. pylori* with a variety of functions ranging from binding to sialylated and sulfated glycoconjugates and receptors containing iron through to the targeted activation of neutrophils. Further studies with mutants in animal models are required to determine the functional role(s) of the HP-NAP protein, and how it contributes to the pathogenic mechanisms of *H. pylori in vivo*.

H. pylori Interactions with Fucosylated Histo-Blood Group Antigen Receptors

The fucosylated blood group antigens typically found on erythrocytes are also expressed on the epithelial cell surfaces of the gastro-intestinal tracts then defined as the histo blood group antigens (Clausen *et al.*, 1989). The H-antigen is a fucosylated carbohydrate structure that defines the blood group O phenotype in the common ABO blood group system. The Lewis b antigen is formed by the addition of another branched fucose residue to the H antigen, making it di-fucosylated. The Lewis a antigen is similar to the Lewis b antigen but lacks the terminal fucose residue and is mono-fucosylated. The Lewis b antigen is the dominant fucosylated blood group related antigen expressed on gastric surface mucous cells in the gastric epithelial lining (Sakamoto *et al.*, 1989).

The fucosylated blood group antigens mediate specific binding of *H. pylori* to the epithelial surface mucous cells in human gastric mucosa (Borén *et al.*, 1993), as shown by *in situ* adherence assays that tested the ability of fluorescently labeled *H. pylori* bacteria to adhere to sections of human gastric mucosa (Falk *et al.*, 1993, 1994a). The involvement of H-type 1 and Lewis b antigens was demonstrated by inhibition of the binding to immobilized receptor conjugates by defined (semi)-synthetic glycoprotein conjugates presenting the different fucosylated blood group antigens. A requirement for the Lewis b antigen for long term colonization could partly explain the restricted host tropism of *H. pylori* because Lewis b is found only in humans and primates.

The fucosylated blood group antigens are also present in the mucins of the gastric mucus layer and in human secretions such as saliva, tears, and milk (discussed in Falk et al., 1994b). Human milk glycoconjugates and oligosaccharides contain a large number of complex structures, many of which are fucosylated. Human milk samples of defined Lewis phenotype were analyzed for their ability to inhibit *H. pylori* adherence *in situ*. The Lewis b positive milk sample (from patients with positive secretor status) was most powerful in reducing *H. pylori* adherence, whereas less fucosylated milk samples from patients with the Lewis a phenotype (non-secretor status) sample could not prevent bacterial adherence (Borén *et al.*, 1993). The carbohydrate compositions of natural glycosylation patterns are genetically regulated and generate many of our blood group systems, such as ABO, Lewis and P. Detailed differences in glycosylation patterns might affect the individual predisposition for infections and secretions such as saliva, tears, and milk may act as natural scavengers or clearance factors (for review see Borén *et al.*, 1994).

Individuals with the blood group O phenotype are at a 1.5-2 fold increased risk of developing peptic ulcer (Rotter, 1983; Hallstone *et al.*, 1994; Borén *et al.*, 1994). Interestingly, no correlation was found between blood group phenotypes and the prevalence of *H. pylori* infections in asymptomatic people or individuals with only mild gastritis (Loffeld *et al.*, 1991; Höök-Nikanne *et al.*, 1990). Nevertheless, for virulent *H. pylori* strains, the triple positive strains, binding of the Lewis b antigens target the microbes to the epithelial cells and potentiates the effect of secretion of virulence factors such as the vacuolating cytotoxin and/or neutrophil activating-recruiting factors (Gerhard *et al.*, 1999).

H. pylori Interactions with Sulfated Receptors and the ECM

Electron microscopy has shown that *H. pylori* is in close contact with intercellular epithelial junctions, possibly facilitated by bacterial secretion of phopholipase C (Weitkamp *et al.*, 1993). At the epithelial junctions, *H. pylori* will come in contact with the glycosaminoglycans of the extra cellular matrix (ECM). Most *H. pylori* strains express the ability to bind to heparan sulfate with high affinity (Ascencio *et al.*, 1993). Heparan sulfate is also present on cell surfaces and has been identified as a potent receptor for *H. pylori* (Hirmo *et al.*, 1995). In a solid phase system where glycolipids were separated by thin layer chromatography, sulfatide was identified as a potential receptor structure (Saitoh *et al.*, 1991; Kamisago *et al.*, 1996). Sulfatide is present both on epithelial and glandular cells, and *H. pylori* interacts with sulfated carbohydrate chains in human gastric mucins (Piotrowski *et al.*, 1991, Hirmo *et al.*, 1999). As described for the binding by HP-NAP above, sulfated carbohydrate chains on salivary mucins provide receptor structures for *H. pylori* adhesion to oral surfaces (Namavar *et al.*, 1998).

Acidic pH and heat shock up-regulate the expression of the *H. pylori* protein Hsp70, a stress induced surface protein that has been suggested to mediate binding to sulfatides (Huesca *et al.*, 1996). However, such treatments also seem to induce binding to a multitude of glycolipids in an unspecific way (S. Teneberg, unpublished observations). Other sulfate-binding proteins have also been described, such as the neutrophil activating protein HP-NAP (Teneberg *et al.*, 1997) and the urease enzyme (Icatlo *et al.*, 1998). *H. pylori* can interact with a multitude of other ECM proteins such as collagen, vitronectin and laminin (Trust *et al.*, 1991; Valkonen *et al.*, 1993; Ringner *et al.*, 1994). The binding of *H. pylori* to laminin was suggested to be initiated through interaction with lipopolysaccharide (LPS) (Valkonen *et al.*, 1994), followed by lectin dependent binding. An adhesin has also been described which confers laminin binding on *H. pylori* (Valkonen *et al.*, 1997).

Binding of *H. pylori* to Lactosylceramide and Gangliotria-/Gangliotetraosyl-Ceramides

Numerous diverse species of bacteria can interact with lactosyl-based glycolipids, such as lactosylceramide (Galβ4Glcβ1Cer) gangliotriaosyl-ceramide (GalNAcβ4Galβ4Glcβ1Cer)) and gangliotetraosyl-ceramide (Galβ3GalNAcβ4Galβ4Glcβ1Cer) (Karlsson, 1989). *H. pylori* also interacts with these glycosphingolipids according to the thin-layer chromatogram binding assay (Ångström *et al.*, 1998). The binding to lactosylceramide was often detected in parallel with binding to gangliotria- and gangliotetraosyl-ceramide. However, chemical derivatization of the glycolipids revealed that the binding to these two latter glycosphingolipids is independent of the binding to lactosylceramide (Ångström *et al.*, 1998). The adhesins responsible for these specificities have not yet been identified.

H. pylori Interactions with Phosphatidylethanolamine Receptors

Charged glycoconjugates such as the phospholipid phosphatidylethanolamine (PE) that is integrated in the cell membrane have also been suggested to represent receptor structures for *H. pylori* (Lingwood *et al.*, 1992). Binding to PE could mediate tight association with the host cell membrane and facilitate signal transduction. Affinity chromatography to PE resulted in the isolation of a 63 kDa adhesin candidate (Lingwood *et al*, 1993). However, this protein was later shown to be a catalase (Odenbreit *et al*, 1996a) and seems unlikely to be a true adhesin.

H. pylori Lipopolysaccharide Lewis Antigens as Adhesins

H. pylori LPS contains Lewis x and Lewis y antigens (Monteiro *et al*, 2000,: Aspinall *et al.*, 1996a; Aspinall *et al.*, 1996b) and the following data suggests that they act as adhesins (Appelmelk, 2000):

A monoclonal antibody that recognizes the Lewis x antigen and *H. pylori* LPS inhibits bacterial adhesion to gastric epithelial cells (Osaki *et al*, 1998; B.J.Appelmelk and H. Yamaguchi, unpublished observatios). Inactivation of the *galE* or *rfbM* genes prevents synthesis of Lewis x LPS antigen and these mutants did not adhere to gastric epithelium (Edwards *et al,* 2000). Moreover, synthetic Lewis x binds to human gastric epithelial cells (Edwards *et al,* 2000). These data suggest that *H. pylori* LPS, and in particular the Lewis x moiety, mediates adhesion to gastric epithelium, and predict the existence of a gastric Lewis x antigen-binding lectin. Previously unknown Lewis x antigen-binding polypeptides of 16-29 kDa were indeed found in gastric epithelial cells (Campbell *et al,* 1999). Other studies have shown that sur-factant protein D, a 560 kDa lectin belonging to the innate defense system and expressed in the stomach (Fisher *et al,* 1995) also binds to *H. pylori* LPS (Eggleton *et al,* 1999). It is also possible that *H. pylori* Lewis x antigens interact with the host via a homotypic (Lewis x-Lewis x) interaction; such a Ca^{++}-dependent homotypic interaction is thought to play an important role in embryogenesis (Hakomori *et al,* 1992). Lewis x-Lewis x-antigen interactions have also been demonstrated to be involved in tumor cell-cell adhesion (Kojima *et al.,* 1994). However, although soluble Lewis x can bind to liposome-incorporated Lewis x (Geyer *et al,* 1999), attempts to demonstrate an interaction of soluble Lewis x antigen with *H. pylori* have failed (Ilver *et al,* 1998). Nevertheless, LPS-presented Lewis antigens could make a distinct difference for cellular and/or bacterial interactions, due to the repetitive display, den-sity and complexity of the antigens presented by *H. pylori.* An important aspect of *H. pylori* LPS Lewis x/y antigens is their ability to display phase variation, i.e., the high frequency, reversible switching of phenotype (Appelmelk *et al,* 1999); for instance, a strain that ex-presses Lewis x antigens can yield variants that express a non-fucosylated polylactosamine chain (the i-antigen). When no other adhesins are expressed, LPS-phase variation would allow detachment of bacteria expressing the i-antigen, followed by transmission to another host. During growth in the new host switch-back variants expressing Lewis x antigens would again become competent for adherence and colonization.

Identification of *H. pylori* Adhesins

The very many different adherence properties and possibilities for cellular crosstalks sug-gest that *H. pylori* requires a series of adhesin molecules with complementary specificities for the successful targeting to various micro-environments and for adaptation to various in-flammatory responses in the gastric epithelial lining.

Phase shifting mechanisms based on repeat motifs in the promoter or the structural gene regions were suggested for some members of the large HOP (Helicobacter Outer membrane Proteins) gene family where expression is under the control of a so-called slipped-strand mispairing mechanisms (SSM) i.e., minor changes in the repetitive sequence can confer ex-pression of the protein, by translation in the correct reading frame. Some of the HOP-genes seem to be silent or incomplete, due to lack of functional translational start sites. Thus, the expression of such genes most likely requires mechanisms of recombination and transfer of the silent genes into an expression locus. In *H. pylori* the expression of several OMP encod-ing genes is predicted to be controlled by CT-dinucleotide repeats in the gene sequence en-coding the leader peptide of the pre-secretory proteins (Tomb *et al.,* 1997; Alm *et al.,* 1999). One example is the recently published *hopZ* gene (Peck *et al.,* 1999). HopZ-producing strains show a number of CT repeats in the leader sequence compatible with a continuous in-frame reading. Interestingly, the expression of HopZ protein could be correlated with adherence to the gastric carcinoma cell line AGS (Peck *et al.,* 1999).

The recombination mechanisms and the regulated expression of adhesins with different binding properties are probably activated by specific environmental signals, such as pH, receptor-recognition, PMN-cell interactions and gradients of inflammatory mediators. Alter-natively, in some situations the signals might rather make the microbes non-adherent, to facilitate the rapid escape to the protective mucus layer. The genetic framework for such proficiencies in adherence properties could most likely have a correspondence in the large number of HOPs, suggesting complex interrelations with the host.

Adhesin Identification Methods

There are established techniques for how to identify adhesive activities of normal bacteria. A most common way would be to isolate a spontaneous non-adherent mutant (or phase shift variant) for comparison with the wild type strain. Unfortunately, adhesin molecules are often minor components and identification procedures usually turn out difficult.

The genome sequences of *H. pylori* strains 26695 and J99 revealed some 30 genes encoding putative outer membrane proteins (OMPs), many of which share conserved sequence motifs in the N- and C-terminal regions. The OMP proteins make up a blend of transport proteins, porins and adhesins (Tomb *et al.*, 1997; Alm *et al.*, 1999). The blood group antigen binding adhesin, BabA (Ilver *et al*, 1998), the adherence-associated proteins, AlpA and AlpB (Odenbreit *et al.*, 1999), the family of HOP-porins (Exner *et al.*, 1995; Hancock *et al.*, 1998), and the recently described HopZ adhesin protein (Peck *et al.*, 1999), all belong to the family of HOP-proteins (Tomb *et al.* 1997). The large number of distinguishable *H. pylori* adhesion specificities requires careful functional analysis and detailed evaluation by both biochemical and genetic approaches. Reverse genetics is possible in *H. pylori* by homologous recombination after natural transformation (Nedenskov-Sorenson *et al.*, 1990; Hofreuter *et al.*, 1998), allowing knock-out (KO) mutations and the insertion of antibiotic resistance cassettes (Wang *et al.*, 1990). KO-mutants can be tested for the loss of specific adherence traits in order to identify the specific genes involved. The use of these techniques will be useful to characterize the role of the HOP proteins.

In the following sections we will focus on two examples of the identification of specific *H. pylori* adhesins, namely BabA and AlpA/B.

The BabA Adhesin

BabA was identified on the basis of its high specificity and affinity for the Lewis b antigen (Ilver *et al.*, 1998). The majority of 100 clinical isolates possessed Lewis b antigen binding properties. Immunogold electron microscopy technique, using gold particles coupled to the Lewis b antigen showed that the BabA adhesin was distributed over the entire bacterial outer membrane, but not on the flagella. The BabA adhesin was then detected as a 74 kDa band on immunoblots by receptor overlay analysis with biotinylated (soluble) Lewis b conjugates and streptavidin. BabA from *H. pylori* strains from different geographical areas are of similar molecular weight.

Conventional protein purification is difficult with an organism such as *H. pylori* due to its fastidious growth and poor yields. We therefore developed the novel ReTagging technique (Receptor activity directed affinity Tagging) for identification and protein purification of the BabA adhesin. ReTagging should prove useful for diverse studies of interactions in general, for both infectious diseases and inflammatory processes. The adhesin on the bacterial surface (or any receptor-bound ligand in general) is specifically tagged with biotin through the power of the receptor-adhesin interaction. A multi-functional crosslinker structure which contained a biotin group and a photo reactive group was covalently coupled to the Lewis b conjugate (Figure 1A). The Lewis b-crosslinker complex is then mixed with a suspension of *H. pylori* cells. The Lewis b oligosaccharides bind tightly to the bacteria bringing the crosslinker into the close vicinity of the adhesin. The bacteria are then exposed to UV-light, which activates

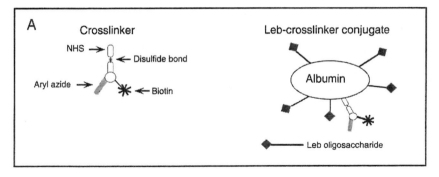

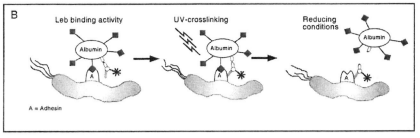

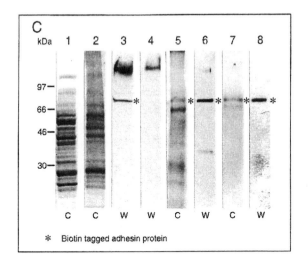

Figure 1. Receptor Activity Directed Affinity Tagging/ReTagging of the BabA adhesin (Ilver *et al.*, 1998). A) A multi-functional crosslinker structure attached to the Lewis b conjugate. B) The Lewis b-crosslinker complex is mixed with *H. pylori*. The Lewis b oligosaccharides bind tightly to the bacterial adhesin. The crosslinker is activated by UV-light, which attaches the biotin residue to the adhesin protein. As a result, the formation of the receptor-adhesin complex has transferred and covalently positioned a biotin handle onto the adhesin protein. The biotin tagged adhesin is then prepared and available for immuno detection and/or protein purification. C) By the ReTagging technique and immunoblot detection with streptavidin of separated proteins on SDS-gel, the 74 kDa BabA adhesin was identified. For preparative protein purification, streptavidin-coated magnetic beads were used for affinity purification of the BabA adhesin. Lanes 1and 2; SDS-PAGE of bacterial protein extract (control) before (1) and after (2) UV-crosslinking. Lanes 3and 4; Streptavidin based immuno-blot detection demonstrating the 74 kDa biotin tagged BabA protein before (3) and after (4) magnetic beads extraction. Lane 5 and 6; BabA protein eluted from the streptavidin coated magnetic beads. Lane 5; visualized with Coomassie blue stain. Lane 6; The biotin tagged BabA protein detected by streptavidin. Lane 7 and 8; The protein preparation after final size fractionation. The eluted BabA band was subjected to N-terminal sequencing. C= Coomassie blue stained gel. W= Western blot immuno detection by streptavidin and chemoluminescence.

the crosslinker and transfers the biotin residue to the adhesin protein. The biotin tagged adhesin can then be recognized, and analyzed by standard procedures (Figure 1B). The ReTagging technique revealed an adhesin with an apparent molecular weight of 74 kDa (Figure 1C). The biotin-tagged adhesin was purified using streptavidin-coated magnetic beads (Figure 1C).

The purified BabA-adhesin was N-terminally sequenced and oligonucleotide primers were constructed for PCR amplification of the corresponding DNA. The amplified DNA fragment was cloned and used as a probe for screening a plasmid library. Two genes were identified, *babA* (encoding a 78 kDa protein) and *babB* (encoding a 75.5 kDa protein) with identical N-terminal sequences. Further ReTagging and purification of the adhesin provided peptide material for an extended N-terminal amino acid sequence (41 aa). The gene matching the adhesin protein N-terminal peptide sequence was identified as the *babA* gene. Interestingly, the *babA* and *babB* genes encode two distinctly different proteins, but with almost identical N-terminal and completely identical and conserved 300 amino acids C-terminal domains.

Interestingly, the *babA* and *babB* genes have exchanged their positions in the genomes of 26695 and J99, most likely by recombination mechanisms. In both *H. pylori* genomes sequenced so far there are only one *babA* and one *babB* gene present. However, sequence analyses of the *bab*-genes in strain CCUG17875 revealed the presence of two *babA* genes. The coding region of the two genes are identical with the exception of 10 additional nucleotides in the *babA2* gene, resulting in the formation of a translational initiation codon, while the *babA1* gene is not translated (Ilver *et al.*, 1998).

As discussed, the clinical relevance of the expressed *babA2* gene was recently demonstrated, as defined by the virulent triple-positive strains (Gerhard *et al.*, 1999). These results suggest that cell adhesion and intimate adherence mediated by the Lewis b antigens may provoke intense inflammatory responses with a potential to develop severe disease.

The AlpA and AlpB Proteins

In an initial approach to investigate the interaction of *H. pylori* with epithelial cells, we attempted to complement *Campylobacter jejuni* with a *H. pylori* cosmid library that had been established in *E. coli*. Gene transfer into *C. jejuni* was inefficient and deletions occurred in the cosmid clones, suggesting that the heterologous expression of *H. pylori* genes may be difficult in *C. jejuni* (Odenbreit and Haas, unpublished observations). It is also unclear whether *H. pylori* promoters would be transcriptionally active in *Campylobacter* species and whether the functional presentation of active adhesins is dependent on genes located at different sites in the chromosome. Thus, the genetic tools for this approach are not yet available, but the concept of heterologous expression of *H. pylori* outer membrane associated genes with accompanying bio-panning of adhesin activities in *Campylobacter* species holds enough potential that it might be worth trying again in the future.

Construction of a H. pylori Mutant Library and Screening for Adherence Defects

In a second approach, saturating transposon mutagenesis was used to generate knock-out mutants of single genes involved in adhesion. No functional transposon element for *H. pylori* was initially available and therefore, the transposon shuttle mutagenesis method (Seifert *et al.*, 1986; Haas *et al.*, 1993; Brune *et al.*, 1999) was adapted to *H. pylori*. *H. pylori* genes were cloned in *E. coli*, mutated and reintroduced into the chromosome of *H. pylori* to generate knock-out mutants. The *blaM* reporter gene within the Tn*Max9* mini-transposon can be

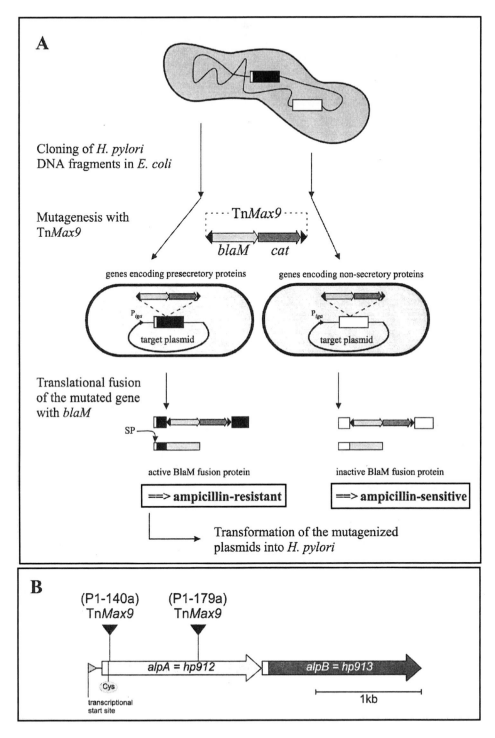

Figure 2. Generation of a library of *H. pylori* strains with selective mutations in genes encoding secretory proteins, followed by isolation of the AlpA and AlpB proteins (Odenbreit *et al.*, 1999). A) SP, Signal peptide, BlaM, mature ß-lactamase, *cat*, chloramphenicol-acetyltransferase, Tn*Max9* mini transposon. B) Schematic representation of the *alpAB* operon involved in *H. pylori* adhesion to gastric epithelial cells. P1-140a and P1-179a are original mutant strains with defects in binding to Kato-III cells and gastric tissue sections (see text). Cys, position of the lipoprotein processing site of AlpA.

used to identify genes encoding secretory proteins (Broome-Smith *et al.*, 1990; Kahrs *et al.*, 1995) because only these hybrid proteins are exported to the periplasm and result in resistance to ampicillin. To guarantee the expression of active *blaM* fusions, *H. pylori* gene fragments in the range of 3-6kb were inserted downstream of a weak but constitutive promoter (P_{iga}) in plasmid pMin2 (Figure 2A). 2400 mutated plasmids yielded 192 active *blaM* fusions. After transformation, 135 *H. pylori* mutant strains were isolated (Odenbreit *et al.*, 1996b). Partial sequencing of the transposon insertion sites revealed a random distribution of Tn*Max9* over the genome, confirming that a multitude of genes had been mutated.

All 135 *H. pylori* mutant strains were tested for the ability to bind to the gastric epithelial cell line Kato-III. Additionally, bacterial adherence was tested *in situ*, i.e., to histological sections of human gastric biopsy material. Two binding-defective mutants (P1-140a, P1-179a) were identified (Figure 2B).

Characterization of the alpAB Operon

In both adhesion-defective mutants the transposon had targeted the same gene. The deduced amino acid sequence of this gene included an N-terminal signal sequence that is specific for lipoproteins (Odenbreit *et al.*, 1999). Thus, the gene was designated as *alpA* (adherence-associated lipoprotein) (Figure 2B). An open reading frame (ORF) located 67bp downstream of *alpA* encodes a related protein (47% identity) and was designated *alpB*. Both genes are organized in an operon (Odenbreit *et al.*, 1999) (Figure 2B). Both proteins belong to the HOP outer membrane protein family (HP0912, HP0913) and are predicted to form a β-barrel structure, consisting of amphipathic, anti-parallel β-pleated sheets, with possible porin function (Tomb *et al.*, 1997). The *alpA* promoter sequence lacks the consensus sequence known as σ-factor binding sites, suggesting that an unusual σ-factor might be necessary for *alpAB* expression. Both genes are constitutively expressed *in vitro* but information on their expression *in vivo* is still lacking.

The functional role of the AlpAB proteins in adherence is still not well defined. The pre-incubation of gastric histological sections with an AlpA-specific antibody abrogates the Alp-dependent binding phenotype, suggesting that AlpA is a structural part of an adhesin complex (Odenbreit *et al.*, 1999). However, the expression of either AlpA or AlpB is insufficient for adherence, suggesting that both gene products are necessary. A direct interaction of one of the proteins with a receptor structure on the epithelial cells has not yet been demonstrated. Thus, the AlpA/B proteins might not confer adhesive properties themselves, but rather act as part of a supra molecular complex on the bacterial surface, such as an usher or platform for the presentation of different adhesins.

Therapeutic Implications

The interaction between bacterial adhesins and host tissue receptors determines in part the tissue tropism of the colonization process and is necessary for triggering the inflammatory responses and concomitant development of disease.

The high prevalence of *H. pylori* infection and the increased frequency of antibiotic resistance emphasize the need for alternative drugs. The possibility to interfere with the adhesin-receptor recognition may represent a new approach for the development of anti-microbial strategies. Detailed knowledge about the receptor-adhesin complexes and their tight interactions, in combination with the combinatorial synthesis of complex carbohydrate analogues, might allow anti-adhesive compounds such as receptor analogues to be used in therapy. Soluble receptor analogues based on the sialylated blood group antigens can provide anti-inflammatory therapy by reducing the rolling and recruitment of inflammatory cells to the endothelial cell lining (Mulligan *et al.*, 1993). Information from X-ray diffraction analyses of complexes of glycosidase-sialylated receptor structures have provided efficient anti-viral drug strategies against the influenza virus via the inactivation of the viral neuraminidase (Service *et al.*, 1997). Efficient inhibition of the binding of fresh clinical *H. pylori* strains to gastrointestinal epithelial cells by sialylated oligosaccharides has recently been demonstrated. Receptor analogues were evaluated in both primate animal models and in clinical trials with *H. pylori* infected patients and have yielded promising result, as mentioned above (Simon *et al.*, 1997; Mysore *et al.*, 1999). These promising results in these sets of experiments suggest that receptor analogues could be used to treat chronic infectious diseases such as peptic ulcer disease. In addition the carbohydrate binding domains (CBD) of adhesins have evolved to specific adherence motifs with regard to receptor specificities and affinity levels. As a consequence to a requirement for biologically conserved receptor-binding motifs, micro-organisms have acquired a potential "Achilles' heel". The carbohydrate-binding domains of adhesins that bind to sialylated glycoconjugate receptors exhibit consensus sequence motifs consisting of positively charged amino acids (Morschhauser *et al.*, 1990). Immunization with the type 1 fimbrial FimH adhesin protein of *E. coli* protects against urinary tract infection (Langermann *et al.*, 1997). The immune response to the bacterial adhesin protein provides antibodies that can activate the complement system and antibodies that bind directly to the CBD of the adhesin. Antibodies targeted to the adhesin will thereby sterically interfere with adherence. The presence of biologically conserved CBD motifs provides targets for vaccine-based strategies against *H. pylori*. The BabA protein, and in particular the Lewis b antigen binding domain, together with the AlpA/B and the HopZ adhesin proteins form a group of potential vaccine candidates against the diseases caused by adherent *H. pylori* strains.

References

Ångström, J., Teneberg, S., Abul Milh, M., *et al.* 1998. The lactosylceramide binding specificity of *Helicobacter pylori*. Glycobiology. 8: 297-309.

Alm, R.A., Ling, L.S., Moir, D.T., *et al.* 1999. Genomic-sequence comparison of two unrelated isolates of the human gastric pathogen *Helicobacter pylori*. Nature. 397: 176-180.

Appelmelk, B.J., Martin, S.L., Monteiro, M.A., *et al.* 1999. Phase variation in *Helicobacter pylori* lipopolysaccharide due to changes in the lengths of poly(C) tracts in (3-fucosyltransferase genes. Infect. Immun. 67: 5361-5366.

Appelmelk, B.J., Monteiro, M.A., Martin, S.L.,Moran, A.P., and Vandenbroucke-Grauls, C.M.J.E. 2000. Why *Helicobacter pylori* has Lewis antigens. Trends Microbiol. In Press.

Ascencio, F., Fransson, L.-Å., and Wadström, T. 1993. Affinity of the gastric pathogen *Helicobacter pylori* for the N-sulphated glycosaminoglycan heparan sulphate. J. Med. Microbiol. 38: 240-244.

Aspinall, G.O., Monteiro, M.A., Pang, H., *et al.* 1996a. Lipopolysaccharide of the *Helicobacter pylori* type strain NCTC 11637 (ATCC 43504): structure of the O antigen chain and core oligosaccharide regions. Biochemistry. 35: 2489-2497.

Aspinall, G.O., Monteiro, M.A. 1996b. Lipopolysaccharides of *Helicobacter pylori* strains P466 and MO19: structures of the O antigen and core oligosaccharide regions. Biochemistry. 35: 2498-2504.

Blaser, M.J. 1992. Hypotheses on the pathogenesis and natural history of Helicobacter-induced inflammation. Gastroenterology. 102: 720-727.

Blaser, M.J. 1993. *Helicobacter pylori:* microbiology of a 'slow' bacterial infection. Trends Microbiol. 1: 255-260.

Blaser, M.J. 1996. The bacteria behind ulcers. Sci. Amer. 274: 92-97.

Borén, T., Falk, P., Roth, K.A. *et al.* 1993. Attachment of *Helicobacter pylori* to human gastric epithelium mediated by blood group antigens. Science. 262: 1892-1895

Borén, T., Normark, S., and Falk, P. 1994. *Helicobacter pylori* ; Molecular Basis for Host Recognition and Bacterial Adherence. Trends in Microbiology. 2: 221-228.

Broome-Smith, J.K., Tadayyon, M. and Zhang, Y. 1990. β-lactamase as a probe of membrane protein assembly and protein export. Mol. Microbiol. 4: 1637-1644.

Brune, W., Menard, C., Hobom, U., Odenbreit, S., Messerle, M. and Koszinowski, U.H. 1999. Rapid identification of essential and nonessential herpesvirus genes by direct transposon mutagenesis. Nat. Biotechnol. 17: 360-364.

Campbell, B.J., Rogerson, K.A., Rhodes, J.M. 1999. Adherence of *Helicobacter pylori* to human gastric epithelium mediated by an epithelial cell-surface Lewis x carbohydrate-binding protein. Gastroenterology. 116: A131.

Chmiela, M., Lelwala-Guruge, J., and Wadström, T. 1994. Interaction of cells of *Helicobacter pylori* with human polymorphonuclear leukocytes: possible role of hemagglutinins. FEMS Immunol. Med. Microbiol. 9: 41-48.

Clausen, H., Hakomori, S., 1989. ABH and related histo-blood group antigens: immunochemical differences in carrier isotypes and their distribution. Vox. Sang. 56: 1-20.

Clyne, M., Drumm, B., 1993. Adherence of *Helicobacter pylori* to primary human gastrointestinal cells. Infect. Immun. 61: 4051-4057.

Covacci, A., Rappuoli, R. 2000. Tyrosine-phosphorylated bacterial proteins: Trojan horses for the host cell. J. Exp. Med. 191: 587-592.

Davril, M., Degroote, S., Humbert, P., Galabert, C., Dumur, V., Lafitte, J.J., Lamblin, G. and Roussel P. 1999. The sialylation of bronchial mucins secreted by patients suffering from cystic fibrosis or from chronic bronchitis is related to the severity of airway infection. Glycobiology. 9: 311-321.

Dubois, A., Tamawski, A., Newell, D.G., *et al.* 1991. Gastric injury and invasion of parietal cells by spiral bacteria in Rhesus monkeys. Are gastritis and hyperchlorhydria infectious diseases? Gastroenterology. 100: 884-891.

Edwards, N.J., Monteiro, M.A., Walsh, E.J., *et al.* 2000. The Lewis x O-antigen side chain promotes adhesion of *Helicobacter pylori* to gastric epithelium. Mol. Microbiol. 35: 1530-1539.

Eggleton, P., Murray, E., Dodds, A., *et al.* 1999. Surfactant protein D binding to *Helicobacter pylori* lipopolysaccharide. *Gut* 45, Suppl. 3: A35.

Enders, G., Brooks, W., von Jan, N., Lehn, N., Bayerdorffer, E. and Hatz, R. 1995. Expression of adhesion molecules on human granulocytes after stimulation with *Helicobacter pylori* membrane proteins: comparison with membrane proteins from other bacteria. Infect. Immun. 63: 2473-2477.

Evans, D.G., Evans Jr, D.J., Molds, J.J., and Graham, D.Y. 1988. N-acetylneuraminyllactose-binding fibrillar hemagglutinin of *Campylobacter pylori*: a putative colonization factor antigen. Infect. Immun. 56: 2896-2906.

Evans, D.G., Karjalainen, T.K., Evans, D.J., Graham, D.Y., and Lee, C.-H. 1993. Cloning, nucleotide sequence, and expression of a gene encoding an adhesion subunit protein of *Helicobacter pylori*. J. Bacteriol. 175: 674-683.

Evans, D.G., and Evans Jr, D.J. 1995. Adhesion properties of *Helicobacter pylori*. Methods Enzymol. 253: 336-360.

Evans Jr., D.J., Evans, D.G., Takemura, T., Nakano, H., Lampert, H.C., Graham, D.Y., Granger, D.N., and Kvietys, P.R. 1995. Characterization of *Helicobacter pylori* neutrophil-activating protein. Infect. Immun. 63: 2213-2220.

Exner, M.M., Doig, P., Trust, T.J., and Hancock R.E.W. 1995. Isolation and characterization of a family of porin proteins from *Helicobacter pylori*. Infect. Immun. 63: 1567-1572.

Falk, P., Roth, K.A., Borén, T. *et al*. 1993. An in vitro adherence assay reveals that *Helicobacter pylori* exhibits cell lineage-specific tropism in the human gastric epithelium. Proc. Natl. Acad. Sci. USA. 90: 2035-2039.

Falk, P., Borén, T., Haslam, D. *et al*. 1994a. Bacterial adhesion and colonization assays. Methods Cell. Biol. 45: 165-192.

Falk, P., Borén, T., and Normark, S. 1994b. Strategies for characterization of microbial host receptors. Methods Enzymol. 236: 353-374.

Filipe, M.I. 1979. Mucins in the human gastrointestinal epithelium: a review. Invest. Cell. Pathol. 2: 195-216.

Fisher, J.H., and Mason, R. 1995. Expression of pulmonary surfactant protein D in rat gastric mucosa. Am. J. Respir. Cell. Mol. Biol. 12:13-8.

Fox, J.G., Dangler, C.A., Taylor, N.S. King, A., Koh, T.J., Wang, T.C. 1999. High-salt diet induces gastric epithelial hyperplasia and parietal cell loss, and enhances *Helicobacter pylori* colonization in C57BL/6 mice. Cancer. Res. 59: 4823-8.

Frazier, B.A., Pfeifer, J.D., Russell, D.G. *et al*. 1993. Paracrystalline inclusions of a novel ferritin containing nonheme iron, produced by the human gastric pathogen *Helicobacter pylori:* evidence for a third class of ferritins. J. Bacteriol. 175: 966-972.

Gerhard, M., Lehn, N., Neumayer, N., *et al*. 1999. Clinical relevance of the *Helicobacter pylori* gene for blood-group antigen-binding adhesin. Proc. Natl. Acad. Sci. USA. 96: 12778-12783

Geyer, A., Gege, C., and Schmidt, R.R., 1999. Carbohydrate-carbohydrate recognition between Lewis x glycoconjugates. Angew. Chemie. Int. Ed. 38: 1466-1468.

Guruge, J.L., Falk, P.G., Lorenz, R.G., Dans, M., Wirth, H.P., Blaser, M.J., Berg, D.E. and Gordon, J.I. 1998. Epithelial attachment alters the outcome of *Helicobacter pylori* infection. Proc. Natl. Acad. Sci. USA 95: 3925-3930.

Hakomori, S., 1992. Le(X) and related structures as adhesion molecules. Histochem. J. 24: 771-776.

Hallstone, A.E., Perez, E.A., Borén, T., *et al*. 1994. Blood type and the risk of gastric disease. Science. 264: 1386-1388.

Hancock, R.E.W., Alm, R., Bina, J., and Trust, T. 1998. *Helicobacter pylori*: a surprisingly conserved bacterium. Nature Biotechnol. 16: 216-217.

Hessey, S.J., Spencer, J. Wyatt, J.I., *et al*. 1990. Bacterial adhesion and disease activity in Helicobacter associated chronic gastritis. Gut. 31: 134-138.

Hirmo, S., Utt, M., Ringner, M., and Wadström, T. 1995. Inhibition of heparan sulphate and other glycosaminoglycans binding to *Helicobacter pylori* by various polysulphated carbohydrates. FEMS Immunol. Med. Microbiol. 10: 301-6.

Hirmo, S., Kelm, S., Schauer, R., *et al*. 1996. Adhesion of *Helicobacter pylori* strains to α-2,3-linked sialic acids. Glycoconj. J. 13: 1005-1011.

Hirmo, S., Kelm, S., Iwersen, M., *et al*. 1998. Inhibition of *Helicobacter pylori* sialic acid-specific haemagglutination by human gastrointestinal mucins and milk glycoproteins. FEMS Immunol. Med. Microbiol. 20: 275-81.

Hirmo, S., Artursson, E., Puu, G., *et al*. 1999. *Helicobacter pylori* interactions with human gastric mucin studied with a resonant mirror biosensor. J. Microbiol. Methods. 37: 177-82.

Hofreuter, D., Odenbreit, S., Henke, G., and Haas, R. 1998. Natural competence for DNA transformation in *Helicobacter pylori*: identification and genetic characterization of the *comB* locus. Mol. Microbiol. 28: 1027-1038.

Huang, J., Keeling, P.W.N., and Smyth, C.J. 1992. Identification of erythrocyte-binding antigens in *Helicobacter pylori*. J. General. Microbiol. 138: 1503-1513.

Huesca, M., Borgia, S., Hoffman, P., and Lingwood, C.A. 1996. Acidic pH changes receptor binding specificity of *Helicobacter pylori*: a binary adhesion model in which surface heat shock (stress) proteins medfiate sulfatide recognition in gastric colonization. Infect. Immun. 64: 2643-2648.

Höök-Nikanne, J., Sistonen, P., and Kosunen, T.U. 1990. Effect of ABO blood group and secretor status on the frequency of *Helicobacter pylori* antibodies. Scand. J. Gastroenterol. 25: 815-818.

Icatlo Jr, F.C., Kuroki, M., Kobayashi, C., Yokoyama, H., Ikemori, Y., Hashi, T., Kodama, Y. 1998. Affinity purification of *Helicobacter pylori* urease. Relevance to gastric mucin adherence by urease protein. J. Biol. Chem. 273: 18130-8.

Ilver, D., Arnqvist, A., Ögren, J., *et al.* 1998. *Helicobacter pylori* adhesin binding fucosylated histo-blood group antigens revealed by retagging. Science. 279: 373-377.

Johansson, L., Johansson, P., and Miller-Podraza, H. 1999. NeuAca3Gal is part of the *Helicobacter pylori* binding epitope in polyglycosylceramides of human erythrocytes. Eur. J. Biochem. 266: 559-565.

Jones, A.C., Logan, R.P.H., Foynes, S., Cockayne, A., Wren, B.W., and Penn, C.W. 1997. A flagellar sheath protein of *Helicobacter pylori* is identical to HpaA, a putative N-acetylneuraminyllactose-binding hemagglutinin, but is not an adhesin for AGS cells. J. Bacteriol. 179: 5643-5647.

Kahrs, A.F., Odenbreit, S., Schmitt, W., Heuermann, D., Meyer, T.F. and Haas, R. 1995. An improved TnMax mini-transposon system suitable for sequencing, shuttle mutagenesis and gene fusions. Gene. 167: 53-57.

Kamisago, S., Iwamori, M., Tai, T., Mitamura, K., Yazaki, Y., Sugano, K. 1996. Role of sulfatides in adhesion of *Helicobacter pylori* to gastric cancer cells. Infect. Immun. 64: 624-8.

Karlsson, K.A. 1989. Animal glycosphingolipids as membrane attachment sites for bacteria. Annu. Rev. Biochem. 58: 309-350.

Karlsson, K.-A. 1998. Meaning and therapeutic potential of microbial recognition of host glycoconjugates. Mol. Microbiol. 29: 1-11.

Karlsson, H., Johansson, L., Miller-Podraza H., and Karlsson, K.-A. 1999. Fingerprinting of large oligosaccharides linked to ceramide by matrix-assisted laser desorption/ionization time-of-flight mass spectrometry: highly heterogeneous polyglycosylceramides of human erythrocytes with receptor activity for *Helicobacter pylori*. Glycocobiology. 9: 765-778.

Kelm, S., and Schauer, R. 1997. Sialic acids in molecular and cellular interactions. Int. Rev. Cytol. 175: 137-240.

Kirschner, E. and Blaser, M. 1995. The dynamics of *Helicobacter pylori* infection of the human stomach. J. Theor. Biol. 176: 281-290.

Kobayashi, Y., Okazaki, K., and Murakami, K. 1993. Adhesion of *Helicobacter pylori* to gastric epithelial cells in primary cultures obtained from stomachs of various animals. Infect. Immun. 61: 4058-4063.

Kojima, N., Fenderson, B.A., Stroud, M.R., *et al.* 1994. Further studies on cell adhesion based on Le(x)-Le(x) interaction, with new approaches: embryoglycan aggregation of F9 teratocarcinoma cells, and adhesion of various tumourcells based on Le(x) expression. Glycoconj. J. 1 1: 238-248.

Langermann, S., Palaszynski, S., Barnhart, M., *et al.* 1997. Prevention of mucosal *Escherichia coli* infection by FimH-adhesin-based systemic vaccination. Science. 276: 607-610.

Lelwala-Guruge, J., Ljungh, Å., and Wadström, T. 1992. Haemagglutination patterns of *Helicobacter pylori*. Frequency of sialic acid-specific and non-sialic acid-specific haemagglutinins. APMIS. 100: 908-913.

Lelwala-Guruge, J., Ascencio, F., Kreger, A.S., Ljungh, Å., and Wadström, T. 1993. Isolation of a sialic acid-specific surface haemagglutinin of *Helicobacter pylori* strain NCTC 11637. Zbl. Bakt. 280: 93-106.

Lingwood, C.A., Huesca, M., and Kuksis, A. 1992. The glycerolipid receptor for *Helicobacter pylori* (and exoenzyme S) is phosphatidylethanolamine. Infect. Immun. 60: 2470-2474.

Lingwood, C.A., Wasfy, G., Han, H., *et al.* 1993. Receptor affinity purification of a lipid-binding adhesin from *Helicobacter pylori*. Infect. Immun. 61: 2474-2478.

Loffeld, R.J., and Stobberingh, E. 1991. *Helicobacter pylori* and ABO blood groups. J. Clin. Pathol. 44: 516-517.

Madrid, J.F., Ballesta, J., Castells, M.T. *et al.* 1990. Glycoconjugate distribution in the human fundic mucosa revealed by lectin- and glycoprotein-gold cytochemistry. Histochemistry. 95: 179-187.

Miller-Podraza, H., Abul Milh, M., Bergström, J. and Karlsson, K.-A. 1996. Recognition of glycoconjugates by *Helicobacter pylori*: an apparently high affinity binding of human polyglycosylceramides, a second sialic acid-based specificity. Glycoconjugate J. 13: 453-460.

Miller-Podraza, H., Bergström, J., Abul Milh, M. and Karlsson, K.-A. 1997. Recognition of glycoconjugates by *Helicobacter pylori*. Comparison of two sialic acid-dependent specificities based on haemagglutination and binding to human erythrocyte glycoconjugates. Glycoconjugate J. 14: 467-471.

Miller-Podraza, H., Larsson, T., Nilsson, J., Teneberg, S., Matrosovich, M., and Johansson, L. 1998. Epitope dissection of receptor-active gangliosides with affinity for *Helicobacter pylori* and influenza virus. Acta. Biochim. Pol. 45: 439-449.

Miller-Podraza, H., Bergström, J., Teneberg, S., Abul Milh, M., Longard, M., Olsson, B.-M. Uggla, L. and Karlsson, K.-A. 1999. *Helicobacter pylori* and neutrophils: Sialic acid-dependent binding to various isolated glycoconjugates. Infect. Immun. 67: 6309-6313.

Monteiro, M.A., Appelmelk, B.J., Rasko, D.A., Moran, A.P., Hynes, S.O., MacLean, L.L., Chan, K.H., Michael, F.S., Logan, S.M., O'Rourke, J., Lee, A., Taylor, D.E., Perry, M.B. 2000. Lipopolysaccharide structures of *Helicobacter pylori* genomic strains 26695 and J99, mouse model *H. pylori* Sidney strain, *H. pylori* P466 carrying sialyl Lewis X, and *H. pylori* UA915 expressing Lewis B classification of *H. pylori* lipopolysaccharides into glycotype families. Eur. J. Biochem. 267: 305-20.

Morschhauser, J., Hoschutzky, H., Jann, K. *et al.* 1990. Functional analysis of the sialic acid binding adhesin SfaS of pathogenic *Escherichia coli* by site directed mutagenesis. Infect. Immun. 58: 2133-2138.

Mullen, P.J., Carr, N., Milton, J.D., and Rhodes, J.M. 1995. Immunohistochemical detection of *O*-acetylated sialomucins in intestinal metaplasia and carcinoma of the stomach. Histopathology. 27: 161-7.

Mulligan, M.S., Paulson, J.C., De-Frees, S., *et al.* 1993. Protective effects of oligosaccharides in P-selectin dependent lung injury. Nature. 364: 149-151.

Mysore, J.V., Wigginton, T., Simon, P.M., Zopf, D., Heman-Ackah, L.M., and Dubois, A. 1999. Treatment of *Helicobacter pylori* infection in rhesus monkeys using a novel antiadhesion compound. Gastroenterology 117: 1316-1325.

Namavar, F., Sparrius, M., Veerman, E.C.I., Appelmelk, B.J., Vand andenbroucke-Grauls, C.M.J.E. 1998. Neutrophil-activating protein mediates adhesion of *Helicobacter pylori* to sulfated carbohydrates on high-molecular-weight salivary mucin. Infect. Immun. 66: 444-447.

Nedenskov-Sorenson, P., Bukholm, G., and Bovre K., 1990. Natural competence for genetic transformation in *Campylobacter pylori*. J. Inf. Dis. 161: 365-366.

Odenbreit, S., Wieland, B., and Haas, R. 1996a. Cloning and genetic characterization of *Helicobacter pylori* catalase and construction of a catalase-deficient mutant strain. J. Bacteriol. 178: 6960-6967.

Odenbreit, S., Till, M. and Haas, R. 1996b. Optimized BlaM-transposon shuttle mutagenesis of *Helicobacter pylori* allows the identification of novel genetic loci involved in bacterial virulence. Mol. Microbiol. 20: 361-373.

Odenbreit, S., Till, M., Hofreuter, D., Faller, G. and Haas, R. 1999. Genetic and functional characterisation of the *alpAB* gene locus essential for adhesion of *Helicobacter pylori* to human gastric tissue. Mol. Microbiol. 31: 1537-1548.

Odenbreit, S., Puls, J., Sedlmaier,B., Gerland, E., Fischer, W., and Haas, R. 2000: Translocation of *Helicobacter pylori* CagA into gastric epithelial cells by type IV secretion. Science 287: 1497-1500.

Osaki, T., Yamaguchi, H., and Taguchi, H. 1998. Establishment and characterization of a monoclonal antibody to inhibit adhesion of *Helicobacter pylori* to gastric epithelial cells. J. Med. Microbiol. 47:505-12.

Ota, H., Nakayama, J., Momose, M., Hayama, M., Akamatsu, T., Katsuyama, T., Graham, D.Y., Genta, R.M. 1998. *Helicobacter pylori* infection produces reversible glycosylation changes to gastric mucins. Virchows. Arch. 433: 419-26.

O´Toole, P.W., Janzon, L., Doig, P., Huang, J., Kostrzynska, M., and Trust, T.J. 1995. The putative neuraminyllactose-binding hemagglutinin HpaA of *Helicobacter pylori* CCUG 17874 is a lipoprotein. J. Bacteriol. 177: 6049-6057.

Peck, B., Ortkamp, M., Diehl, K.D., Hundt, E. and Knapp, B. 1999. Conservation, localization and expression of HopZ, a protein involved in adhesion of *Helicobacter pylori*. Nucleic. Acids. Res. 27: 3325-3333.

Piotrowski, J., Slomiany, A., Murty, V.L., Fekete, Z., and Slomiany, B.L. 1991. Inhibition of *Helicobacter pylori* colonization by sulfated gastric mucin. Biochem. Int. 24: 749-56

Rautelin, H., Blomberg, B., Fredlund, H., Järnerot, G., and Danielsson, D. 1993. Incidence of *Helicobacter pylori* strains activating neutrophils in patients with peptic ulcer disease. Gut. 34: 599-603.

Ringner, M., Valkonen, K.H., and Wadström, T. 1994. Binding of vitronectin and plasminogen to *Helicobacter pylori*. FEMS Immunol. Med. Microbiol. 9: 29-34.

Roberts, J.A., Marklund, B.I., Ilver, D. *et al.* 1994. The Gal((1-4)Gal-specific tip adhesin of *Escherichia coli* P-fimbriae is needed for pyelonephritis to occur in the normal urinary tract. Proc. Natl. Acad. Sci. USA. 91: 11889-11893.

Rotter, J.I. 1983. Peptic ulcer. In: The Principles and Practice of Medical Genetics. Emery AEH, Rimoin DL, eds. New York: Churchill Livingstone. 863-878.

Saitoh, T., Natomi, H., Zhao, W., Okuzumi, K., Sugano, K., Iwamori, M., and Nagai, Y. 1991. Identification of glycolipid receptors for *Helicobacter pylori* by TLC immunostaining. FEBS Lett. 282: 385-387.

Sakamoto, S., Watanabe, T., Tokumaru, T., *et al.* 1989. Expression of Lewis a, Lewis b, Lewis x, Lewis y, sialyl-Lewis a and sialyl-Lewis x blood group antigens in human gastric carcinoma and in normal gastric tissue. Cancer. Res 49: 745-752.

Schreiber, S., Stüben, M., Josenhans, C., *et al.* 1999. *In vivo* distribution of Helicobacter felis in the gastric mucus of the mouse: Experimental method and results. Infect. Immun. 67: 5151-5156.

Segal, E.D., Falkow, S., and Tompkins, L.S. 1996. *Helicobacter pylori* attachment to gastric cells induces cytoskeletal rearrangements and tyrosine phosphorylation of host cell proteins. Proc. Natl. Acad. Sci. USA. 93: 1259-1264.

Segal, E.D., Lange, C., Covacci, A., Tompkins, L.S. and Falkow, S. 1997. Induction of host signal transduction pathways by *Helicobacter pylori*. Proc. Natl. Acad. Sci. USA. 94: 7595-7599.

Segal, E.D., Cha, J., S. Lo, Falkow, S., and Tompkins, L.S. 1999. Altered states: Involvement of phosphorylated CagA in the induction of host cellular growth changes by *Helicobacter pylori* Proc. Natl. Acad. Sci USA. 96: 14559-14564.

Seifert, H.S., Chen, E.Y., So, M. and Heffron, F. 1986 Shuttle mutagenesis: A method of transposon mutagenesis for Saccharomyces cerevisiae. Proc. Natl. Acad. Sci. USA. 83: 735-739.

Service, R.F. 1997. Researchers seek new weapon against the flu. Science. 275:756.

Simon, P.M., Goode, P.L., Mobasseri, A. *et al.* 1997. Inhibition of *Helicobacter pylori* binding to gastrointestinal epithelial cells by sialic acid-containing oligosaccharides. Infect. Immun. 65: 750-757.

Slomiany, B.L., Piotrowski, J., Samanta, A., VanHorn, K., Murty, V.L.N., and Slomiany, A. 1989. *Campylobacter pylori* colonization factor shows specificity for lactosylceramide sulfate and GM3 ganglioside. Biochem. Int. 19: 929-936.

Smoot, D.T., Resau, J.H., Naab, T., *et al.* 1993. Adherence of *Helicobacter pylori* to cultured human gastric epithelial cells. Infect Immun. 61: 350-355.

Stein, M., Rappuoli, R., and Covacci, A. 2000. Tyrosine phosphorylation of the *Helicobacter pylori* CagA antigen after *cag*-driven host cell translocation Proc. Natl.Acad. Sci. USA. 97: 1263-1268.

Strömberg, N., Nyholm, P.-G., Pascher, I. *et al.* 1991. Saccharide orientation at the cell surface affects glycolipid receptor function. Proc. Natl. Acad. Sci. USA. 88: 9340-9344.

Syder, A.J., Guruge, J.L., Li, Q., Hu, Y., Oleksiewicz, C.M., Lorenz, R.G., Karam, S.M., Falk, P.G., and Gordon, J.I. 1999. *Helicobacter pylori* attaches to NeuAcα2,3Galβ1,4 glycoconjugates produced in the stomach of transgenic mice lacking parietal cells. Mol. Cell. 3: 263-274.

Taylor, N.S., Hasubski, A.T., Fox, J.G., *et al.* 1992. Haemagglutination profiles of *Helicobacter* species that cause gastritis in man and animals. J. Med. Microbiol. 37: 299-303.

Teneberg, S., Miller-Podraza, H., Lampert, H.C., Evans Jr, D.J., Evans, D.G., Danielsson, D., and Karlsson, K.-A. 1997. Carbohydrate-binding specificity of the neutrophil-activating protein of *Helicobacter pylori*. J. Biol. Chem. 272: 19067-19071.

Tomb, J.-F., White, O., Kerlavage, A.R., *et al.*, 1997. The complete genome sequence of the gastric pathogen *Helicobacter pylori*. Nature. 388: 539-547.

Tonello, F., Dundon, W.G., Satin, B., Molinari, M., Tognon, G., Grandi, G., Del Giudice, Rappuoli, R., and Montecucco, C. 1999. The *Helicobacter pylori* neutrophil-activating protein is an iron-binding protein with dodecameric structure. Mol. Microbiol. 34: 238-246.

Trust, T.J., Doig, P., Emody, L., *et al.* 1991. High-affinity binding of the basement membrane proteins collagen type IV and laminin to the gastric pathogen *Helicobacter pylori*. Infect. Immun. 59: 4398-4404.

Tzouvelekis, L.S., Mentis, A.F., Makris, A.M., Spiliadis, C., Blackwell, C., and Weir, D.M. 1991. In vitro binding of *Helicobacter pylori* to human gastric mucin. Infect. Immun. 59: 4252-425.

Valkonen, K.H., Ringner, M., Ljungh, Å., *et al.* 1993. High-affinity binding of laminin by *Helicobacter pylori*:   evidence for a lectin-like interaction. FEMS Immunol. Med. Microbiol. 7: 29-37.

Valkonen, K.H., Wadström, T., and Moran, A.P. 1994. Interaction of lipopolysaccharides of *Helicobacter pylori* with basement membrane protein laminin. Infect. Immun. 62: 3640-3648.

Valkonen, K., Wadström, T., and Moran, A.P. 1997. Identification of the *N*-acetylneuraminyllactose-specific laminin-binding protein of *Helicobacter pylori*. Infect. Immun. 65: 916-923.

Varki, A. 1994. Selectin ligands. Proc. Natl. Acad. Sci. USA. 91: 7390-7397.

Wadström, T., Hirmo, S., and Borén, T. 1996. Biochemical aspects of *Helicobacter pylori* colonization of the human gastric mucosa. Aliment. Pharmacol. Ther. 10(Suppl I): 17-27.

Wang, Y., and Taylor, D.E., 1990. Chloramphenicol resistance in *Campylobacter coli*: nucleotide sequence, expression, and cloning vector construction. Gene 94:23-8

Weitkamp, J.H., Perez-Perez, G.I., Bode, G., *et al.* 1993. Identification and characterization of *Helicobacter pylori* phospholipase C activity. Zentralbl. Bakteriol. 280: 11-27.

Winberg, J., Mollby, R., Bergström, J., *et al.* 1995. The PapG-adhesin at the tip of P-fimbriae provides *Escherichia coli* with a competitive edge in experimental bladder infections of cynomolgus monkeys. J. Exp. Med. 182: 1695-1702.

From: *Helicobacter pylori: Molecular and Cellular Biology*
ISBN 1-898486-25-5 © 2001 Horizon Scientific Press, Wymondham, UK.

13

Helicobacter pylori Lipopolysaccharides

Anthony P. Moran[*]

Abstract

Lipopolysaccharides (LPS), a family of toxic phosphorylated glycolipids in the outer membrane of Gram-negative bacteria are composed of a lipid moiety (termed lipid A), a core oligosaccharide, and a polymeric O-specific polysaccharide chain. Compared with LPS of other bacteria, *H. pylori* LPS has, in general, low immunological activity which may aid the persistence of infection in the gastric mucosa. The molecular basis for this lower activity resides in the under-phosphorylated and unusally acylated lipid A component of LPS. The core oligosaccharide of *H. pylori* LPS contributes to the binding of the bacterium to the host glycoprotein laminin, and hence interferes with gastric cell receptor-laminin interaction in the basement membrane. The core sugars of certain *H. pylori* strains, particularly those associated with gastric ulceration, have been implicated in pepsinogen induction, which may result also in reduced mucosal integrity. Of particular interest, the O-chains of most, but not all, *H. pylori* strains mimic Lewis (Le) antigens. Although investigations have focussed on the role of these antigens in *H. pylori*-associated autoimmunity, which remains to be unequivocally established, other pathogenic consequences of Lewis mimicry are becoming apparent. Expression of Lewis antigens may be crucial for *H. pylori* colonisation and adherence and, by aiding bacterial interaction with the gastric mucosa, potentiate delivery of secreted products, and hence influence the inflammatory response.

Introduction

Helicobacter pylori has the typical trilaminar cell envelope of Gram-negative bacteria, of which the outermost layer is the outer membrane whose major component is lipopolysaccharide (LPS). These molecules, also termed endotoxins, are a family of toxic phosphorylated glycolipids (Moran, 1994). In general, LPS are the main surface antigens (O-antigens) of Gram-negative bacteria, and are essential for the physical integrity and function of the bacterial outer membrane (Rietschel *et al.*, 1990). Although LPS are bound to the cell wall, and hence the synonym endotoxins, they can be released by multiplying or disintegrating bacteria to interact with the host.

The LPS of various bacterial species share a common architecture of three principal domains (Figure 1), which have different structural and functional properties. The outer polysaccharide moiety, called the O-specific polysaccharide chain is a polymer of repeating

[*]email: anthony.moran@nuigalway.ie

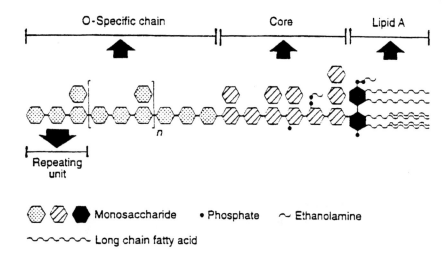

Figure 1. General structure of LPS showing the three major domains of the molecule; O-specifc chain, core oligosaccharide, and lipid A.

units that may contain up to seven different sugars (Rietschel *et al.*, 1990). The second region is the core oligosaccharide composed of a short series of about 10 to 15 sugars (Rietschel *et al.*, 1990; Holst *et al.*, 1996). The innermost component of LPS, which anchors the molecule in the outer membrane, is a lipid moiety termed lipid A (Moran, 1994; Holst *et al.*, 1996). In a given bacterial species, the structure of lipid A is highly conserved, that of the core oligosaccharide is more variable, whereas that of the O-specific chain can be highly variable.

Macromolecular Structure of *H. pylori* LPS

Fresh clinical isolates of *H. pylori* produce high-molecular-weight, smooth-form LPS (S-LPS) but, unlike enterobacterial LPS, produce O-chains of relatively constant chain length (Moran *et al.*, 1992; Moran 1995). Despite the relative homogeneity of O-chain length, interstrain variations in the electrophoretic banding pattern, mobility and staining properties of *H. pylori* LPS have been observed and indicate potential structural differences in the LPS (Moran *et al.*, 1992). Antigenic and structural differences in the LPS of different strains have been detected by immunoblotting, passive hemagglutination, lectin binding, and enzyme-linked immunosorbent assays and indicate sufficient diversity for use as the basis of a strain typing scheme (Mills *et al.*, 1992; Simoons-Smit *et al.*, 1996; Hynes *et al.*, 1999a).

In contrast to fresh clinical isolates of *H. pylori*, culture collection strains that have been subcultured many times on solid media *in vitro* produce low-molecular-weight, rough-form LPS (R-LPS) without O-polysaccharide chains (Moran *et al.*, 1992). This suggests that changes may be induced in LPS expression by growth conditions which would have major implications for biological studies on LPS *in vitro*. Supporting this, it has been demonstrated that strains may undergo a phenotypic change from S- to R-LPS when grown on conventional solid media (Moran *et al.*, 1992; Moran and Walsh, 1993b). Therefore, care should be taken with bacterial growth conditions when performing LPS studies in order not to induce production of an aberrant R-LPS molecular form. This phase shift can be reversed, and the expression of S-LPS stabilized *in vitro*, when strains are grown in liquid media (Moran and Walsh, 1993a; Walsh and Moran, 1997).

Table 1. Immunological and endotoxic assays in which *H. pylori* LPS exhibits significantly lower activities than enterobacterial LPS

Property	References
Pyrogenicity, mitogenicity, lethality	Muotiala *et al.* (1992)
LAL assay	Pece *et al.* (1995)
IL-1, IL-6 and TNF-α induction	Birkholz *et al.* (1993), Pece *et al.* (1995)
IL-8 from neutrophils	Crabtree *et al.* (1994)
IL-8 from epithelial cell lines	Crabtree *et al.* (1995)
Priming neutrophils to release toxic oxygen radicals	Nielsen *et al.* (1994)
E-selectin expression on endothelial cells	Darveau *et al.* (1995)
Induction of prostaglandin E_2 and nitric oxide	Perez-Perez *et al.* (1995)
Abolition of suppressor T-cell activity	Baker *et al.* (1994)
NK cell activity	Tarkkanen *et al.* (1993)

Immunological Activity of *H. pylori* LPS and the Host-Parasite Relationship

In general, LPS of Gram-negative bacteria strongly activate B-lymphocytes, granulocytes and mononuclear cells and are, therefore, potent immunostimulators (Rietschel *et al.*, 1990; Moran, 1994; Holst *et al.*, 1996). Also, LPS molecules are endowed with a broad spectrum of endotoxic properties, such as pyrogenicity and lethal toxicity, which contribute to the pathogenic potential of Gram-negative bacteria (Rietschel *et al.*, 1990; Moran, 1994). It is well established that lipid A is responsible for the immunological and endotoxic properties of LPS, although the degree of bioactivity may be modulated by the saccharide component (Rietschel *et al.*, 1990, 1991; Holst *et al.*, 1996).

Immunological Properties of *H. pylori* LPS

H. pylori LPS have much lower immunobiological activities than enterobacterial LPS (for extensive reviews, see Moran, 1995, 1996a; Moran and Aspinall, 1998). For example, the pyrogenicity and mitogenicity of *H. pylori* LPS is 1000-fold lower, and lethal toxicity in mice, 500-fold lower, compared with *Salmonella typhimurium* LPS (Muotiala *et al.*, 1992). The induction of cytokines such as interleukin- (IL-) 1, IL-6 and tumour necrosis factor alpha (TNF-α) from activated human mononuclear cells by *H. pylori* LPS is significantly lower than that by *Escherichia coli* LPS (Birkholz *et al.*, 1993; Pece *et al.*, 1995). In particular, induction of IL-8 from neutrophils is 1000-fold lower when induced by *H. pylori* LPS than by enterobacterial LPS (Crabtree *et al.*, 1994) and, likewise, *H. pylori* stimulates only low-level secretion of IL-8 from epithelial cell lines (Crabtree *et al.*, 1995). Consistently, compared with enterobacterial LPS, *H. pylori* LPS exhibits significantly lower activities in a number of endotoxic and immunological assays that have been used (Table 1).

Molecular Basis for Low Immunological Activities

We have hypothesised that *H. pylori* LPS, and its lipid A component in particular, have evolved their present structure as a consequence of adaptation of the bacterium to its ecological niche in the gastric mucosa (Moran, 1995, 1996a). LPS, which is an essential component of the bacterial outer membrane, by inducing a low immunological response, may prolong *H. pylori* infection for longer than that by more aggressive and short-lived pathogens. This hypothesis is supported by data from studies using a range of immunological, biochemical and analytical chemical approaches.

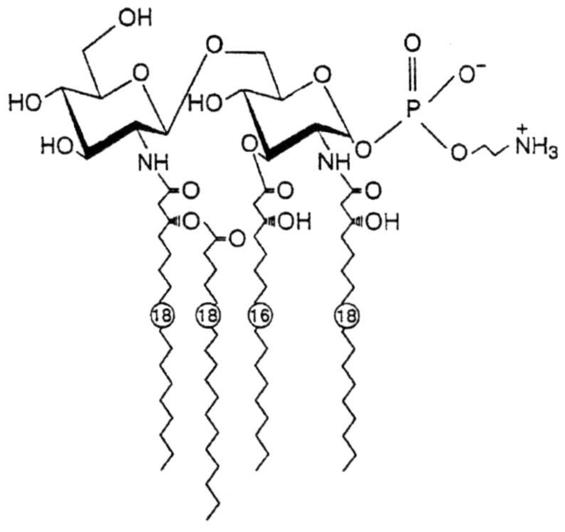

Figure 2. Proposed chemical structure of the predominant lipid A molecular species found in *H. pylori* LPS. Numbers in circles refer to the number of carbon atoms in acyl chains. Note the presence of only four fatty acids attached to the D-glucosamine disaccharide backbone and absence of phosphate from the non-reducing end of the backbone.

Chemical Modification and Fractionation of H. pylori LPS

Modification and isolation of *H. pylori* LPS components with subsequent testing in immunological assays has given insights into the molecular basis for the observed lower immunological activities. Using this approach, the phosphorylation pattern in lipid A has been shown to influence induction of TNF-α, and the core oligosaccharide modulated this effect (Pece *et al.*, 1995). A similar phenomenon was observed with induction of other mediators released from mononuclear leucocytes (Semeraro *et al.*, 1996) and dephosphorylation of *H. pylori* LPS influences activity in the *Limulus* amoebocyte lysate (LAL) assay (Pece *et al.*, 1995). In contrast, dephosphorylation does not alter the priming activity of *H. pylori* LPS on neutrophils to release toxic oxygen radicals (TOR), which suggests the lesser importance of phosphorylation compared with acylation pattern in TOR priming (Nielsen *et al.*, 1994). Moreover, the lack of abolition of suppressor T-cell activity has been attributed to the presence of long chain fatty acids in *H. pylori* lipid A (Baker *et al.*, 1994). Therefore, depending on the particular immunological activity examined, the phosphorylation or acylation pattern assumes importance.

Structural Analysis of H. pylori Lipid A

The structure of *H. pylori* lipid A has been investigated in two studies (Suda *et al.*, 1997; Moran *et al.*, 1997). Both studies found under-phosphorylation and under-acylation of the lipid A compared with enterobacterial lipid A (Rietschel *et al.*, 1990, 1991; Holst *et al.*, 1996; Moran, 1994), in agreement with preliminary reports (Moran, 1995; Moran *et al.*, 1996b). However, one study reported a lipid A with only three fatty acid substituents (Suda *et al.*, 1997), whereas when milder isolation conditions were used a lipid A backbone carrying four fatty acids was observed (Moran *et al.*, 1997). Thus, the latter represents the more intact form of the molecule and is shown in Figure 2. A structural comparison of *H. pylori* lipid A with that of *E. coli* has been performed (Moran, 1998) and, collectively, the data allow the following conclusions: (i) There is one less phosphate group (4'-phosphate is absent) on the backbone of *H. pylori* lipid A in comparison to *E. coli* lipid A. (ii) Only four fatty acids are present in *H. pylori* lipid A (two fatty acids are absent at postion 3') compared to six in *E.coli*. (iii) The average length of the fatty acid chains in *H. pylori* lipid A is longer (16 to 18 carbons) than in *E. coli* (12 to 14 carbons). Thus, there are major differences between the structures of these two lipid A molecules which, based on established, structure-bioactivity relationships of lipid A molecules (Rietschel *et al.*, 1990, 1991; Moran, 1994; Holst *et al.*, 1996), are likely to translate into differences in immunological activities (Moran, 1998).

Binding of H. pylori LPS to LPS-Binding Protein

In serum, LPS-binding protein (LBP) acts as a catalytic protein to present LPS to the monocyte/macrophage cell surface, where the newly formed LPS-LBP complex interacts with the CD14 surface receptor which, independently or in association with a second receptor, lead to the induction of proinflammatory cytokines (Moran, 1994). The comparative binding of *H. pylori* LPS and *E. coli* LPS to LBP has been examined, and the former shown to bind more poorly with slower binding kinetics to LBP, but also with poorer binding to CD14 (Cunningham *et al.*, 1996). Collectively, as LBP binds LPS through its lipid A component, these findings are consistent with a proportionately lower ability of *H. pylori* LPS to activate monocytes and reflect the unusual phosphorylation and fatty acid substitution in lipid A (Moran *et al.*, 1997; Moran, 1998).

Comparison with Commensal LPS

The long-term human commensal bacterium *Bacteroides fragilis*, a member of the normal gut microbiota, produces a lipid A bearing some structural similarity to *H. pylori* lipid A, particularly under-phosphorylation and an unusual acylation pattern, which is accompanied by low immunological activities (Weintraub *et al.*, 1989). Since the primary role of LPS is to provide a functional macromolecular matrix in the outer membrane through which a bacterium can interact with its environment (Muotiala *et al.*, 1992; Moran, 1994), the LPS of commensal bacteria apparently have evolved to retain this role but also to reduce the immunological response, hence aiding persistence of the bacterium. In an analogous manner, the low immunological activity of *H. pylori* LPS, particularly its lipid A, may aid persistence and development of chronic infection.

Long-term Consequences of *H. pylori* LPS-induced Immunoactivity

Despite the low immunological activity of *H. pylori* LPS, *H. pylori* colonisation of the human antral mucosa is associated with inflammation which can be induced by other *H. pylori*-derived molecules (Mai *et al.*, 1991; Moran, 1998). Nevertheless, as a consequence of enzymatic degradation of LPS by phagocytes, some LPS and/or lipid A partially modified structures can be excreted by exocytosis. These compounds, retaining some immunological activity, could play a role as subliminal, low grade, persistent stimuli involved in *H. pylori* pathogenesis during long-term chronic infection (Pece *et al.*, 1995) contributing to gastric damage, and potentially to extragastric sequelae. Consistent with this, *H. pylori* LPS has been shown to induce nitric oxide synthase in an *in vivo* animal model and hence to contribute to gastric damage (Lamarque *et al.*, 1997; Szepes *et al.*, 1997). Furthermore, we have studied the ability of *H. pylori* LPS to induce the production of procoagulant activity (PCA), identified as tissue factor, and to induce plasminogen activator inhibitor type 2 (PAI-2) by human mononuclear leucocytes (Semeraro *et al.*, 1996). Although both effects were potentiated by a CD14-mediated mechanism, the potency of *H. pylori* LPS was about 1000-fold lower than that of *S. typhimurium* LPS, and was influenced by the phosphorylation pattern in lipid A and modulated by the core oligosaccharide. Nevertheless, the structural prerequisites for

A

B

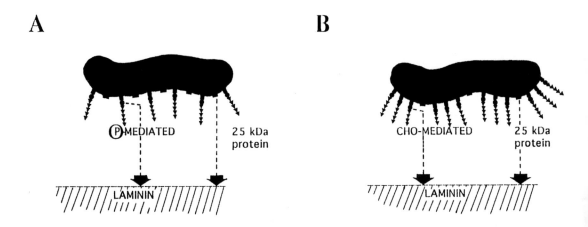

Figure 3. Representation of the dual mechanisms of interaction of (A) haemagglutinating and (B) poorly haemagglutinating *H. pylori* strains with laminin. Binding of the haemagglutinating strain is mediated by a phosphorylated structure (P) in the core oligosaccharide of LPS and a lectin-like 25 kDa protein, whereas binding of the poorly haemagglutinating strain is mediated by a conserved carbohydrate structure (CHO) in the core oligosaccharide of LPS and the same 25 kDa protein.

low-grade expression of tissue factor (leading to fibrin formation) and PAI-2 (leading to persistence of the formed fibrin) are present within the *H. pylori* molecule and may contribute to the inflammatory response during chronic infection with *H. pylori*. For more detailed discussion of this topic, see reviews elsewhere (Moran and Aspinall, 1998; Moran, 1999).

LPS Lectin-like Interactions and Laminin Binding

Laminin is a glycoprotein component of the extracellular matrix, and plays a role in the structure of the basement membrane. Bacterial interactions with extracellular and basement membrane proteins play an important role in the pathogenesis and virulence of a number of infections (Ljungh *et al.*, 1996). *H. pylori* strains have been shown to exhibit high affinity binding to laminin (Trust *et al.*, 1991; Moran *et al.*, 1993; Valkonen *et al.* 1993), but this interaction is unlikely to be involved in the initial colonisation of gastric mucosal cells, since putative primary adhesins recognise receptors in the mucus layer and on the epithelial cell surface (Moran, 1995, 1996b; Moran and Wadström 1994, 1998). Nonetheless, *H. pylori* is observed associating with intercellular junctions, and laminin binding may explain its association with this microenvironment (Moran, 1996a).

Preliminary studies indicated the involvement of a lectin-like interaction (Valkonen *et al.*, 1993) and LPS (Moran *et al.*, 1993) in laminin binding by *H. pylori*. We characterised the LPS-mediated interaction of *H. pylori* with laminin and found two mechanisms operating (Valkonen *et al.*, 1994). Laminin binding by haemagglutinating strains differs from that of non-haemagglutinating strains; a phosphorylated structure in the core oligosaccharide of LPS mediates the interaction of a haemagglutinating strain, whereas a conserved non-phosphorylated structure in the core oligosaccharide mediates the interaction of a poorly haemagglutinating strain (Figure 3). Although both these interactions are mediated by the core of LPS, these structures can be exposed on the bacterial surface as a serological response against the core occurs in gastritis patients, but is more developed in duodenal ulcer patients (Pece *et al.*,

1997). The lectin-like adhesin that recognizes the sugar chain of laminin is sialic acid-specific, recognizes α(2,3)-sialyllactose, is conserved in haemagglutinating and poorly hemagglutinating strains, and has been identified as a 25 kDa protein (Valkonen *et al.*, 1997) (Figure 3). Also, studies of *H. pylori*-infected individuals have shown a serological response against this protein, indicating that the adhesin is produced *in vivo* (A.P. Moran *et al.*, unpublished). However, the binding domain of LPS on the laminin molecule has not, to date, been identified. Nevertheless, the binding of *H. pylori* to laminin illustrates a dual recognition system, whereby sugars in the core of LPS interact with a region in laminin, probably a peptide region, and a protein on *H. pylori* (with lectin-like properties) recognises a stretch of sugars in laminin. It has been proposed that the initial binding of laminin by *H. pylori* is mediated by LPS and that this is followed by the specific lectin binding (Ljungh *et al.*, 1996; Valkonen *et al.*, 1997).

Once adherent to gastric epithelium, laminin binding by *H. pylori* may have pathological consequences (Moran, 1996a). A 67 kDa protein receptor (an integrin) for laminin on gastric epithelial cells has been isolated, and *H. pylori* LPS has been shown to interfere with the specific interaction between the integrin and laminin *in vitro* (Slomiany *et al.*, 1991). Therefore, the binding of *H. pylori* to laminin mediated by LPS may disrupt epithelial cell-basement membrane interactions contributing to the disruption of gastric mucosal integrity and the development of gastric leakiness associated with the bacterium. Significant penetration between cells by *H. pylori*, and a weakening of adhesion of the tight junctions between cells, has been observed by microscopy and could be the origin of gastric leakiness. Although certain cytoprotective anti-ulcer drugs have been reported to counteract the anti-adhesive effect of LPS on intergrin action (Slomiany *et al.*, 1991), further studies are required to verify this hypothesis, since other soluble factors of *H. pylori* have been identified contributing to gastric leakiness (Terrés *et al.*, 1998).

Pepsinogen Stimulation by *H. pylori* LPS

Elevated concentrations of serum pepsinogen I, the endocrine component of pepsinogen secreted by chief cells, are considered a major risk factor for both the development and recurrence of duodenal ulcers, and decreased levels are associated with *H. pylori* eradication. Using isolated rabbit gastric glands, stimulation of pepsinogen has been induced by cell sonicates of *H. pylori* (Cave and Cave, 1991). Another study, using gastric mucosa from guinea pigs in Ussing chambers, reported a 50-fold stimulation of pepsinogen secretion with *H. pylori* LPS compared with only a 12-fold increase with *E. coli* LPS, but no stimulation of acid secretion (Young *et al.*, 1992). Microscopic examination confirmed the structural integrity of chief cells, indicating the absence of a non-specific toxic effect, but showed degranulation of zymogen granules in *H. pylori* LPS-treated tissue suggesting a specific mechanism. These observations have been confirmed in separate studies (Young *et al.*, 1996; Moran and Aspinall, 1998). However, the pepsinogen stimulatory effect observed in rabbit gastric glands was comparatively lower (Cave and Cave, 1991), but this may reflect the stimulatory effect of a lower concentration of LPS in the tested cell sonicates (Moran, 1996a). Comparison of the stimulatory effect of LPS derived from two duodenal ulcer patients and two asymptomatic individuals showed that the former preparations induced 50-fold stimulation, whereas the latter induced levels like those of controls (Moran *et al.*, 1998). Hence, strain differences in pepsinogen induction can occur. Moreover, experiments using fractionated LPS, dephosphorylated LPS, and polymyxin B inhibition studies have indicated the involvement of structures in the core of LPS from *H. pylori* duodenal ulcer strains (Young *et al.*, 1996; Moran *et al.*, 1998).

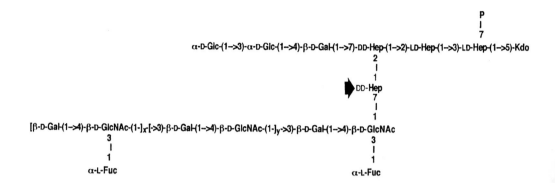

Figure 4. Proposed structure of the polysaccharide component of LPS of *H. pylori* NCTC 11637. The arrow indicates the point at which further substitution by heptan and glucan chains occurs in some other *H. pylori* strains. Abbreviations for sugars: Gal, galactose; Glc, glucose; GlcNAc, *N*-acetylglucosamine; Fuc, fucose; DD-Hep, D-*glycero*-D-*manno*-heptose; LD-Hep, L-*glycero*-D-*manno*-heptose; Kdo, 3-deoxy-D-*manno*-2-octulosonic acid.

Compared with other bacterial species, the core regions of *H. pylori* strains exhibit an unusual conformation as determined by chemical structural studies. As exemplified by the core oligosaccharide of *H. pylori* NCTC 11637 (Figure 4), a branching occurs from the D-*glycero*-D-*manno*-heptose (DD-Hep) residue in the inner core through a second such residue to which the first repeating unit of the O-polysaccharide chain is attached (Aspinall *et al.*, 1996). However, strain differences in substitution of the intervening DD-Hep between the core oligosaccharide and O-chain occur, whereby heptan and glycan chains may be present adding to structural complexity (Aspinall and Moran, 1997). Moreover, preliminary serological analysis has shown the presence of epitopes in the core that are common with other bacterial species, whereas others are specific to *H. pylori* (Walsh and Moran, 1996). Structural investigations of the LPS cores of pepsinogen-inducing and non-inducing strains have identified differences in substitution in the outer core-O-chain domain (Aspinall *et al.*, 1999). Thus, potential structures are present which explain the activation of pepsinogen by *H. pylori* LPS *in vitro*, but these require further examinination in an *in vivo* animal model of pepsinogen induction.

Expression of Lewis Antigens in *H. pylori* LPS

The Lewis blood group antigen system is composed of type 1 antigens, Lewis a (Le[a]) and Lewis b (Le[b]) and type 2 antigens, Lewis x (Le[x]) and Lewis y (Le[y]), and is biochemically related to the ABO blood groups (Green, 1989). In 1994, the expression of Le[x] in the O-polysaccharide chain of the *H. pylori* type strain NCTC 11637 was first reported (Aspinall *et al.*, 1994).

Lewis Antigens in *H. pylori* LPS

Several structural studies on LPS of certain *H. pylori* strains have shown mimicry of Le[x] and/ or Lewis y (Le[y]) blood group determinants (Aspinall *et al.*, 1996, 1997; Aspinall and Monteiro, 1996; Monteiro *et al.*, 1998a, 1998b; Knirel *et al.*, 1999). These are formed by mono- or di-fucosylated *N*-acetyl-β-lactosamine units attached to the LPS core (see Figure 4). Typically, the O-chains have a poly-(N-acetyl-β-lactosamine) chain decorated with multiple lateral α-L-fucose residues (Aspinall *et al.*, 1994, 1996, 1997; Aspinall and Monteiro, 1996; Knirel *et al.*, 1999) or, in some strains, with additional glucose or galactose residues (Monteiro *et al.*, 1998b; Aspinall *et al.*, 1999). Furthermore, Lewis[a] (Le[a]), Lewis[b] (Le[b]), and H type I antigenic determinants (Monteiro *et al.*, 1998a), and sialyl-Le[x] (Monteiro *et al.*, 2000) have been found in the O-chains of other *H. pylori* strains. Based on serological screening, 80-90% of *H. pylori* strains express Le[x] and/or Le[y] antigens (Simoons-Smit *et al.*, 1996; Wirth *et al.*, 1996; Marshall *et al.*, 1999; Heneghan *et al.*, 2000), but such analysis may under-estimate Le[x] and Le[y] expression since strains non-typable with anti-Lewis antibodies have been shown by chemical analysis to express these antigens (Knirel *et al.*, 1999). Nevertheless, detailed chemical structural analysis has shown that some strains possesss O-chains composed of 3-*C*-methyl-D-mannose, a novel sugar not found previously in nature (Moran *et al.*, 1999; Kocharova *et al.*, 2000), thereby confirming that not all *H. pylori* strains express Lewis antigen mimicry.

Lewis Antigen Mimicry in *H.pylori* Pathogenesis

Although Le[x], Le[y], and related blood group antigens are present in the human gastric mucosa (Sakamoto *et al.*, 1989), the pathogenic implications of Lewis antigen mimicry by *H. pylori* remain unclear. Various implications and consequences of Lewis antigen mimicry have been hypothesised and are being systematically investigated.

Lewis Antigens and Camouflage

It has been speculated that expression of Le[x] and Le[y] antigens may camouflage the bacterium in the gastric mucosa upon initial infection (Moran 1996a, 1996b; Moran *et al.*, 1996c). Surface and foveolar epithelia co-express either Le[a] and Le[x] in Le(a+,b-) individuals (non-secretors), or Le[b] and Le[y] in Le(a-,b+) individuals (secretors), whereas glandular epithelium lacks type 1 antigens (Le[a] and Le[b]) but expresses Le[x] and Le[y] (type 2 antigens) irrespective of the secretor phenotype. Extending this hypothesis, and because Le[a] and Le[b] which are expressed in the foveolar epithelium of the gastric mucosa are isoforms of Le[x] and Le[y], Wirth *et al.* (1997) examined the relationship between bacterial and host gastric Lewis expression, as determined by erythrocyte Lewis (a,b) phenotype, and found that the relative proportion of bacterial expression of Le[x] and Le[y] corresponded to the host Le(a+,b-) and Le(a-,b+) blood group phenotypes, respectively. In contrast, similar studies in Irish and Canadian patient populations did not find this correlation (Heneghan *et al.*, 1998, 2000; Taylor *et al.*, 1998). The discrepancies between these results may reflect the study populations. Of considerable importance is that 26% of the study population of Wirth *et al.* (1997) were of the recessive Lewis phenotype, Le(a-,b-), which is higher than that usually observed for a Caucasian or European white population, and reflects the heterogenous origin of their patients. Distribution of this recessive phenotype group of patients to their true secretor/non-secretor phenotype by salivary testing, would influence the outcome of the obtained results. Importantly, in the other studies, no patients of a recessive phenotype were observed (Heneghan *et al.*, 1998).

Lewis Antigens and Autoimmunity

In prolonged chronic infection by *H. pylori*, expression of Le[x] and Le[y] antigens has been implicated in the pathogenesis of atrophic gastritis by the induction of autoreactive antibodies (Appelmelk *et al.*, 1996, 1997; Moran *et al.*, 1996a, 1996c), but this pathogenic mechanism has not been unequivocally established (Appelmelk *et al.*, 1998) (see Chapter 4). The presence of autoantibodies in *H. pylori*-infected individuals has been correlated with the degree of gastric infiltration, with the numbers of inflammatory cells, and with glandular atrophy (Negrini *et al.*, 1996; Faller *et al.*, 1997). Moreover, *H. pylori* isolates from patients with severe atrophy expressed Le^x and Le^y antigens, and upon immunisation of mice yielded a strong gastric autoantibody response (Negrini *et al.*, 1996). In contrast, *H. pylori* isolates from individuals with near-normal mucosa often lacked Lewis antigen expression, and were less able to induce autoantibodies in mice. Furthermore, growth in mice of a *H. pylori*-induced, anti-Le^y-secreting hybridoma resulted in gastric histopathological changes consistent with those of gastritis (Negrini *et al.*, 1991; Appelmelk *et al.*, 1996). The oligosaccharide of the β-chain of the H$^+$,K$^+$-ATPase gastric proton pump of parietal cell canaliculi, which is similar in chemical structure to Lewis antigens, has been implicated in pathogenic autoimmune responses, and anti-Le^y *H. pylori*-induced autoreactive antibodies can react with the β-chain (Appelmelk *et al.*, 1996). In addition, anti-Lewis and related antigen–parietal cell antibodies were induced in a transgenic mouse model of *H. pylori* infection in which gastric pathology developed (Guruge *et al.*, 1998). Collectively, these data have been taken to indicate that Lewis antigen mimicry plays a role in the induction of autoimmunity in *H. pylori*-associated disease. Nevertheless, this interpretation remains controversial.

Experiments attempting to abolish the binding of anti-canalicular serum autoantibodies by absorption with *H. pylori* have been contradictory. Using Lewis-positive lysates of *H. pylori*, a decrease was seen in one such series (Negrini *et al.*, 1991), but no significant reduction was observed in another (Faller *et al.*, 1998), and anti-H$^+$,K$^+$-ATPase serum autoantibodies were not absorbed with *H. pylori* in another study (Ma *et al.*, 1994). Moreover, sera reacted with recombinant H$^+$,K$^+$-ATPase expressed in *Xenopus* oocytes which was deduced not to express Lewis antigens (Claeys *et al.*, 1998). Because of these findings, it has been speculated that *H. pylori*-associated autoimmunity parallels the classical model for induction of organ-specific autoimmunity, with a central role for increased autoantigen presentation resulting in loss of tolerance, and anti-Lewis antibodies only reflecting gastric damage (Appelmelk *et al.*, 1998; Faller *et al.*, 1998). However, the outer membrane of *H. pylori* undergoes blebbing, producing vesicles. The membranes of these blebs contain LPS, and the associated Lewis antigens, which can be demonstrated by immuno-electronmicroscopy (Moran, 1999), and also other virulence factors (Fiocca *et al.*, 1999). Thus, the Lewis antigens of *H. pylori* can be presented to the immune system in this format, rather than on whole bacterial cells, and hence the use of whole cells or lysates of *H. pylori* may not be optimal to remove *H. pylori*-induced anti-Lewis antibodies. In addition, it is unclear as to what type of glycosylation occurs in the *Xenopus* oocytes. Therefore, further studies are required to resolve these outstanding issues.

Lewis Antigens and the Inflammatory Response

As reviewed elsewhere (Moran *et al.*, 1996a, 1996c; Moran, 1998), it can be proposed that recognition and cross-linking of CD15 by *H.pylori*-induced anti-Le^x autoantibodies may potentially affect polymorph adhesiveness to the endothelium, hence influencing neutrophil infiltration, and furthermore, expression of Le^x on the bacterial surface may influence the T-cell response, shifting immunity from cell-mediated to antibody-mediated which is ineffective in combating *H. pylori* chronic infection. Nevertheless, these putative mechanisms of pathogenesis have not received extensive attention and require further investigation. In patient studies, although it has become apparent that bacterial colonisation density and the ensuing inflammatory response can be influenced by host expression of ABO and Le^a blood group determinants (Heneghan *et al.*, 1998), bacterial Le^x expression is associated with peptic ulcer disease (Marshall *et al.*, 1999), and is statistically related to neutrophil infiltration (Heneghan *et al.*, 2000).

On the other hand, it has been suggested that expression of host Lewis antigens by the bacterium could aid the persistence of *cagA*-positive *H. pylori* proinflammatory strains by counterbalancing their proinflammatory effects. Consistent with this, Wirth *et al.* (1996) showed that *H. pylori* isolates positive for both Le^x and Le^y were predominantly *cagA*-positive and that a *cagA*-ablated strain had diminished expression of Le^x. Although small numbers of isolates from different countries were examined in the study, predominantly isolates were from North America. Our investigations of *H. pylori* isolates from a geographically distinct Irish population has indicated the lack of an association between *cagA* status and Lewis antigen expression (Marshall *et al.*, 1999). This contrasting lack of an association could be attributable to the adaptation of *H. pylori* strains with differing attributes to differing human populations (Marshall *et al.*, 1999), and remains a question for further investigation.

Lewis Antigen Mimicry and Bacterial Adherence

Based on knowledge from cell biology where homologous and heterologous Lewis antigen interactions have been documented, it was hypothesised that Lewis antigen mimicry in LPS could play a role in bacterial adhesion (Moran, 1996b). Consistent with this, loss of O-polysaccharide chain and Lewis antigen expression by *H. pylori* results in lack of ability to establish infection in a variety of mouse colonisation models (Hynes *et al.*, 1999b). Construction of an isogenic *H. pylori galE* mutant expressing low-molecular-weight LPS without an O-chain (Kwon *et al.*, 1998), resulted in loss of ability to colonise compared with the Le^x-expressing parental strain (Moran,1999; Moran *et al.*, 2000). Also, a *H. pylori galE* mutant and another mutant in which *rfbM* had undergone insertional mutagenesis, yielding an LPS with an O-chain but without fucosylation (and hence without Le^x expression), were constructed and were shown in bacterial adherence studies to gastric antral mucosa *in situ* not to exhibit tropic binding to the mucosa compared with the parental strain (Edwards *et al.*, 1999). Thus, Lewis antigens may be crucial for bacterial colonisation, and by aiding bacterial interaction with the gastric mucosa through adherence phenomena, aid delivery of secreted products (Moran, 1996b; Rieder *et al.*, 1997) and, thereby, influence the inflammatory response.

Other Putative Mechanisms of LPS-induced Gastric Damage

H. pylori LPS has been reported not to inhibit mucin synthesis in segments of rat stomach *in vitro* (Slomiany *et al.*, 1992), consistent with observations on the effect of *H. pylori* on mucin synthesis in human antral biopsy specimens and in a mucus-secreting human cell line (Micots *et al.*, 1993; Moran, 1996a), but did induce a profound inhibitory effect on the process of mucus glycosylation and sulphation (Slomiany *et al.*, 1992). Furthermore, prolonged exposure to the LPS caused a significant change in the macromolecular nature of the mucin, and induced a shift from a high- to a low-molecular-weight form. A mucin receptor, a protein of 97 kDa, from gastric epithelial cells has been isolated and *H. pylori* LPS reported to inhibit mucin binding *in vitro* (Piotrowski *et al.*, 1993). Both these phenomena could interfere with the mucus perimeter *in vivo* and require further investigation. Moreover, *H. pylori* LPS has been reported to inhibit acid secretion in pylorus-ligated conscious rats (Ootsuba *et al.*, 1997), and the LPS has been shown to influence enterochromaffin-like cell secretion and proliferation which may contribute to the abnormalities in gastric acid secretion associated with *H. pylori* (Kidd *et al.*, 1997). Also, induction of apoptosis in gastric epithelia of rats induced by *H. pylori* has been reported contributing to the gastric mucosal injury (Piotrowski *et al.*,

1997). Despite the potential importance of the above pathogenic mechanisms, these observations await independent verification, and the putative molecular structures within *H. pylori* LPS need to be identified.

LPS Biosynthesis Genes and Biological Implications

The first LPS biosynthesis gene to be identified and cloned from *H. pylori* was a *rfaC* homologue encoding heptosyl transferase I involved in core oligosaccharide synthesis (Goldberg *et al.* 1996). Attempts to produce knockout mutants of this gene in *H. pylori* have, to date, proven unsuccessful reflecting technical difficulties and the possibility that such mutations are lethal, underlining the importance of the unusual conformation found in the *H. pylori* core oligosaccharide. Subsequently, due to the potential biological importance of Lewis antigen mimicry in *H. pylori* O-chains, attention has focussed on cloning and seqencing first the $\alpha(1,3)$-fucosyltransferase gene involved in Le^x synthesis (Martin *et al.*, 1997; Ge *et al.*, 1997), then the $\alpha(1,2)$-fucosyltransferase gene involved in Le^y synthesis (Wang *et al.*, 1999) as well as the $\beta(1,4)$-galactosyltransferase gene required for synthesis of *N*-acetyllactosamine in the backbone of Lewis antigens (Logan *et al.*, 2000). Despite this interest, mutants in these genes have not been tested extensively in animal models or other biological systems. However, mutants in two other LPS genes affecting O-chain synthesis (*galE* and *rfbM*) have been generated and tested biologically as described above.

The availability of two complete genome sequences of *H. pylori* strains (26695 and J99), and their comparative analysis (Tomb *et al.*, 1997; Alm *et al.*, 1999), has given further insights into putative LPS biosynthesis genes in *H. pylori*. At least 27 genes likely to be involved in LPS biosynthesis have been found in *H. pylori* (Tomb *et al.*, 1997), but unlike other bacteria these genes are scattered throughout the genome, not clustered at one locus. This may reflect the synthetic mechanism of *H. pylori* LPS, particularly of the O-chain, by sequential addition of monosaccharides to the growing chain rather than the polymerisation of repeating units (Aspinall and Moran, 1997; Berg *et al.*, 1997). *H. pylori* has homologues to all enzymes required for lipid A synthesis, including *lpxA*, *lpxB*, *lpxD* and *envA* orthologues (Marais *et al.*, 1999; Doig *et al.*, 1999). Also, orthologues of genes involved in core oligosaccharide synthesis were identified: *rfaF*, *rfaE*, *rfaD*, and *rfaC* for synthesis of the inner core; *kdsA*, *kdsB* and *rpe* for the 3-deoxy-D-*manno*-2-octulsonic acid (Kdo) region of the inner core; and three copies of *rfaJ*, *pgi*, and *galU* for outer core synthesis (Marais *et al.*, 1999). With respect to O-chain synthesis, each genome has two $\alpha(1,3)$-fucosyltransferase genes which differ in the number of a 7 amino acid sequence repeat (Tomb *et al.*, 1997; Alm *et al.*, 1999). DNA motifs near the 5'-end of these genes at two distinct polynucleotide repeats have been deduced to indicate regulation through slipped-strand repair (Alm *et al.*, 1999). Phase variation in Lewis antigen expression by *H. pylori* LPS has been reported based largely on serological findings and requires further chemical structural validation (Appelmelk *et al.*, 1999). No putative gene for $\alpha(1,2)$-fucosyltransferase was identified in the genomes, but a truncated gene with a C14 tract was found and *in silico* insertion of a C-G pair yielded a full-length protein with strong homology to $\alpha(1,2)$-fucosyltransferases (Berg *et al.*, 1997). Three genes which are homologues of the $\alpha(1,2)$-glucosyltransferase gene (*rfaJ*) have been found but no $\alpha(1,2)$-linked glucose has been described in *H. pylori* LPS (Berg *et al.*, 1997; Marais *et al.*, 1999). It has been suggested that these genes encode $\beta(1,4)$-galactosyltransferase and/or $\beta(1,3)$-*N*-acetylglucosaminyltransferase functions required for Lewis antigen synthesis (Berg *et al.*, 1997). Although of low homology, an orthologue of *neuA*, that codes for CMP-*N*-

acetylneuraminic acid synthetase (Tomb *et al.*, 1997), was suggested to be involved in sialyl-Le[x] expression (Berg *et al.*, 1997) which has been verified chemically to occur in *H. pylori* LPS (Monteiro *et al.*, 2000). Furthermore, a homologue of GDP-D-mannose dehydratase and homologues of galactosyltransferases have been suggested to be involved in O-chain synthesis (Marais *et al.*, 1999).

Despite the above deductions, the assignment of roles to putative open reading frames in the sequenced *H. pylori* genomes must be performed with caution, particularly since the putative functions have not been subjected to mutational and biochemical analyses (see Chapter 17). For future studies, mutational analysis of the putative genes, followed by chemical verification of the LPS structures present, and testing of the validated mutants in animal models and other relevant biological test systems promises to give deeper insights into the role of LPS in *H. pylori* pathogenesis.

Acknowledgements

This work was supported by grants from the Irish Health Research Board and the Millennium Research Fund. The author thanks his many colleagues for their continued support and encouragement.

References

Alm, R.A., Ling, L.-S.L., Moir, D.T., King, B.L., Brown, E.D., Doig, P.C., Smith, D.R., Noonan, B., Guild, B.C., de Jonge, B.L., Carmel, G., Tummino, P.J., Caruso, A., Uria-Nickelsen, M., Mills, D.M., Ives, C., Gibson, R., Merberg, D., Mills, S.D., Jiang, Q., Taylor, D.E., Vovis, G.F., and Trust, T.J. 1999. Genomic-sequence comparison of two unrelated isolates of the human gastric pathogen *Helicobacter pylori*. Nature 397: 176-180.

Appelmelk, B.J., Faller, G., Claeys, D., Kirchner, T., and Vandenbroucke-Grauls, C.M.J.E. 1998. Bugs on trial: the case of *Helicobacter pylori* and autoimmunity. Immunol. Today. 19: 296-299.

Appelmelk, B.J., Martin, S.L., Monteiro, M.A., Clayton, C.A., McColm, A.A., Zheng, P., Verboom, T., Maaskant, J.J., van den Eijnden, D.H., Hokke, C.H., Perry, M.B., Vandenbroucke-Grauls, C.M.J.E., and Kusters, J.G.1999. Phase variation in *Helicobacter pylori* lipopolysaccharide due to changes in the lengths of poly(C) tracts in α3-fucosyltransferase genes. Infect. Immun. 67: 5361-5366.

Appelmelk, B.J., Negrini, R., Moran, A.P., and Kuipers, E.J. 1997. Molecular mimicry between *Helicobacter pylori* and the host. Trends Microbiol. 5: 70-73.

Appelmelk, B.J., Simoons-Smit, I.M., Negrini, R., Moran, A.P., Aspinall, G.O., Forte, J.G., de Vries, T., Quan, H., Verboom, T., Maaskant, J.J., Ghiara, P., Kuipers, E.J., Bloemena, E., Tadema, T.M., Townsend, R.R., Tyagarajan, K., Crothers, J.M., Monteiro, M.A., Savio, A., and de Graaf, J. 1996. Potential role of molecular mimicry between *Helicobacter pylori* lipopolysaccharide and host Lewis blood group antigens in autoimmunity. Infect. Immun. 64: 2031-2040.

Aspinall, G.O., Mainkar, A.S., and Moran, A.P. 1999. A structural comparison of lipopolysaccharides from two strains of *Helicobacter pylori*, of which one strain (442) does and the other strain (471) does not stimulate pepsinogen secretion. Glycobiology 9: 1235-1245.

Aspinall, G.O., and Monteiro, M.A. 1996. Lipopolysaccharides of *Helicobacter pylori* strains P466 and MO19: structures of the O antigen and core oligosaccharide regions. Biochemistry. 35: 2498-2504.

Aspinall, G.O., Monteiro, M.A., Pang, H., Walsh, E.J., and Moran, A.P. 1996. Lipopolysaccharide of the *Helicobacter pylori* type strain NCTC 11637 (ATCC 43504): Structure of the O antigen chain and core oligosaccharide regions. Biochemistry. 35: 2489-2497.

Aspinall, G.O., Monteiro, M.A., Pang, H., Walsh, E.J., and Moran, A.P. 1994. O antigen chains in the lipopolysaccharide of *Helicobacter pylori* NCTC 11637. Carbohydr. Lett. 1: 151-156.

Aspinall, G.O., Monteiro, M.A., Shaver, R.T., Kurjanczyk, L.A., and Penner, J.L. 1997. Lipopolysaccharides of *Helicobacter pylori* serogroups O:3 and O:6. Structures of a class of lipopolysaccharides with reference to the location of oligomeric units of D-*glycero*-(-D-*manno*-heptose residues. Eur. J. Biochem. 248: 592-601.

Aspinall, G.O., and Moran, A.P. 1997. *Helicobacter pylori* lipopolysaccharide structure and mimicry of Lewis blood group antigens. In: Pathogenesis and Host Response in *Helicobacter pylori* Infections, A.P. Moran, and C.A. O'Morain, eds. Normed Verlag, Bad Homburg, Germany. p. 34-42.

Baker, P.J., Hraba, T., Taylor, C.E., Stashak, P.W., Fauntleroy, M.B., Zähringer, U., Takayama, K., Sievert, T.R., Hronowski, X., Cotter, R.J., and Perez-Perez, G. 1994. Molecular structures that influence the immunomodulatory properties of the lipid A and inner core region oligosaccharides of bacterial lipopolysaccharides. Infect. Immun. 62: 2257-2269.

Berg, D.E., Hoffman, P.S., Appelmelk, B.J., and Kusters, J.G. 1997. The *Helicobacter pylori* genome sequence: genetic factors for long life in the gastric mucosa. Trends Microbiol. 5: 468-474.

Birkholz, S., Knipp, U., Nietzki, C., Adamek, R.J., and Opferkuch, W. 1993. Immunological activity of lipopolysaccharide of *Helicobacter pylori* on human peripheral mononuclear blood cells in comparison to lipopolysaccharides of other intestinal bacteria. FEMS Immunol. Med. Microbiol. 6: 317-324.

Cave, T.R., and Cave, D.R. 1991. *Helicobacter pylori* stimulates pepsin secretion from isolated rabbit gastric glands. Scand. J. Gastroenterol. 26 (Suppl. 181): 9-14.

Claeys D., Faller G., Appelmelk B.J., Negrini R., and Kirchner T. 1998. The gatric H^+,K^+-ATPase is a major autoantigen in chronic *Helicobacter pylori* gastritis with body mucosa atrophy. Gastroenterology. 115: 340-347.

Crabtree, J.E., Perry, S., Moran, A., Peichl, P., Tompkins, D.S., and Lindley, I.J.D. 1994. Neutrophil IL-8 secretion induced by *Helicobacter pylori*. Am. J. Gastroenterol. 89: 1337.

Crabtree, J.E., Rembacken, B., Lindley, I.J.D., and Moran, A.P. 1995. Preferential induction of interleukin-8 from gastric epithelial cell lines by *Helicobacter pylori* lipopolysaccharide. Gut. 37 (Suppl. 1): A31.

Cunningham, M.D., Seachord, C., Ratcliffe, K., Bainbridge, B., Aruffo, A., and Darveau, R.P. 1996. *Helicobacter pylori* and *Porphyromonas gingivalis* lipopolysaccharides are poorly transferred to recombinant soluble CD14. Infect. Immun. 64: 3601-3608.

Darveau, R.P., Cunningham, M.D., Bailey, T., Seachord, C., Ratcliffe, K., Bainbridge, B., Dietsch, M., Page, R.C., and Aruffo, A. 1995. Ability of bacteria associated with chronic inflammatory disease to stimulate E-selectin expression and promote neutrophil adhesion. Infect. Immun. 63: 1311-1317.

Doig, P., de Jonge, B.L., Alm, R.A., Brown, E.D., Uria-Nickelsen, M., Noonan, B., Mills, S.D., Tummino, P., Carmel, G., Guild, B.C., Moir, D.T., Vovis, G.F., and Trust, T.J. 1999. *Helicobacter pylori* physiology predicted from genomic comparison of two strains. Microbiol. Mol. Biol. Rev. 63: 675-707.

Edwards, N.J., Monteiro, M., Walsh, E., Moran, A., Roberts, I.S., and High, N.J. 1999. The role of the lipopolysaccharide in adhesion of *Helicobacter pylori* to gastric mucosa. In: Abstracts of the 10th International Workshop on *Campylobacter, Helicobacter* and Related Organisms. H.L.T. Mobley, I. Nachamkin, and D. McGee, eds. University of Maryland, Baltimore. p.148.

Faller, G., Steininger, H., Appelmelk, B.J., and Kirchner, T. 1998. Evidence of novel pathogenic pathways for the formation of antigastric autoantibodies in *Helicobacter pylori* gastritis. J. Clin. Pathol. 51: 244-245.

Faller, G., Steininger, H., Kränzlein, J., Maul, H., Kerkau, T., Hensen, J., Hahn, E.G., and Kirchner, T. 1997. Antigastric autoantibodies in *Helicobacter pylori* infection: implications of the histological and clinical parameters of gastritis. Gut 41: 619-623.

Fiocca, R., Necchi, V., Sommi, P., Ricci, V., Telford, J., Cover, T.L., and Solcia, E. 1999. Release of *Helicobacter pylori* vacuolating cytotoxin by both a specific secretion pathway and budding of outer membrane vesicles. Uptake of released toxin and vesicles by gastric epithelium. J. Pathol. 188: 220-226.

Ge, Z., Chan, N.W.C., Palcic, M.M., and Taylor, D.E. 1997. Cloning and heterologous expression of an α1,3-fucosyltransferase gene from the gastric pathogen *Helicobacter pylori*. J. Biol. Chem. 272: 21357-21363.

Goldberg, J.B., Wright, A., Ramakrishna, J., Moran, A., Kersulyte, D. and Berg, D.E. 1996. Cloning of the lipopolysaccharide (LPS)-core gene, *rfaC*, from *Helicobacter pylori*. Gut 39 (Suppl. 2): A69.

Green, C. 1989. The ABO, Lewis and related blood group antigens: a review of structure and biosynthesis. FEMS Microbiol. Immunol. 47: 321-330.

Guruge, J.L., Falk, P.G., Lorenz, R.G., Dans, M., Wirth, H.-P., Blaser, M.J., Berg, D.E., and Gordon, J.I. 1998. Epithelial attachment alters the outcome of *Helicobacter pylori* infection. Proc. Natl. Acad. USA. 95: 3925-3930.

Heneghan, M.A., McCarthy, C.F., and Moran, A.P. 1998. Is colonisation of gastric mucosa by *Helicobacter pylori* strains expressing Lewis x and y epitopes dependent on host secretor status? In: Campylobacter, Helicobacter and Related Organisms. A.J. Lastovica, D.G. Newell, and E.E. Lastovica, eds. Institute of Child Health and University of Cape Town, Cape Town, South Africa. p. 473-477.

Heneghan, M.A., McCarthy, C.F., and Moran, A.P. 2000. Relationship of blood group determinants on *Helicobacter pylori* lipopolysaccharide with host Lewis phenotype and inflammatory response. Infect. Immun. 68: 937-941.

Heneghan, M.A., Moran, A.P., Feeley, K.M., Egan, E.L., Goulding, J., Connolly, C.E., and McCarthy, C.F. 1998. Effect of host Lewis blood group antigen expression on *Helicobacter pylori* colonisation density and the consequent inflammatory response. FEMS Immunol. Med. Microbiol. 20: 257-266.

Holst, O., Ulmer, A.J., Brade, H., Flad, H.-D., and Rietschel, E.T. 1996. Biochemistry and cell biology of bacterial endotoxins. FEMS Immunol. Med. Microbiol. 16: 83-104.

Hynes, S.O., Hirmo, S., Wadström, T., and Moran, A.P. 1999a. Differentiation of *Helicobacter pylori* isolates based on lectin binding of cell extracts in an agglutination assay. J. Clin. Microbiol. 37: 1994-1998.

Hynes, S.O., Sjunnesson, H., Sturegård, E., Wadström, T., and Moran, A.P. 1999b. A relationship between the expression of high molecular weight lipopolysaccharide by *Helicobacter pylori* and colonizing ability in mice. In: Abstracts of the 10th International Workshop on *Campylobacter, Helicobacter* and Related Organisms. H.L.T. Mobley, I. Nachamkin, and D. McGee, eds. University of Maryland, Baltimore. p. 158.

Kidd, M., Miu, K., Tang, L.H., Perez-Perez, G.I., Blaser, M.J., Sandor, A. and Modlin, I.M. 1997. *Helicobacter pylori* lipopolysaccharide stimulates histamine release and DNA synthesis in rat enterochromaffin-like cells. Gastroenterology 113: 1110-1117.

Knirel, Y.A., Kocharova, N.A., Hynes, S.O., Widmalm, G., Andersen, L.P., Jansson, P.-E., and Moran, A.P. 1999. Structural studies on lipolysaccharides of serologically non-typable strains of *Helicobacter pylori*, AF1 and 007, expressing Lewis antigenic determinants. Eur. J. Biochem. 266: 123-131.

Kocharova, N.A., Knirel, K.A., Widmalm, G., Jansson, P.-E., and Moran, A.P. 2000. Structure of an atypical O-antigen polysaccharide of *Helicobacter pylori* containing a novel monosaccharide 3-*C*-methyl-D-mannose. Biochemistry. 39: 4755-4760.

Kwon, D.H., Woo, J.-S., Perng, C.-L., Go, M.F., Graham, D.Y., and El-Zaatari, F.A.K. 1998. The effect of *galE* gene inactivation on lipopolysaccharide profile of *Helicobacter pylori*. Curr. Microbiol. 37: 144-148.

Lamarque, D., Szepes, Z., Moran, A.P., Delchier, J.C., and Whittle, B.J.R. 1997. A purified lipopolysaccharide (LPS) from *Helicobacter pylori* (Hp) causes epithelial cell damage and induction of nitric oxide synthase in rat duodenum and colon. Ir. J. Med. Sci. 166 (Suppl. 3): 11.

Ljungh, Å., Moran, A.P., and Wadström, T. 1996. Interactions of bacterial adhesins with extracellular matrix and plasma proteins: pathogenic implications and therapeutic possibilities. FEMS Immunol. Med. Microbiol. 16: 117-126.

Logan, S.M., Conlan, J.W., Monteiro, M.A., Wakarchuk, W.W., and Altman, E. 2000. Functional genomics of *Helicobacter pylori*: identification of a beta-1,4-galactosyltransferase and generation of mutants with altered lipopolysaccharide. Mol. Microbiol. 35: 1156-1167.

Ma, J.Y., Borch, K., Sjöstrand, S.E., Janzon, L., and Mårdh, S. 1994. Positive correlation between H,K-adenosine triphosphate autoantibodies and *Helicobacter pylori* antibodies in patients with pernicious anaemia. Scand. J. Gastroenterol. 29: 961-965.

Mai, U.E.H., Perez-Perez, G.I., Wahl, L.M., Wahl, S.M., Blaser, M.J., and Smith, P.D. 1991. Soluble surface proteins from *Helicobacter pylori* activate monocytes/macrophages by lipopolysaccharide-independent mechanism. J. Clin. Invest. 87: 894-900.

Marais, A., Mendz, G.L., Hazell, S.L., and Mégraud, F. 1999. Metabolism and genetics of *Helicobacter pylori*: the genome era. Microbiol. Mol. Biol. Rev. 63: 642-674.

Marshall, D.G., Hynes, S.O., Coleman, D.C., O'Morain, C.A., Smyth, C.J., and Moran, A.P. 1999. Lack of a relationship between Lewis antigen expression and *cagA*, CagA, *vacA* and VacA status of Irish *Helicobacter pylori* isolates. FEMS Immunol. Med. Microbiol. 24: 79-90.

Martin, S.L., Edbrooke, M.R., Hodgman, T.C., van den Eijnden, D.H., and Bird, M.I. 1997. Lewis X biosynthesis in *Helicobacter pylori*. Molecular cloning of an α(1,3)-fucosyltransferase gene. J. Biol. Chem. 272: 21349-21356.

Micots, I., Augeron, C., Laboisse, C.L., Muzeau, F., and Mégraud, F. 1993. Mucin exocytosis: a major target for *Helicobacter pylori*. J. Clin. Pathol. 46: 241-245.

Mills, S.D., Kurjanczyk, L.A., and Penner, J.L. 1992. Antigenicity of *Helicobacter pylori* lipopolysaccharides. Infect. Immun. 30: 3175-3180.

Monteiro, M.A., Appelmelk, B.J., Rasko, D.A., Moran, A.P., McLean, L.L., Chan, K.H., St. Michael, F.S, Logan, S.M., O'Rourke, J., Lee, A., Hynes, S.O., Taylor, D.E., and Perry, M.B. 2000. Lipopolysaccharide structures of *Helicobacter pylori* genomic strains 26695 and J99, mouse model *H. pylori* Sydney strain, *H. pylori* P466 carrying sialyl-Lewis X, and *H. pylori* UA915 expressing Lewis B. Classification of *H. pylori* lipopolysaccharides into glycotype families. Eur. J. Biochem. 267: 305-320.

Monteiro, M.A., Chan, K.H.N., Rasko, D.A., Taylor, D.E., Zheng, P.Y., Appelmelk, B.J., Wirth, H.P., Yang, M.Q., Blaser, M.J., Hynes, S.O., Moran, A.P., and Perry, M.B. 1998a. Simultaneous expression of type 1 and type 2 Lewis blood group antigens by *Helicobacter pylori* lipopolysaccharides. Molecular mimicry between *H. pylori* lipopolysaccharides and human gastric epithelial cell surface glycoforms. J. Biol. Chem. 273: 11533-11543.

Monteiro, M.A., Rasko, D., Taylor, D.E., and Perry, M.B. 1998b. Glucosylated *N*-acetyllactosamine O-antigen chain in the lipopolysaccharide from *Helicobacter pylori* strain UA861. Glycobiology. 8: 107-112.

Moran, A.P. 1994. Structure-bioactivity relationships of bacterial endotoxins. J. Toxicol. - Toxin Rev. 14: 47-83.

Moran, A.P. 1995. Cell surface characteristics of *Helicobacter pylori*. FEMS Immunol. Med. Microbiol. 10: 271-280.

Moran, A.P. 1996a. The role of lipopolysaccharide in *Helicobacter pylori* pathogenesis. Aliment Pharmacol. Ther. 10 (Suppl. 1): 39-50.

Moran, A.P. 1996b. Pathogenic properties of *Helicobacter pylori*. Scand. J. Gastroenterol. 31 (Suppl. 215): 22-31.

Moran, A. 1998. The products of *Helicobacter pylori* that induce inflammation. Eur. J. Gastroenterol. Hepatol. 10 (Suppl. 1): S3-S8.

Moran, A.P. 1999. *Helicobacter pylori* lipopolysaccharide-mediated gastric and extragastric pathology. J. Physiol. Pharmacol. 50: 787-805.

Moran, A.P., Appelmelk, B.J., and Aspinall, G.O. 1996a. Molecular mimicry of host structures by lipopolysaccharides of *Campylobacter* and *Helicobacter* spp.: implications in pathogenesis. J. Endotoxin Res. 3: 521-531.

Moran, A.P., and Aspinall, G.O. 1998. Unique structural and biological features of *Helicobacter pylori* lipopolysaccharides. Prog. Clin. Biol. Res. 397: 37-49.

Moran, A.P., Helander, I.M., and Kosunen, T.U. 1992. Compositional analysis of *Helicobacter pylori* rough-form lipopolysaccharides. J. Bacteriol. 174: 1370-1377.

Moran, A.P., Knirel, Y.A., Kocharova, N.A., Hynes, S.O., Widmalm, G., Andersen, L.P., and Jansson, P.-E. 1999. Lipopolysaccharides of serologically non-typable strains of *Helicobacter pylori*. Gut. 45 (Suppl. III): A7.

Moran, A.P., Kuusela, P., and Kosunen, T.U. 1993. Interaction of *Helicobacter pylori* with extracellular matrix proteins. J. Appl. Bacteriol. 75: 184-189.

Moran, A.P., Lindner, B., and Walsh, E.J. 1997. Structural characterization of the lipid A component of *Helicobacter pylori* rough- and smooth-form lipopolysaccharides. J. Bacteriol. 179: 6453-6463.

Moran, A.P., McGovern, J.J., Lindner, B., and Walsh, E.J. 1996b. Structural analysis of lipid A from *Helicobacter pylori* rough- and smooth-form LPS. J. Endotoxin Res. 3 (Suppl. 1): 35.

Moran, A.P., Prendergast, M.A., and Appelmelk, B.J. 1996c. Molecular mimicry of host structures by bacterial lipopolysaccharides and its contribution to disease. FEMS Immunol. Med. Microbiol. 16: 105-115.

Moran, A.P., Sturegård, E., Sjunnesson, H., Wadström, T., and Hynes, S.O. 2000. The realtionship between O-chain expression and colonisation ability of *Helicobacter pylori* in a mouse model. FEMS Immunol. Med. Microbiol. In press.

Moran, A.P., and Wadström, T. 1994. Bacterial pathogenic factors. Curr. Opin. Gastroenterol. 10 (Suppl. 1): 17-21.

Moran, A.P., and Wadström, T. 1998. Pathogenesis of *Helicobacter pylori*. Curr. Opin. Gastroenterol. 14 (Suppl. 1): S9-S14.

Moran, A.P., and Walsh, E.J. 1993a. *In vitro* production of endotoxin by the gastroduodenal pathogen *Helicobacter pylori* - relevance to serotyping. Ir. J. Med. Sci. 162: 384.

Moran, A.P., and Walsh, E.J. 1993b. Expression and electrophoretic characterization of smooth-form lipopolysaccharides of *Helicobacter pylori*. Acta Gastro-Enterol. Belgica. 56 (Suppl): 98.

Moran, A.P., Young, G.O., and Lastovica, A.J. 1998. Pepsinogen induction by *Helicobacter pylori* lipopolysaccharides. Gut. 43 (Suppl. 2): A15.

Muotiala, A., Helander, I.M., Pyhälä, L., Kosunen, T.U., and Moran, A.P. 1992. Low biological activity of *Helicobacter pylori* lipopolysaccharide. Infect. Immun. 60: 1714-1716.

Negrini, R., Lisato, L., Zanella, I., Cavazzini, L., Gullini, S., Villanacci, V., Poiesi, C., Albertini, A., and Ghielmi, S. 1991. *Helicobacter pylori* infection induces antibodies cross-reacting with human gastric mucosa. Gastroenterology. 101: 437-445.

Negrini, R., Savio, A., Poiesi, C., Appelmelk, B.J., Buffoli, F., Paterlini, A., Cesari, P., Graffeo, M., Vaira, D., and Franzin, G. 1996. Antigenic mimicry between *Helicobacter pylori* and gastric mucosa in the pathogenesis of body atrophic gastritis. Gastroenterology. 111: 655-665.

Nielsen, H., Birkholz, S., Andersen, L.P., and Moran, A.P. 1994. Neutrophil activation by *Helicobacter pylori* lipopolysaccharides. J. Infect. Dis. 170: 135-139.

Ootsuba, C., Okumura, T., Takahashi, N., Wakebe, H., Imagawa, K., Kikuchi, M., and Kohgo, Y. 1997. *Helicobacter pylori* lipopolysaccharide inhibits acid secretion in pylorus-ligated conscious rats. Biochem. Biophys. Res. Commun. 236: 532-537.

Pece, S., Fumarola, D., Giuliani, G., Jirillo, E., and Moran, A.P. 1995. Activity in the *Limulus* amebocyte lysate assay and induction of tumour necrosis factor by diverse *Helicobacter pylori* lipopolysaccharide preparations. J. Endotoxin Res. 2: 455-462.

Pece, S., Messa, C., Caccavo, D., Giuliani, G., Greco, B., Fumarola, D., Berloco, P., Di Leo, A., Jirillo, E. and Moron, A.P. 1997. Serum antibody response against *Helicobacter pylori* NCTC 11637 smooth- and rough-lipopolysaccharide phenotypes in patients with *H. pylori*-related gastropathy. J. Endotoxin Res. 4: 383-390.

Perez-Perez, G.I., Shepherd, V.L., Morrow, J.D., and Blaser, M.J. 1995. Activation of human THP-1 cells and rat bone marrow-derived macrophages by *Helicobacter pylori* lipopolysaccharide. Infect. Immun. 63: 1183-1187.

Piotrowski, J., Piotrowski, E., Skrodzka, D., Slomiany, A., and Slomiany, B. 1997. Induction of acute gastritis and epithelial apoptosis by *Helicobacter pylori* lipopolysaccharide. Scand. J. Gastroenterol. 32:203-211.

Piotrowski, J., Slomiany, A., and Slomiany, B. 1993. Inhibition of gastric mucosal mucin receptor by *Helicobacter pylori* lipopolysaccharide. Biochem. Mol. Biol. Int. 31: 1051-1058.

Rieder, G., Hatz, R.A., Moran, A.P., Walz, A., Stolte, M., and Enders, G. 1997. Role of adherence in interleukin-8 induction in *Helicobacter pylori*-associated gastritis. Infect. Immun. 65: 3622-3630.

Rietschel, E.T., Brade, L., Holst, O., Kulshin, V.A., Lindner, B., Moran, A.P., Schade, U.F., Zähringer, U., and Brade, H. 1990. Molecular structure of bacterial endotoxin in relation to bioactivity. In: Endotoxin Research Series Vol. 1. Cellular and Molecular Aspects of Endotoxin Reactions. A. Nowotny, J.J. Spitzer, and E.J. Ziegler, eds. Elsevier Science, Amsterdam, The Netherlands. p.15-32.

Rietschel, E.T., Brade, L., Schade, U., Seydel, U., Zähringer, U., Holst, O., Kuhn, H.-M., Kulschin, V.A., Moran, A.P., and Brade, H. 1991. Bacterial endotoxins: relationships between chemical structure and biological activity of the inner core-lipid A domain. In: Microbial Surface Components and Toxins in Relation to Pathogenesis. E.Z. Ron and S. Rottem, eds. Plenum, New York. p. 209-217.

Sakamoto, J., Watanabe, T., Tokumaru, T., Takagi, H., Nakazato, H., and Lloyd, K.O. 1989. Expression of Lewis[a], Lewis[b], Lewis[x], Lewis[y], sialyl-Lewis[a], and sialyl-Lewis[x] blood group antigens in human gastric carcinoma and in normal gastric tissue. Cancer Res. 49: 745-752.

Semeraro, N., Montemurro, P., Piccoli, C., Muolo, V., Colucci, M., Giuliani, G., Fumarola, D., Pece, S., and Moran, A.P. 1996. Effect of *Helicobacter pylori* lipopolysaccharide (LPS) and LPS-derivatives on the production of tissue factor and plasminogen activator inhibitor type 2 by human blood mononuclear cells. J. Infect. Dis. 174: 1255-1260.

Simoons-Smit, I.M., Appelmelk, B.J., Verboom, T., Negrini, R., Penner, J.L., Aspinall, G.O., Moran, A.P., Fei Fei, S, Bi-Shan, S., Rudnica, W., Savio, A. and de Graaff, J. 1996. Typing of *Helicobacter pylori* with monoclonal antibodies against Lewis antigens in lipopolysaccharide. J. Clin. Microbiol. 34: 2196-2200.

Slomiany, B.L., Liau, Y.H., Lopez, R.A., Piotrowski, J., Czajkowski, A., and Slomiany, A. 1992. Effect of *Helicobacter pylori* lipopolysaccharide on the synthesis of sulfated gastric mucin. Biochem. Int. 27: 687-697.

Slomiany, B.L., Piotrowski, J., Rajiah, G., and Slomiany, A. 1991. Inhibition of gastric mucosal laminin receptor by *Helicobacter pylori* lipopolysaccharide: effect of nitecapone. Gen. Pharmacol. 22: 1063-1069.

Suda, Y., Ogawa, T., Kashihara, W., Oikawa, M., Shimoyama, T., Hayashi, T., Tamura, T., and Kusumoto, S. 1997. Chemical structure of lipid A from *Helicobacter pylori* strain 206-1 lipopolysaccharide. J. Biochem. 121: 1129-1133.

Szepes, Z., Lamarque, D., Moran, A,. Delchier, J.C., and Whittle, B.J.R. 1997. Purified lipopolysaccharide from *Helicobacter pylori* (Hp) provokes epithelial cell injury and induction of nitric oxide synthase in rat duodenum: inhibition of cytotoxicity by superoxide dismutase. Gastroenterology. 112 (Suppl): A303.

Tarkkanen, J., Kosunen, T.U., and Saksela, E. 1993. Contact of lymphocytes with *Helicobacter pylori* augments natural killer cell activity and induces production of gamma interferon. Infect. Immun. 61: 3012-3016.

Taylor, D.E., Rasko, D.A., Sherburne, R., Ho, C., and Jewell, L.D. 1998. Lack of correlation between Lewis antigen expression by *Helicobacter pylori* and gastric epithelial cells in infected patients. Gastroenterology. 115: 1113-1122.

Terrés, A.M., Pajares, J.M., Hopkins, A.M., Murphy, A., Moran, A., Baird, A.W., and Kelleher, D. 1998. *Helicobacter pylori* disrupts epithelial barrier function in a process inhibited by protein kinase C activators. Infect. Immun. 66: 2943-2950.

Tomb, J.-F., White, O., Kerlavage, A.R., Clayton, R.A., Sutton, G.G., Fleischmann, R.D., Ketchum, K.A., Klemk, H.P., Gill, S., Dougherty, B.A., Nelson, K., Quackenbush, J., Zhou, L., Kirkness, E.F., Peterson, S., Loftus, B., Richardson, D., Dodson, R., Khalak, H.G., Glodek, A., McKenney K., Fitzgerald, L.M., Lee, N., Adams, M.D., Hickey, E.K., Berg, D.E., Gocayne, J.D., Utterback, T.R., Peterson, J.D., Kelley, J.M., Cotton, M.D., Weidman, J.M., Fujii, C., Bowman, C., Watthey, L., Wallin, E., Hayes, W.S., Borodovsky, M., Karp, P.D., Smith, H.O., Fraser, C.M., and Venter, J.C. 1997. The complete genome sequence of the gastric pathogen *Helicobacter pylori*. Nature 388: 539-547.

Trust, T.J., Doig, P., Emödy, L., Kienle, Z., Wadström, T., and O'Toole, P. 1991. High affinity binding of the basement membrane proteins collagen type IV and laminin to the gastric pathogen *Helicobacter pylori*. Infect. Immun. 59: 4398-4404.

Valkonen, K.H., Ringner, M., Ljungh, Å., and Wadström, T. 1993. High affinity binding of laminin by *Helicobacter pylori*: evidence for a lectin-like interaction. FEMS Immunol. Med. Microbiol. 7: 29-38.

Valkonen, K.H., Wadström, T., and Moran, A.P. 1994. Interaction of lipopolysaccharides of *Helicobacter pylori* with basement membrane protein laminin. Infect. Immun. 62: 3640-3648.

Valkonen, K.H., Wadström, T., and Moran, A.P. 1997. Identification of the *N*-acetylneuraminyllactose-specific laminin-binding protein of *Helicobacter pylori*. Infect. Immun. 65: 916-923.

Walsh, E.J., and Moran, A.P. 1996. Serological analysis of the core region of *Helicobacter pylori* lipopolysaccharide. Gut. 39 (Suppl. 2): A71-A72.

Walsh, E.J., and Moran, A.P. 1997. Influence of medium composition on the growth and antigen expression of *Helicobacter pylori*. J. Appl. Microbiol. 83: 67-75.

Wang, G., Rasko, D.A., Sherburne, R., and Taylor, D.E. 1999. Molecular genetic basis for the variable expression of Lewis Y in *Helicobacter pylori*: analysis of the α(1,2)-fucosyltransferase gene. Mol. Microbiol. 31: 1265-1274.

Weintraub, A., Zähringer, U., Wollenweber, H.W., Seydel, U., and Rietschel, E.T. 1989. Structural characterization of the lipid A component of *Bacteroides fragilis* strain NCTC 9343 lipopolysaccharide. Eur. J. Biochem. 183:425-431.

Wirth, H.P., Yang, M., Karita, M., and Blaser, M.J. 1996. Expression of the human cell surface glycoconjugates Lewis X and Lewis Y by *Helicobacter pylori* isolates is related to *cagA* status. Infect. Immun. 64: 4598-4605.

Wirth, H.P., Yang, M., Peek, R.M., Tham, K.T., and Blaser, M.J. 1997. *Helicobacter pylori* Lewis expression is related to the host Lewis phenotype. Gastroenterology. 113: 1091-1098.

Young, G.O., Lastovica, A.J., Brown, S., McGovern, J.J., and Moran, A. 1996. Effect of pepsinogen release of various sub-fractions of *H. pylori* lipopolysaccharide. In: Campylobacters, Helicobacters and Related Organisms. D. Newell, J.M. Ketley, and R.A. Feldman, eds. Plenum, New York. p. 697-700.

Young, G.O., Stemmet, N., Lastovica, A., van der Merwe, E.L., Louw, J.A., Modlin, I.M., and Marks, I.N. 1992. *Helicobacter pylori* lipopolysaccharide stimulates gastric mucosal pepsinogen secretion. Aliment. Pharmacol. Ther. 6: 169-177.

From: *Helicobacter pylori: Molecular and Cellular Biology*
ISBN 1-898486-25-5 © 2001 Horizon Scientific Press, Wymondham, UK.

14

The *cag* Type IV Secretion System of *Helicobacter pylori* and CagA Intracellular Translocation

Markus Stein, Rino Rappuoli and Antonello Covacci[*]

Abstract

The evolution and dissemination of virulence traits between bacteria is based on mobilization of DNA by a) phage conversion, b) distant events of horizontal transfer that promoted the pathogenicity island assembly and c) plasmid mobilization. Type III and Type IV secretory apparatuses are often located on mobilized DNA and are associated with an increased virulence of the bacterial specialist. Type I strains of *Helicobacter pylori* possess the *cag* pathogenicity island that encodes a specialized Type IV secretion machinery activated during infection. The *cag* organelle is involved in cellular responses like induction of pedestals, secretion of interleukin-8 and phosphorylation of proteic targets. Co-cultivation of epithelial cells with *Helicobacter pylori* triggers signal transduction and tyrosine-phosphorylation of a 145 kDa putative host cell protein. We and others have recently demonstrated that this protein is not derived from the host but rather is the bacterial immunodominant antigen CagA, a virulence factor commonly expressed during infection in patients of peptic ulcer disease and thought to be orphan of a specific biological function. CagA is delivered into the epithelial cells by the *cag* type IV secretion system where it is phosphorylated on tyrosine residues by an as yet unidentified host cell kinase and wired to eukaryotic signal transduction pathways and cytoskeletal plasticity.

Helicobacter pylori is Associated with Disease

In 1600, Giordano Bruno was burned alive in Rome for the heresy of describing an infinite universe, with infinite planets rotating around infinite suns and infinite forms of life populating this worlds. Today, the universe is still infinite and infinite is the variety of microbes, of which at most 1% have been classified. Human, plant and animal diseases are often of infectious origin, sometimes sustained by unrecognized and/or uncultivated etiologic agents. Until 1982, when it was isolated by accidental extended incubation, *Helicobacter pylori* (*Hp*) (Dundon *et al.*, 1998) was part of the unknown microbial world. Today it is a well recognized pathogen, associated with gastritis, peptic ulcer and gastric cancer lymphoma (Parsonnet *et al.*, 1997; Blaser 1998). It is a Gram-negative and microaerophilic rod that lives for decades in the extreme environment of the human stomach - an equilibrium that permanently binds the parasite to the host.

[*]corresponding author, email: antonello_covacci@biocine.it

It is thought that some virulent strains evolved from non-pathogenic commensal progenitors after the rapid acquisition of new genetic traits that were acquired from unrelated organisms by horizontal DNA transfer. The newly acquired DNA, often derived from phages or plasmids, can be integrated into the chromosome (pathogenic *Escherichia coli, Salmonella, Helicobacter*) or have become part of large plasmids (*Yersinia, Shigella, Agrobacterium*) (Groisman and Ochman, 1996; Lee, 1996; Hacker *et al.*, 1997). In various pathogens, these regions encode blocks of virulence determinants, one reason why they were designated as pathogenicity islands (PAIs). PAIs are large DNA elements (10 to >200 kbp) which are often characterized by instability, flanking direct repeats, and a G+C content different from the chromosome, features that are associated with their foreign origin. However, PAIs are not static regions at the end of their evolutionary development and have continued to undergo evolutionary modifications. Some have lost mobilization factors, acquired cryptic genes by point mutations and deletions and have adjusted their G+C content and codon usage to that of the host chromosome. Altogether, it seems that the acquisition of PAIs provided certain microorganisms with selective advantage against their commensal parents and enabled them to colonize new and less competitive niches.

Structural Organization of the *cag*- Pathogenicity Island

Most clinical isolates of *Hp* possess the *cag* PAI. *cag* stands for *cytotoxin associated gene*, because the first identified protein in the island, CagA, is frequently coexpressed with the vacuolating cytotoxin, VacA. However, their genes are 300 kb apart and VacA expression is independent of the presence of the *cag*A gene; *cagA* null mutants still produce the VacA protein (Crabtree *et al.*, 1995; Tummuru *et al.*, 1994; Xiang *et al.*, 1995).

The nucleotide sequence of the 40 kb PAI of *Hp* strain CCUG 17874 was released in 1996 (Censini *et al.*, 1996; Akopyanz *et al.*, 1998; Tummuru *et al.*, 1995), and possibly represents the first full length PAI accessible to computer analyses. The G+C content of *cag* is slightly different from the overall G+C value of the chromosome (35% vs 39%). Based on the presence or absence of *cag, Hp* strains have been grouped into two families called Type I, which possess the *cag*-island, and Type II, which do not. The presence of two 31-bp direct repeats at the end of *cag* indicates that it was originally acquired as a single unit which was inserted at the end of the chromosomal glutamate racemase gene (*glr*) (Censini *et al.*, 1996; Ritter *et al.*, 1995). The lack of glutamate racemase is lethal and integration of *cag* does not disrupt the coding sequence of the glutamate racemase gene. Instead a 31 bp module, which contains the recombination site and corresponds to the last nucleotides of the *glr* gene is simply duplicated and flanks *cag* on both sides (Censini *et al.*, 1996). This module also forms the core of the left and right ends of a *Helicobacter* insertion sequence, the IS*605* element.

In the reference strain CCUG 17874 [also called NCTC 11638], an intervening sequence splits the *cag* PAI into the *cag*I and the *cag*II domains (Censini *et al.*, 1996; Akopyanz *et al.*, 1998). The intervening segment is flanked on both sides by an IS*605* element and encodes two putative transposases. Both enzymes, TnpA and TnpB, resemble analogous genes found in the insertion sequences IS*200* of Gram-negative bacteria and IS*1341* of the thermophilic bacterium *PS3* (Censini *et al.*, 1996).

Southern analysis showed that the structure of the *cag* locus varies as if intermediate strains had emerged following IS*605* insertion and homologous recombination from the Type I lineage (Censini *et al.*, 1996, Akopyanz *et al.*, 1998). For example, 10 out of 40 different clinical isolates contained *cag*I and *cag*II fused without IS*605*, while other strains with multiple copies of IS*605* exhibited partial or total deletion of *cag*I or total deletion of *cag*I and

Table 1.

Nomenclature of *cag* Genes According To:			Properties of Single Gene Inactivation in the *cag* Region		
Tomb et al.	*Akopyants et al.*	*Censini et al.*	IL-8 secretion	NF-κB activation	CagA-Ptyr translocation ion
		G27wt,*cag*+	+	+	+
		G27Δcag	-	-	-
HP0520 *cag1*	ORF6	cagζ			
HP0521 *cag2*	ORF7	cagε	-		
HP0522 *cag3*	ORF8	cagδ			
HP0523 *cag4*	ORF9	cagγ			
HP0524 *cag5*	ORF10	cagβ(virD4)	++?		-
HP0525	ORF11	cagα virB11)	-		-
HP0526 *cag6*	ORF12	cagZ			
HP0527 *cag7*	ORF13	*cag*Y(virB10)	-		-
HP0527 *cag7*	ORF14		-		
HP0528 *cag8*	ORF15	cagX(virB9)	-		-
HP0529 *cag9*	ORF16	*cag*W(virB8)	-		
HP0530 *cag10*	ORF17	cagV			
HP0531 *cag11*	ORF18	cag U	-		
HP0532 *cag12*	ORF19	cagT (virB7)	-		
HP0533					
HP0534 *cag13*	ORF20	cagS	-		
	ORF21	tnpA			
	ORF22	tnpB			
HP0535 *cag14*		cagQ			
		cagR			
HP0536 *cag15*		cagP			
		cagO			
HP0537 *cag16*		cagM	-	-	-
HP0538 *cag17*		cagN	+	+	+
HP0539 *cag18*		cagL	-	-	-
HP0540 *cag19*		cagI	-	-	-
HP0541 *cag20*		cagH	-	-	-
HP0542 *cag21*		cagG	-	-	-
HP0543 *cag22*		cagF	-	+	-
HP0544 *cag23*		cagE(virB4), picB(*Tummuru*)	-	-	-
HP0545 *cag24*		*cag*D	-*		-
HP0546 *cag25*		cagC	-*		-
		cagB	-*		-
HP0547 *cag26*		cagA	+	+	-
HP0548		cagΩ			
HP0549 *glr*		glr			

*cag*II. Maeda *et al.* (1999) provided data about the deletions and structural variants of *cag* in Japanese isolates. 59 of 63 clinical isolates contained the complete *cag* region. In the remaining four strains, *cag* was partially deleted and *cag*A was not transcribed even though the *cag*A gene was present. The size of the cag deletion varied but always started from *cag*B and included the *cag*A promotor region.

The *cag* PAI can also be deleted due to DNA transfer of an empty site from a *cag*- strain followed by homologous recombination (Kersulyte *et al.*, 1999). Instability of the *cag* PAI due to inversions or deletions may be important for virulence. Deletions, possibly in response to an inductive signal or due to selective forces, might promote the adaptation of the

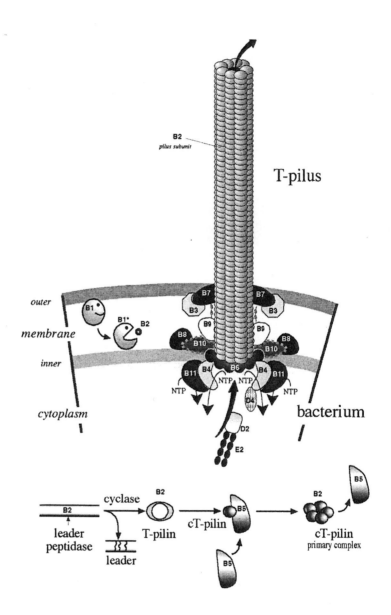

Figure 1. VirB complex of *Agrobacterium tumefaciens* as a type IV model system. The localization and interactions of the individual VirB proteins within the secretion system are shown. In the lower part of the figure the steps leading to the formation of cyclic prepilin (cT-pilin) aggregates (with the molecular chaperon VirB5) are indicated. CagM corresponds to a periplasmic chaperone (Pacman-like character). Three levels of energizer are also sketched: the VirB11 homologue provides energy for the assembly of the conduit; the VirB4 homologue energizes the injection process and the VirD4 ATPase is required for interactions with injected components (e.g. the CagA protein). The VirB2 protein is the major pilus subunit.

pathogen at different stages of infection and chronic persistence of infection might reflect continuous fluctuations in the proportions of *cag*⁺ and *cag*⁻ bacteria (Atherton *et al.*, 1996; Blum *et al.*, 1994; Covacci *et al.*, 1997; Figura *et al.*, 1998, Hamlet *et al.*, 1999). In addition to these dramatic rearrangement, single point mutations can also dramatically abrogate the function of the *cag* PAI because most of the genes within *cag* encode components of a secretion engine.

cag Encodes A Type IV Secretion Apparatus

Stucture of the Type IV System

Computer analysis of the approximately 30 *cag* genes and their gene products revealed that most of the proteins have one or more predicted transmembrane helices, suggesting their association with the bacterial inner membrane. Database homology searches revealed that several *cag* proteins are homologous to membrane associated proteins of the *tra* conjugation system of *E. coli* and other type IV secretion systems (Christie, 1997) (Table 1). Type IV secretion systems probably evolved after an ancestral duplication event from bacterial conjugation systems and specialized into protein secretion structures that target virulence factors to the bacterial surface, the supernatant or even into host cells (Stein *et al.*, 2000; Winans *et al.*, 1996; Covacci and Rappuoli, 1993; Weiss *et al.*, 1993).

Type IV systems build channels, that serves as conduits through which unfolded molecules pass through the cell envelope, or pili, that are believed to mediate the physical contact between the donor bacterium and a recipient cell (Figure 1). The core structural components of type IV secretion systems seem to be quite conserved among different species, but each also encodes unique proteins without known homologies. These unique proteins may modify and complement the core components specifically for each of the different functions of these systems. In *Helicobacter* the conserved core structure of the type IV system includes homologues of VirB4 (*cag*E), VirB7 (CagT), VirB8 (*cag*W, HP529), VirB9 (HP528), VirB10 (HP527), and VirB11 (HP525) proteins (Anersson *et al.*, 1998; Christie, 1997; Censini *et al.*, 1996; Tummuru *et al.*, 1995) (Table 2). In *A. tumefaciens,* VirB4 and VirB11 hydrolyze ATP *in vitro*. The energy of ATP hydrolysis is then utilized to drive transporter assembly or substrate translocation (Baker *et al.*, 1997; Christie, 1997). Both ATPases are tightly associated with the inner membrane. Each forms a homodimer, which associates with other transporter components during the early phase of assembly, although the final oligomeric structure of these proteins in the fully assembled transporter is unknown. The VirB11 homologue is probably localized exclusively on the cytoplasmic face of the inner membrane, whereas VirB4 might be an integral protein extending into the periplasmic space. VirB7 and VirB9 are candidates for components of an outer membrane pore (Christie, 1997).

The localization of the remaining components of the *Helicobacter cag* system is currently unknown. Some may be structural subunits of a (putative) trans-envelope channel, or they may transiently assist in the assembly of the type IV structure. Studies of *cag* mutants revealed that some of the proteins, most notably the lipoprotein CagT (a VirB7 homologue), and the VirB9 homologue CagX, provide important stabilizing functions for other Cag proteins with homologies to VirB proteins. The high number of genes in *cag* suggests that at least some of the Cag proteins fullfill a chaperon-like function. In most secretion systems, macromolecular export is accomplished by the help of specific chaperons, which keep the exported proteins in the unfolded state, protect them from degradation, and target them to their secretion site. For exemple, the dependence of CagT on CagM may be functionally

Table 2. List of bacterial species containing recognized type III and type IV secretion systems

Bacterial Species	Type III	Type IV	References and Nomenclature
Actinobacillus actinomycetemcomitans		+	+
Agrobacterium tumefaciens		+	***virB, D, E*** regions of Ti plasmid (Baker *et al*, 1997; Fullner *et al*. 1996)
Bordetella bronchiseptica	+	+	+
Bordetella pertussis	+	+	***ptl,*** pertussis toxin liberation genes, (Covacci and Rappuoli, 1993; Weiss *et al*., 1993)
Brucella suis		+	***vir*** (O'Callaghan *et al*., 1999)
Chlamydia spp	+		
Citrobacter rodentium	+		
Escherichia coli	+	+	+
Enteropathogenic *E.coli*	+		
Enterohemorrhagic *E.coli*	+		
Erwinia amylovora	+		
Erwinia chrysanthemi	+		
Erwinia herbicola pv. *gysophila*	+		
Erwinia stewartii	+		
Hafnia alveii	+		
Helicobacter pylori		+	***Cag,*** see text
Legionella pneumophila		+	***icm/dot*** and ***lvh*** (Kirby and Isberg, 1998; Segal and Shuman, 1998; Segal *et al*., 1999)
Pseudomonas aeruginosa	+		
Pseudomonas syringae	+		
Ralstonia solanacearum	+		
Rickettsia prowazekii		+	(Andersson *et al*., 1998)
Rhizobium spp.	+		
Salmonella enterica	+		
Shigella spp.	+		
Xantomonas spp.	+		
Yersinia spp.	+		

related to periplasmic chaperons (C. Lange *et al*., Personal Commun.). Interestingly, there is no obvious homologue of the VirB2 pilus of *Agrobacterium* present in *cag* (Lai and Kado, 1998). However, this does not exclude the possibility that a pilus structure is encoded by one of the *cag* genes unique for *Helicobacter*. Another intriguing feature of *cag* is the presence of a VirD4 homologue (HP524) that couples protein-protein interaction with conjugative DNA transfer in *Agrobacterium* and which is localized in the inner membrane. It has been hypothesized that the *Helicobacter* secretion system can export proteins and even nucleoproteins across kingdom boundaries (Christie, 1997). Alternatively, the VirD4 homologue may have an altered function in *Helicobacter* such as the control of inflammation as discussed below.

Other Members of the Type IV Secretion Family

Bordetella pertussis, the causative agent of whooping cough, uses the Ptl transporter to export the 6-subunit pertussis toxin across the bacterial envelope into the supernatant. The functional toxin then binds its receptor on the epithelial cell surface and is internalized whereupon it manifests its toxic activity (Covacci and Rappuoli, 1993; Weiss *et al*., 1993). This has led to the hypothesis that the Ptl proteins do not elaborate a pilus. However, PtlA is related to the VirB2 pilin subunit (Covacci and Rappuoli, 1993; Christie, 1997; Weiss *et al*., 1993; Winans *et al*., 1996), indicating that the Ptl system does assemble at least a vestige of a pilus, or that VirB2-type proteins supply a function that is critical to transporter assembly and unrelated to pilus assembly.

Further gene clusters encoding homologues of the type IV family are involved in virulence of *Legionella pneumophila*, the causative agent of Legionnaire's Disease and Pontiac Fever (Kirby and Isberg, 1998; Segal and Shuman, 1998; Segal *et al.*, 1999). Mutational studies to identify genes essential for intracellular growth and macrophage killing resulted in the identification of the *icm/dot* genes. Both intracellular growth and macrophage killing are thought to result from export of a toxin effector via the *icm/dot* system. Products of two of the genes, *icmE* and *dotB*, are related to VirB10 and VirB11, and products of fourteen other *icm* or *dot* genes have homologues in bacterial conjugation systems. In addition the Icm/Dot system has been shown to direct the interbacterial movement of plasmid RSF1010. This observation suggests that this system too has retained a functional vestige of the ancestral DNA conjugation system from which it evolved. Recently, a type IV secretion system, Lvh, was localized on a chromosomal island in *L. pneumophila* and encodes for 11 VirB homologues (all but VirB1) (Segal *et al.*, 1999). Unlike the Icm/Dot system, the Lvh system is not required for intracellular replication, but is partially required for RSF1010 conjugation. The presence of the *lvh* island can also restore conjugation in *icmE* and *dotB* mutants, thus showing that both systems can interact with one another (Segal *et al.*, 1999).

Whether other type IV export systems, including the *B. pertussis* Ptl system or the *Hp* Cag system, are also capable of directing conjugative DNA transfer to recipient bacteria or even to eukaryotic cells remains to be tested.

Homologues of most of the VirB proteins, including a VirB2 homologue, were recently discovered *in Brucella suis*. Mutants in *virB5*, *virB9* or *virB10* were highly attenuated in an *in vitro* infection model with human macrophages, indicating that the *virB* region is essential for the intracellular survival and multiplication of *B. suis* (O'Callaghan *et al.*, 1999) Two homologues of VirB4 and one homologue each of VirB8, VirB9, VirB10, VirB11 and VirD4 were detected within the genome sequence of *Rickettsia prowazekii*, suggesting that this microorganism encodes still another type IV transport system similar to *Hp* and *L. pneumophila* (Andersson *et al.*, 1998).

Together, these findings raise the intriguing possibility that this subset of homologues corresponds to a minimal ancestral protein subassembly upon which bacteria have designed transporters for novel purposes ranging from intercellular DNA transfer and toxin export to the direct injection of virulence factors into eukaryotic cells. It should be mentioned that the secretion of autotransporting proteins such as IgA protease has also been referred to as type IV secretion. However, this secretion process involves a mechanism that is completely different from as the type IV secretion systems exhibited by *Hp* and numerous other pathogens.

Substrates of The Type IV Machineries

The flagellar apparatus, or a simplified version of it, may be the ancestor of the type III secretion machines (Kubori *et al.*, 1998; Macnab, 1992). Type III systems subsequently specialized as extracellular, tubular protrusions or as intracellular gated complexes involved in substrate transfer between different sub-cellular compartments (Cormack *et al.*, 1996; Valdivia and Falkow, 1997). This analogy can be extended to the type IV family, which apparently originated from a conjugative apparatus that later acquired entirely new functions by association with other classes of genes (Figure 2). Type III and IV systems are both related to proteins involved in filamentous phage assembly/secretion (Russel, 1995, Salmond *et al.*, 1995). Type III secretion systems have been used as injection machineries by various human pathogens, including *Salmonella*, *Yersinia*, *Shigella*, enteropathogenic *E. coli* (EPEC), and various plant pathogens that deliver virulence proteins directly into the host cell cytoplasm (Hueck, 1998; Lee 1997). Inside the cell, the injected molecules then act as cytotoxins, kinases,

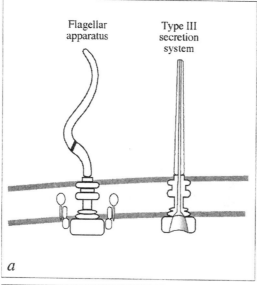

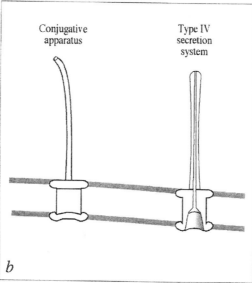

Figure 2. Schematic representation of the similarities of type III secretion systems with bacterial flagellar system (a) and of type IV secretion systems with conjugative pili (b). The authors suggest that they originated from duplication events followed by functional specialization.

phosphatases, or activate enzymatic processes. The outcome of these events differs with the pathogen and can include the induction of bacterial internalization into the host cell, anti-phagocytic activity, rearrangements of cytoskeletal proteins plus pedestal formation, cyto-toxic effects, or the induction of apoptosis. Common to all effector molecules is that they perturb or interfere with cellular functions and modify host-signaling pathways, thus creating conditions that favour the survival of the pathogen.

In contrast to type III systems, only few substrates have been described for proteins secreted by type IV systems. In *A. tumefaciens*, the VirD2 protein binds to the single-stranded T-DNA and is injected into the plant cell as a nucleoproteic particle. The transferred DNA molecule enters the nucleus, is incorporated into the plant chromosome and leads to the expression of genes that alter the cellular physiology. *B. pertussis* uses a type IV system to secrete active pertussis toxin, which then acts on the host cell. In *L. pneumophila*, the Icm/Dot proteins export virulence factor(s) targeted to the intracellular environment to enhance

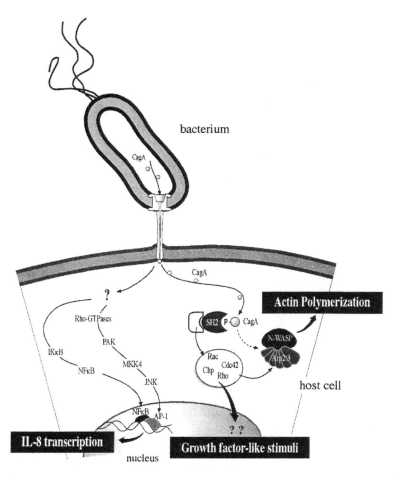

Figure 3. Two independent signalling pathways are induced in the host cell following contact between *Hp* cells and host cells and CagA translocation. The fact that globular actin polymerizes suggests an interaction of activated CagA with N-WASP (expressed in epithelial cells) and the Arp2/3 machinery (directly or indirectly via small GTP-binding proteins). The X-factor is required for IL-8 induction and can be identified i) with an injected component or ii) by physical interaction of the type IV secretion system with the cellular surface. Signal tranduction pathways leading to Growth factors-like stimuli are indicated.

survival by preventing phagosome-lysosome fusions and killing host macrophages (Kirby and Isberg, 1998; Segal and Shuman, 1998). Tyrosine phosphorylated proteins were recently detected in the host cell close to the vacuoles in which *L. pneumophila* resides (Susa and Marre, 1999). If these proteins were of bacterial origin, it is conceivable that they are injected into the host cell via the Icm/Dot or the Lvh type IV secretion. However, this is very speculative and it was unclear until recently whether, similarly to type III systems, type IV systems can inject virulence effectors into the host cell cytoplasm. Therefore, the discovery that the *cag* PAI-mediated translocation of the CagA molecule of *Helicobacter* was highly intriguing (Segal *et al.*, 1999, Asahi *et al.*, 2000, Odenbreit *et al.*, 2000; Stein *et al.*, 2000). The *cagA* gene is localized at the 3´-end of the PAI and its molecular size can vary between strains from 128 to 142 kDa, depending on the number of internal repeats. Animal models and clinical investigations have demonstrated that 70-80% of all clinical isolates express CagA. CagA positive strains are more virulent and are associated with more severe disease than strains lacking CagA expression. However, a specific function for CagA remained elusive and the molecule was primarily used as a marker to differentiate between type I strains containing a complete PAI and less pathogenic type II strains.

cag-mediated Signal Transduction Events

cag-induced signaling events offer new insights into the pathogenic effects triggered by the *Hp cag* PAI and may help explain the relationships between colonization of the host, inflammation, chronic infection and disease.

Inflammatory Response

The first identified *cag*-mediated host-signaling event was the secretion of IL-8 by gastric epithelial cells. IL-8 is an important chemoattractant of neutrophils. The infiltrating neutrophils then activate processes that lead to a strong inflammatory response in the gastric mucosa of the human stomach (see Chapter 5). In consequence, persistent infections with *Helicobacter* result in chronic inflammatory effects, which contribute to disease. Other cytokines (IL-1, IL-6), chemokines (RANTES, GRO, MIP-1a, ENA-78, MCP-1) and tumor necrosis factor alpha, which are activated by *Helicobacter* contribute to increase the inflammation. Expression of IL-8, IL-6 and MCP-1, depend on the activation of the transcription factor nuclear factor (B (NFκB) (Figure 3). All the *cag* mutants that have been tested fail to promote the secretion of IL-8 and do not activate NFκB with the exceptions of *cag*A, *cag*F, *cag*N, and the *vir*D4 homologue (Censini *et al.*, 1996; Glocker *et al.*, 1998; Munzenmaier *et al*, 1997; Tummuru *et al.*, 1995). Interestingly, IL-8 induction is increased 3.6-fold by the *vir*D4 mutant, suggesting that the VirD4-homologue might downregulate IL-8 secretion induced by other *cag* genes in order to control host inflammatory responses (Crabtree *et al.*, 1999). The *cag* molecule(s) responsible for the induction of IL-8 *via* NFκB are not defined. One speculative possibility might be that the *cag* type IV system injects an inducer protein into the host cells to elicit the inflammatory response. Alternatively, a *cag* encoded protein or even a complex of Cag proteins, such as the type IV structure or a pilus, might act as activator.

A second signaling pathway leading to the activation of IL-8 secretion was recently described. *cag*-positive strains of *Hp,* but not individual *cag* mutants, induce expression of c-fos and c-jun and activate the transcription factor AP-1 (Naumann *et al.*, 1999; Meyer-ter-Vehn *et al.*, 2000) (Figure 3). Similarly to NFκB, AP-1 complexes are able to positively regulate the promoter activity of the IL-8 gene. Both cell responses are activated by the MAP-kinases. ERK1/2, MEK1/2, and JNK, but not p38 ERK kinase, induce the transcription of c-fos *via* phosphorylation of the transcription factor Elk-1. c-Jun activation depends on phosphorylation by JNK kinase. c-Fos and c-Jun are able to form heterodimers that participate in stable AP-1 complexes, resulting in prolonged AP-1 activation. AP-1 is an important regulator of cell proliferation and differentiation. It is therefore not surprising that deregulation of c-Fos and c-Jun expression can cause neoplastic transformations and increase the rate of tumor formation. This circumstance may account for the hyper-proliferation observed in epithelial cells during chronic infections with *Hp* and may in the long term contribute to the induction of gastric carcinoma.

The signaling pathways described in this section are triggered by activation of transcription factors. Although these events depend on the presence of an intact *cag*-PAI, transcription factors can also be activated by a variety of different molecules. Among these are chemokines, growth factors or stress responses. Thus, Type I strains may exhibit a closer *cag* mediated host cell contact than Type II strains and somehow trigger AP-1 and NFκB activation by triggering a non-specific stress response. In fact, induction of IL-8 has been described for various pathogenic bacteria which exchange signals with host cells through structures involved in a direct pathogen-host cell interaction.

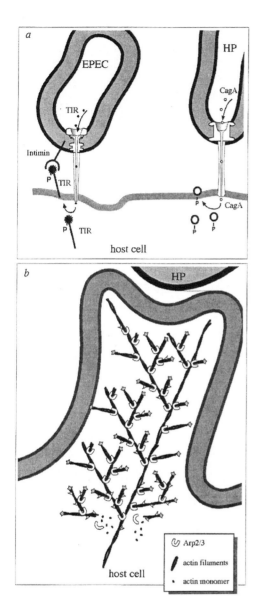

Figure 4. Type III and Type IV secretion systems converge functionally. a) EPEC and *Hp* inject Tir and CagA effector proteins by Type III or Type IV systems, respectively. Tir, a product of the LEE pathogenicity island, is the translocated EPEC receptor for Intimin. b) cellular responses in both systems: cortical actin polymerization and pedestal protrusions are induced after phosphorylation of a tyrosine residue by a host cell kinase.

Translocation and Tyrosine-phosphorylation of CagA

The only *cag* molecule which is currently known to be injected into the host cell through the Type IV secretion system is CagA (Segal *et al.*, 1999, Asahi *et al.*, 2000, Odenbreit *et al.*, 2000; Stein *et al.*, 2000). A *cag*A knock-out mutant induces wild-type levels of IL-8 and NF(B in tissue culture cells and therefore the CagA-induced signals are clearly independent and separate processes from the inflammatory signaling events. This is also evident from the observation that *cag*F and *vir*D4 mutants can still induce IL-8 secretion while the same mutants are deficient for tyrosine-phosphorylation and translocation of CagA.

After infection of host cells with *Hp* wild-type, but not with type IV mutants, CagA was detected in the membrane fraction of gastric epithelial cells. After translocation, tyrosine residues in CagA are phosphorylated by a host cell kinase, followed by the appearance of surface protrusions or pedestals (Figure 4). These structures are a consequence of rearrangements of actin and other cytoskeletal proteins, including the profilin binding protein WASP, beneath the attaching bacterium (Segal *et al.*, 1996). This model is supported by the electron-micrographic detection of immuno-gold labeled tyrosine-phosphorylated proteins (most likely CagA) at the basis of the pedestals that cup the bacterium. A similar model was described for the translocated intimin receptor (Tir) from EPEC (Kenny *et al.*, 1997) (Figure 4). Whether CagA, similar to Tir, is a receptor mediating the intimate contact between the *Helicobacter* and the host cell warrants future investigation. However, Tir and CagA lack sequence homologies and these proteins are associated with unrelated secretion mechanisms (type III and type IV secretion, respectively). Despite their differences, both secretion systems can translocate virulence effectors into host cells, suggesting the functional convergence of secretion engines during the evolution of pathogens.

The nucleation of actin filaments at the cell cortex depends on activated Arp 2/3 complexes that interact with Rho-family proteins (i.e. Cdc42) *via* WASP-like family proteins (Welch, 1999) (Figure 3). The bacterial pathogens *Listeria monocytogenes* and *Shigella flexneri* trigger actin polymerization by a WASP-like bacterial product (Listeria ActA) or by stimulating N-WASP recruitment (Shigella IcsA) (Loisel *et al.*, 1999; Egile *et al.*, 1999). EPEC-induced pedestals involve activation of the Chp small GTPase and WASP recruitment (Kalman *et al.*, 1999). By analogy, CagA might mimic WASP, CagA might recruit WASP through a specific small GTPase, or CagA might activate a system inducing conformational changes in the Arp 2/3 complex after binding the N-WASP-like protein CA region (Figure 3). Alternatively, CagA effects might proceed by a novel pathway that has not yet been described for other pathogens.

Open Questions

The *cag* Type IV secretion system and the connection with the host cells through translocated effector(s) reflect an important aspect of the virulence of *Hp*. Many issues remain unresolved: i) the evolutionary origin and mechanism of acquisition of the *cag* PAI; ii) the functional convergence between Type III and Type IV secretion systems; iii) the functional analogies between Tir and CagA; iv) the intracellular targets. The *cag* PAI unifies principles and mechanisms from different pathogens for subversion and control of host cells. Further studies are likely to provide a framework to unify apparently unrelated bacterial pathogens.

Acknowledgements

We thank B.B. Finlay, S. Falkow, C. Montecucco and J. Hacker for helpful suggestions and M. Achtman and S. Suerbaum for comments on the manuscript. We acknowledge P. Christie for comments and N. Lange, S. Censini, S. Guidotti, M. Marchetti and H. Pahl for granting access to experimental data. We gratefully acknowledge G. Corsi for the illustrations and C. Mallia for editorial assistance.

References

Akopyanz, N., Bukanov, N.O., Westblom, T.U., and Berg, D.E. 1992. PCR-based RFLP analysis of DNA sequence diversity in the gastric pathogen *Helicobacter pylori*. Nucleic Acids Res. 20: 6221-6225.

Akopyanz, N., Clifton, S.W. , Kersulyte, D. , Crabtree, J.E., Youree, B.E., Reece, C.A.,

Bukanov, N.O., Drazek, E.S., Roe, B.A., and Berg, D.E. 1998. Analyses of the *cag* pathogenicity island of *Helicobacter pylori*. Mol. Microbiol. 28: 37-53.

Andersson, S.G., Zomorodipour, A., Andersson, J.O., Sicheritz-Ponten, T., Alsmark, U.C., Podowski, R.M., Naslund, A.K., Eriksson, A.S., Winkler, H.H., and Kurland, C.G. 1998. The genome sequence of *Rickettsia prowazekii* and the origin of mitochondria. Nature 396: 133-140.

Asahi, M., Awuma, T., Ito, S., Ito, Y., Suto, H., Nagai, Y., Tsubokawa, M., Tohyama, Y., Maeda, S., Omata, M., Suzuki, T, and Sasakawa C. 2000. *Helicobacter pylori* CagA Protein can be tyrosine phosphoylated in the gastric epithelial cells. J. Exp. Med. 191: 593-602

Atherton, J.C., Tham, K.T., Peek Jr, R.M., Cover, T.L., and Blaser, M.J. 1996. Density of *Helicobacter pylori* infection *in vivo* as assessed by quantitative culture and histology. J. Infect. Dis. 174: 552-556.

Baker, B., Zambryski, P., Staskawicz, B., and Dinesh-Kumar, S.P. 1997. Signaling in plant-microbe interactions. Science 276: 726-733.

Blaser, M. J. 1998. *Helicobacter pylori* and gastric diseases. B.M.J. 316: 1507-1510.

Blum, G., Ott, M., Lischewski, A., Ritter, A., Imrich, H., Tschape, H., and Hacker, J. 1994. Excision of large DNA regions termed pathogenicity islands from tRNA-specific loci in the chromosome of an *Escherichia coli* wild-type pathogen. Infect. Immun. 62: 606-614.

Censini, S., Lange, C., Xiang, Z., Crabtree, J.E., Ghiara, P., Borodovsky, M., Rappuoli, R., and Covacci, A. 1996. *cag*, a pathogenicity island of *Helicobacter pylori*, encodes type I-specific and disease-associated virulence factors. Proc. Natl. Acad. Sci. USA. 93: 14648-14653.

Christie, P. 1997. *Agrobacterium tumefaciens* T-complex transport apparatus: a paradigm for a new family of multifunctional transporters in eubacteria. J. Bacteriol. 179: 3085-3094.

Christie, P.J. 1997. The *cag* pathogenicity island: mechanistic insights. Trends Microbiol. 5: 264-265.

Cormack, B.P., Valdivia, R.H., and Falkow, S. 1996. FACS-optimized mutants of the green fluorescent protein (GFP). Gene 173: 33-38.

Covacci, A., and Rappuoli, R. 1993. Pertussis toxin export requires accessory genes located downstream from the pertussis toxin operon. Mol. Microbiol. 8: 429-434.

Covacci A., Censini, S., Bugnoli, M., Petracca, R., Burroni, D., Macchia, G., Massone, A., Papini, E., Xiang, Z., Figura, N., and Rappuoli, R. 1993. Molecular characterization of the 128-kDa immunodominant antigen of *Helicobacter pylori* associated with cytotoxicity and duodenal ulcer. Proc. Natl. Acad. Sci. USA. 90: 5791-5795.

Covacci, A., Falkow, S., Berg, D.E., and Rappuoli, R. 1997. Did the inheritance of a pathogenicity island modify the virulence of *Helicobacter pylori*? Trends Microbiol. 5: 205-208.

Covacci, A., and Rappuoli, R. 1998. *Helicobacter pylori*: molecular evolution of a bacterial quasi-species. Curr. Opin. Microbiol. 1: 96-102.

Covacci, A., and Rappuoli, R. 2000. Tyrosine-phosphorylated bacterial proteins: trojan horses for the host cell. J. Exp. Med. 191: 587-592

Cover, T.L., Cao, P., Lin, C.D., Tham, K.T., and Blaser, M.J. 1993. Correlation between vacuolating cytotoxin production by *Helicobacter pylori* isolates *in vitro* and *in vivo*. Infect. Immun. 61: 5008-5012.

Cover, T.L. 1996. The vacuolating cytotoxin of *Helicobacter pylori*. Mol. Microbiol. 20: 241-246.

Crabtree, J.E., Xiang, Z., Lindley, I.J., Tompkins, D.S., Rappuoli, R. and Covacci, A. 1995. Induction of interleukin-8 secretion from gastric epithelial cells by a cagA negative isogenic mutant o of *Helicobacter pylori*. J. Clin. Pathol. 48: 967-969.

Crabtree, J.E., Kersulyte, D., Li, S.D., Lindley, I.J., and Berg, D.E. 1999. Modulation of *Helicobacter pylori* induced interleukin-8 synthesis in gastric epithelial cells mediated by *cag* PAI encoded VirD4 homologue. J. Clin. Pathol. 52: 653-6577

Deibel, C., Kramer, S., Chakraborty, T., and Ebel, F. 1998. EspE, a novel secreted protein of attaching and effacing bacteria, is directly translocated into infected host cells, where it appears as a tyrosine-phosphorylated 90 kDa protein. Mol. Microbiol. 30: 147-161.

Dundon W.G., Beesley, S.M., and Smyth, C.J. 1998. *Helicobacter pylori*-a conundrum of genetic diversity. Microbiology 144: 2925-2939.

Egile, C., Loisel, T.P., Laurent, V., Li, R., Pantaloni, D., Sansonetti, P.J., and Carlier, M.F. 1999. Activation of the CDC42 effector N-WASP by the *Shigella flexneri* IcsA protein promotes actin nucleation by Arp2/3 complex and bacterial actin-based motility. J. Cell. Biol. 146: 1319-1332

Euler, A.R., Zurenko, G.E., Moe, J.B., Ulrich, R.G., and Yagi, Y. 1990. Evaluation of two monkey species (Macaca mulatta and Macaca fascicularis) as possible models for human *Helicobacter pylori* disease. J. Clin. Microbiol. 28: 2285-2290

Falkow, S. 1998. Who speaks for the microbes? Emerg. Infect. Dis. 4: 495-497.

Figura, N., Vindigni, C., Covacci, A., Presenti, L., Burroni, D., Vernillo, R., Banducci, T., Roviello, F., Marrelli, D., Biscontri, M., Kristodhullu, S., Gennari, C., and Vaira, D. 1998. *cagA* positive and negative *Helicobacter pylori* strains are simultaneously present in the stomach of most patients with non-ulcer dyspepsia: relevance to histological damage. Gut 42: 772-778.

Finlay, B.B., and Falkow, S. 1997. Common themes in microbial pathogenicity revisited. Microbiol. Mol. Biol. Rev. 61: 136-169.

Finlay, B.B., and Cossart, P. 1997. Exploitation of mammalian host cell functions by bacterial pathogens. Science 276: 718-725.

Fullner, K.J., Lara, J.C., and Nester, E.W. 1996. Pilus assembly by Agrobacterium T-DNA transfer genes. Science 273: 1107-1109.

Glocker, E., Lange, C., Covacci, A., Bereswill, S., Kist, M., and Pahl, H.L. 1998. Proteins encoded by the *cag* pathogenicity islandof *Helicobacter pylori* are required for NF-kappaB activation. Infect. Immun. 66: 2346-2348.

Go, M.F., Kapur, V., Graham, D.Y., and Musser, J.M. 1996. Population genetic analysis of *Helicobacter pylori* by multilocus enzyme electrophoresis: extensive allelic diversity and recombinational population structure. J. Bacteriol. 178: 3934-3938.

Groisman, E.A., and Ochman, H. 1996. Pathogenicity islands: bacterial evolution inquantum leaps. Cell 87: 791-794.

Hacker, J., Blum-Oehler, G., Muhldorfer, I., and Tschape, H. 1997. Pathogenicity islands of virulent bacteria: structure, function and impact on microbial evolution. Mol. Microbiol. 23: 1089-1097.

Hamlet A., Thoreson A.C., Nilsson O., Svennerholm A.M., and Olbe L. 1999. Duodenal *Helicobacter pylori* infection differs in *cag*A genotype between asymptomatic subjects and patients with duodenal ulcers. Gastroenterology 116: 259-68

Hazell, S.L., Andrews, R.H., Mitchell, H.M., and Daskalopoulous, G. 1997. Genetic relationship among isolates *of Helicobacter pylori*: evidence for the existence of a *Helicobacter pylori* species-complex. FEMS Microbiol. Lett. 150: 27-32.

Hueck, C.J. 1998. Type III protein secretion systems in bacterial pathogens of animals and plants. Microbiol. Mol. Biol. Rev. 62: 379-433.

Jiang, Q., Hiratsuka, K., Taylor, D.E. 1996. Variability of gene order in different *Helicobacter pylori* strains contributes to genome diversity. Mol. Microbiol. 20: 833-842.

Kalman, D., Weiner, O.D., Goosney, D.L., Sedat, J.W., Finlay, B.B., Abe, A., and Bishop,

J.M. 1999. Enteropathogenic *E. coli* acts through WASP and Arp2/3 complex to form actin pedestals. Nature Cell. Biol. 1: 389-391.

Kenny, B., DeVinney, R., Stein, M., Reinscheid, D.J., Frey, E.A., and Finlay, B.B. 1997. Enteropathogenic *E. coli* (EPEC) transfers its receptor for intimate adherence into mammalian cells. Cell 91: 511-520.

Kersulyte, D., Chalkauskas, H., and Berg, D.E. 1999. Emergence of recombinant strains of *Helicobacter pylori* during human infection. Mol.Microbiol. 31: 31-43

Kirby, J.E., and Isberg, R.R.1998. Legionnaires' disease: the pore macrophage and the legion of terror within. Trends Microbiol. 6: 256-258.

Kubori T, Matsushima, Y., Nakamura, D., Uralil, J., Lara-Tejero, M., Sukhan, A., Galan, J.E., and Aizawa, S.I. 1998. Supramolecular structure of the *Salmonella typhimurium* type III protein secretion system. Science 280: 602-605.

Lai, E.M., and Kado, C.I. 1998. Processed VirB2 is the major subunit of the promiscuous pilus of *Agrobacterium tumefaciens*. J. Bacteriol. 180: 2711-2717.

Lange, C., Covacci, A., and Rappuoli, R. 1999. The CagT of *cag*, the pathogenicity island of *Helicobacter pylori*, is a lipoprotein that assembles into a core system and requires CagM for stabilization. Submitted.

Lee, C.A. 1996. Pathogenicity islands and the evolution of bacterial pathogens. Infect. Agents. Dis. 5: 1-7.

Lee, C.A. 1997. Type III secretion systems: machines to deliver bacterial proteins into eukaryotic cells? Trends Microbiol. 5: 148-156.

Loisel, T.P., Boujemaa, R., Pantaloni, D., and Carlier, M.F. 1999. Reconstitution of actin-based motility of Listeria and Shigella using pure proteins. Nature 401: 613-616.

Marchetti, M., Arico, B., Burroni, D., Figura, N., Rappuoli, R., and Ghiara, P. 1995. Development of a mouse model of *Helicobacter pylori* infection that mimics human disease. Science 267: 1655-1658

Macnab, R.M. 1992. Genetics and biogenesis of bacterial flagella. Annu. Rev. Genet. 26: 131-158.

Meyer-Ter-Vehn, T., Covacci, A., Kist, M., and Pahl, H.L. 2000. *Helicobacter pylori* activates mitogen-activated protein kinase cascades and induces expression of the proto-oncogenes c-fos and c-jun. J. Biol. Chem. 275: 16064-16072.

Munzenmaier, A., Lange, C., Glocker, E., Covacci, A., Moran, A., Bereswill, S., Baeuerle, P.A., Kist, M., and Pahl, H.L. 1997. A secreted/shed product of *Helicobacter pylori* activates transcription factor nuclear factor-kappa B. J. Immunol. 159: 6140-6147.

Naumann, M., Wessler, S., Bartsch, C., Wieland, W., Covacci, A., Haas, R. and Meyer, T.F. 1999. Activation of activator protein 1 and stress response kinases in epithelial cells colonized by *Helicobacter pylori* Encoding the *cag* pathogenicity island. J. Biol. Chem. 274: 31655–31662

O'Callaghan, D., Cazevieille, C., Allardet-Servent, A., Boschiroli, M.L., Bourg, G., Foulongne, V., Frutos, P., Kulakov, Y. and Ramuz, M. 1999. A homologue of the *Agrobacterium tumefaciens* VirB and *Bordetella pertussis* Ptl type IV secretion systems is essential for intracellular survival of *Brucella suis*. Mol. Microbiol. 33: 1210-1220

Odenbreit, S., Puls, J., Sedlmaier, B., Gerland, E., Fischer, W., and Haas, R. 2000. Translocation of *Helicobacter pylori* CagA into gastric epithelial cells by type IV secretion. Science 287: 1497-1500

Parsonnet, J., Friedman, G.D., Orentreich, N., and Vogelman, H. 1997. Risk for gastric cancer in people with CagA positive or CagA negative *Helicobacter pylori* infection. Gut 40: 297-301.

Phadnis, S.H., Ilver, D., Janzon, L., Normark, S., and Westblom, T.U. 1994. Pathological significance and molecular characterization of the vacuolating toxin gene of *Helicobacter pylori*. Infect. Immun. 62: 1557-1565.

Ritter, A., Blum, G., Emody, L., Kerenyi, M., Bock, A., Neuhierl, B., Rabsch, W., Scheutz, F., and Hacker, J. 1995. tRNA genes and pathogenicity islands: influence on virulence and metabolic properties of uropathogenic *Escherichia coli*. Mol. Microbiol. 17: 109-121.

Rossi, G., Rossi, M., Vitali, C.G., Fortuna, D., Burroni, D., Pancotto, L., Capecchi, S., Sozzi, S., Renzoni, G., Braca, G., Del Giudice, G., Rappuoli, R., Ghiara, P., and Taccini, E. 1999. A conventional beagle dog model for acute and chronic infection with *Helicobacter pylori*. Infect. Immun. 67: 3112-20

Russel, M. 1995. Moving through the membrane with filamentous phages. Trends Microbiol. 3: 223-228.

Russel, M. 1998. Macromolecular assembly and secretion across the bacterial cell envelope: type II protein secretion systems. J. Mol. Biol. 279: 485-499.

Salmond, G.P., Bycroft, B.W., Stewart, G.S., and Williams, P. 1995. The bacterial 'enigma': cracking the code of cell-cell communication. Mol. Microbiol. 16: 615-624 .

Segal, E.D., Falkow, S., and Tompkins, L.S. 1996. *Helicobacter pylori* attachment to gastric cells induces cytoskeletal rearrangements and tyrosine phosphorylation of host cell proteins. Proc. Natl. Acad. Sci. USA.93: 1259-1264.

Segal, E.D., Lange, C., Covacci, A., Tompkins, L.S., and Falkow, S. 1997. Induction of host signal transduction pathways by *Helicobacter pylori*. Proc. Natl. Acad. Sci. USA. 94: 7595-7599.

Segal, E.D., Cha, J., Lo, J., Falkow, S., and Tompkins, L.S. 1999. Altered states: involvement of phosphorylated CagA in the induction of host cellular growth changes *by Helicobacter pylori*. Proc. Natl. Acad. Sci. U S A. 96: 14559-64.

Segal, G., and Shuman, H.A. 1998. How is the intracellular fate of the *Legionella pneumophila* phagosome determined? Trends Microbiol. 6: 253-255.

Segal, G., Russo J.J., and Shuman, H.A. Relationships between a new type IV secretion system and the *icm/dot* virulence system of *Legionella pneumophila*. Mol.Microbiol. 34: 799-809

Schmitt, W., and Haas, R. 1994. Genetic analysis of the *Helicobacter pylori* vacuolating cytotoxin: structural similarities with the IgA protease type of exported protein. Mol. Microbiol. 12: 307-319.

Stein, M., Rappuoli, R., and Covacci, A. 2000. Tyrosine-phosphorylation of the *Helicobacter pylori* CagA antigen after *cag*-driven host cell translocation. Proc. Natl. Acad.Sci. U S A. 97: 1263-1268..

Susa, M. and Marre, R. 1999. *Legionella pneumophila* invasion of MRC-5 cells induces tyrosine protein phosphorylation. Infect. Immun. 67: 4490-4498

Telford J.L., Ghiara, P., Dell'Orco, M., Comanducci, M., Burroni, D., Bugnoli, M., Tecce, M.F., Censini, S., Covacci, A., Xiang, Z., and Rappuoli, R. 1994. Gene structure of the *Helicobacter pylori* cytotoxin and evidence of its key role in gastric disease. J. Exp. Med. 179: 1653-1658.

Tummuru, M.K., Cover, T.L., and Blaser, M.J. 1993. Cloning and expression of a high-molecular-mass major antigen of *Helicobacter pylori*: evidence of linkage to cytotoxin production. Infect. Immun. 61: 1799-1809.

Tummuru, M.K., Cover, T.L., and Blaser, M.J. 1994. Mutation of the cytotoxin-associated *cag*A gene does not affect the vacuolating cytotoxin activity of *Helicobacter pylori*. Infect. Immun. 62: 2609-2613.

Tummuru, M., Sharma, S.A., and Blaser, M.J. 1995. *Helicobacter pylori pic*B, a homologue of the *Bordetella pertussis* toxin secretion protein, is required for induction of IL-8 in gastric epithelial cells. Mol. Microbiol. 18: 867-876.

Valdivia, R.H., and Falkow, S. 1997. Fluorescence-based isolation of bacterial genes expressed within host cells. Science 277: 2007-2011.

Valdivia, R.H., and Falkow, S. 1997. Probing bacterial gene expression within host cells. Trends Microbiol. 5: 360-363.

Xiang, Z., Censini, S., Bayeli, P.F., Telford, J.L., Figura, N., Rappuoli, R., and Covacci, A. 1995. Analysis of expression of CagA and VacA virulence factors in 43 strains of *Helicobacter pylori* reveals that clinical isolates can be divided into two major types and that CagA is not necessary for expression of the vacuolating cytotoxin. Infect. Immun. 63: 94-98.

Weiss, A.A., Johnson, F.D., and Burns, D.L. 1993. Molecular characterization of an operon required for pertussis toxin secretion. Proc. Natl. Acad. Sci. USA. 90: 2970-2974.

Welch, M.D. (1999) The world according to arp: regulation of actin nucleation by the Arp2/3 complex. Trends Cell. Biol. 9: 423-427

Winans, S.C., Burns, D.L., and Christie, P.J. 1996. Adaptation of a conjugal transfer system for the export of pathogenic macromolecules. Trends Microbiol. 4: 64-68.

Yokota, K., Kurebayashi,Y., Takayama, Y., Hayashi, S., Isogai, H., Isogai, E., Imai, K. Yabana, T., Yachi, A. and Oguma, K. 1991. Microbiol. Immunol. 35: 475-480

From: *Helicobacter pylori: Molecular and Cellular Biology*
ISBN 1-898486-25-5 © 2001 Horizon Scientific Press, Wymondham, UK.

15

Helicobacter pylori VacA Vacuolating Cytotoxin and HP-Nap Neutrophil Activating Protein

Cesare Montecucco[*], Emanuele Papini, Marina de Bernard,
William G. Dundon, Mario Zoratti, Jean-Marc Reyrat,
John L. Telford, Giuseppe del Giudice and Rino Rappuoli

Abstract

The vacuolating toxin (VacA) is a major determinant of *Helicobacter pylori*-associated gastric disease. Unlike former beliefs, this toxin is active in almost all strains of *H. pylori*. In non-polarised cells, VacA alters the endocytic pathway, resulting in the release of acid hydrolases and the reduction of both extracellular ligand degradation and antigen processing. In polarized monolayers, VacA reduces the transepithelial electrical resistance and forms trans-membrane anion-specific channels. Localization of the VacA channels in acidic intracellular compartments causes osmotic swelling which, together with membrane fusion, leads to vacuole formation. The neutrophil activating protein of *H. pylori* (HP-Nap) induces the production of oxygen radicals in human neutrophils via a cascade of intracellular activation events which may contribute to the damage of the stomach mucosa. The activities of both VacA and HP-Nap are discussed in relation to *H. pylori's* nutrient requirements in the hostile environment of the human stomach .

Introduction

The entire sequence of the circular genome of two *H. pylori* strains is presently available (Tomb *et al.,* 1997; Alm *et al.,* 1999) (see Chapter 17) together with portions of the genome of several other strains. These genomes contain between 1,495 (strain J99) and 1,590 (strain 26695) open reading frames (ORFs). Most of these ORFs encode structural proteins or proteins necessary for the growth and division of the bacterial cell (Doig *et al.,* 1999; Ge and Taylor, 1999; Marais *et al.,* 1999). Some of them, including urease and the proteins involved in its biosynthesis, are necessary for the colonization and survival of *H. pylori* in its particular ecological niche, i.e. bound to the apical surface of stomach epithelial cells and within the protective mucus layer. This environment has peculiar and unique features, which are worth recalling here, due to their relevance to the activities of the virulence factor discussed below. To reach this particular site, the bacterium must pass through the oral cavity and the oesophageal tract, which are characterized by the presence of hydrolytic enzymes and a neutral pH.

[*]corresponding author, email: cesare@civ.bio.unipd.it

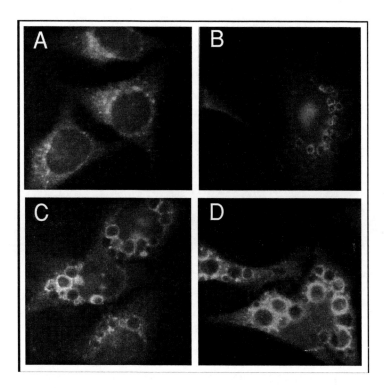

Figure 1. Vacuole formation and growth in HeLa cells after exposure to the vacuolating cytotoxin of *H. pylori*. The picture shows four consecutive and progressive stages of vacuolization at 0 (A), 160 min (B), 260 (C) and 540 min (D) after addition of 50 nM of VacA in the presence of 5 mM ammonium chloride at 37° C. The vacuolar membranes were stained with anti-rab7 antibodies prior to photography.

Once through the pylorus, the pH in the stomach lumen is much lower, with values in the range of 1 to 4. Such acid conditions denature most proteins, thus promoting their hydrolysis by stomach proteases. *H. pylori* can survive these harsh conditions by buffering the pH in its immediate vicinity through the action of urease, which catalyses the hydrolysis of urea to produce ammonia (see Chapter 10). *H. pylori* is believed to rapidly leave the stomach lumen to enter the mucus layer which covers and protects the mucosa against ulceration due to the gastric fluid.

The mucus layer has unique barrier properties and is relatively impermeable even to small molecules. In addition, hydrogen ions can freely diffuse from the apical portion of the mucosal epithelial cells into the gastric lumen, but not *vice versa*. In contrast, bicarbonate anions (and probably iron and nickel ions which are necessary for *H. pylori* growth) are poorly permeable. Thus, a pH gradient exists inside the mucus film resulting in slightly acidic pH near the apical cell membrane under normal conditions, but acid pH if the thickness of the mucus layer is reduced.

H. pylori can enter the mucus because it releases mucus hydrolyzing enzymes, has a spiral shape and is propelled by powerful flagella. Many bacteria adhere strongly to the apical cell membrane via adhesins and by inducing a reorganization of the plasma membrane of the host gastric epithelial cells. The actual volume available to large solutes below the mucus film and above the cells is not known, but it can be safely assumed that it is very limited. As a consequence, molecules released by *H. pylori* into its environment may reach high concentrations even if they are released in limited amounts.

A second introductory consideration is that the supply of nutrients may be very limited, including essential ions necessary for bacterial growth. Within this framework, it is conceivable that several *H. pylori* genes encode for proteins or for the biosynthesis of other molecules which cause changes in the physiological state of the stomach cells and tissues in order to promote growth and diffusion of *H. pylori*. Urease is discussed in Chapter 10 and others probably remain to be discovered. In the present chapter, we focus on two such proteins which are also major virulence factors: the vacuolating cytotoxin (VacA) and the neutrophil activating protein (HP-Nap).

Vacuolating Cytotoxin (VacA)

A Protein Responsible for Cytotoxicity

The first description of a protein cytotoxin by *H. pylori* showed that the supernatants of half of the *H. pylori* isolates could induce the formation of large cytoplasmic vacuoles in cultured eukaryotic cells (Leunk *et al.*, 1988). As shown in Figure 1, these vacuoles arise in the perinuclear area and grow in size to fill the entire cytosol, leading eventually to death by necrosis (Figura *et al.*, 1989). The single protein responsible for cell vacuolization was later purified and found to be a protein of 95 kDa, designated VacA (Cover and Blaser, 1992). Alternative purification protocols have since been developed (Manetti *et al.*, 1995; Icatlo *et al.*, 1998; Reyrat *et al.*, 1998). Antibodies raised against the purified VacA protein prevented the cell vacuolization induced by *H. pylori* supernatants as did some antisera derived from *H. pylori* infected patients (Cover and Blaser, 1992; Manetti *et al.*, 1995).

Gene Structure and Biosynthesis

The partial amino acid sequence of VacA (Cover and Blaser, 1992) allowed its cloning with degenerate primers (Cover *et al.*, 1994; Schmitt and Haas, 1994; Phadnis *et al.*, 1994; Telford *et al.*, 1994). Similar to other genes in *H. pylori*, *vacA* is characterized by considerable sequence variability. The VacA protein is not homologous to any other known protein but four different domains have been identified by secondary structure prediction methods, as depicted in Figure 2. Beginning at the amino-terminus, there is a 33 residue N-terminal signal sequence, which permits the export of the toxin from the cytosol to the periplasm. This is followed by a 37 kDa region predicted to be rich in pleated β- sheets. This region begins with a 32 residue hydrophobic segment and ends with a protease sensitive segment. Indeed part of the VacA toxin released into the medium is nicked at this point (Telford *et al.*, 1994). The following 58 kDa part is predicted to consist of two domains, which are separated by a flexible segment of variable length (Figure 2): the first domain is highly conserved whereas the second one exhibits considerable genetic diversity.

The C-terminal end of the molecule is highly conserved and has three distinctive features: a pair of cysteines separated by ten residues, followed by a 35 kDa region rich in amphipathic pleated beta segments and ending with a stretch of alternating hydrophobic residues that terminates in a C-terminal phenylalanine. These structural elements characterize a domain capable of translocating the portion of the polypeptide chain present at its amino-terminus across the outer membrane of Gram-negatives (Wandersman, 1992, Pohlner *et al.*, 1987). After translocation, part of the 95 kDa protein is released following proteolysis by yet unidentified proteases, but 40-60% of the mature VacA is still associated with the outer membrane of *H. pylori*. After release into the culture medium, the toxin can be additionally cleaved

A

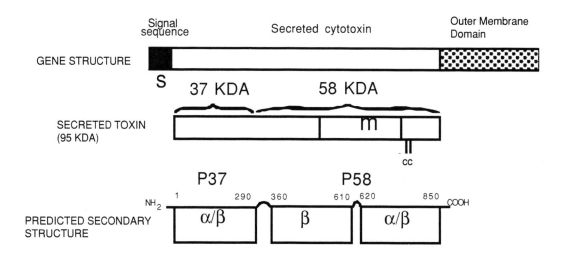

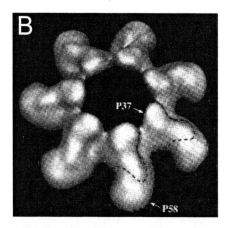

Figure 2. A) Schematic structure of the vacA gene and the secreted VacA toxin of *H. pylori*. The VacA toxin is translated as a 140 kDa protein that can be recognized by a inner membrane protein transport apparatus via the amino-terminal signal sequence (black), which is then removed by periplasmic proteases. The carboxyl-terminal 35 kDa domain (dotted) mediates the translocation of the secreted part of the toxin across the outer membrane to the cell surface. Here, the toxin is released from the bacterium after proteolysis at a site located after the cysteine doublet indicated in the figure (middle panel). Secondary structure methods predict of three different domains within the secreted toxin (middle panel). B) Low resolution electron microscopic structure of VacA. The VacA toxin was deposited on a metal grid and rotary shadowed. The image was generated as detailed by Reyrat *et al.* (1999).

within the segment connecting the p37 and p58 domains, which remain bound to each other via non covalent forces (Telford *et al.*, 1994).

Genetic Variability in the *vacA* Gene

There is considerable strain-dependent sequence variation in the VacA protein (Atherton *et al.*, 1995; Cover, 1996). Signal sequences have been grouped into three different types (s1a, s1b and s2). A second highly divergent segment, referred to as the m-region, is present within p58. The various m sequences were divided into two main types, m1 and m2 (Atherton *et al.*, 1995). Strains with s1/m1 generally secrete high levels of VacA protein and are highly toxigenic in the standard HeLa cell vacuolation assay. Most strains expressing the s2 type fail to release the toxin (Atherton *et al.*, 1995). Strains which express an s1/m2 toxin produce significant quantities of toxin that assembles into the correct structure but is not toxic for HeLa cells. There are no differences in virulence between m1 and m2 strains (Go *et al.*, 1998). Recent data shows that s1/m2 type VacA toxins vacuolate cell lines other than HeLa cells and that m1 and m2 toxins differ in their ability to bind to different cell lines (Pagliaccia *et al.*, 1998). This suggests that the m-region is involved in target cell interaction and the two alleles may have evolved to bind different receptors. *H. pylori* strain 95-54, producing an s1/m2 type toxin, is as active as strain 26695 (s1/m1) in lowering the trans-epithelial resistance (TER) of a polarized cell monolayer (Pelicic *et al.*, 1999). This model system mimics bacterial adhesion to the apical surface of polarized epithelial cells in the gastric tissue, and may reflect important *in vivo* biological properties of VacA (see below).

Strains containing *vacA* genes with the s2 signal peptide are relatively rare and their biological significance is not clear. In marked contrast, m2 strains are found frequently and reflect functional polymorphism. Whereas approximately 80% of strains found in western populations are s1/m1 and 20% are s1/m2, in Chinese populations this figure is reversed (Pan *et al.*, 1998). The high incidence of s1/m2 strains in China may reflect the geographical partitioning of populations of *H. pylori* (see Chapter 19).

Based on the identification of multiple combinations of the divergent s and m regions, it has been suggested that horizontal DNA transfer from other species is at the origin of the mosaic organization of the *vacA* gene, similarly to what has been previously observed for the IgA protease locus of *N. gonorrhoeae* (Halter *et al.*, 1989).

Structure

The secreted VacA toxin has a strong tendency to oligomerize into rosettes (Lupetti *et al.*, 1996; Cover *et al.*, 1997; Lanzavecchia *et al.*, 1998; Reyrat *et al.*, 1999) (Figure 2B). When the 58 kDa domain is expressed in the absence of the 37 kDa subunit, it folds into a soluble structure (Reyrat *et al.*, 1999) but does not form hexamers or heptamers. Images of the free 58 kDa subunit obtained by computer analysis of electron micrographs resemble the peripheral petals of the oligomer. Hence, the monomer organization in the native toxin is probably as indicated in Figure 2B, namely that the toxin consists of a ring of 58 kDa subunits upon which a smaller ring of 37 kDa subunits is superimposed.

The oligomeric form of VacA shows little activity until it is exposed to pH values below 5.5, and is not denatured by media of pH value as low as 1.5 (de Bernard *et al.*, 1995). VacA also shows the remarkable and very unusual property of resisting pepsin digestion at pH 2 at 37°C, but does not resist digestion by proteases at neutral pH (de Bernard *et al.*, 1995). VacA is also rather temperature resistant (Leunk *et al.*, 1988; Yahiro *et al.*, 1997). These properties of the toxin might be relevant for the pathogenesis of duodenal ulcers. In fact, VacA molecules present in the stomach lumen may reach the duodenum through the pylorus, and cause epithelial damage in the intestine region close to the pylorus, but not in the following intestine portions because of toxin degradation in the neutral/slightly alkaline and highly proteolytic intestinal environment.

The structure of VacA changes at acidic pH. The oligomer is dissociated into monomers and surface hydrophobic patches become exposed that mediate binding of the toxin to hydrophobic dyes and, more importantly, its insertion into model biological membranes (Cover *et al.*, 1997; Molinari *et al.*, 1998a). This structural transition takes place in a narrow pH range centered around pH 5.2 (de Bernard *et al.*, 1995; Molinari *et al.*, 1998a). Upon neutralization, VacA does not regain its oligomeric structure for hours, as shown by spectroscopy and sensitivity to proteolysis (de Bernard *et al.*, 1995). These findings suggest that after release from bacteria, VacA can exist in different forms whose biological activities and capabilities of interacting with membranes depend on pH.

Little is known about the structure of VacA bound to the outer membrane of *H. pylori* and its biological relevance remains to be elucidated. Recent work showed that the bacterial associated toxin does not require acid exposure to be active on monolayers of epithelial cells, suggesting that the outer membrane associated VacA is already active and, possibly, predominantly monomeric (Pelicic *et al.,* 1999). In addition, similarly to many gram-negative bacteria, *H. pylori* releases vesicles derived from outer membrane blebs, referred to as Outer Membrane Vesicles (OMV). OMV are rich in membrane associated VacA both *in vitro* and *in vivo* (Fiocca *et al.*, 1999). They can bind to cultured gastric cells and trigger cell vacuolation, and it has been proposed that blebbing of the outer membrane is an additional secretion system responsible for the delivery of VacA, and possibly other virulence factors, to host cells (Fiocca *et al.*, 1999).

Cell Vacuolation

VacA-induced vacuoles are acidic and can accumulate membrane-permeable weak bases, including dyes such as neutral red or acridine orange (Cover *et al.*,1992). This provides a simple and quantitative assay of the total internal volume of these compartments. Vacuolization requires the presence of membrane-permeable amines in the cell culture medium (Ricci *et al.*, 1997). In the presence of 5 mM ammonium and 100 nM VacA, small translucid vacuoles appear within half an hour in the perinuclear area of HeLa cells (Figure 1). These vacuoles contain membrane protein markers of late endosomes and lysosomes (Papini *et al.*, 1994; Molinari *et al.*, 1997) and are capable of incorporating fluid phase markers of the extracellular medium, including BSA-gold (Cover *et al.*, 1992; Catrenich *et al.*, 1992; Papini *et al.*, 1994). Acidification of vacuoles is essential for their formation, enlargement and preservation (Papini *et al.*, 1993; Cover *et al.*, 1993). Acidification is due to the activity of a vacuolar ATPase proton pump (V-ATPase) on the limiting vacuolar membrane (Papini *et al.*, 1996) and vacuole biogenesis requires active rab7, a small GTPase of late endosomes (Papini *et al.*, 1997).Vacuoles in biopsies of stomach mucosa from *H. pylori* infected patients and of HeLa cells in culture contains some electron dense material, but are largely devoid of the large array of multivesicular bodies that is characteristic of late endosomes (LE) and lysosomes (LY) (Leunk *et al.*, 1988; Cover *et al.*, 1992; Ricci *et al.*, 1997; C. Montecucco, unpublished). It appears that VacA induces considerable rearrangement of the organization of LE and LY, with extensive membrane fusion and swelling.

Metabolic and Pathological Effects of Vacuoles Induced by VacA

VacA induced vacuolization has several consequences for cellular physiology that may contribute to pathogenesis and to *H. pylori* survival. The process of cell vacuolation is accompanied by alterations that are likely to be directly relevant for infection and disease pathogenesis: i) marked decrease of the proteolytic activity within the endocytic pathway, including the proteolysis of antigens in the antigen processing compartment of antigen presenting cells (APC) that is needed to generate peptide epitopes and ii) extensive alteration of protein trafficking from the Trans-Golgi Network (TGN) to LE, as indicated by the mistargeting of acid hydrolases to the extracellular medium instead of the lysosomes (Satin *et al.*, 1997; Molinari *et al.*, 1998b). Such effects are detectable even before macroscopic vacuolation and probably correspond to both early alterations of the endocytic pathway and partial neutralization of its late stages (Satin *et al.*, 1997). Protein degradation is an essential cellular function, allowing for the removal of non-functional cell membrane proteins and extracellular ligands with reutilization of amino acids (Mukherjee *et al.*, 1997). The processing of protein antigens by APC cells takes place mainly inside the antigen processing compartment, a specialized form of LE/LY compartment which is capable of fusing with the plasma membrane (Watts, 1997). VacA inhibits the degradation of tetanus toxoid epitopes in the LE processing compartment of tetanus toxoid-specific APC cells (Molinari *et al.*, 1998 b). Consequently, the stimulation of T cell clones specific for epitopes generated in the antigen processing compartment was strongly inhibited by VacA, while that of T cell clones specific for epitopes generated in early endosomes was unaffected (Molinari *et al.*, 1998b). The inhibition of antigen processing and presentation by VacA could be part of a strategy of survival for *H. pylori* because the depression of antigen processing within APC cells of the mucosa could significantly contribute to the long lasting infection with *H. pylori* in the human stomach.

A second consequence of this alteration of the LE/LY compartments is the release of lysosomal acid hydrolases, which are made in the ER as pre-pro-enzymes and are tagged in the Golgi in such a way as to be recognized by tag-specific receptors, which drive them from the TGN to LE and LY. Within these latter compartments, the pre and pro segments are removed and the hydrolase domains are activated (Kornfeld and Mellman, 1989). If pre-pro-acid hydrolases are released on the acidic apical domain of stomach epithelial cells intoxicated by VacA, they could be converted into the active form that is capable of degrading the protective mucus film of the stomach. These host cell derived hydrolytic activities, perhaps in cooperation with hydrolases released by the bacterium (Slomiany and Slomiany, 1991), would loosen the meshwork and thickness of the mucin film, thus increasing its permeability to ions and nutrients that support *H. pylori* growth.

VacA increases the Permeability of Polarized Epithelial Monolayers

Strong adhesion of *H. pylori* to the apical surface of epithelial cells is followed by rearrangement of the plasma membrane and of the underlying actin meshwork which leads to binding that is almost irreversible (Dytoc *et al.*, 1993; Smoot *et al.*, 1993; Boren *et al.*, 1994; Segal *et al.*, 1996). At least part of this activity is believed to be due to the injection of bacterial proteins such as cagA via a putative type IV secretion system (see Chapter 14).

Polarized epithelial monolayers develop trans-epithelial resistance (TER) by sealing the borders of cells via tight junctions and other intercellular structures. The TER level indicates the degree of cell sealing (Kraehenbuhl and Neutra, 1992; Eaton and Simons, 1995). When VacA is activated by low pH and added apically, the TER drops rapidly to 1000-1500 Ohm x cm^2 and maintains this low value for days (Papini *et al.*, 1998). In parallel, the paracellular permeability between cells to small organic molecules and to ions such as Fe^{3+}

and Ni^{2+} is increased. *H. pylori* strains grown on the apical surface of the epithelial cell monolayers cause the same increase in permeability while VacA- mutants do not (Pelicic *et al.*, 1999). These results indicate that VacA increases the supply of essential nutrients that are necessary for bacterial growth on the mucosa (Papini *et al.*, 1998; Montecucco *et al.*, 1999a, 1999b). This supply of nutrients precedes the additional supply discussed above that is derived from the gastric lumen through a damaged mucus layer.

VacA Biochemical Activity

Bacterial protein toxins generally bind to protein or lipid cell surface receptors. No data is available on possible receptor(s) for VacA. High affinity binding sites for VacA would not be necessary for its activity *in vivo* because of the very limited volume into which it is released (Pelicic *et al.*, 1999). Similarly, if the toxin is delivered directly into the host cell by adherent bacteria, high affinity binding sites would also not be necessary.

VacA can promote vacuole formation if expressed within the cytosol of transfected cells (de Bernard *et al.*, 1997) and such activity requires the whole N-terminal domain plus a contiguous region of the C-terminal domain (de Bernard *et al.*, 1998; Ye *et al.*, 1999). In addition, VacA forms ion channels at low pH in potassium-filled liposomes (Moll *et al.*, 1995). Studies with planar lipid bilayers showed that VacA forms anion-selective and voltage-dependent channels only at low pH or after low-pH pre-activation (Tombola *et al.*, 1999a,b; Iwamoto *et al.*, 1999). The VacA channel in the membrane is hexameric (Iwamoto *et al.* 1999). With patch-clamp techniques, Szabò *et al.* (1999) have shown that the electrophysiologial properties of the anion-specific channels formed by VacA on the plasma membrane of HeLa cells closely resemble those of toxin channels in lipid bilayers.

Once VacA has entered eukaryotic cells, it is internalized by endocytosis (Garner and Cover, 1996; Szabò *et al.*, 1999). Hence the toxin anion channel might be active on endocytic compartments as well. In addition, VacA expressed in the cytosol was able to assemble similar channels on intracellular membranes. However, the activity of such intracellular anion-specific channels is only expected to be of importance to organelles endowed with the vacuolar ATPase (v-ATPase) proton pump. This pump is electrogenic; as it acidifies the lumen it also builds up a proton gradient that depresses its further activity. The presence of an anion channel should strongly promote the v-ATPase activity leading to an enhanced accumulation of protons which in turn would drive the uptake of weak bases that can permeate through the membrane. The relevance of the VacA channel activity to the proton pumping of v-ATPase is supported by the finding that cystic-fibrosis cells, which are chloride channel defective, have fewer acidic intracellular organelles (Al-Awqati *et al.*, 1992). An increased lumenal concentration of anions and cations would then cause an osmotically driven swelling with generation of vacuoles (Tombola *et al.*, 1999a). Accordingly, anion-channel inhibitors inhibit as well vacuolation in Hela cells exposed to VacA (Szabò *et al.*, 1999; Tombola *et al.*, 1999b). The importance of the anion-channel activity of VacA is also confirmed by the fact that the anion channels blocker NPPB prevents and reverses the VacA induced decrease of TER of polarized epithelia (Szabò *et al.*, 1999). Also the VacA induced apical anion secretion found in rat intestine (Guarino *et al.*, 1998) can be ascribed to its anion-channel activity.

Because VacA can induce vacuoles when expressed in the cytosol (de Bernard *et al.*, 1997), it became conceivable that VacA interacts with cytosolic protein(s) of the cell and that such interaction may contribute to vacuole formation. The yeast two-hybrid technique has recently emerged as a powerful tool to screen for protein-protein interactions taking place within cells (Kolanus, 1999). Using this approach with VacA as a bait for potential interacting partners expressed by a HeLa cell library a novel protein was identified (de Bernard *et al.*, 2000). The cellular role of this novel protein is currently being investigated.

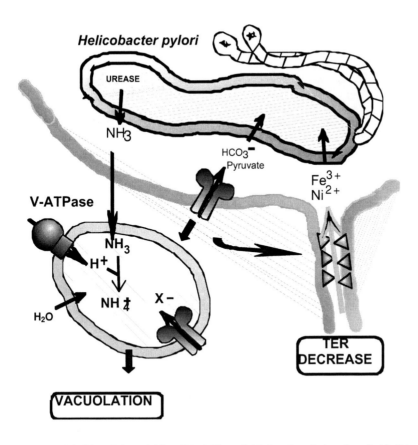

Figure 3. Current model of the cellular activities of VacA. The toxin binds to the apical portion of epithelial cells and inserts into the plasma membrane via hydrophobic protein-lipid interactions. This insertion leads to the formation of anion-selective channels of low conductance, which are capable of releasing bicarbonate and organic anions from the cell cytosol to support bacterial growth. Endocytosis and transport of the VacA toxin channel to endosomes increases their permeability to anions. This enhances the vacuolar ATPase proton pumping activity. In the presence of weak bases, and in particular of the ammonia generated by the *H. pylori* urease, osmotically active acidotropic ions (NH_4^+) will accumulate in the endosomes. This leads to water influx and vesicle swelling, an essential step in vacuole formation. Somehow, the VacA toxin is also capable of altering cell-cell junctions and of modifying the trans-epithelial electrical resistance, with concomitant increased fluxes of iron and nickel ions from the mucosa to the bacterium thus providing it with essential nutrients.

H. pylori Neutrophil Activating Protein

Neutrophils and mononuclear inflammatory cells infiltrate the *H. pylori* infected stomach mucosa (Warren and Marshall, 1983; Marshall *et al.*, 1985; Bayerdorffer *et al.*, 1992; Fiocca *et al.*, 1994; Goodwin, 1997; Suzuki *et al.*, 1997; Yamamura *et al.*, 1999). Moreover, the degree of mucosal damage is correlated with neutrophil infiltration (Warren and Marshall, 1983; Davies *et al.*, 1992, 1994; Fiocca *et al.*, 1994). Protein component(s) in *H. pylori* extracts can attract and activate neutrophils and other inflammatory cells (Karttunen *et al.*, 1990; Mai *et al.*, 1991,1992; Craig *et al.*, 1992; Nielsen and Andersen, 1992a,1992b; Kozol *et al.*, 1993; Reymunde *et al.*, 1993; Evans *et al.*, 1995a; Marchetti *et al.*, 1995) and *H. pylori* strains capable of neutrophil activation are isolated more frequently from peptic ulcer disease than active chronic gastritis (Rautelin *et al.*, 1993). A 150 kDa protein oligomer composed of identical 15 kDa subunits was identified from *H. pylori* water extracts that is capable of

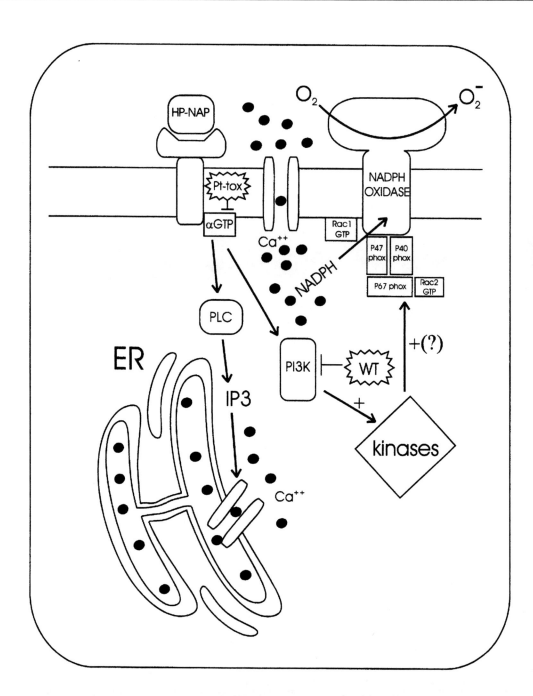

Figure 4. Scheme of the intracellular events involved in HP-Nap activation of human leukocytes. HP-Nap binds to a specific receptor which is coupled to a trimeric G protein sensitive to pertussis toxin. Binding triggers the entry of calcium via plasma membrane calcium channels and via the activation of channels located on the endoplasmic reticulum, which are opened by IP3, produced by activated phospholipase C. The activated HP-Nap receptor also activates a PI3Kinase, which is inhibited by wortmannin. The rise in the cytosolic calcium concentration and the activity of PI3Kinase leads to phosphorylation of the cytosolic subunits of the NADPH oxidase and to their migration to the plasma membrane, causing enzyme activation with production of superoxide anions.

promoting neutrophil adhesion to endothelial cells (Yoshida *et al.*, 1993; Evans *et al.*,1995a). This protein was designated HP-Nap because of its ability to induce neutrophils to produce reactive oxygen radicals (Evans *et al.*,1995a). However, the degree of neutrophil adhesion to endothelial cells is stimulated to various extents by different *H. pylori* strains, suggesting different level of expression of the HP-Nap protein, similarly to VacA (Evans *et al.*, 1995a).

Purified recombinant HP-Nap has been produced in *B. subtilis* and purified. This purified material is also chemotactic for human neutrophils and monocytes (B. Satin, unpublished), suggesting that HP-Nap plays a role in the accumulation of these cells at the site of infection with *H. pylori*. HP-Nap is a powerful stimulant of the production of reactive oxygen radicals (Evans *et al.*, 1995a; Satin *et al.*, 2000). HP-Nap acts via a cascade of intracellular activation events, including increase of cytosolic calcium ion concentration and phosphorylation of proteins leading to the assembly of functional NADPH oxidase on the neutrophil plasma membrane (Figure 4). The stimulation of reactive oxygen radicals by HP-Nap is completely inhibited by pertussis toxin, wortmannin and other specific inhibitors of intracellular molecules involved in cell activation and signalling (Satin *et al.*, 2000).

The nucleotide sequence of the *napA* gene, encoding HP-Nap, was highly conserved within twelve different isolates of *H. pylori* and did not resemble the mosaic structure seen for *vacA* (W.G. Dundon, unpublished). This would indicate the lack of immune selection for diversification of HP-Nap. A spectroscopic and electron microscopic study of HP-Nap revealed that it is a four-helix bundle protein which oligomerizes to a form a dodecamer with a central hole, structurally similar to the *Escherichia coli* DNA binding protein Dps (Grant *et al.*, 1998; Tonello *et al.*, 1999). Dps proteins are a diverse family of bacterial stress proteins that are induced during periods of nutrient limitation (Almirón *et al.*, 1992; Chen and Helmann, 1995; Peña and Bullerjahn, 1995). However, unlike Dps which binds DNA, HP-Nap binds iron and can bind up to 500 atoms of iron per dodecamer (Tonello *et al.*, 1999). In addition, HP-Nap is resistant to thermal and chemical denaturation, similar to ferritins. Indeed, Evans *et al.* (1995b) originally described HP-Nap as a bacterial ferritin based on the nucleotide sequence homologies. The *nap* gene has several A+T-rich regions of dyad symmetry immediately upstream of its start codon which could function as Fur-regulated promoters for iron regulation. Similar A+T rich regions exists upstream of the *H. pylori* ferritin *pfr* gene (Frazier *et al.*, 1993). All these data indicated that HP-Nap is important for iron uptake by *H. pylori*. Iron is an essential nutrient for bacterial growth. The unique niche that *H. pylori* inhabits is limited in available iron and *H. pylori* possesses several mechanisms for iron uptake (Frazier *et al.*, 1995; Housson *et al.*, 1993; Dhaenens *et al.*, 1997; Bereswill *et al.*, 1998; Worst *et al.*, 1999). However, despite the evidence suggesting a role in iron binding, recent data from our laboratory reveal that HP-Nap is constitutively expresssed under iron-depletion conditions, that it is not regulated by the presence or absence of iron and that it does not play a part in the metal resistance of *H. pylori* (W. G. Dundon, unpublished).

Therefore, we hypothesize that HP-Nap was originally an iron binding/iron regulated protein which has evolved to function as a neutrophil activator. Activating neutrophils induces a moderate inflammatory reaction, leading to alteration of the epithelial tight junctions and basal membranes and possibly promotes the release of nutrients from the mucosa to support the growth of *H. pylori* (Blaser, 1993).

Acknowledgements

Work carried out in the authors' laboratories is supported by the European Community grants BMH4-CT97-2410, by the Progetto Finalizzato CNR Biotecnologie, by the CNR-MURST 5% Project and by MURST 40% Project on Inflammation.

References

Al-Awqati, Q., Barasch, J., and Landry, D. 1992. Chloride channels of intracellular organelles and their potential role in cystic fibrosis. J. Exp. Biol. 172: 245-266.

Alm, R.A., Ling, L.-S.L., Moir, D.T., King, B.L., Brown, E.D., Doig, P.C., Smith, D.R., Noonan, B. and 15 other authors. 1999. Genomic-sequence comparison of two unrelated isolates of the human gastric pathogen *Helicobacter pylori*. Nature. 397: 176-180.

Almirón, M., Link, A.J., Furlong, D., and Kolter, R. 1992. A novel DNA-binding protein with regulatory and protective roles in starved *Escherichia coli*. Genes Dev. 6: 2646-2654.

Atherton, J.C., Cao, P., Peek, R.M. Jr, Tummuru, M.K., Blaser, M.J., and Cover, T.L. 1995. Mosaicism in vacuolating cytotoxin alleles of *Helicobacter pylori*. Association of specific *vacA* types with cytotoxin production and peptic ulceration. J. Biol. Chem. 270: 17771-17777.

Bayerdorffer, E., Lehn, N., Hatz, R., Mannes, G.A., Oertel, H., Sauerbruch, T. and Stolte, M. 1992. Difference in expression of *Helicobacter pylori* gastritis in antrum and body. Gastroenterol. 102: 1575-1582.

Bereswill; S., Waidner, U., Odenbreit, S., Lichte, F., Fassbinder, F., Bode, G., and Kist, M. 1998. Structural, functional and mutational analysis of the *pfr* gene encoding a ferritin from *Helicobacter pylori*. Microbiol. 144: 2505-2516.

Blaser, M.J. 1993. *Helicobacter pylori*: microbiology of a "slow" bacterial infection. Trends Microbiol. 1: 255-259.

Boren, T., Normark, S., and Falk, P. 1994. *Helicobacter pylori*: molecular basis for host recognition and bacterial adherence. Trends Microbiol. 2: 221-228.

Catrenich, C.E. and Chestnut, M.H. 1992. Character and origin of vacuoles induced in mammalian cells by the cytotoxin of *Helicobacter pylori*. J. Med. Microbiol. 37: 389-395.

Chen, L., and Helmann, J.D. 1995. *Bacillus subtilis* MrgA is a Dps (PexB) homologue: evidence for metalloregulation of an oxidative-stress gene. Mol. Microbiol. 18: 295-300.

Covacci, A., Telford, J.L., Del Giudice, G., Parsonnet, J., and Rappuoli, R. 1999. *Helicobacter pylori* virulence and genetic geography. Science. 284:1328-1333.

Cover, T.L. and Blaser, M.J. 1992. Purification and characterization of the vacuolating toxin from *Helicobacter pylori*. J. Biol. Chem. 267: 10570-10575.

Cover, T.L., Susan, A.H., and Blaser, M.J. 1992. Characterization of HeLa cell vacuoles induced by *Helicobacter pylori* broth culture supernatant. Human Pathol. 23: 1004-1010.

Cover, T.L., Reddy, L.Y. and Blaser, M.J. 1993. Effects of ATPase inhibitors on the response of HeLa cells to *Helicobacter pylori* vacuolating toxin. Infect. Immun. 61: 1427-1431.

Cover, T.L., Tummuru, M.K.R., Cao, P., Thompson, S.A., and Blaser M.J. 1994. Divergence of genetic sequences for the vacuolating cytotoxin among *Helicobacter pylori* strains. J. Biol. Chem. 269: 10566-10573.

Cover, T.L. 1996. The vacuolating cytotoxin of *Helicobacter pylori*. Mol. Microbiol. 20: 241-246.

Cover, T.L., Hanson, P.I., and. Heuser, J.E. 1997. Acid-induced dissociation of VacA, the *Helicobacter pylori* vacuolating toxin, reveals its pattern of assembly. J. Cell Biol. 138: 759-769.

Craig, P.M., Territo, M.C., Karnes, W.E., and Walsh, J.H. 1992. *Helicobacter pylori* secretes a chemotactic factor for monocytes and neutrophils. Gut. 33: 1020-1023.

Davies, G.R., Simmonds, N.J., Stevens, T.R., Grandison, A., Blake, D.R. and Rampton, D.S. 1992. Mucosal reactive oxygen metabolite production in duodenal ulcer disease. Gut. 33: 1467-1472.

Davies, G.R., Banatvala, N., Collins, C.E., Sheaff, M.T., Abdi, Y., Clements, L., and Rampton, D.S. 1994. Relationship between infective load of *Helicobacter pylori* and reactive oxygen metabolite production in antral mucosa. Scand. J. Gastroenterol. 29: 419-424.

de Bernard, M., Papini, E., de Filippis, V., Gottardi, E., Telford., J., Manetti, R., Fontana, A., Rappuoli, R., and Montecucco, C. 1995. Low pH activates the vacuolating toxin of *Helicobacter pylori*, which becomes acid and pepsin resistant. J. Biol. Chem. 270: 23937-23940.

de Bernard, M., Aricò, B., Papini, E., Rizzuto, R., Grandi, G., Rappuoli, R., and Montecucco, C. 1997. *Helicobacter pylori* toxin VacA induces vacuol formation by acting in the cell cytosol. Mol. Microbiol. 26: 665-674.

de Bernard, M., Burroni, D., Papini, E., Rappuoli, R., Telford, J.L., and Montecucco, C. 1998. Identification of the *Helicobacter pylori* VacA toxin domain active in the cell cytosol. Infect. Immun. 66: 6014-6016.

De Bernard, M., Moschioni, M., Napolitani, G., Rappuoli, R. and Montecucco, C. 2000. The VacA Toxin of *Helicobacter pylori* identifies a new intermediate filament interacting protein. EMBO J., in press

Dhaenens, L., Szczebara, F. and Husson, O. 1997. Identification, characterization, and immunogenicity of the lactoferrin-binding protein of *Helicobacter pylori*. Infect. Immun. 65: 514-518.

Doig, P., De Jonge, B.L., Alm, R.A., Brown, E.D., Uria-Nickelsen, M., Noonan, B., Mills, S.D., Tummino, P., Carmel, G., Guild, B.C., Moir, D.T., Vovis, G.F. and Trust, T.J. 1999. *Helicobacter pylori* physiology predicted from genomic comparison of two strains. Microbiol. Mol. Biol. Rev. 63: 675-707.

Dytoc, M., Gold, B., Louie, M., Huesca, M., Fedorko, L., Crowe, S., Lingwood, C., Brunton, J., and Sherman, P. 1993. Comparison of *Helicobacter pylori* and attaching-effaching *Escherichia coli* adhesion to eukaryotic cells. Infect. Immun. 61: 448-456.

Eaton, S. and Simons, K. 1995. Apical, basal, and lateral cues for epithelial polarization. Cell. 82: 5-8.

Evans, D.J. Jr., Evans, D.G., Takemura, T., Nakano, H., Lampert, H.C., Graham, D.Y., Granger, D.N., and Kvietys, P.R. 1995a. Characterization of a *Helicobacter pylori* neutrophil-activating protein. Infect. Immun. 63: 2213-2220.

Evans, D.J. Jr., Evans, D.G., Lampert, H.C., and Nakano, H. 1995b. Identification of four new prokaryotic bacterioferritins, from *Helicobacter pylori*, *Anabaena variabilis*, *Bacillus subtilis* and *Treponema pallidum*, by analysis of gene sequences. Gene. 153: 123-127.

Figura, N., Guglielmetti, P., Rossolini, A., Barberi, A., Cusi, G., Musmanno, R.A., Russi, M., and Quaranta, S. 1989. Cytotoxin production by *Campylobacter pylori* strains isolated from patients with peptic ulcers and from patients with chronic gastritis only. J. Clin. Microbiol. 27: 225-226.

Fiocca, R., Luinetti, O., Villani, L., Chiaravalli, A.M., Capella, C., and Solcia, E. 1994. Epithelial cytotoxicity, immune responses, and inflammatory components of *Helicobacter pylori* gastritis. Scand. J. Gastroenterol. 205: 11-21.

Fiocca, R., Necchi, V., Sommi, P., Ricci, V., Telford, J.L., Cover, T.L., and Solcia, E. 1999. Release of *Helicobacter pylori* vacuolating cytotoxin by both a specific secretion pathway and budding of outer membrane vesicles. Uptake of released toxin and vesicles by gastric epithelium. J. Pathol.188: 220-226.

Frazier, B.A., Pfeifer, J.D., Russell, D.G., Falk, P., Olsen, A.N., Hammar, M., Westblom, T.U., and Normark, S.J. 1995. Paracrystalline inclusions of a novel ferritin containing nonheme iron, produced by the human gastric pathogen *Helicobacter pylori*: evidence for a third class of ferritins. J. Bacteriol. 175: 966-972.

Garner, J.A. and Cover, T.L. 1996. Binding and internalization of the *Helicobacter pylori* vacuolating cytotoxin by epithelial cells. Infect. Immun. 64: 4197-4203.

Ge, Z., and Taylor, D.E. 1999. Contributions of genome sequencing to understanding the biology of *Helicobacter pylori*. Annu. Rev. Microbiol. 53: 353-387.

Go, M.F., Cissell, L., and Graham, D.Y. 1998. Failure to confirm association of *vacA* gene mosaicism with duodenal ulcer disease. Scand. J. Gastroenterol. 33: 132-136.

Goodwin, C.S. 1997. *Helicobacter pylori* gastritis, peptic ulcer, and gastric cancer: clinical and molecular aspects. Clin. Infect. Dis. 25: 1017-1019.

Grant, R.A., Filman, D.J., Finkel, S.E., Kolter R, and Hogle, J.M. 1998. The crystal structure of Dps, a ferritin homolog that binds and protects DNA. Nat. Struct. Biol. 5: 294-303.

Guarino, A., Bisceglia, M., Canani, R.B., Boccia, M.C., Mallardo, G., Bruzzese, E., Massari, P., Rappuoli, R., and Telford, J.L. 1998. Enterotoxic effect of the vacuolating toxin produced by *Helicobacter pylori* in Caco-2 cells. J. Infect. Dis. 178: 1373-1378.

Halter, R., Pohlner, J., and Meyer, T.F. 1989. Mosaic-like organization of IgA protease genes *in Neisseria gonorrhoeae* generated by horizontal genetic exchange *in vivo*. EMBO J. 8: 2737-2744.

Housson, M.-O., Legrand, D., Spik, G., and LeClerc, H. 1993. Iron acquistion by *Helicobacter pylori:* importance of human lactoferrin. Infect Immun. 61: 2694-2697.

Icatlo, F.C., Kuroki, M., Kobayashi, C., Yokoyama, H., Ikemori, Y., Hashi, T., and Kodama, Y. 1998. Affinity purification of *Helicobacter pylori* urease. Relevance to gastric mucin adherence by urease protein. J. Biol. Chem. 273: 18130-18138.

Iwamoto, H., Czajkowsky, D.M., Cover, T.L., Szabo, G., and Shao, Z. VacA from *Helicobacter pylori*: a hexameric chloride channel. 1999. FEBS Lett.: 450:101-104.

Ji, X., Fernandez, T., Buroni, D., Pagliaccia, C., Atherton, J.C., Reyrat, J.M., Rappuoli, R., and Telford, J.L. 2000. Cell specifity of *Helicobacter pylori* cytotoxin is determined by a short region in the polymorphic midregion. Infect. Immun. 68: 3754-3757.

Karttunen, R., Andersson, G., Poikonen, K., Kosunen, T.U., Karttunen, T., Juutinen, K., and Niemela, S. 1990. *Helicobacter pylori* induces lymphocyte activation in peripheral blood cultures. Clin. Exp. Immunol. 82: 485-488.

Kolanus, W. 1999. The two hybrid toolbox. Curr. Top. Microbiol. Immunol. 243:37-54.

Kornfeld, S. and Mellman, I. 1989. The biogenesis of lysosomes. Annu. Rev. Cell Biol. 5: 483-525.

Kozol, R., McCurdy, B., and Czanko, R. 1993. A neutrophil chemotactic factor present in *H. pylori* but absent in *H. mustelae*. Dig. Dis. Sci. 38: 137-141.

Kraehenbuhl, J.P. and Neutra, M.R. 1992. Molecular and cellular basis of immune protection of mucosal surfaces. Physiol. Rev. 72: 853-879.

Lanzavecchia, S., Bellon, P.L., Lupetti, P., Dallai, R., Rappuoli, R., and Telford, J.L. 1998. Three-dimensional reconstruction of metal replicas of the *Helicobacter pylori* vacuolating cytotoxin. J. Struct. Biol. 121: 9-18.

Leunk, R.D., Johnson, P.T., David, B.C., Kraft, W.G., and Morgan, D.R. 1988. Cytotoxin activity in broth-culture filtrates of *Campylobacter pylori*. J. Med. Microbiol. 26: 93-99.

Lupetti, P., Heuser, J.E., Manetti, R., Lanzavecchia, S., Bellon, P.L., Dallai, R., Rappuoli, R., and Telford, J.L. 1996. Oligomeric and subunit structure of the *Helicobacter pylori* vacuolating cytotoxin. J. Cell Biol. 133: 801-807.

Mai, U.E., Perez-Perez, G.I., Allen, J.B., Wahl, S.M., Blaser, M.J., and Smith, P.D. 1992. Surface proteins from *Helicobacter pylori* exhibit chemotactic activity for human leukocytes and are present in gastric mucosa. J. Exp. Med. 175: 517-525.

Mai, U.E., Perez-Perez, G.I., Wahl, L.M., Wahl, S.M., Blaser, M.J., and Smith, P.D 1991. Soluble surface proteins from *Helicobacter pylori* activate monocytes/macrophages by lipopolysaccharide-independent mechanism. J. Clin. Invest. 87: 894-900.

Manetti, R., Massari, P., Burroni, D., de Bernard, M., Marchini, A., Olivieri, R., Papini, E., Montecucco, C., Rappuoli, R., and Telford, J.L. 1995. *Helicobacter pylori* cytotoxin: importance of native conformation for induction of neutralizing antibodies. Infect. Immun. 63: 4476-4480.

Marais, A., Mendz, G.L., Hazell, S.L., and Mégraud, F. 1999. Metabolism and genetics of *Helicobacter pylori*: the genome era. Microbiol. Mol. Biol. Rev. 63: 642-674.

Marchetti, M., Aricò, B., Burroni, D., Figura, N., Rappuoli, R., and Ghiara, P. 1995. Development of a mouse model of *Helicobacter pylori* infection that mimics human disease. Science. 265: 1656-1658.

Marshall, B.J., Armstrong, J.A., McGeche, D.B., and Glancy, R.J. 1985. Attempt to fulfil Koch's postulates for pyloric Campylobacter. Med. J. Aust. 142: 436-439.

Molinari, M., Galli, C., Norais, N., Telford, J.L., Rappuoli, R., Luzio, J.P., and. Montecucco, C. 1997. Vacuoles induced by *Helicobacter pylori* toxin contain both late endosomal and lysosomal markers. J. Biol. Chem. 272: 25339-25344.

Molinari, M., Galli, C., de Bernard, M., Norais, N., Ruysschaert, J.M., Rappuoli, R., and Montecucco, C. 1998a. The acid activation of *Helicobacter pylori* toxin VacA: structural and membrane binding studies. Biochem. Biophys. Res. Comun. 248: 334-340.

Molinari, M., Salio, M., Galli, C., Norais, N., Rappuoli, R., Lanzavecchia, A., and Montecucco, C. 1998b. Selective inhibition of Li-dependent antigen presentation by *Helicobacter pylori* toxin VacA. J. Exp. Med. 187: 135-140.

Moll, G., Papini, E., Colonna, R., Burroni, D., Telford, J.L., Rappuoli, R., and Montecucco, C. 1995. Lipid interaction of the 37-kDa and 58-kDa fragments of the *Helicobacter pylori* cytotoxin. Eur. J. Biochem. 234: 947-952.

Montecucco, C., Papini, E., de Bernard, M., and Zoratti, M. 1999a. Molecular and cellular activities of *Helicobacter pylori* pathogenic factors. FEBS Lett. 452:16-21

Montecucco, C., Papini, E., de Bernard, M., Telford, J.L., and Rappuoli, R. 1999b. *Helicobacter pylori* vacuolating cytotoxin and associated pathogenic factors. In: The Comprehensive Sourcebook of Bacterial Protein Toxins. Ed. Alouf, J.E. and Freer, J.H. Academic Press., San Diego, CA 92101.

Mukherjee, S., Richik, N.G., and Maxfield, F.R. 1997. Endocytosis. Physiol. Rev. 77: 759-803.

Nielsen, H. and Andersen, L.P. 1992a. Activation of human phagocyte oxidative metabolism by *Helicobacter pylori*. Gastroenterol. 103: 1747-1753.

Nielsen, H. and Andersen, L.P. 1992b. Chemotactic activity of *Helicobacter pylori* sonicate for human polymorphonuclear leucocytes and monocytes. Gut. 33: 738-742.

Pagliaccia, C., de Bernard, M., Lupetti, P., Ji, X., Burroni, D., Cover, T.L., Papini, E., Rappuoli, R., Telford, J.L., and Reyrat, J.M. 1998. The m2 form of the *Helicobacter pylori* cytotoxin has cell type-specific vacuolating activity. Proc. Natl. Acad. Sci. USA. 95: 10212-10127.

Pan, Z.J., Berg, D.E., van der Hulst, R.W., Su, W.W., Raudonikiene, A., Xiao, S.D., Dankert, J., Tytgat, G.N., and van der Ende, A. 1998. Prevalence of vacuolating cytotoxin production and distribution of distinct *vacA* alleles in *Helicobacter pylori* from China. J. Infect. Dis. 178: 220-226.

Papini, E. Bugnoli, M., de Bernard, M., Figura, N., Rappuoli, R., and Montecucco, C. 1993. Bafomycin A1 inhibits *Helicobacter pylori*-induced vacuolization of HeLa cells. Mol. Microbiol. 7, 323-327.

Papini, E., de Bernard, M., Milia, E., Zerial, M., Rappuoli, R., and Montecucco, C. 1994. Cellular vacuoles induced by *Helicobacter pylori* originate from late endosomal compartments. Proc. Natl. Acad. Sci. USA. 91: 9720-9724.

Papini, E., Gottardi, E., Satin, B., de Bernard, M., Telford, J., Massari, P., Rappuoli, R., Sato S.B., and Montecucco, C. 1996. The vacuolar ATPase proton pump on intracellular vacuoles induced by *Helicobacter pylori*. J. Med. Microbiol. 44, 1-6.

Papini, E., Satin, B., Bucci, C., de Bernard, M., Telford, J.L., Manetti, R., Rappuoli, Zerial, M., and Montecucco, C. 1997. The small GTP binding protein rab7 is essential for cellular vacuolation induced by *Helicobacter pylori* cytotoxin. EMBO J. 16: 15-24.

Papini, E., Satin, B., Norais, N., de Bernard, M., Telford, J.L., Rappuoli, R., and Montecucco, C. 1998. Selective increase of the permeability of polarized epithelial cell monolayers by *Helicobacter pylori* vacuolating toxin. J. Clin. Invest. 102: 813-820.

Pelicic, V., Reyrat, J.M., Sartori, L., Pagliaccia, C., Rappuoli, R., Telford, J.L. Montecucco, C., and Papini, E. 1999. *Helicobacter pylori* VacA cytotoxin associated with the bacteria increases epithelial permeability independently of its vacuolating activity. Microbiol. 145: 2043-2050.

Peña, M.O., and Bullerjahn, G.S. 1995. The DpsA protein of *Synechococcus sp.*strain PCC7942 is a DNA binding hemoprotein. J. Biol. Chem. 270: 22478-22482.

Phadnis, S.H., Ilver, D., Janzon, L., Normark, S., and Westblom, T.U. 1994. Pathological significance and molecular characterization of the vacuolating toxin gene of *Helicobacter pylori*. Infect. Immun. 62: 1557-1565.

Pohlner, J., Halter, R., Beyreuther, K., and Meyer, T.F. 1987. Gene structure and extracellular secretion of *Neisseria gonorrhoeae* IgA protease. Nature. 325: 458-462.

Rautelin, H., Blomberg, B., Fredlund, H., Jarnerot, G., and Danielsson, D. 1993. Incidence of *Helicobacter pylori* strains activating neutrophils in patients with peptic ulcer disease. Gut. 34: 599-603.

Reymunde, A., Deren, J., Nachamkin, I., Oppenheim, D., and Weinbaum, G. 1993. Production of chemoattractant by *Helicobacter pylori*. Dig. Dis. Sci. 38: 1697-1701.

Reyrat, J.M., Charrel, M., Pagliaccia, C., Burroni, D., Lupetti, P., de Bernard, M., Xi, J, Norais, N., Papini, E., Dallai, R., Rappuoli, R., and Telford, J.L. 1998. Characterization of a monoclonal antibody and its use to purify the cytotoxin of *Helicobacter pylori*. FEMS Lett. 165: 79-84.

Reyrat, J.M., Lanzavecchia, S., Lupetti, P., de Bernard, M., Pagliaccia, C., Pelicic, V., Charrel, M., Ulivieri, C., Norais, N., Ji, X., Cabiaux, V., Papini, E., Rappuoli, R., and Telford, J.L. 1999 3D imaging of the 58 kDa cell binding subunit of the *Helicobacter pylori* cytotoxin. J. Mol. Biol. 290: 459-470.

Ricci, V., Sommi, P., Fiocca, R., Romano, M., Solcia, E., and Ventura, U. 1997. *Helicobacter pylori* vacuolating toxin accumulates within the endosomal-vacuolar compartment of cultured gastric cells and potentiates the vacuolating activity of ammonia. J. Pathol. 183: 453-459.

Satin, B., Norais, N., Telford, J.L., Rappuoli, R., Murgia, M., Montecucco, C., and Papini, E. 1997. Vacuolating toxin of *Helicobacter pylori* inhibits maturation of procathepsin D and degradation of Epidermal Growth Factor in HeLa cells through a partial neutralization of acidic intracellular compartments. J. Biol. Chem. 272: 25022-25028.

Satin, B., Del Giudice, G., Della Bianca, V., Dusi, S., Laudanna, C., Tonello, F., Kelleher, D., Rappuoli, R., Montecucco, C., and Rossi, F. 2000. The neutrophil activating protein (HP-NAP) of *Helicobacter pylori* is a protective antigen and a major virulence factor. J. Exp. Med. 191-1467-1476.

Schmitt, W. and Haas, R. 1994. Genetic analysis of the *Helicobacter pylori* vacuolating cytotoxin: structural similarities with the IgA protease type of exported protein. Mol. Microbiol. 12: 307-319.

Segal, E.D., Falkow, S., and Tompkins, L.S. 1996. *Helicobacter pylori* attachment to gastric cells induces cytoskeletal rearrangements and tyrosine phosphorylation of host cell proteins. Proc. Natl. Acad. Sci. USA. 93: 1259-1264.

Slomiany, B.L. and Slomiany, A. 1991. Role of mucus in gastric mucosal protection. J. Physiol. Pharmacol. 42: 147-161.

Smoot, D.T., Resau, J.H., Naab, T., Desbordes, B.C., Gilliam, T., Bull-Henry, K., Curry, S.B., Nidiry, J., Sewchand, J., Mills-Robertson, K., Frontin, K., Abebe, E., Dillon, M., Chippendale, G.R., Phelps, P.C., Scott, V.F., and Mobley, H.L.T. 1993. Adherence of *Helicobacter pylori* to cultured human gastric epithelial cells. Infect. Immun. 61: 350-355.

Suzuki, M. Mori, M., Miyayama, A., Iwai, N., Tsunematsu, N., Oonuki, M., Suzuki, H., Hibi, T., and Ishii, H. 1997. Enhancement of neutrophil infiltration in the corpus after failure of *Helicobacter pylori* eradication. J. Clin Gastroenterol. 25: S222-S228

Szabò, I., Brutsche, S., Tombola, F., Moschioni, M., Satin, B., Telford, J.L., Rappuoli, R., Montecucco, C., Papini, E, and Zoratti, M. 1999. Formation of anion-selective channels in the cell plasma membrane by the toxin VacA of *Helicobacter pylori* is required for its biological activity. EMBO J. 18: 5517-5527.

Telford, J.L., Ghiara, P., Dell'Orco, M., Comanducci, M., Burroni, D., Bugnoli, M., Tecce, M.F., Censini, S., Covacci, A., Xiang, Z., Papini, E., Montecucco, C., Parente, L., and Rappuoli, R. 1994. Purification and characterization of the vacuolating toxin from *Helicobacter pylori*. J. Exp. Med. 179: 1653-1658.

Tomb, J.F., White, O., Kerlavage, A.R., Clayton, R.A., Sutton, G.G., Fleischmann, R.D., Ketchum, K.A., Klenk, H.P., Gill, S., Dougherty, B.A., Nelson, K., Quackenbush, J., Zhou, L., Kirkness, E.F., Peterson, S., Loftus, B., Richardson, D., Dodson, R. Khalak, H.G., Glodek, A., McKenney, K., Fitzegerald, L.M., Lee, N., Adams, M.D., Venter, J.C., *et al.* 1997. The complete genome sequence of the gastric pathogen *Helicobacter pylori*. Nature. 388: 539-547.

Tombola, F., Carlesso, C., Szabò, I., de Bernard, M., Reyrat, J.M., Telford, J.L., Rappuoli, R., Montecucco, C., Papini, E., and Zoratti, M. 1999a. *Helicobacter pylori* vacuolating toxin forms anion-selective channels in planar lipid bilayers: possible implications for the mechanism of cellular vacuolation. Biophys. J. 96: 1401-1409.

Tombola, F., Oregna, F., Brutsche, S., Szabò, I., Del Giudice, G., Rappuoli, R., Montecucco, C., Papini, E., and Zoratti, M. 1999b. Inhibition of the vacuolating and anion channel activities of the VacA toxin of *Helicobacter pylori*. FEBS Lett. 460: 221-225

Tonello, F., Dundon, W.G., Satin, B., Molinari, M., Tognon, G., Grandi, G., Del Giudice, G., Rappuoli, R., and Montecucco, C. 1999. The *Helicobacter pylori* neutrophil-activating protein is an iron-binding protein with dodecameric structure. Mol. Microbiol. 34: 238-246.

van der Ende, A., Pan, Z.J., Bart, A., van der Hulst, R.W., Feller, M., Xiao, S.D., Tytgat, G.N., and Dankert, J. 1998. CagA-positive *Helicobacter pylori* populations in China and the Netherlands are distinct. Infect. Immun. 66:1822-1826.

Wandersman, C. 1992. Secretion across the bacterial outer membrane. Trends Genet. 8: 317-322.

Warren, J.R. and Marshall, B.J. 1983. Unidentified curved bacilli on gastric epithelium in active chronic gastritis. Lancet 1: 1273-1275.

Watts, C. 1997. Capture and processing of exogenus antigen for presentation on MHC molecules. Ann. Rev. Immunol. 15: 821-850.

Worst, D.J., Maaksant, J., Vandenbrouke-Grauls, C.M.J.E., and Kusters, J.G. 1999. Multiple haem-utilization in *Helicobacter pylori*. Microbiol. 145: 681-688.

Yahiro, K., Niidome, T., Hatakeyama, T., Aoyagi, H., Kurazono, H., Padilla, P.I., Wada, A., and Hirayama, T. 1997. *Helicobacter pylori* vacuolating cytotoxin binds to the 140-kDa protein in human gastric cancer cell lines, AZ-521 and AGS. Biochem. Biophys. Res. Commun. 238: 629-632.

Ye, D., Willhite, D.C., and Blanke, S.R. 1999. Identification of the minimal intracellular vacuolating domain of the *Helicobacter pylori* vacuolating toxin. J. Biol. Chem. 274: 9277-9282.

Yamamura, F., Yoshikawa, N., Akita, Y., Mitamura, K., and Miyasaka, N. 1999. Relationship between *Helicobacter pylori* infection and histologic features of gastritis in biopsy specimens in gastroduodenal diseases, including evaluation of diagnosis by polymerase chain reaction assay. J. Gastroenterol. 34: 461-466

Yoshida, N., Granger, D.N., Evans, D.J. Jr., Evans, D.G., Graham, D.Y., Anderson, D.C., Wolf, R.E., and Kvietys, P.R. 1993. Mechanisms involved in *Helicobacter pylori*-induced inflammation. Gastroenterol. 105: 1431-1440.

From: *Helicobacter pylori: Molecular and Cellular Biology*
ISBN 1-898486-25-5 © 2001 Horizon Scientific Press, Wymondham, UK.

16

Development of a *Helicobacter pylori* Vaccine

Subhas Banerjee and Pierre Michetti*

Abstract

Novel strategies are needed to control *H. pylori* infection on a global scale and the potential value of a vaccine is increasingly recognized. Studies in rodents have demonstrated the feasibility of prophylactic and therapeutic mucosal immunization against *Helicobacter* infections. Initial human trials showed that oral immunization with *H. pylori* urease is safe, immunogenic, and can result in a decreased gastric bacterial load. However, more potent vaccines will be needed to protect against or cure *H. pylori* infection in humans. To achieve this goal, our knowledge of the mechanisms of immune protection in the stomach needs to be improved. In rodents, the MHC II-restricted CD4$^+$ T cell response plays a prominent role whereas antibodies are not necessary for protection. In humans, mechanisms that mediate protection against noninvasive gastric pathogens are largely unknown. Multivalent vaccines, safe adjuvants, and improved vaccine delivery systems all need to be studied. Divalent combinations of antigens are superior to single antigen vaccines in rodents. The availability of two genomic sequences of *H. pylori* will certainly help identify additional candidate vaccine antigens. Detoxified bacterial toxin adjuvants, live vaccine vectors and alternate routes of immunization are currently being actively evaluated.

Introduction

Helicobacter pylori colonizes the gastric mucosa of approximately half the world's adult human population, with near universal adult infection in some developing countries (see chapter 3). *H. pylori* is now accepted as being the major etiological factor for chronic gastritis and peptic ulcer disease. It is also associated with the development of distal gastric adenocarcinoma and gastric lymphoma, and is therefore now deemed a WHO class I carcinogen (Anonymous, 1994). *H. pylori* may also play a role in subgroups of patients with non ulcer dyspepsia and in childhood diarrhea in developing countries (Sullivan *et al.,* 1990; McColl *et al.,* 1998). An estimated 20% of the infected population develop symptoms of gastric disease. *H pylori* is therefore a worldwide public health problem of considerable magnitude.

Eradication regimes for *H. pylori* based on multiple antibiotic therapy typically yield cure rates of 85-90%. These cure rates will probably decrease with time due to the selection

*corresponding author, email: pmichett@caregroup.harvard.edu

of antibiotic resistant bacteria. In parts of the world such as Hong Kong and India, 70-80% of *H. pylori* isolates now exhibit resistance to metronidazole (Ling *et al.,* 1996, Fock, 1997), and resistance to clarithromycin, although currently still low, is increasing. The high cost of antibiotic therapy is a major financial burden for many developing countries. Moreover, reinfection rates in developing countries are extremely high, e.g. 73% within 8 months in Peru (Ramirez-Ramos *et al.,* 1997), making antibiotic therapy an ineffectual approach (Ramirez-Ramos *et al.,* 1997). Alternative strategies are therefore required to combat *H. pylori* on a large scale. Effective prophylactic and therapeutic vaccination would be an attractive global solution to this widespread health problem.

The main argument against diverting scarce public financial resources to *H. pylori* vaccine research is that the prevalence of *H. pylori* infection is falling, at least in the developed world, and the incidence of gastric cancer is decreasing worldwide at rates of over 10% a decade (Coleman *et al.,* 1993). However, a recent study using the methodology of the American Institute of Medicine confirmed that the development and use of a *H. pylori* vaccine in the USA could be expected to provide public health benefits considerably superior to those provided by several other vaccines, including vaccines directed against hepatitis B, respiratory syncitial virus, influenza, varicella, rotavirus and herpes simplex virus (Tsugawa *et al.,* 1998). The Institute of Medicine has itself recently carried out a cost-effectiveness analysis of the benefits of investing in the development of potential new vaccines (Stratton *et al.,* 1999). A *H. pylori* vaccine was grouped in level II, on a scale of I to IV, where level I has the most favorable cost:benefit ratio.

Vaccination Against a Persisting Pathogen

An early concern was that a vaccine approach to eradicating *H. pylori* was unlikely to be successful, as the natural host immune response is ineffective in clearing infection, which is usually lifelong in the absence of treatment. The mechanisms by which *H. pylori* persists in immunocompetent hosts remain unclear. Although both humoral and cellular responses occur in infected individuals, they are associated with anomalies that may contribute to the persistence of *H. pylori* infection.

H. pylori induces circulating IgG and IgA, and is associated with the production of secretory IgA (sIgA) in the gastric mucosa. However, a recent study indicated that *H. pylori* specific IgA in the gastric juice of infected humans is predominantly of the non-secretory type (Birkholz *et al.,* 1998). The stability of non-secretory IgA is poor because it is susceptible to acid hydrolysis and proteolytic cleavage (Berdoz *et al.,* 1999). The lack of the secretory component on *H. pylori*-specific IgA may thus contribute to the persistence of the infection.

The mucosal T cell response in *H. pylori* infection may in fact contribute to tissue damage. Mucosal and systemic T helper cell responses to *H. pylori* are predominantly of the Th1 type, characterized by a predominance of interferon γ (INF-γ) producing CD4$^+$ T cells and resultant low IgG1 and high IgG2a levels (Fan *et al.,* 1994; Karttunen *et al.,* 1995; Mohammadi *et al.,* 1996; D'Elios *et al.,* 1997; Bamford *et al.,* 1998). The severity of gastritis in mice infected with *H. felis* correlates with INF-γ levels and can be reversed by blocking Th1 responses with anti-INF-γ antibodies, with no effect on bacterial load (Mohammadi *et al.,* 1996, 1997). B cell and antibody deficient mice infected with *H. felis* develop gastritis identical to infected wild type mice, indicating that the Th1 response is sufficient to induce gastritis in mice (Blanchard *et al.,* 1999). The mucosal damage associated with gastritis may play a role in the altered sIgA production described above.

Despite these observations, numerous studies in mice and ferrets have demonstrated the feasibility of prophylactic and therapeutic vaccination against *Helicobacter* infections, achieving protection rates averaging 60-80% (Chen *et al.*, 1993; Czinn *et al.*, 1993; Doidge *et al.*, 1994; Michetti *et al.*, 1994; Corthesy-Theulaz *et al.*, 1995; Cuenca *et al.*, 1996). Morever, a recent abstract suggests that subinfectious doses of *H. pylori* can induce protective immunity in mice against subsequent infection, rather than resulting in colonization (Radcliff *et al.*, 1999). These studies indicate that an efficacious immune response can occur in the gastric lumen, and that gastric *Helicobacter spp* are susceptible to such an immune response.

Mechanisms of Protection Induced by Immunization

Early vaccination studies in animals suggested a correlation between antibody levels and protection (Lee *et al.*, 1995; Ferrero *et al.*, 1997; Goto *et al.*, 1999). The role of antibodies against gastric *Helicobacter* infections has been partially clarified by several recent studies. Intranasal vaccination with urease in mice did not result in protection although high levels of specific circulating and secretory antibodies, were induced (Weltzin *et al.*, 1997; Ermak *et al.*, 1998). Co-administration of a mucosal adjuvant and a specific antigen was necessary for protection, suggesting that mechanisms other than simply an antibody response need to be induced. Passive transfer of sera containing urease specific IgG antibodies from immunized to naïve mice also failed to protect against *H. pylori* infection (Ermak *et al.*, 1998). Thus, circulating antibodies alone are insufficient mediators of protection (Ermak *et al.*, 1998). Recent studies with antibody deficient gene-targeted mice also confirmed that antibodies are not essential for protection in this animal (see below). Thus, although a protective role for antibodies has not yet been ruled out in larger animals, it is possible that antibody levels may merely serve as surrogate markers for other immune mechanisms that are responsible for protection.

Early studies also suggested the importance of cell-mediated responses in protection against *Helicobacter* infections. Passive transfer of T cells from donor mice immunized against *H. felis* led to reductions in bacterial load in recipient mice (Mohammadi *et al.*, 1997). These reductions were noted upon transfer of interleukin 4 (IL-4) producing clones (Th2) but not with INF-γ producing clones (Th1). A switch from a Th1 to a Th2 response has also been reported during therapeutic immunization in mice (Saldinger *et al.*, 1998). These observations are compatible with the IgA and IgG1 isotypes that were associated with protection (Lee *et al.*, 1995; Ferrero *et al.*, 1997). These istoypes are induced by Th2 cytokines (Kunimoto *et al.*, 1989; Sonoda *et al.*, 1989). These observations led to the proposal that widening the host response from a purely Th1 type to a balanced Th2/Th1 type, as reflected by an IgG1/IgG2a ratio close to unity, may provide the best protection against *Helicobacter* infection (Ermak *et al.*, 1998).

Three recent studies have elegantly confirmed that antibody-independent mechanisms are sufficient to provide protection against *H. pylori* infection (Ermak *et al.*, 1998; Blanchard *et al.*, 1999; Pappo *et al.*, 1999). Blanchard *et al.* (1999) were able to induce protective immunity against *H. felis* in IgA-knockout mice and B cell/antibody deficient mice, confirming that antibody-independent mechanisms are sufficient to induce protection against *H pylori* infection. Similarly, Ermak *et al.* (1998) have used gene targeted mice deficient in either MHC I or MHC II receptors, or in the constant Ig chain μIg gene, to show that protection against *H. pylori* is mediated by MHC II-mediated mechanisms. In this study too, antibody responses were unnecessary for protection. The level of protection correlated with the density of gastric lamina propria T cells. Pappo *et al.* (1999) recently confirmed that MHC II gene-targeted mice are unable to mount effective protective responses against *H. pylori* infection following vaccination. MHC II knockout animals failed to generate serum IgG and gastric IgA antibodies to *H. pylori*, and also lacked CD4$^+$ T cells in the gastric mucosa. In addition, although INF-γ upregulates MHC II mechanisms, studies in knockout mice confirmed that INF-γ is not necessary for protection against *H. pylori* (Sawai *et al.,* 1999).

Human Studies

To date, only three trials of *H. pylori* vaccination in humans have been reported. An initial study on 12 healthy subjects with asymptomatic *H. pylori* infection confirmed the safety of orally administered recombinant *H. pylori* urease (Kreiss *et al.,* 1996). No adjuvant was used in this study, which was primarily designed to test the safety of urease in humans, and as with studies with mice in the absence of adjuvant, no immune response to urease was observed. A second study used a vaccine strain of *Salmonella typhi* modified to express *H. pylori* urease subunits A and B (DiPetrillo *et al.,* 1999). All 8 volunteers in the study developed immune responses to *S. typhi* antigens, but none developed immune responses against *H. pylori,* for reasons that remain unclear. Possible explanations include decreased immunogenicity of bacteria-bound urease, low urease expression by the innoculum, or a rapid plasmid copy number decrement *in vivo*. In the third study, urease was administered to 26 *H. pylori* infected volunteers together with *E. coli* heat labile enterotoxin (LT) as a mucosal adjuvant. The volunteers possessed specific IgA and IgG antibodies against *H. pylori* upon recruitment to the study. Following immunization, a dose-related rise in anti-urease IgA titers was seen, and an increase in circulating anti-urease IgA producing cells was observed in all *H. pylori*-infected volunteers (Michetti *et al.,* 1999). Despite an increase in the numbers of urease specific IgG antibody secreting cells (ASC), no increase was seen in IgG titers. A trend towards higher ASC responses was seen in volunteers receiving higher doses of LT, suggesting that the magnitude of the response may depend on the amount of adjuvant used. Eradication of pre-existing *H. pylori* infection was not observed and the degree of gastritis remained unchanged following vaccination. However, quantitative cultures of gastric biopsy specimens showed an encouraging decrease in bacterial densities following vaccination, not seen in the placebo group. All volunteers were previously infected with *H pylori*, preventing a test of the prophylactic potential of this vaccine. The major adverse effect was diarrhea, seen in two thirds of the volunteers. The incidence and severity of the diarrhea declined with subsequent immunizations, presumably due to development of antibodies to LT. No post-immunization gastritis was seen in any of the three human studies.

One explanation for the disappointing results in human studies given the success of therapeutic vaccination in rodents, is that the mouse model, although convenient, is not a natural system for *Helicobacter* infection. Infection may be tenuous in mice, resulting in high success rates with prophylactic and therapeutic vaccination. Indeed, tinidazole monotherapy results in 100% eradication of *H. felis* infection in mice (Hook-Nikanne *et al.,* 1996) whereas even triple antibiotic therapy rarely achieves >90% eradication in humans. Furthermore, the histopathology and urease assays used in several studies to estimate protection in mice, are insensitive in detecting small numbers of bacteria in gastric tissue, and may have overestimated true protection rates in many early mouse studies. In a recent study, complete sterilizing protection, as defined by the absence of *H. pylori,* on culture was obtained in only 10-20% of mice following mucosal immunization, although all other animals showed significant reductions in bacterial densities (Kleanthous *et al.,* 1998).

Natural models of *Helicobacter* infection such as ferrets and Rhesus monkeys may reflect human infection more closely, and vaccination studies in these animals are perhaps more predictive of results that may be expected in humans. Oral immunization of ferrets naturally infected with *H. mustelae* with urease plus cholera toxin led to the eradication of infection in only 30% of immunized animals (Cuenca *et al.*, 1996). Rhesus monkeys (*Macaca mulatta*) may be the best animal model of *H. pylori* infection. Seventy percent of colony-raised animals are naturally infected by one year of age, infected animals develop chronic active gastritis, and the eradication of infection requires triple therapy. Following immunization of 9 month old rhesus monkeys with urease plus LT, naturally acquired *H. pylori* infection was prevented in 31% of animals at a 10 month follow up, compared with a 7% *H. pylori* negative rate in placebo treated controls (Dubois *et al.*, 1998). In a further study, naturally infected monkeys vaccinated with urease plus LT showed no decrease in *H. pylori* by histology or quantitative culture counts, despite a >2 fold increase in specific serum IgG and IgA and an increase in urease specific ASCs in the gastric mucosa (Lee *et al.*, 1998). Prophylactic vaccination was also studied in monkeys after eradication of pre-existing natural infection with antibiotic therapy. Upon challenge, only two of five animals were protected against reinfection, although all five vaccinated animals had lower quantitative culture counts (Lee *et al.*, 1998). A recent study compared mucosal vaccination alone, parenteral vaccination alone and mucosal priming followed by parenteral vaccination in 15 *H. pylori* negative rhesus monkeys (5 in each group). Following challenge with *H. pylori*, only animals vaccinated with the mucosal prime-parenteral boost strategy showed a modestly reduced colonization level on quantitative culture, compared with sham immunized animals (Lee *et al.*, 1999). No animals were free of infection. Thus, in natural models of *Helicobacter* infection, only a limited prophylactic effect of vaccination has be demonstrated in primates and therapeutic immunization has only been demonstrated in ferrets.

Vaccine Antigens

Vaccination with several antigens, including the urease B subunit, heat shock proteins HspA and HspB, VacA, CagA, and catalase, has been shown to confer protection against *Helicobacter* infection in mice and ferrets. Logically, the ideal antigen should be a strongly immunogenic surface antigen, without intrinsic toxicity or significant antigenic variation in all strains of *H. pylori*. VacA and CagA are produced only by type I strains, and VacA induces gastric erosions in mice (Telford *et al.*, 1994). HspB has homologies to the GroEL family of proteins involved in autoimmune reactions (Ferrero *et al.*, 1995). The *H. pylori* blood group binding antigen (BabA) which mediates adherence of *H. pylori* to gastric mucosal cells has recently been purified and characterized, and has also been proposed as a potential vaccine antigen (Ilver *et al.*, 1998). However, BabA is also expressed mainly by type I strains. In contrast, urease and HspA are expressed by all *H. pylori* strains, and the safety of both antigens has been established in mice, and urease has been shown to be safe in humans.

Immunization of mice with a dual antigen preparations comprising either HspA or VacA plus the urease B subunit conferred increased protection rates compared to any of the antigens used alone (Ferrero *et al.*, 1995; Ghiara *et al.*, 1997). It is possible that multivalent vaccines containing more than one antigen may be required for successful protection against *H. pylori* in humans. The continued search for alternative protective antigens therefore remains an important thrust of the vaccine effort. The availability of two fully sequenced *H. pylori* genomes (Tomb *et al.*, 1997; Lee, 1998; Alm *et al.*, 1999) is likely to provide further suitable candidate antigens for vaccine development (Tomb *et al.*, 1997; Lee, 1998; Alm *et al.*, 1999). Two dimensional gel electrophoresis could identify conserved *H. pylori* proteins that are recognized by the sera of *H. pylori* infected subjects, and may be a useful technique in this search (McAtee *et al.*, 1998). *H. pylori* genomic expression libraries screened by plaque immunoblot assays using *H. pylori* antisera may offer an additional approach to identifying novel candidate antigens (Hocking *et al.*, 1999).

Adjuvants

Mucosal adjuvants enhance the antigenicity of poorly immunogenic antigens and can help disrupt oral tolerance. They also broaden the width of the immune response by inducing the recruitment of T cells in the gastric mucosa. Adjuvants can also be used to modulate $CD4^+$ T cell subset differentiation, and different adjuvants induce Th1 and Th2 responses to varying extents. Systemic vaccination with urease and one of four adjuvants which stimulate Th1 and Th2 responses to differing extents confirmed that the best protection followed immunization with an adjuvant promoting both strong Th1 and Th2 responses (Guy *et al.*, 1998). Vaccination trials in animal have used mainly cholera toxin (CT) and LT. These adjuvants are toxic in humans and stimulate diarrhea. Genetically modified toxins retaining their adjuvant properties have been developed in order to overcome this problem. A mutant LT protein (LTK63), lacking toxic ADP-ribosylating activity, was as effective as native LT and equal numbers of immunized mice were protected against infection with either LT or LTK63 (Marchetti *et al.*, 1998). The nontoxic B subunit of cholera toxin (CT-B) has also been proposed as an adjuvant for mucosal immunization against *H. felis* (Lee *et al.*, 1994). However, its observed efficacy may reflect contamination with small amounts of CT-A, as no protection has been observed after vaccination using recombinant CT-B as adjuvant (Blanchard *et al.*, 1998).

Routes of Immunization

Mucosal immunization is superior to systemic immunization in inducing protection against enteric infections. It results in more sustained mucosal humoral and cellular immune responses, and has shown better efficacy in field trials in humans (Levine *et al.*, 1993). Oral immunization induces $\alpha4\beta7^+$ ASCs, with preferential homing into the gut lamina propria, in contrast to systemic immunization (Kantele *et al.*, 1999). In mice, $\alpha4\beta7$ function has been shown to be required for protection against *H. felis* infection after oral immunization, in association with high numbers of gastric mucosal $\alpha4\beta7^+$ $CD4^+$ T cells (Michetti *et al.*, 1998). Oral vaccination may not, however, be the optimal route for inducing gastric protection. Kleanthous *et al.* (1998) showed that mice were equally well protected against *H. pylori* infection whether they were immunized by either oral, intranasal or rectal routes. The rectal route alone resulted in a balanced Th1/Th2 response, and induced the highest levels of gastric IgA (Kleanthous *et al.*, 1998).

Some studies suggest that systemic immunization could represent a viable alternative to mucosal immunization against *H. pylori* (Guy *et al.*, 1998; Blanchard *et al.*, 1999), but these results differ from those found by other investigators. In a well designed study, mucosal vaccination was superior to parenteral immunization, which resulted in only modest declines in bacterial load (Ermak *et al.*, 1998). Furthermore, even if mucosal priming with urease plus LT was followed by parenteral boosting with urease plus alum, less protection was stimulated than by mucosal vaccination (Ermak *et al.*, 1998). Similarly, mucosal priming and parenteral boosting in rhesus monkeys which were subsequently challenged with *H. pylori* resulted in a modestly reduced level of bacterial colonization compared with sham immunized animals, whereas parenteral vaccination alone did not induce protection (Lee *et al.*, 1999). The site of systemic immunization may also modulate the response. Guy *et al.*,(1999) found that *Helicobacter* infection was preferentially eradicated following subcutaneous systemic immunization in the lumbar region rather than the neck.

Alternative Vaccine Delivery Systems

Biodegradable poly(D,L-lactide-co-glycolide) (PLG) microspheres have recently been evaluated in mice as a non toxic adjuvant plus oral delivery system for *H. pylori* lysates (Kim *et al.,* 1999). Supposedly, this system should induce more efficient phagocytosis with transport to lymph nodes, where sustained antigen release would result in prolonged immune stimulation. However, the initial immunogenicity results were disappointing and lower antibody titers were achieved than with *H. pylori* lysate plus CT. Studies assessing the prophylactic and therapeutic ability of this system have not yet been carried out.

An encouraging alternative approach is to use attenuated live bacteria to express and deliver *H. pylori* antigens. There are several theoretical advantages to this approach: potentially toxic adjuvants are not required, purification of the antigen is not necessary, antigen denaturation and degradation in the stomach does not occur, and only one to two immunizations are needed, all of which should contribute to lower costs. Bacterial systems should also be able to deliver more than one vaccine antigen, and if deemed necessary to modulate the immune reponse, recombinant adjuvants could be included in the system. Both mucosal and systemic humoral responses can be stimulated, and mixed Th1 and Th2 responses can be elicited. *Salmonella typhimurium* strains expressing the A and B subunits of *H. pylori* urease have been developed (Corthesy-Theulaz *et al.,* 1998; Gomez-Duarte *et al.,* 1998). *Salmonella* can persist in infected intestinal tissues for several weeks post immunization, and prolonged antigen stimulation should be possible. Immunization in mice using this system resulted in high levels of serum IgA, IgG1 and IgG2a and of mucosally secreted IgA, with protection rates of 60% to 100% (Corthesy-Theulaz *et al.,* 1998; Gomez-Duarte *et al.,* 1998). However, the first human study utilizing a bacterial expression system (with a different antigen expression system) yielded disappointing results (DiPetrillo *et al.,* 1999). Lactic acid bacteria (*Lactococcus lactis*) and poliovirus replicons expressing *H. pylori* urease subunit B have also recently been decribed (Lee *et al.,* 1999; Novak *et al.,* 1999).

Direct DNA vaccination might offer still another alternative but has not been studied for *Helicobacter* infection. Plasmid DNA packaged in a delivery system such as the polysaccharide chitosan can be administered orally and is capable of generating both systemic and mucosal immunity (Roy *et al.,* 1999).

Conclusion

Global approaches to the control of *H. pylori* infection are needed, and a vaccine may be cost-effective in this context. Vaccination studies in animals have established a proof of principle and have helped identify vaccine antigens. Unfortunately, the available animal models have only limited relevance to the development of a human *H. pylori* vaccine. Initial trials in volunteers are encouraging, but further human studies will be required to define vaccine antigens, adjuvants, and delivery systems. Clarifying the mechanisms of protection will certainly constitute an important step in this development.

References

Anonymous. 1994. Schistosomes, liver flukes and *Helicobacter pylori*. IARC Working Group on the Evaluation of Carcinogenic Risks to Humans. IARC Monogr Eval Carcinog Risks Hum. 61: 1-241.

Alm, R.A., Ling, L.S., Moir, D.T., King, B.L., Brown, E.D., Doig, P.C., Smith, D.R., Noonan, B., Guild, B.C., deJonge, B.L., Carmel, G., Tummino, P.J., Caruso, A., Uria-Nickelsen, M., Mills, D.M., Ives, C., Gibson, R., Merberg, D., Mills, S.D., Jiang, Q., Taylor, D.E., Vovis, G.F., and Trust, T.J.. 1999. Genomic-sequence comparison of two unrelated isolates of the human gastric pathogen *Helicobacter pylori*. Nature. 397: 176-80.

Bamford, K.B., Fan, X., Crowe, S.E., Leary, J.F., Gourley, W.K., Luthra, G.K., Brooks, E.G., Graham, D.Y., Reyes, V.E., and Ernst, P.B. 1998. Lymphocytes in the human gastric mucosa during *Helicobacter pylori* have a T helper cell 1 phenotype. Gastroenterology. 114: 482-92.

Berdoz, J., Blanc, C.T., Reinhardt, M., Kraehenbuhl, J.P., and Corthesy, B. 1999. *In vitro* comparison of the antigen-binding and stability properties of the various molecular forms of IgA antibodies assembled and produced in CHO cells. Proc. Natl. Acad. Sci. USA. 96: 3029-34.

Birkholz, S., Schneider, T., Knipp, U., Stallmach, A., and Zeitz, M. 1998. Decreased *Helicobacter pylori*-specific gastric secretory IgA antibodies in infected patients. Digestion. 59: 638-45.

Blanchard, T.G., Czinn, S.J., Redline, R.W., Sigmund, N., Harriman, G., and Nedrud, J.G. 1999. Antibody-independent protective mucosal immunity to gastric *Helicobacter* infection in mice. Cellular Immunology. 191: 74-80.

Blanchard, T.G., Gottwein, J.M., Targoni, O.S., Eisenberg, J.C., Zagorski, B.M., Trezza, R.P., Redline, R., Nedrud, J.G., Tary-Lehmann, M., Lehman, P.V. 1999. Systemic vaccination inducing either TH1 or TH2 immunity protects mice from challenge with *H. pylori*. Gastroenterology. 116: A695.

Blanchard, T.G., Lycke, N., Czinn, S.J., and Nedrud, J.G. 1998. Recombinant cholera toxin B subunit is not an effective mucasal adjuvant for oral immunization of mice against *Helicobacter felis*. Immunology. 94: 22-27

Chen, M., Lee, A., Hazell, S., Hu, P., and Li, Y. 1993. Immunization against gastric infection with *Helicobacter* species: first step in the prophylaxis of gastric cancer? Zentrabl Bakteriol. 280: 155-165

Coleman, M.P., Esteve, J., Damiecki, P., Arslan, A., and Renard, H. 1993. Trends in cancer incidence and mortality. Internat. Agency Res. Cancer. 121: 193-224

Corthesy-Theulaz, I., Porta, N., Glauser, M., Saraga, E., Vaney, A.C., Haas, R., Kraehenbuhl, J.P., Blum, A.L., and Michetti, P. 1995. Oral immunization with *Helicobacter pylori* urease B subunit as a treatment against *Helicobacter* infection in mice. Gastroenterology. 109: 115-121.

Corthesy-Theulaz, I.E., Hopkins, S., Bachmann, D., Saldinger, P.F., Porta, N., Haas, R., Zheng-Xin, Y., Meyer, T., Bouzourene, H., Blum, A.L., and Kraehenbuhl, J.P. 1998. Mice are protected from *Helicobacter pylori* infection by nasal immunization with attenuated *Salmonella typhimurium* phoPc expressing urease A and B subunits. Infect. Immun. 66: 581-586.

Cuenca, R., Blanchard, T.G., Czinn, S.J., Nedrud, J.G., Monath, T.P., Lee, C.K., and Redline, R.W. 1996. Therapeutic immunization against *Helicobacter mustelae* in naturally infected ferrets. Gastroenterology. 110: 1770-1775.

Czinn, S.J., Cai, A., and Nedrud, J.G. 1993. Protection of germ free mice from infection by *Helicobacter felis* after active oral or passive IgA immunization. Vaccine. 139: 637-642.

D'Elios, M.M., Manghetti, M., De Carli, M., Costa, F., Baldari, C.T., Burroni, D., Telford, J., Romagnani, S., and Del Prete, G. 1997. T helper 1 effector cells specific for *Helicobacter pylori* in gastric antrum of patients with peptic ulcer disease. J. Immunol. 158: 962-967.

DiPetrillo, M.D., Tibbetts, T., Kleanthous, H., Killeen, K.P., and Hohmann, E.L. 1999. Safety and immunogenicity of phoP/phoQ-deleted *Salmonella typhi* expressing *Helicobacter pylori* urease in adult volunteers. Vaccine. 18: 449-459.

Doidge, C., Gust, I., Lee, A., Buck, F., Hazell, S., and Manne, U. 1994. Therapeutic immunization against *Helicobacter* infection. Lancet. 343: 913-914.

Dubois, A., Lee, C.K., Fiala, N., Kleanthous, H., Mehlman, P.T., and Monath, T.P. 1998. Immunization against natural *Helicobacter pylori* infection in nonhuman primates. Infect. Immun. 66: 4340-6.

Ermak, T.H., Giannasca, P.J., Nichols, R., Myers, G.A., Nedrud, J.G., Weltzin, R.A., Lee, C.K., Kleanthous, H., and Monath, T.P. 1998. Immunization of mice with urease vaccine affords protection against *Helicobacter pylori* infection in the absence of antibodies and is mediated by MHC class II-restricted responses. J. Experim. Med. 188: 1-12.

Fan, X.J., Chua, A., Shahi, C.N., McDevitt, J., Keeling, P.W., and Kelleher, D. 1994. Gastric T lymphocyte responses to *Helicobacter pylori* in patients with *H. pylori* colonization. Gut. 35: 1379-1384.

Ferrero, R.L., Thiberge, J.M., Kansau, I., Wuscher, N., Huerre, M., and Labigne, A. 1995. The GroES homolog of *Helicobacter pylori* confers protective immunity against mucosal infection in mice. Proc. Natl. Acad. Sci. USA. 92: 6499-6505.

Ferrero, R.L., Thiberge, J.M., and Labigne, A. 1997. Local immunoglobulin G antibodies in the stomach may contribute to immunity against *Helicobacter* infection in mice. Gastroenterology. 113: 185-94.

Fock, K.M. 1997. Peptic ulcer disease in the 1990s: an Asian perspective. J. Gastroenterol Hepatol. 12: S23-8.

Ghiara, P., Rossi, M., Marchetti, M., Di Tommaso, A., Vindigni, C., Ciampolini, F., Covacci, A., Telford, J.L., De Magistris, M.T., Pizza, M., Rappuoli, R., and Del Giudice, G. 1997. Therapeutic intragastric vaccination against *Helicobacter pylori* in mice eradicates an otherwise chronic infection and confers protection against reinfection. Infect. Immun. 65: 4996-5002.

Gomez-Duarte, O.G., Lucas, B., Yan, Z.X., Panthel, K., Haas, R., and Meyer, T.F. 1998. Urease subunits A and B delivered by attenuated *Salmonella typhimurium* vaccine strain protects mice against gastric colonization by *Helicobacter pylori*. Vaccine. 16: 460-471.

Goto, T., Nishizono, A., Fujioka, T., Ikewaki, J., Mifune, K., and Nasu, M. 1999. Local secretory immunoglobulin A and postimmunization gastritis correlate with protection against *Helicobacter pylori* infection after oral vaccination of mice. Infect. Immun. 67: 2531-9.

Guy, B., Hessler, C., Fourage, S., Haensler, J., Vialon-Lafay, E., Rokbi, B., and Millet, M.J. 1998. Systemic immunization with urease protects mice against *Helicobacter pylori* infection. Vaccine. 16: 850-6.

Guy, B., Hessler, C., Fourage, S., Rokbi, B., and Quentin Millet, M. 1999. Comparison between targeted and untargeted systemic immunization with adjuvanted urease to cure *Helicobacter pylori* infection in mice. Vaccine. 17: 1130-1135.

Hocking, D., Webb, E., Radcliff, F., Rothel, L., Taylor, S., Pinczower, G., Kapouleas, C., Braley, H., Lee, A., and Doidge, C. 1999. Isolation of recombinant protective *Helicobacter pylori* antigens. Infect. Immun. 67: 4713-9.

Hook-Nikanne, J.P., Aho, P., Karkkainen, P., Kosunen, T.U., and Salaspuro, M. 1996. The *Helicobacter* felis mouse model in assessing anti-*Helicobacter* therapies and gastric mucosal prostaglandin E2 levels. Scand. J. Gastroenterol. 31: 334-338.

Ilver, D., Arnqvist, A., Ogren, J., Frick, I.M., Kersulyte, D., Incecik, E.T., Berg, D.E., Covacci, A., Engstrand, L., and Borén, T. 1998. *Helicobacter pylori* adhesin binding fucosylated histo-blood group antigens revealed by retagging. Science. 279: 373-7.

Kantele, A., Westerholm, M., Kantele, J.M., Mäkelä, P.H., and Savilahti, E. 1999. Homing potentials of circulating antibody-secreting cells after administration of oral or parenteral protein or polysaccharide vaccine in humans. Vaccine. 17: 229-36.

Karttunen, R., Karttunen, T., Ekre, H.P., and MacDonald, T.T. 1995. Interferon gamma and interleukin 4 secreting cells in the gastric antrum in *Helicobacter pylori* positive and negative gastritis. Gut. 36: 341-345

Kim, S.Y., Doh, H.J., Ahn, J.S., Ha, Y.J., Jang, M.H., Chung, S.I., and Park, H.J. 1999. Induction of mucosal and systemic immune response by oral immunization with H. *pylori* lysates encapsulated in poly(D,L-lactide-co-glycolide) microparticles. Vaccine. 17: 607-16.

Kleanthous, H., Myers, G.A., Georgakopoulos, K.M., Tibbits, T.J., Ingrassia, J.W., Gray, H.L., Ding, R., Zhang, Z.X., Lei, W., Nichols, R., Lee, C.K., Ermak, T.E., and Monath, T.P. 1998. Rectal and intranasal immunization with recombinant urease induce distinct local and serum immune responses in mice and protect against *Helicobacter pylori* infection. Infect. Immun. 66: 2879-2886.

Kreiss, C., Buclin, T., Cosma, M., Corthesy-Theulaz, I., and Michetti, P. 1996. Safety of oral immunization with recombinant urease in patients with *Helicobacter pylori* infection. Lancet. 347: 1630-1631.

Kunimoto, D.Y., Nordan, R.P., and Strober, W. 1989. IL-6 is a potent cofactor of IL-1 in IgM synthesis and of IL-5 in IgA synthesis. J. Immunol. 143: 2230-2235.

Lee, A. 1998. The *Helicobacter pylori* genome - new insights into pathogenesis and therapeutics. New Engl. J. Med. 338: 832-833.

Lee, A., Chen, M.H. 1994. Successful immunization against gastric infection against infection with *Helicobacter* species: use of a cholera toxin B-subunit-whole cell vaccine. Infect. Immun. 62: 3594-3597.

Lee, C.K., Soike, K., Giannasca, P., Hill, J., Weltzin, R., Kleanthous, H., Blanchard, J., and Monath, T.P. 1999. Immunization of rhesus monkeys with a mucosal prime, parenteral boost strategy protects against infection with *Helicobacter pylori*. Vaccine. 17: 3072-82.

Lee, C.K., Soike, K., Hill, J., Georgakopoulos, K., Tibbitts, T., Ingrassia, J., Gray. H., Boden, J., Kleanthous, H., Giannasca, P.J., Emark, T., Weltzin, R., Blanchard, J., Monath, T.P. 1998. Immunization with recombinant *Helicobacter pylori* urease decreases colonization levels following experimental infection of rhesus monkeys. Vaccine. 17: 1493-1505.

Lee, C.K., Weltzin, R., Thomas, W.D., Jr., Kleanthous, H., Ermak, T.H., Soman, G., Hill, J.E., Ackerman, S.K., and Monath, T.P. 1995. Oral immunization with recombinant *Helicobacter pylori* urease induces secretory IgA antibodies and protects mice from challenge with *Helicobacter felis*. J. Infect. Dis. 172: 161-72.

Lee, M.H., Roussel, Y., Wilks, S., and Tabaqchali, S. 1999. Expression of *Helicobacter pylori* urease subunit B gene in *Lactobacillus lactis* MG1363: A novel approach to vaccine development. Gut. 45: A58.

Levine, M.M., and Noriega, F. 1993. Vaccines to prevent enteric infections. Baillieres.Clin.Gastroenterol. 7: 501-517.

Ling, T.K., Cheng, A.F., Sung, J.J., Yiu, P.Y., and Chung, S.S. 1996. An increase in *Helicobacter pylori* strains resistant to metronidazole: a five-year study. Helicobacter. 1: 57-61.

Marchetti, M., Rossi, M., Giannelli, V., Giuliani, M.M., Pizza, M., Censini, S., Covacci, A., Massari, P., Pagliaccia, C., Manetti, R., Telford, J.L., Douce, G., Dougan, G., Rappuoli, R., and Ghiara, P. 1998. Protection against *Helicobacter pylori* infection in mice by intragastric vaccination with *H. pylori* antigens is achieved using a non-toxic mutant of *E. coli* heat-labile enterotoxin (LT) as adjuvant. Vaccine. 16: 33-7.

McAtee, C., Lim, M., Fung, K., Velligan, M., Fry, K., Chow, T., and Berg, D. 1998. Identification of potential diagnostic and vaccine candidates of *Helicobacter pylori* by two-dimensional gel electrophoresis, sequence analysis, and serum profiling. Clinical and Diagnostic Laboratory Immunology. 5: 537-542.

McColl, K., Murray, L., El-Omar, E., Dickson, A., El-Nujumi, A., Wirz, A., Kelman, A., Penny, C., Knill-Jones, R., and Hilditch, T. 1998. Symptomatic benefit from eradicating *Helicobacter pylori* infection in patients with nonulcer dyspepsia. N Engl J Med. 339: 1869-74.

Michetti, M., Kelly, C.P., Kraehenbuhl, J.P., Bouzourene,H., and Michetti, P. 2000. Gastric mucosal a4b7 integrin-positive lymphocytes and immune protection against helicobacter infection in mice. Gastroenterology. 119: 109-118.

Michetti, P., Corthésy-Theulaz, I., Davin, C., Haas, R., Vaney, A.C., Heitz, M., Bille, J., Kraehenbuhl, J.P., Saraga, E., and Blum, A.L. 1994. Immunization of BALB/c mice against *Helicobacter* felis infection with H. *pylori* urease. Gastroenterology. 107: 1002-1011.

Michetti, P., Kreiss, C., Kotloff, K.L., Porta, N., Blanco, J.L., Bachmann, D., Herranz, M., Saldinger, P.F., Corthésy-Theulaz, I., Losonsky, G., Nichols, R., Simon, J., Stolte, M., Ackerman, S., Monath, T.P., and Blum, A.L. 1999. Oral immunization with urease and Escherichia coli heat-labile enterotoxin is safe and immunogenic in *Helicobacter pylori*-infected adults. Gastroenterology. 116: 804-12.

Mohammadi, M., Czinn, S., Redline, R., and Nedrud, J. 1996. *Helicobacter* specific cell mediated responses display a predominant Th1 phenotype and promote a delayed type hypersensitivity response in the stomachs of mice. J Immunol. 156: 4729-4738.

Mohammadi, M., Nedrud, J., Redline, R., Lycke, N., and Czinn, S.J. 1997. Murine CD4 - cell response to *Helicobacter* infection: TH1 cells enhance gastritis and TH2 cells reduce bacterial load. Gastroenterology. 113: 1848-1857.

Novak, M.J., Smythies, L.E., McPherson, S.A., Smith, P.D., and Morrow, C.D. 1999. Poliovirus replicons encoding the B subunit of *Helicobacter pylori* urease elicit a Th1 associated immune response. Vaccine. 17: 2384-91.

Pappo, J., Torrey, D., Castriotta, L., Savinainen, A., Kabok, Z., and Ibraghimov, A. 1999. *Helicobacter pylori* infection in immunized mice lacking major histocompatibility complex class I and class II functions. Infect. Immun. 67: 337-41.

Radcliff, F.J., Labigne, A., Ferrero, R.L. 1999. Mice exposed to a low inoculum dose of H. *pylori* are abl eto resist infection upon rechallenge. Gut. 45: A57.

Ramirez-Ramos, A., Gilman, R.H., Leon-Barua, R., Recavarren-Arce, S., Watanabe, J., Salazar, G., Checkley, W., McDonald, J., Valdez, Y., Cordero, L., and Carrazco, J. 1997. Rapid recurrence of *Helicobacter pylori* infection in Peruvian patients after successful eradication. Clin Infect Dis. 25: 1027-1031.

Roy, K., Mao, H.Q., Huang, S.K., and Leong, K.W. 1999. Oral gene delivery with chitosan—DNA nanoparticles generates immunologic protection in a murine model of peanut allergy. Nat Med. 5: 387-91.

Saldinger, P.F., Porta, N., Launois, P., Louis, J.A., Waanders, G.A., Bouzourene, H., Michetti, P., Blum, A.L., and Corthésy-Theulaz, I.E. 1998. Immunization of BALB/c mice with *Helicobacter* urease B induces a T helper 2 response absent in *Helicobacter* infection. Gastroenterology. 115: 891-7.

Sawai, N., Kita, M., Kodama, T., Tanahashi, T., Yamaoka, Y., Tagawa, Y., Iwakura, Y., and Imanishi, J. 1999. Role of gamma interferon in *Helicobacter pylori*-induced gastric inflammatory responses in a mouse model. Infect. Immun. 67: 279-85.

Sonoda, E., Matsumoto, R., Hitoshi, Y., Ishii, T., Sugimoto, M., Araki, S., Tominaga, A., Yamaguchi, N., and Takatsu, K. 1989. Transforming growth factor beta induces IgA production and acts additively with interleukin 5 for IgA production. J. Exp. Med. 170: 1415-20.

Stratton, K.R., Durch, J.S., and Lawrence, R.S. 2000. Vaccines for the 21st century: a tool for decision making. Washington, DC, National Academy Press. In press. Prepublication copy can be viewed on the internet at: http://stills.nap.edu/html/vac21

Sullivan, P.B., Thomas, J.E., Wight, D.G., Neale, G., Eastham, E.J., and Corrah, T., Lloyd-Evans, N., Greenwood, B.M. 1990. *Helicobacter pylori* in Gambian children with chronic diarrhea and malnutrition. Arch Dis Child. 65: 189-191.

Telford, J.L., Ghiara, P., Dell'Orco, M., Comanducci, M., Burroni, D., Bugnoli, M., Tecce, M.F., Censini, S., Covacci, A., and Xiang, Z. 1994. Gene structure of the *Helicobacter pylori* cytotoxin and evidence of its key role in gastric disease. J. Exp. Med. 179: 1653-58.

Tomb, J.F., White, O., Kerlavage, A.R., Clayton, R.A., Sutton, G.G., Fleischmann, R.D., Ketchum, K.A., Klenk, H.P., Gill, S., Dougherty, B.A., Nelson, K., Quackenbush, J., Zhou, L., Kirkness, E.F., Peterson, S., Loftus, B., Richardson, D., Dodson, R., Khalak, H.G., Glodek, A., McKenney, K., Fitzegerald, L.M., Lee, N., Adams, M.D., Venter, J.C., *et al.* 1997. The complete genome sequence of the gastric pathogen *Helicobacter pylori*. Nature. 388: 539-547.

Tsugawa, M., Shachter, R., Owens, D.K., and Parsonnet, J. 1998. Cost-effectiveness of vaccine development for H. *pylori* using the Institute of Medicine model. Gastroenterology. 114: A313.

Weltzin, R., Kleanthous, H., Guirakhoo, F., Monath, T.P., and Lee, C.K. 1997. Novel intranasal immunization techniques for antibody induction and protection of mice against gastric *Helicobacter felis* infection. Vaccine. 15: 370-376.

From: *Helicobacter pylori: Molecular and Cellular Biology*
ISBN 1-898486-25-5 © 2001 Horizon Scientific Press, Wymondham, UK.

17

Comparative Analysis of the *Helicobacter pylori* Genomes

Richard A. Alm* and Trevor J. Trust

Abstract

Helicobacter pylori is the first bacterial species to have two completed genomes sequences available for comparison. The availability of the genomic sequence from two unrelated *H. pylori* isolates, J99 and 26695, has enabled a detailed analysis on the overall level of genetic diversity that has been previously reported for this organism. The arrangement of the genes in the two chromosomes was more similar than predicted, being disrupted by a few organizational differences and the presence of genes that were unique to only one of the strains, which comprised between 6% and 7% of the total coding capability of each strain. Half of the unique genes to each strain were contained within a hypervariable region of the chromosome. Analysis of the orthologous genes and their respective proteins displayed a high degree of allelic differences, and these differences were greater in the *H. pylori* specific genes, assisting in the understanding of the high discriminatory power of nucleotide-dependent typing techniques. As with other bacterial genomes sequenced to date, only approximately 60% of the identified genes encoded a product that could be assigned a function. These results will promote analyses on the population genetics and the evolution of the association of *H. pylori* with the human host. The genome sequences also provide direct access to all of the potential therapeutic targets allowing for the development of either novel anti-bacterial drugs or vaccines against *H. pylori* for people suffering from *H. pylori* related diseases.

Introduction

In recent years there has been a massive increase in the amount of genomic sequence generated by both the private and public research arenas. Since the completion in 1995 of the first bacterial genome sequence, *Haemophilus influenzae*, the genomes of twenty additional bacterial genomes have been published with several others in the final stages of analysis. A significant impact on world health has been recognized since the discovery of *Helicobacter pylori* in the early 1980's (Marshall and Warren, 1984) and the association of *H. pylori* infection with several gastroduodenal diseases including chronic gastritis, duodenal ulcers, MALT lymphoma and gastric adenocarcinoma (Cover and Blaser, 1992; Hunt, 1996; Labigne and de Reuse, 1996). This gastric pathogen, which infects approximately half of the world's population, has subsequently been met with a substantial level of research on all facets of its

*corresponding author, email: richard.alm@astrazeneca.com

biology. This has included the completion of the genome sequences of two unrelated isolates (Alm *et al.*, 1999; Tomb *et al.*, 1997) which afforded the first comparison of completed genome sequences from two isolates of the same bacterial species. There have been numerous reports on the level of genomic diversity, using several distinct techniques, observed among *H. pylori* isolates (Blaser, 1997a; Logan and Berg, 1996; Marshall *et al.*, 1998). The role that this genetic variation plays is unclear, but has been suggested to be involved in the chronicity of *H. pylori* infection (Logan and Berg, 1996) as well as the ability of *H. pylori* infection to progress to a diverse panel of diseases, and even then only in a fraction of infected people (Atherton *et al.*, 1997; Dubois *et al.*, 1996). Further, it has been suggested that such genetic variation may result in some *H. pylori* strains playing a beneficial role towards the host (Blaser, 1997b), perhaps by lowering intragastric acidity in patients producing high levels of gastric acid and thus providing a protective effect against gastroesophageal reflux disease (Richter *et al.*, 1998).

The availability of genome sequence information from the two independent strains has enabled an initial comparison to address some of these diversity issues. Further, analysis of the two genomes allowed the identification of a set of genes that are likely to be common to the majority of *H. pylori* strains, and a comparison of the metabolic capabilities encoded by each strain (Doig *et al.*, 1999). This information will play an important role in the selection of novel targets to allow development of novel anti-*H. pylori* therapeutic agents as well as potential antigens to be included in either a prophylactic or therapeutic vaccine.

Comparative Features of the Two *H. pylori* Genomes

General Features

The two *H. pylori* strains that have been completely sequenced, J99 and 26695, are likely to be completely unrelated. *H. pylori* J99 was isolated in the United States in 1995 from a patient diagnosed with duodenal ulcer and was subjected to minimal passage in the laboratory before its genome was sequenced. In contrast, *H. pylori* 26695 was isolated in the early 1980's in the United Kingdom from a patient with gastritis of 'unknown severity' (Eaton *et al.*, 1989). Further, this strain had been passaged extensively in the laboratory prior to its sequence being completed in 1997. The *H. pylori* J99 chromosome of 1,643,831 base pairs is 24,036 base pairs smaller than the size of the *H. pylori* strain 26695 genome, but both strains are in the average range of *H. pylori* genome sizes estimated by pulsed-field gel electrophoresis (Jiang *et al.*, 1996; Taylor *et al.*, 1992). The overall (G+C)% content of both *H. pylori* genomes is 39%. However, analysis of the (G+C)% distribution across the genomes identified 9 and 8 regions of DNA with different (G+C)% content DNA for strains J99 and 26695, respectively. Three of these regions, including the *cag*PAI, contain the same genes and are found in the same relative location in both J99 and 26695 (Alm *et al.*, 1999). There are two large regions of low (G+C)% content DNA in strain 26695 that encode proteins unique to this strain, whereas J99 contains a single unique low (G+C)% region at one of these relative locations. In addition, both genomes contain several other small regions of low (G+C)% content DNA, many of which contain strain-specific sequences including numerous DNA restriction/modification systems and these regions are often found in the same physical location in both J99 and 26695.

Both genomes contain two copies of the 16S and 23S-5S rRNA loci in the same two relative physical locations, but the strain 26695 genome differs in having an orphan 5S rRNA copy in addition to those associated with each 23S-5S locus. In both *H. pylori* genomes, the 23S-5S and the 16S rRNA loci are not contiguous, contrary to the organization seen in the

Table 1. Sequence similarity of orthologous genes between two *H. pylori* strains

Functional group	Number of genes in			Average identity		Genes of different length	
	J99	26695	both[a]	nt%	aa%	Number	Av Diff (aa)
Amino acid biosynthesis	44	44	44	94.3	95.9	4	3
Biosynthesis of cofactors	60	59	59	93.6[b]	94.4[b]	8[b]	3.9
Cell envelope	160	164[cde]	156	92.9	93.9	52	18.8
Cellular processes[hi]	97	114[fg]	93	94.1	95.1	22	22.8
DNA replication	23	23	21	94.1	95.1	8	5.5
DNA restriction, modification, recombination and repair	66	68	51	90.6	90.6	25	21.1
Energy metabolism	104	104[j]	102	95.0	96.9	9	3
Fatty acid and phospholipid metabolism	28	29	28	95.0	96.6	3	3.7
Purines, pyrimidine, nucleoside and nucleotide biosynthesis	34	34	34	95.0	96.7	0	0
Regulatory functions	32	32[k]	31[k]	94.4	96.4	5	5.2
Transcription	13	13	13	95.4	96.9	1	2
Translation	128	128	128	95.0	96.9	16	3.1
Transport and binding proteins	88[l]	87	87	94.5	96.6	9	4.9
Total for predicted function	877	899	846	94.0	95.4	162	13.9
Conserved with no function	275	290	267	92.5	93.3	62	21.0
H. pylori specific with no function	344	364	288	88.4	87.3	95	34.9
Total for no known function	618	654	555	90.4	90.2	157	29.4
Total	**1496[h]**	**1553[h]**	**1402[m]**	**92.6**	**93.4**	**319**	**21.5**

[a] relative to the number of JHP genes

[b] does not include JHP862 which represents a partial duplication of the *folE* gene

[c] includes 8 HP genes which make up 4 JHP genes (73, 238, 851, 1138).

[d] includes 2 HP genes which make up JHP86

[e] includes 2 HP genes which make up JHP625

[f] includes 2 HP genes which make up JHP556

[g] includes 4 HP genes which make up JHP35 and JHP36

[h] Includes the recently identified JHP1126A(HP1203A) gene which displays similarity to the functionally required domains of the *E. coli* SecE protein (Doig *et al.*, 1999).

[i] see text, J99 has 1 IS606 copy, while 26695 has 2 IS606 copies and 5 IS605 copies. The orthologous transposase genes in IS606 were not included in this analysis.

[j] includes 2 HP genes which make up JHP841

[k] includes 2 HP genes which make up JHP151, J99 has two duplicate orthologs of HP1365

[l] includes an additional partial duplication of HP0818 in J99

[m] The 94 gene difference represents the 89 J99-specific genes, 3 duplications and *tnpA/tnpB* from IS606

majority of the organisms whose genomes have been completely sequenced, including *E. coli* (Blattner *et al.*, 1997) and *Bacillus subtilis* (Kunst *et al.*, 1997). Separation of the 23S-5S and 16S rRNA loci likely requires more complex regulation to effectively coordinate the expression of these three rRNA species than is required in organisms in which the three rRNA genes are adjacent to each other. The genome of *H. pylori* 26695 contains nearly identical copies of approximately 6 kb of DNA that encodes *H. pylori*-specific ORFs, on one side of both 23S-5S regions. In contrast, in the J99 genome, one 23S-5S copy is flanked by an orthologous 6 kb piece of DNA, but the second 23S-5S locus is adjacent to a DNA that encodes J99-specific genes (Doig *et al.*, 1999).

Both the J99 and 26695 chromosomes contain 36 tRNA species, and all of them map to the same relative locations in both strains (Doig *et al.*, 1999). Neither *H. pylori* strain contains Asn or Gln tRNA synthetases although they both contain homologues of the *B. subtilis* *gatABC* genes (Alm *et al.*, 1999; Doig *et al.*, 1999) which have been shown to amidate Glu-charged tRNAs to provide the necessary Gln-charged tRNA (Curnow *et al.*, 1997). The possibility exists that these genes or similar such modification genes are responsible for the amidation of appropriate Asp-charged tRNAs to provide Asn-charged tRNAs.

Certain genes, including *dnaA, dnaN* and *gyrA*, are often found near the origin of replication in prokaryotes (Salzberg *et al.*, 1998). These genes are not in close proximity either to each other or to the repeated heptamer that was designated as nucleotide number one in the *H. pylori* 26695 sequence. For ease of comparison, nucleotide number one in J99 was also designated at the start of the heptamer repeat (AGTGATT) which produces translational stops in all reading frames (Alm *et al.*, 1999). Unlike strain 26695 that has twenty-six tandem copies of this heptamer repeat, the J99 genome has only two tandem copies at this location. However, the same heptamer sequence is repeated in thirteen tandem copies approximately 2.5 kb upstream in J99, whereas in the same location, strain 26695 contains only two tandem copies. An algorithm based on the analysis of the skewed distribution of repeated octamers in the *H. pylori* 26695 genome sequence has identified a putative origin of replication in the area between 1.566Mb and 1.640Mb (Salzberg *et al.*, 1998).

H. pylori J99 possesses 1,496 predicted ORFs, with the recent identification of a *secE* ortholog (Doig *et al.*, 1999), whereas strain 26695 was predicted to contain 1,590 ORFs (Tomb *et al.*, 1997). During the comparative annotation process, this number was reduced to 1,552 (Alm *et al.*, 1999) due to the removal of 38 predicted ORFs that encoded products shorter than 50 amino acids in length that either contained in-frame stop codons in J99 or displayed significant nucleotide drift similar to other intergenic regions and were deemed unlikely to encode protein products. In both strains, only approximately 58% of the predicted proteins could be assigned a putative function based on sequence similarity (Table 1). The remaining gene products can be separated into those that contain an ortholog in another bacterial species but the function is unknown (~18%) or those that remain *H. pylori*-specific (~23%). Initial analysis of the recently released *Campylobacter jejuni* NCTC 11168 sequence released by the Sanger Center has identified approximately 90 gene products that are, to date, only contained within these two closely related species (Doig *et al.*, 1999). This would reduce the number of *H. pylori*-specific genes to a level where they represent approximately 17% of the coding capacity of the *H. pylori* genome.

There have been two insertion sequences identified in *H. pylori*, IS605 and IS606. These elements can be located, often in multiple copies, at different locations on the chromosomes of *H. pylori* strains (Hook-Nikanne *et al.*, 1998; Kersulyte *et al.*, 1998). Although *H. pylori* J99 lacks a complete copy of the IS605 element, strain 26695 contains 5 complete copies, 3 of which lie in the hypervariable plasticity zone and the remaining two are associated with organizational differences between the two strains (Figure 1, see below). Both J99 and 26695 have complete copies of the IS*606* element, although the single copy in J99 is present in a different location than either of the two copies in strain 26695 and none of these are associated with any organizational differences (Figure 1). Both J99 and 26695 also contain vestiges of these insertion elements (Alm *et al.*, 1999). Furthermore, the location of the complete and partial IS elements in both strains appears to be biased. Only one of the eight full-length IS elements (an IS605 element) and six of eighteen IS fragments in both strains are located in that half of the genome that surrounds the heptamer repeats designated as nucleotide number one. In addition, there are five cases of coincidence in the location of partial or complete copies of insertion elements in these two strains, suggesting that certain loci are more receptive to insertion sequences or that these elements have been maintained from a common progenitor strain.

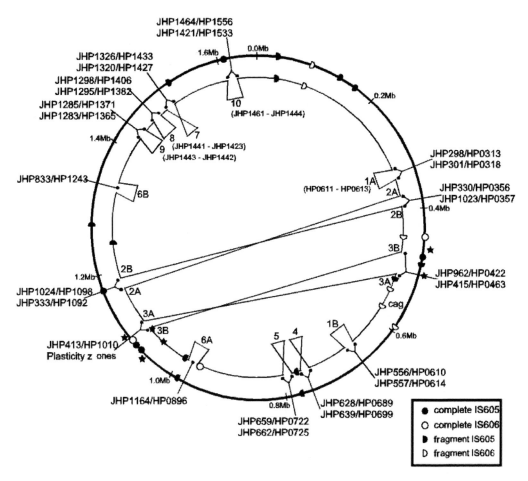

Figure 1. Schematic representation of the alignment of the *H. pylori* J99 and 26695 chromosomes. The outer circle represents the 1.6Mb *H. pylori* 26695 chromosome. The orthologous genes in *H. pylori* J99 are represented by the inner circle and were mapped onto the 26695 chromosome; with artificial breaks being introduced where necessary although the scale is slightly different due to the size difference of the genomes. The organizational rearrangements that were required are numbered 1 through 10 but their size is not drawn to scale. At each of these rearrangements, the orthologous gene pairs that flank the endpoints are listed, except for 6A and 6B where only the single gene that has been transposed is indicated. The location of the complete (circles) or partial (half-circles) copies of the IS*605* (solid) and IS*606* (open) insertion elements in both genomes are indicated. The *cag* Pathogenicity Island is at the same relative location in both genomes and is indicated between the genomes. In J99, the *cag* island lies within two IS*606* fragments. The hypervariable plasticity zone is associated with region 3B, and spans the areas between the solid stars. In the case of translocations 1A, 8, 9 and 10, the affected genes are indicated in parentheses in the center of the inner circle.

Nucleotide Diversity

Several studies have been performed to examine the nucleotide diversity of selected genes among different *H. pylori* isolates. Variation has been detected in the *cagA, vacA, flaA, flaB, cysS* and *ureC/glmM* genes, as well as the *tnpA/B* genes from IS*605* by sequencing these loci from panels of strains (Evans Jr *et al.*, 1998; Garner and Cover, 1995; Hook-Nikanne *et al.*, 1998; Kansau *et al.*, 1996; Suerbaum *et al.*, 1998b). Recently, Achtman *et al.* (1999) assessed the sequence structure of seven housekeeping genes from a geographically diverse panel of strains. These analyses identified two weak clonal groupings based on geographical separation albeit that recombination was extremely common at most loci. A common theme in all of

Table 2. Nucleotide and amino acid identity between orthologous *H. pylori* genes.

| | Number (%) of predicted ORFs | |
| | Nucleotide | Amino acid |
Percent Identity		
100	0 (0)	41 (2.9)
99.0-99.9	0 (0)	89 (6.4)
98.0-98.9	8 (0.6)	180 (12.9)
97.0-97.9	64 (4.6)	192 (13.7)
96.0-96.9	185 (13.2)	168 (12.0)
95.0-95.9	276 (20.0)	163 (11.7)
94.0-94.9	291 (20.8)	116 (8.3)
93.0-93.9	192 (13.7)	90 (6.4)
92.0-92.9	114 (8.2)	79 (5.7)
91.0-91.9	54 (3.9)	48 (3.4)
90.0-90.9	35 (2.5)	38 (2.7)
<90	178 (12.7)	193 (13.8)
Total[a]	1397 (100[b])	1397 (100[b])

[a] This total does not include strain specific genes. Genes that appear "split" by putative frameshifts in either strain have been classed as the larger ORF in the above analysis.
[b] The total is different due to rounding.

these sequencing studies is that the presence of identical orthologous gene sequences between two *H. pylori* strains is extremely rare and that recombination between alleles is common. Numerous methods that would detect nucleotide diversity, other than direct sequencing, have also been used to measure the genetic diversity of *H. pylori* isolates, including RAPD-PCR and PCR-based RFLP (Akopyanz *et al.*, 1992; Berg *et al.*, 1997; Han *et al.*, 1997). These data, some of which sample the whole genome, also demonstrate that it is rare for two *H. pylori* isolates to possess the same pattern.

The availability of two complete genome sequences has allowed a complete analysis of the level of nucleotide and corresponding amino acid diversity among the orthologous gene pairs. The comparison of orthologous genes at the nucleotide level indicated that there was a silent drift in the amino acid coding triplets that failed to result in a change in the predicted protein sequence. While there are no two orthologous genes that are 100% identical, 41 proteins share 100% identity between the two strains. More significant however, is that only 8 gene pairs (0.6%) share a level of identity that is greater than 98%, while 269 pairs of proteins (19.3%) share that level of identity (Table 2). This analysis was extended to determine whether the level of nucleotide conservation correlated with functional prediction. Indeed, the orthologous genes with a predicted function possessed an average of 94.0% identity compared to the average of 92.6% over all the orthologous genes (Table 1; Alm and Trust, 1999). The expected increase in the predicted protein similarity due to 'silent' nucleotide changes is also evident in both classes, being 93.4% and 95.4% for all the orthologous proteins and those with a predicted function, respectively (Table 1). The *H. pylori*-specific genes demonstrate the lowest level of identity, and is the only group where the average identity of the corresponding proteins is lower (Table 1). The possibility exists that some of these predicted coding sequences may either not represent true genes or that some of them are slowly collapsing into redundancy. The orthologous proteins were also analyzed for any differences in their length (Table 1). The orthologous proteins with no known function, either conserved in other species or *H. pylori* specific were significantly more disparate, with almost 30% of the proteins being different in length, and the average difference being 29.4 amino acid residues. In contrast, only 19% of the orthologous proteins with a predicted function differed in length, and then only by an average of 13.9 residues (Table 1). However, the majority of the differences in the functional categories came from the DNA restriction/modification enzymes, the outer membrane proteins and other cell envelope associated proteins and the Vir and VacA homologues (Alm and Trust, 1999). Analysis of the length divergence in the 'metabolic' functional groups demonstrated that only 49 orthologous protein pairs (6%) differed in length, and then only by an average of 3.6 residues.

Codon usage tables for the genes of both *H. pylori* J99 and 26695 were constructed to assess whether the observed differences in the third position of the amino acid coding triplet affected the codon bias (Table 3). These data suggest that *H. pylori* does possess a codon bias, and that both strains adhere to that bias essentially equivalently, independent of the drift in individual genes. Between *H. pylori* J99 and 26695, the sixty-one nucleotide triplets that encode amino acids are all ±0.1% with respect to their proportion of all amino acids. Further, in only four cases is there greater than a 1% differential when assessing the proportion of each codon to a specific amino acid (Table 3). It appears that while the 'wobble' position in the coding triplet changes frequently among orthologous genes, the overall genome bias does not change significantly. This may suggest that the nucleotide differences may drift randomly, possibly as a result of poor DNA mismatch during replication (Alm and Trust, 1999). Both *H. pylori* genomes lack an identifiable *mutH* or *mutL* ortholog (Alm and Trust, 1999), the mutator genes that, in combination with MutS, are responsible for efficient correction of purine/pyrimidine mismatches during replication (Horst *et al.*, 1999). A high frequency of transition mutations has been noted between orthologous genes in *H. pylori* and the lack of the complete MutHLS system may represent the underlying mechanism (Wang *et al.*, 1999).

H. pylori proteins are rich in lysine which represents ~8.8% of the total amino acids encoded by the organism (Table 3). Although this represents a large increase with respect to the abundance of lysine seen in *E. coli* (4.4%) the overall bias for basic amino acids is modified by a lower percentage of arginine (3.4%) than *E. coli* (5.5%). However, the result is that a majority of the *H. pylori* proteins have a predicted isoelectric point (p*I*) that is basic (> 7.0) when compared to *E. coli*.

The third 'wobble' position in the coding triplet is often where organisms shows distinct bias and is usually influential with respect to the overall (G+C)% content of the genome. For example, the prevalence of (G+C) nucleotides at this position in a high (G+C)% organism (~65%) such as *Pseudomonas aeruginosa* is very high, at ~82%. Similarly, in organisms containing lower (G+C)% content DNA such as *Borrelia burgdorferi* (29%) or *Campylobacter jejuni* (31%), the (G+C) bias in the wobble position is considerably lower at around 20%. Analysis of the (G+C) preference in the wobble position of the coding triplets in *H. pylori* indicated that the bias at the third position is not as strong as would be expected, having a (G+C)% of 42.7 and 42.0% for J99 and 26695, respectively which is higher than that of the whole genome.

Physical Arrangement of Coding Regions

The physical arrangement of the relative gene order on the *H. pylori* chromosome was predicted, on the basis of pulsed-field gel electrophoresis and several genetic probes, to be quite distinct between different *H. pylori* isolates (Jiang *et al.*, 1996; Takami *et al.*, 1993; Taylor *et al.*, 1992). Comparative analysis of the gene order between the two sequenced strains was performed by comparing each of the 1,496 *H. pylori* J99 genes to the orthologous gene in 26695 (if present in that strain) with respect to the adjacent gene on both sides. This analysis indicated that the immediate genetic neighbor of 1,268 genes (84.8%) was the same in both strains. The presence of genes that were specific to either J99 or 26695 resulted in 161 of the J99 genes having their relative gene order being disrupted on one side. A total of forty genes

Table 3. Codon Usage of *H. pylori* genes.

AA	Codon	J99 % of all aa	J99 % of type	26695 % of all aa	26695 % of type	AA	Codon	J99 % of all aa	J99 % of type	26695 % of all aa	26695 % of type
Ala	GCA	0.7	9.9	0.7	10.9	Ile	AUA	0.9	12.6	0.9	12.9
	GCC	1.5	21.4	1.4	20.3		AUC	2.8	38.9	2.7	37.7
	GCG	2.1	30.4	2.0	29.5		AUU	3.5	48.5	3.6	49.4
	GCU	2.6	38.3	2.7	39.3		total[a]	7.1	100	7.2	100
	total	6.9	100	6.8	100	Lys	AAA	6.8	76.6	6.9	77.1
Arg	AGA	0.9	25.4	0.9	26.9		AAG	2.1	23.4	2.0	22.9
	AGG	0.9	25.3	0.8	24.6		total[a]	8.8	100	8.9	100
	CGA	0.3	7.5	0.2	7.0	Met	AUG	2.2	100	2.2	100
	CGC	0.8	24.5	0.8	24.3		total	2.2	100	2.2	100
	CGG	0.1	3.2	0.1	3.0	Phe	UUC	1.2	21.8	1.1	21.2
	CGU	0.5	14.1	0.5	14.1		UUU	4.2	78.2	4.3	78.8
	total[a]	3.4	100	3.5	100		total	5.4	100	5.4	100
Asn	AAC	2.6	43.7	2.5	43.2	Pro	CCA	0.5	14.6	0.5	14.9
	AAU	3.3	56.3	3.3	56.8		CCC	0.9	26.5	0.9	26.6
	total[a]	5.9	100	5.9	100		CCG	0.3	9.8	0.3	9.8
Asp	GAC	1.3	28.4	1.3	27.2		CCU	1.6	49.2	1.6	48.7
	GAU	3.4	71.6	3.5	72.8		total[a]	3.3	100	3.3	100
	total	4.7	100	4.8	100	Ser	AGC	2.7	39.8	2.7	39.5
Cys	UGC	0.7	66.5	0.7	65.7		AGU	1.0	14.2	1.0	14.3
	UGU	0.4	33.5	0.4	34.3		UCA	0.6	8.9	0.6	8.9
	total	1.1	100	1.1	100		UCC	0.6	8.4	0.6	8.2
Gln	CAA	3.1	85.0	3.2	85.2		UCG	0.4	5.6	0.4	5.6
	CAG	0.5	15.0	0.6	14.8		UCU	1.6	23.1	1.6	23.5
	total[a]	3.7	100	3.7	100		total[a]	6.8	100	6.8	100
Glu	GAA	5.1	73.6	5.0	73.4	Thr	ACA	0.7	15.0	0.7	15.8
	GAG	1.8	26.4	1.8	26.6		ACC	1.4	32.0	1.4	31.5
	total[a]	6.9	100	6.9	100		ACG	1.0	22.5	1.0	22.2
Gly	GGA	0.6	10.2	0.6	10.9		ACU	1.3	30.5	1.3	30.5
	GGC	2.1	35.5	2.0	35.2		total	4.4	100	4.4	100
	GGG	2.2	37.8	2.1	37.3	Trp	UGG	0.7	100	0.7	100
	GGU	1.0	16.5	1.0	16.6		total	0.7	100	0.7	100
	total[a]	5.8	100	5.8	100	Tyr	UAC	1.1	31.1	1.1	30.4
His	CAC	0.7	31.8	0.7	31.1		UAU	2.5	68.9	2.6	69.6
	CAU	1.4	68.2	1.5	68.9		total[a]	3.7	100	3.7	100
	total[a]	2.1	100	2.1	100	Val	GUA	0.6	10.4	0.6	10.3
Leu	CUA	0.8	7.2	0.8	7.1		GUC	0.8	14.1	0.8	13.7
	CUC	1.0	8.9	1.0	8.7		GUG	2.8	48.6	2.7	48.5
	CUG	0.4	3.8	0.4	3.9		GUU	1.5	26.9	1.5	27.4
	CUU	1.6	14.5	1.6	14.7		total[a]	5.7	100	5.6	100
	UUA	4.3	38.8	4.4	39.0	Stop	UAA	0.2	56.1	b	b
	UUG	3.0	26.9	3.0	26.6		UAG	0.1	16.7	b	b
	total[a]	11.2	100	11.2	100		UGA	0.1	27.2	b	b
							total[a]	0.3	100		

a; Rounding affects these totals, all are correct to one decimal place.
b; These values were not calculated.

were flanked by J99-specific (not present in 26695) genes on both sides. Importantly, only 27 genes (1.8%) were flanked on one side by a gene that was common to both strains but in a different genetic order, representing a gene-shuffling event. There were 9 strings of conserved genes that were over 50 genes in length that represents 46% of the total number of common genes, with the longest being 133 genes. The limited gene shuffling observed is consistent with a low level of evolutionary divergence within *H. pylori* (Tatusov *et al.*, 1996). While comparison of these two genomes demonstrated very limited gene order rearrangements, the possibility exists that the gene order in other *H. pylori* isolates may be disrupted to a larger degree by insertion or deletion of additional genes or other organizational rearrangements.

Ten segments of the J99 genomic sequence, ranging in size from 1 kb to 83 kb, had to be artificially inverted and/or transposed to enable the regions containing the orthologous genes to be mapped onto the *H. pylori* 26695 circular chromosome (Figure 1). Two of these (rearrangements 2 and 3) are large inversions and translocations and result in the movement of a large number of genes to a different quadrant of the circular chromosome, as well as the inversion of their relative transcriptional direction. The remaining eight organizational differences are smaller in size and thus represent the movement of a smaller number of genes. Insertion elements, repeated sequences, or restriction/modification genes are associated with at least one of the endpoints in all of these regions in one or the other of the two chromosomes (Alm *et al.*, 1999). These findings are consistent with a role for such elements in the mechanisms giving rise to these organizational differences between these two chromosomes, although the precise role is presently unknown.

The genes included in translocation 1 in both strains (JHP299-JHP300 and HP0611-HP0613) are flanked by DNA with a lower (G+C)% content and are associated in J99 with a 169 base pair repeat, copies of which are found in varying lengths with approximately 80% identity at several locations in both genomes. Inversion 4 is associated with the presence of a type II restriction/modification system unique to the J99 genome. One endpoint of inversion 5 lies adjacent to a left end repeat of IS605 in both strains. However, this short inversion could result from a simple recombination across the inverted, approximately 1 kb of nearly identical nucleotide sequence coding for the C-terminal domain of the two genes encoding outer membrane proteins (omp) located there in both strains (Alm *et al.*, 1999). Rearrangement 6 is the equivalent of a reciprocal exchange of two additional omp genes which share a similar highly conserved C-terminal domain and have been proposed to be the Lewis binding adhesins babA and babB (Ilver *et al.*, 1998). In addition, the sequences upstream of the babA and babB initiation codons share some sequence homology, which may allow allelic exchange within or between strains. Interestingly, sequences with similar homologies are also located upstream of other members of this outer membrane superfamily (JHP7 and JHP1261). A 184 base pair inverted repeat is contained partially within the C-terminal coding region of two divergently transcribed histidine-rich genes (HP1427 and HP1432) that are associated with inversion 7 in *H. pylori* 26695. Due to the organizational difference, these sequences appear as a direct repeat and the two histidine-rich genes are adjacent to each other (JHP1320 and JHP1321) and are in the same transcriptional orientation in strain J99 (Alm *et al.*, 1999; Doig *et al.*, 1999). Rearrangement 8 is an inversion and translocation that is associated with the presence of DNA restriction/modification genes that are unique to each strain (JHP1296-1297 in J99 and HP1404-1405 in 26695). The inversion and translocation that represents rearrangement 9 is associated with a DNA restriction/modification gene JHP1442, HP1366) as well as the precise duplication of a response regulator gene (JHP1283 and JHP1443 are two orthologs of HP1365) in *H. pylori* J99 (Alm *et al.*, 1999). The endpoint of rearrangement 10 is adjacent to an IS605 element in strain 26695 but not in J99 (Figure 1).

Rearrangement 2 is a large inversion and translocation that was required to align 75kb of orthologous sequence between J99 and 26695 (Alm *et al.*, 1999). An IS605 sequence is located at one endpoint of this rearrangement in 26695, encoded by genes HP1095 and HP1096. There is also the insertion of unique genes in both J99 (JHP331-332) and 26695 (HP1094) associated with this rearrangement. The DNA in *H. pylori* 26695 that encodes the 31 amino acid residue protein HP1097 is located at a similar location in J99, albeit with a frameshift and no initiation codon which suggests that it does not represent a true ORF (see above). Rearrangement 3 is the most intriguing of the rearrangements required to align the two genomes and represents an inversion and translocation of 83kb of DNA and is associated in both genomes with a large region of lower (G+C)% DNA (Alm *et al.*, 1999). These regions are called the plasticity zones as they contain almost half of the unique genes in each strain, many of which are also *H. pylori*-specific as they have no similarity to any gene product in the databases. The presence of low (G+C)% content DNA clustered in discrete regions is suggestive of horizontal DNA transfer, and the strain-specific sequence differences are consistent with different origins for this apparent foreign DNA. The plasticity zone in J99 is present as one contiguous sequence, whereas the strain 26695 plasticity zone is split into two segments separated by approximately 600 kb (Figure 1). The two segments of the strain 26695 plasticity zone contain two copies of repeat 7, an orphan copy of the 5S RNA (not linked to the 23S rRNA), three full length IS605 elements, one full length IS606 element, and one IS605 fragment. In contrast, the J99 plasticity zone lacks any insertion elements and contains only one copy of repeat 7. A simple recombination event across two inverted copies of repeat 7 in a progenitor strain resembling strain 26695 would yield an arrangement similar to that of J99 at one endpoint of this organizational difference. However, a simple reciprocal event in a progenitor strain could not account for the complex changes required to produce an arrangement similar to that of J99 at the other endpoint (Alm *et al.*, 1999). Importantly, two copies of the IS605 insertion element and neighboring strain-specific chromosomal DNA from *H. pylori* 26695 (HP0999, HP1000, HP1001) are present in the same genetic configuration on a *H. pylori* plasmid isolated from an unrelated strain. Plasmid DNA from some *H. pylori* strains does possess a (G+C)% content in this lower range (Lee *et al.*, 1997). These data suggest that a mechanism of plasmid integration may exist in *H. pylori* for the introduction of novel DNA into the chromosome, lending strength to the hypothesis that horizontal DNA transfer gives rise to much of the diversity seen within *H. pylori*.

Strain-specific Genes and Metabolic Analyses

The variable clinical outcome of infections caused by *H. pylori* has been associated with strain to strain differences in its genetic content. For example the presence of the putative pathogenicity island, *cag* (*cag*PAI), has been connected to an enhanced risk for the development of duodenal ulcers and adenocarcinoma of the distal stomach (Akopyants *et al.*, 1998b; Figura, 1997). Both of the two sequenced *H. pylori* strains contained a set of specific genes absent from the other genome. *H. pylori* J99 contains 89 genes that are not present in strain 26695, of which only 25 had an assignable function based on sequence similarity. In contrast, *H. pylori* 26695 contained 117 genes that were absent from strain J99, with only 26 having a putative function assigned to them.

The availability of two completely annotated *H. pylori* genomes has allowed an analysis of the physiology of this bacterium and the metabolic capabilities of both J99 and 26695 and how the unique genes may alter the metabolic phenotype of each particular strain Doig *et al.*, 1999). Although only approximately 60% of the genes in *H. pylori* could be assigned a function, the predicted metabolic capabilities are generally in excellent agreement with those predicted from experimental observations (Doig *et al.*, 1999). *H. pylori* appears to use fermentation to generate energy, and can undergo both anaerobic and aerobic respiration and has quite an extensive repertoire of genes encoding nitrogen metabolism enzymes (Doig *et al.*, 1999). Both *H. pylori* strains possess relatively few enzymes for the biosynthesis of amino acids, but possess a large number of putative transporter uptake systems (Doig *et al.*, 1999). It is likely that these amino acids will be used as carbon sources, as both strains appear to have a limited ability to acquire sugars from the surrounding milieu.

The exact role that many of the genes that are unique to either *H. pylori* J99 or 26695 play in altering the metabolic diversity was difficult to assess as 64 (72%) and 91 (78%) of the unique genes could not be assigned a function in J99 or 26695, respectively. The large majority of these genes of unknown function (56 in J99 and 70 in 26695) were also *H. pylori*-specific and share no sequence similarity to information in the public databases. The metabolic impact of the remaining unique gene products is minimal, and it is predicted that both *H. pylori* J99 and 26695 are extremely similar physiologically (Doig *et al.*, 1999). Approximately 60% of the gene products that had a predicted function and were unique to either J99 or 26695 were DNA restriction/modification enzymes, consistent with the fact that each *H. pylori* isolate appears to contain its own complement of these enzymes (Akopyants *et al.*, 1998a). The plasticity zones of both strains contain all of the unique genes assigned to the DNA Replication (*topA* genes) and Cellular Processes (*virB4* and *virD4* genes) categories, with the exception of HP0548 in 26695 that encodes the cag-omega putative helicase (Alm *et al.*, 1999). The 5' end of this gene is deleted in *H. pylori* J99, and as such was not called a coding sequence. Several of the genes unique to either J99 or 26695 encode products that are likely to be able to alter the appearance of the bacterial cell envelope such as outer membrane proteins or lipopolysaccharide biosynthetic enzymes and subsequently may alter the interaction with the host immune system (Alm *et al.*, 1999). The remaining four unique genes two in each strain) are predicted to be involved in energy and phospholipid metabolism. *H. pylori* J99 contains two additional genes that encode dehydrogenases, whereas strain 26695 possess an acetyl-CoA synthetase and a second, larger acyl-carrier protein (Alm *et al.*, 1999; Doig *et al.*, 1999). The presence of the former gene in *H. pylori* 26695 allows the direct conversion of acetate to acetyl-CoA whereas strain J99 appears to perform this by running the fermentative pathway in the reverse direction. Strain 26695 has a frameshift that has been independently verified in its phosphotransacetylase gene (*pta*) which is required for this reaction and likely converts acetate directly (Doig *et al.*, 1999).

The genomic locations of the J99- or 26695-unique genes were very similar. Almost half of the genes unique to either J99 or 26695 are located in their respective hypervariable plasticity zone(s) (Figure 1). Furthermore, in approximately half of the loci where one strain contained a unique gene(s), the other strain also contained a unique gene(s), with orthologous flanking genes (Alm *et al.*, 1999), suggesting that the *H. pylori* genome may possess limited regions where these genes can be located.

Gene Expression

Analysis of the *H. pylori* genomes for clues about gene expression suggests that regulation of gene expression in this organism may be less complex than in some other bacterial pathogens. Both *H. pylori* J99 and 26695 possess only three identifiable sigma factors, which provide promoter recognition specificity to the DNA polymerase III holoenzyme. These are the general housekeeping sigma factor, σ^{70}, encoded by the *rpoD* gene, the flagellar-specific sigma factor, σ^{28}, encoded by the *fliA* gene and the σ^{54} subunit encoded by the *rpoN* gene. Of these, the σ^{54} factor, which controls gene expression of several different gene classes, has been shown to be susceptible to environmental regulation from numerous stimuli (Alm *et al.*, 1993; Merrick, 1993). Regulatory networks also appear limited in *H. pylori*. There are three histidine kinases in addition to CheA, although only two of these are adjacent to transcriptional response regulators. In addition to the chemotaxis response regulator CheY (and the three related CheV molecules with a CheY domain), there are five two-component regulators in a paralogous family.

H. pylori has several genes that contain homopolymeric tracts and dinucleotide repeats that allow the modulation of gene expression by slipped-strand repair (Alm *et al.*, 1999; Saunders *et al.*, 1998; Tomb *et al.*, 1997). Many of the genes that appear regulated in this manner, such as outer membrane proteins and lipopolysaccharide genes, have the ability to alter the antigenic make-up of the surface of the *H. pylori* bacterium. Such a mechanism could be involved in antigenic variation and adaptive evolution and may affect the ability of the organism to colonize and cause disease. The best-studied examples of this phenomenon in *H. pylori* are the α-fucosyltransferase genes that are responsible for the late stages of lipopolysaccharide biosynthesis. The specific lipopolysaccharide serotype of several clinical isolates as well as NCTC 11637 phase variants have been correlated with the varying length of the homopolymeric C tract and the subsequent expression status of the two α-(1,3) fucosyltransferase and the single α-(1,2) fucosyltransferase genes (Appelmelk *et al.*, 1999). Comparative analysis of the genes that encode outer membrane proteins suggests that this slipped-strand mechanism may also be responsible for modulation of the spacing between the initiation codon and the ribosome-binding site of specific genes (Alm *et al.*, 2000). This may represent another subtle mechanism by which *H. pylori* can alter the level of gene expression. Further, both *H. pylori* genomes also contain pseudogenes, or authentic frameshifts in several genes as well as in-frame termination codons (Doig *et al.*, 1999; Tomb *et al.*, 1997). This may also represent another mechanism by which *H. pylori* can regulate gene expression, or may simply represent the initiation of decay of unwanted genetic information.

Analysis of the *H. pylori* genomes revealed the paucity of transcription terminating stem loop structures which suggests that the transcription of the majority of genes in *H. pylori* may be terminated in a rho dependent fashion. Preliminary analysis of the *Campylobacter jejuni* NCTC 11168 completed genome sequence indicates a similar trend. Further, both species appear to have fewer cases of functionally similar genes being adjacent to each other. Rather, genes that encode products of seemingly totally different cellular processes are adjacent to each other, with many of them appearing to be co-transcriptionally linked in an operon based on the spacing (in many cases overlapping) between the termination and initiation codons.

Use of Genomic Information for Development of New Treatment Strategies

The continual advance in high-throughput DNA sequencing technologies combined with the significant improvements in bioinformatic capabilities has enabled complete genomic sequence to be rapidly generated and analyzed from increasingly complex organisms. The post-genomic era has arrived and predicting, from the large repository of data now available, or determining the function of the encoded gene products has become a primary focus of many research efforts. In particular, the study of infectious diseases has been revolutionized by the ability to identify the majority of the potential virulence determinants carried by a pathogenic bacterium. Subsequently, there has been significant interest in the potential use of this sequence data for the generation of novel therapeutics and vaccines.

The availability of the genome sequence of two unrelated *H. pylori* strains has provided an unprecedented opportunity for the development of novel therapies against this organism. The current anti-*H. pylori* therapies involve a complex combination of antimicrobial compounds and acid suppression. However, there are several reasons why therapy aimed at the eradication of *H. pylori* in a given patient may fail. These include patient non-compliance with the complex regimen as well as the occurrence of increasing resistance to the commonly used antibiotics such as metronidazole and clarithromycin (Alarcon *et al.*, 1999; Graham, 1998). With the prediction that antibiotic resistance levels in *H. pylori* will increase, there is a need to develop novel methods to treat patients suffering from *H. pylori*-associated gastrointestinal diseases. Therapeutic strategies under development are a therapeutic vaccine as well as the development of specific anti-*H. pylori* antibiotics.

The comparative analysis of the genomes of *H. pylori* strains J99 and 26695 have identified genes that are unique to each strain (see above), but also a set of genes that are common to both strains. Many of these common orthologs are required for the survival of *H. pylori* in the host, and as such represent potential targets suitable for therapeutic intervention. This knowledge is the basis behind the 'Genomics to Drugs' strategy that is being pursued by several pharmaceutical companies, both for the development of anti-*H. pylori* agents but also for other classes of novel antibiotics. Using the genomic information of *H. pylori*, many of the biochemical capabilities of the organism can be predicted and then experimentally confirmed. This information can then be used to predict those biological functions, and the corresponding enzymes, which are essential for the viability of the organism. The absolute requirement for a particular gene product for viability can be measured *in vitro* by constructing defined isogenic mutants in a laboratory strain of *H. pylori*. These essential proteins can then be characterized and used in target-based screening approaches to identify chemical compounds that can inhibit their respective biochemical activities. Subsequent optimization of these compounds to provide suitable microbiological and pharmacological criteria will enable the development of novel, specific anti-*H. pylori* agents. The increase in prokaryotic genomic information, and the recent completion of the genome sequence of the closely related *C. jejuni* organism, has enabled, through comparative genomics, the identification of many genes that remain *H. pylori*-specific. The products of these *H. pylori*-specific genes represent an excellent source of potential antibiotic targets enabling development of *H. pylori* specific therapeutics.

An alternative to antibiotic therapy is the development of a vaccine that can reduce or eradicate the burden of *H. pylori* in a large population. Therapeutic vaccination has been demonstrated using experimental animal models with recombinant *H. pylori* antigens (Dunkley *et al.*, 1999; Ferrero *et al.*, 1995; Ghiara *et al.*, 1997; Hocking *et al.*, 1999; Kleanthous *et al.*, 1998; Marchetti *et al.*, 1998; Radcliff *et al.*, 1997). The advantages of such an immunotherapy would be its specificity for *H. pylori* and its activity against the emerging antibiotic-resistant organisms. Furthermore, it is likely that vaccination may also provide protection against potential re-infection. The *H. pylori* genomic information provides the sequence of all the potential antigens that can be included in a vaccine. The genome-selected proteins are recombinantly expressed and purified and their efficacy is then evaluated in an experimental *H. pylori* infection model. To date, recombinant *H. pylori* antigens must be administered together with a powerful mucosal adjuvant such as cholera toxin or heat-labile toxin to observe efficacy in the murine model. The requirement to include a mucosal adjuvant in order to either protect the animal from subsequent *H. pylori* infection or reduce the pre-existing bacterial burden was interpreted to mean that the anti-*H. pylori* immunity was likely to be mediated by secretory antibodies. Such a protective mechanism would suggest that antigens that are exposed on the bacterial cell surface would represent the best vaccine candidates. To this end, outer membrane proteins or adhesins would seem to be likely candidates for inclusion in a vaccine. *H. pylori* possesses an outer membrane profile that is distinct from other Gram-negative bacteria such as the Enterobacteriaceae in that it appears to contain a large number of less-abundant proteins (Doig and Trust, 1994). Indeed, the *H. pylori* genomic sequences revealed the presence of a large number of genes that encode outer membrane proteins present in several distinct paralogous families (Alm *et al.*, 2000). The largest of these families, with 32 members in common, contain proteins that have been experimentally shown to possess porin activity (Doig *et al.*, 1995; Exner *et al.*, 1995) as well as adhesive specificity for gastric epithelial cells (Ilver *et al.*, 1998; Odenbreit *et al.*, 1999). However, recent evidence suggests that an antigen-specific antibody response is not required to demonstrate protective efficacy in a murine model of *H. pylori* infection (Ermak *et al.*, 1998; Pappo *et al.*, 1999). These data suggest that cell-mediated immunity may play a role in the protective immune response against *H. pylori* infection, suggesting that a broader range of *H. pylori* proteins may be able to induce protective immunity.

Future Considerations

The reason why only a small percentage of people infected with *H. pylori* progress into serious disease has been the cause for much debate and the variable clinical outcome of infections caused by *H. pylori* has been associated with strain-to-strain differences in genetic content. The comparison of the genomes of these two unrelated *H. pylori* isolates has demonstrated that the allelic diversity of the genes is, as expected, substantial but that the gene order across these two genomes is largely conserved. This supports observations from other laboratories that have demonstrated a smaller number of genes conserved in order across a wide range of isolates (Bina *et al.*, 2000; Ge and Taylor, 1999; Suerbaum *et al.*, 1998a). The genomic information has also allowed an initial look into the 'common set' of genes by identifying and eliminating those genes that appear only in one of the two strains.

Does this genomic comparison give any indication as to why infection with *H. pylori* causes such a wide spectrum of disease in the human population? Differences in gene expression, by mechanisms such as slipped-strand regulation or methylation are likely to play a role, and advances in microarray technology will allow the gene expression patterns of *H. pylori* strains to be examined at a global scale. The strain-unique genes, either present on a plasmid, in the plasticity zone or scattered about the genome are also likely to affect the outcome of infection with a particular *H. pylori* strain. However, none of the strain-specific genes, at least in the two strains whose genomes have been sequenced that came from patients with very different disease states, appear to have homologues known to be virulence factors in other pathogenic bacteria. The genomic comparison suggests that human host factor(s), socio-cultural differences or genetic predispositions as well as environmental factors may play a significant, and perhaps, unappreciated role in the susceptibility, severity and outcome of a *H. pylori* infection (Alm *et al.*, 1999; Alm and Trust, 1999).

With the genomics technology advancing exponentially, it is possible that the genome of another *H. pylori* strain or a related *Helicobacter* species may soon be completed. Analysis of the sequence of another *Helicobacter* species may provide important information as to genes that may be required for host specificity. The genome sequence of another *H. pylori* strain would not only provide additional information regarding the global nucleotide diversity within this species as a model of genetic divergence but would also continue to provide information as to the co-evolution of this pathogen with the human species. Together, all of the *Helicobacter* genomic information should provide new opportunities for discovery and development of novel therapeutic strategies for treatment of diseases associated with this globally important pathogen.

References

Achtman, M., Azuma, T., Berg, D.E., Ito, Y., Morelli, G., Pan, Z.J., Suerbaum, S., Thompson, S.A., van der Ende, A. and van Doorn, L.J. 1999. Recombination and clonal groupings within *Helicobacter pylori* from different geographical regions. Mol. Microbiol. 32: 459-470.

Akopyants, N., Fradkov, A., Diatchenko, L., Hill, J., Siebert, P., Lukyanov, S., Sverdlov, E. and Berg, D. 1998a. PCR-based subtractive hybridization and differences in gene content among strains of *Helicobacter pylori*. Proc. Natl. Acad. Sci. USA. 95: 13108-13113.

Akopyants, N.S., Clifton, S.W., Kersulyte, D., Crabtree, J.E., Youree, B.E., Reece, C.A., Bukanov, N.O., Drazek, E.S., Roe, B.A. and Berg, D.E. 1998b. Analyses of the *cag* pathogenicity island of *Helicobacter pylori*. Mol. Microbiol. 28: 37-53.

Akopyanz, N., Bukanov, N.O., Westblom, T.U. and Berg, D.E. 1992. PCR-based RFLP analysis of DNA sequence diversity in the gastric pathogen *Helicobacter pylori*. Nucl. Acids Res. 20: 6221-6225.

Alarcon, T., Domingo, D. and Lopez-Brea, M. 1999. Antibiotic resistance problems with *Helicobacter pylori*. Int. J. Antimicrob. Agents 12: 19-26.

Alm, R.A., Guerry, P. and Trust, T.J. 1993. The *Campylobacter* sigma 54 *flaB* flagellin promoter is subject to environmental regulation. J. Bacteriol. 175: 4448-4455.

Alm, R.A., Ling, L.L., Moir, D.T., King, B.L., Brown, E.D., Jiang, Q., Doig, P.C., Smith, D.R., Noonan, B., Guild, B.C., deJonge, B.L., Carmel, G., Tummino, P.J., Caruso, A., Uria-Nickelsen, M., Mills, D.M., Ives, C., Merberg, D., Mills, S.D., Taylor, D.E., Vovis, G.F. and Trust, T.J. 1999. Genomic sequence comparison of two unrelated isolates of the human gastric pathogen *Helicobacter pylori*. Nature. 397: 186-190.

Alm, R.A. and Trust, T.J. 1999. Analysis of the genetic diversity of *Helicobacter pylori* - the tale of two genomes. J. Mol. Med. 77: 834-846.

Alm, R.A., Bina, J., Andrews, B.M., Doig, P., Hancock, R.E.W., and Trust, T.J. 2000.Comparative genomics of *Helicobacter pylori:* Analysis of the outer membrane protein families. Infect. Immun. 68: 4155-4168.

Appelmelk, B.J., Martin, S.L., Monteiro, M.A., Clayton, C.A., McColm, A.A., Zheng, P., Verboom, T., Maaskant, J.J., Eijnden, D.H.v.d., Hokke, C.H., Perry, M.B., Vandenbroucke-Grauls, C.M. and Kusters, J.G. 1999. Phase variation in Helicobacter pylori lipopolysaccharide due to changes in the lengths of poly C. tracts in alpha3-fucosyltransferase genes. Infect. Immun. 67: 5361-5366.

Atherton, J.C., Peek Jr., R.M., Tham, K.T., Cover, T.L. and Blaser, M.J. 1997. Clinical and pathological importance of heterogeneity in *vacA*, the vacuolating cytotoxin gene of *Helicobacter pylori*. Gastroenterol. 112: 92-99.

Berg, D.E., Gilman, R.H., Lelwala-Guruge, J., Srivastava, K., Valdez, Y., Watanabe, J., Miyagi, J., Akopyants, N.S., Ramirez-Ramos, A., Yoshiwara, T.H., Recavarren, S. and Leon-Barua, R. 1997. *Helicobacter pylori* populations in Peruvian patients. Clin. Infect. Dis. 25: 996-1002.

Bina, J.E., Alm, R.A., Uria-Nickelsen, M., Thomas, S.R., Trust, T.J. and Hancock, R.E.W. 2000. *Helicobacter pylori* uptake and efflux: Basis for intrinsic susceptibility to antibiotics *in vitro*. Antimicrob. Agents Chemo. 44: 248-254.

Blaser, M.J. 1997a. Heterogeneity of *Helicobacter pylori*. *Eur. J. Gastroenterol. Hepatol.* 9: S3-S6.

Blaser, M.J. 1997b. Not all *Helicobacter pylori* strains are created equal: should all be eliminated? Lancet. 349: 1020-1022.

Blattner, F.R., III, G.P., Bloch, C.A., Perna, N.T., Burland, V., Riley, M., Collado-Vides, J., Glasner, J.D., Rode, C.K., Mayhew, G.F., Gregor, J., Davis, N.W., Kirkpatrick, H.A., Goeden, M.A., Rose, D.J., Mau, B. and Shao, Y. 1997. The complete genome sequence of *Escherichia coli* K-12. Science. 277: 1453-1462.

Cover, T.L. and Blaser, M.J. 1992. *Helicobacter pylori* and gastroduodenal disease. Annu. Rev. Med. 42: 135-145.

Curnow, A.W., Hong, K.W., Yuan, R., Kim, S.I., Martins, O., Winkler, W., Henkin, T.M. and Soll, D. 1997. Glu-tRNAGln amidotransferase: a novel heterotrimeric enzyme required for correct decoding of glutamine codons during translation. Proc. Natl. Acad. Sci. USA. 94: 11819-11826.

Doig, P., deJonge, B.L., Alm, R.A., Brown, E.D., Uria-Nickelsen, M., Noonan, B., Mills, S.D., Tummino, P., Carmel, G., Guild, B.C., Moir, D.T., Vovis, G.F. and Trust, T.J. 1999. *Helicobacter pylori* physiology predicted from genomic comparison of two strains. Microbiol. Mol. Biol. Rev. 63: 675-707.

Doig, P., Exner, M.M., Hancock, R.E.W. and Trust, T.J. 1995. Isolation and characterization of a conserved porin protein from *Helicobacter pylori*. J. Bacteriol. 177: 5447-5452.

Doig, P. and Trust, T.J. 1994. Identification of surface-exposed outer membrane antigens of *Helicobacter pylori*. Infect. Immun. 62: 4526-4533.

Dubois, A., Berg, D.E., Incecik, E.T., Fiala, N., Heman-Ackah, L.M., Perez-Perez, G.I. and Blaser, M.J. 1996. Transient and persistent infection of non-human primates with *Helicobacter pylori*: implications for human disease. Infect. Immun. 64: 2885-2891.

Dunkley, M.L., Harris, S.J., McCoy, R.J., Musicka, M.J., Eyers, F.M., Beagley, L.G., Lumley, P.J., Beagley, K.W. and Clancy, R.L. 1999. Protection against *Helicobacter pylori* infection by intestinal immunisation with a 50/52-kDa subunit protein. FEMS Immunol. Med. Microbiol. 24: 221-225.

Eaton, K.A., Morgan, D.R. and Krakowka, S. 1989. *Campylobacter pylori* virulence factors in gnotobiotic piglets. Infect. Immun. 57: 1119-1125.

Ermak, T.H., Giannasca, P.J., Nichols, R., Myers, G.A., Nedrud, J., Weltzin, R., Lee, C.K., Kleanthous, H. and Monath, T.P. 1998. Immunization of mice with urease vaccine affords protection against *Helicobacter pylori* infection in the absence of antibodies and is mediated by MHC class II-restricted responses. J. Exp. Med. 188: 2277-2288.

Evans Jr, D.J., Queiroz, D.M., Mendes, E.N. and Evans, D.G. 1998. Diversity in the variable region of *Helicobacter pylori cagA* gene involves more than simple repetition of a 102-nucleotide sequence. Biochem. Biophys. Res. Commun. 245: 780-784.

Exner, M.M., Doig, P., Trust, T.J. and Hancock, R.E.W. 1995. Isolation and characterization of a family of porin proteins from *Helicobacter pylori*. Infect. Immun. 63: 1567-1572.

Ferrero, R.L., Thiberge, J.M., Kansau, I., Wuscher, N., Huerre, M. and Labigne, A. 1995. The GroES homolog of *Helicobacter pylori* confers protective immunity against mucosal infection in mice. Proc. Natl. Acad. Sci. USA. 92: 6499-6503.

Figura, N. 1997. *Helicobacter pylori* factors involved in the development of gastroduodenal mucosal damage and ulceration. J. Clin. Gastroenterol. 25: S149-S163.

Garner, J.A. and Cover, T.L. 1995. Analysis of genetic diversity in cytotoxin-producing and non-cytotoxin-producing *Helicobacter pylori* strains. *J. Infect. Dis.* 172: 290-293.

Ge, Z. and Taylor, D.E. 1999. Contributions of genome sequencing to understanding the biology of *Helicobacter pylori*. Annu. Rev. Microbiol. 53: 353-387.

Ghiara, P., Rossi, M., Marchetti, M., Tommaso, A.D., Vindigni, C., Ciampolini, F., Covacci, A., Telford, J.L., Magistris, M.T.D., Pizza, M., Rappuoli, R. and Giudice, G.D. 1997. Therapeutic intragastric vaccination against *Helicobacter pylori* in mice eradicates an otherwise chronic infection and confers protection against reinfection. Infect. Immun. 65: 4996-5002.

Graham, D.Y. 1998. Antibiotic resistance in *Helicobacter pylori*: implications for therapy. Gastroenterol. 115: 1272-1277.

Han, J., Yu, E., Lee, I. and Lee, Y. 1997. Diversity among clinical isolates of *Helicobacter pylori* in Korea. Mol. Cells. 7: 544-547.

Hocking, D., Webb, E., Radcliff, F., Rothel, L., Taylor, S., Pinczower, G., Kapouleas, C., Braley, H., Lee, A. and Doidge, C. 1999. Isolation of recombinant protective *Helicobacter pylori* antigens. Infect. Immun. 67: 4713-4719.

Hook-Nikanne, J., Berg, D.E., Jr, R.M.P., Kersulyte, D., Tummuru, M.K. and Blaser, M.J. 1998. DNA sequence conservation and diversity in transposable element IS605 of *Helicobacter pylori*. Helicobacter. 3: 79-85.

Horst, J.P., Wu, T.H. and Marinus, M.G. 1999. *Escherichia coli* mutator genes. Trends Microbiol. 7: 29-36.

Hunt, R.H. 1996. The role of *Helicobacter pylori* in pathogenesis: the spectrum of clinical outcomes. Scand. J. Gastroenterol. Suppl. 220: 3-9.

Ilver, D., Arnqvist, A., Ogren, J., Frick, I.M., Kersulyte, D., Incecik, E.T., Berg, D.E., Covacci, A., Engstrand, L. and Boren, T. 1998. *Helicobacter pylori* adhesin binding fucosylated histo-blood group antigens revealed by retagging. Science 279: 373-377.

Jiang, Q., Hiratsuka, K. and Taylor, D.E. 1996. Variability of gene order in different *Helicobacter pylori* strains contributes to genome diversity. Mol. Microbiol. 20: 833-842.

Kansau, I., Raymond, J., Bingen, E., Courcoux, P., Kalach, N., Bergeret, M., Braimi, N., Dupont, C. and Labigne, A. 1996. Genotyping of *Helicobacter pylori* isolates by sequencing of PCR products and comparison with the RAPD technique. Res. Microbiol. 147: 661-669.

Kersulyte, D., Akopyants, N.S., Clifton, S.W., Roe, B.A. and Berg, D.E. 1998. Novel sequence organization and insertion specificity of IS605 and IS606: chimaeric transposable elements of *Helicobacter pylori*. Gene. 223: 175-186.

Kleanthous, H., Myers, G.A., Georgakopoulos, K.M., Tibbitts, T.J., Ingrassia, J.W., Gray, H.L., Ding, R., Zhang, Z.Z., Lei, W., Nichols, R., Lee, C.K., Ermak, T.H. and Monath, T.P. 1998. Rectal and intranasal immunizations with recombinant urease induce distinct local and serum immune responses in mice and protect against *Helicobacter pylori* infection. Infect. Immun. 66: 2879-2886.

Kunst, F., Ogasawara, N., Moszer, I., Albertini, A.M., Alloni, G., Azevedo, V., Bertero, M.G., Bessieres, P., Bolotin, A., Borchert, S., Borriss, R., Boursier, L., Brans, A., Braun, M., Brignell, S.C., Bron, S., Brouillet, S., Bruschi, C.V., Caldwell, B., Capuano, V., Carter, N.M., Choi, S.-K., Codani, J.-J., Connerton, I.F., Cummings, N.J., Daniel, R.A., Denizot, F., Devine, K.M., Dusterhoft, A., Ehrlich, S.D., Emmerson, P.T., Entian, K.D., Errington, J., Fabret, C., Ferrari, E., Foulger, D., Fritz, C., Fujita, M., Fujita, Y., Fuma, S., Galizzi, A., Galleron, N., Ghim, S.-Y., Glaser, P., Goffeau, A., Golightly, E.J., Grandi, G., Guiseppi, G., Guy, B.J., Haga, K., Haiech, J., Harwood, C.R., Henaut, A., Hilbert, H., Holsappel, S., Hosono, S., Hullo, M.-F., Itaya, M., Jones, L., Joris, B., Karamata, D., Kasahara, Y., Klaerr-Blanchard, M., Klein, C., Kobayashi, Y., Loetter, P., Koningstein, G., Krogh, S., Kumano, M., Kurita, K., Lapidus, A., Lardinois, S., Lauber, J., Lazarevic, V., Lee, S.-M., Levine, A., Liu, H., Masuda, S., Mauel, C., Medigue, C., Median, N., Mellado, R.P., Mizuno, M., Moestl, D., Nakai, S., Noback, M., Noone, D., O'Reilly, M., Ogawa, K., Ogiwara, A., Oudega, B., Park, S.-H., Parro, V., Pohl, T.M., Portetelle, D., Porwollik, S., Prescott, A.M., Presecan, E., Pujic, P., Purnelle, B., *et al.* 1997. The complete genome sequence of the Gram-positive bacterium *Bacillus subtilis*. Nature. 390: 249-256.

Labigne, A. and de Reuse, H. 1996. Determinants of *Helicobacter pylori* pathogenicity. Infect. Agents Dis. 5: 191-202.

Lee, W.K., An, Y.S., Kim, K.H., Kim, S.H., Song, J.Y., Ryu, B.D., Choi, Y.J., Yoon, Y.H., Baik, S.C., Rhee, K.H. and Cho, M.J. 1997. Construction of a *Helicobacter pylori-Escherichia coli* shuttle vector for gene transfer in *Helicobacter pylori*. Appl. Environ. Microbiol. 63: 4866-4871.

Logan, R.P.H. and Berg, D.E. 1996. Genetic diversity of *Helicobacter pylori.* Lancet. 348: 1462-1463.

Marchetti, M., Rossi, M., Giannelli, V., Giuliani, M.M., M, M.P., Censini, S., Covacci, A., Massari, P., Pagliaccia, C., Manetti, R., Telford, J.L., Douce, G., Dougan, G., Rappuoli, R. and Ghiara, P. 1998. Protection against *Helicobacter pylori* infection in mice by intragastric vaccination with *H. pylori* antigens is achieved using a non-toxic mutant of *E. coli* heat-labile enterotoxin LT. as adjuvant. Vaccine. 16: 33-37.

Marshall, B.J. and Warren, J.R. 1984. Unidentified curved bacilli in the stomach of patients with gastritis and peptic ulceration. Lancet. 1: 1311-1315.

Marshall, D.G., Dundon, W.G., Beesley, S.M. and Smyth, C.J. 1998. *Helicobacter pylori* - a conundrum of genetic diversity. *Microbiol.* 144: 2925-2939.

Merrick, M.J. 1993. In a class of its own—the RNA polymerase sigma factor sigma 54 sigma N. Mol. Microbiol. 10: 903-909.

Odenbreit, S., Till, M., Hofreuter, D., Faller, G. and Haas, R. 1999. Genetic and functional characterization of the *alpAB* gene locus essential for the adhesion of *Helicobacter pylori* to human gastric tissue. Mol. Microbiol. 31: 1537-1548.

Pappo, J., Torrey, D., Castriotta, L., Savinainen, A., Kabok, Z. and Ibraghimov, A. 1999. *Helicobacter pylori* infection in immunized mice lacking major histocompatibility complex class I and class II functions. Infect. Immun. 67: 337-341.

Radcliff, F.J., Hazell, S.L., Kolesnikow, T., Doidge, C. and Lee, A. 1997. Catalase, a novel antigen for *Helicobacter pylori* vaccination. Infect. Immun. 65: 4668-4674.

Richter, J.E., Falk, G.W. and Vaezi, M.F. 1998. *Helicobacter pylori* and gastroesophageal reflux disease: the bug may not be all bad. Am. J. Gastroenterol. 93: 1800-1802.

Salzberg, S.L., Salzberg, A.J., Kerlavage, A.R. and Tomb, J.F. 1998. Skewed oligomers and origins of replication. Gene. 217: 57-67.

Saunders, N.J., Peden, J.F., Hood, D.W. and Moxon, E.R. 1998. Simple sequence repeats in the *Helicobacter pylori* genome. Mol. Microbiol. 27: 1091-1098.

Suerbaum, S., Brauer-Steppkes, T., Labigne, A., Cameron, B. and Drlica, K. 1998a. Topoisomerase I of *Helicobacter pylori*: juxtaposition with a flagellin gene *flaB*. and functional requirement of a fourth zinc finger motif. Gene. 210: 151-161.

Suerbaum, S., Smith, J.M., Bapumia, K., Morelli, G., Smith, N.H., Kunstmann, E., Dyrek, I. and Achtman, M. 1998b. Free recombination within *Helicobacter pylori*. Proc. Natl. Acad. Sci. USA. 95: 12619-12624.

Takami, S., Hayashi, T., Tonokatsu, Y., Shimoyama, T. and Tamura, T. 1993. Chromosomal heterogeneity of *Helicobacter pylori* isolates by pulsed-field gel electrophoresis. Zbl. Bakt. 280: 120-127.

Tatusov, R.L., Mushegian, A.R., Bork, P., Brown, N.P., Hayes, W.S., Borodovsky, M., Rudd, K.E. and Koonin, E.V. 1996. Metabolism and evolution of *Haemophilus influenzae* deduced from a whole-genome comparison with *Escherichia coli*. Curr. Biol. 6: 279-291.

Taylor, D.E., Eaton, M., Chang, N. and Salama, S.M. 1992. Construction of a *Helicobacter pylori* genome map and demonstration of diversity at the genome level. J. Bacteriol. 174: 6800-6806.

Tomb, J.-F., White, O., Kerlavage, A.R., Clayton, R.A., Sutton, G.G., Fleischmann, R.D., Ketchum, K.A., Klenk, H.P., Gill, S., Dougherty, B.A., Nelson, K., Quackenbush, J., Zhou, L., Kirkness, E.F., Peterson, S., Loftus, B., Richardson, D., Dodson, R., Khalak, H.G., Glodek, A., McKenney, K., Fitzegerald, L.M., Lee, N., Adams, M.D., Hickey, E.K., Berg, D.E., Gocayne, J.D., Utterback, T.R., Peterson, J.D., Kelley, J.M., Cotton, M.D., Weidman, J.M., Fujii, C., Bowman, C., Watthey, L., Wallin, E., Hayes, W.S., Borodovsky, M., Karp, P.D., Smith, H.O., Fraser, C.M. and Venter, J.C. 1997. The complete genome sequence of the gastric pathogen *Helicobacter pylori*. Nature. 388: 539-547.

Wang, G., Humayun, M.Z. and Taylor, D.E. 1999. Mutation as an origin of genetic variability in *Helicobacter pylori*. Trends Microbiol. 7: 488-493.

From: *Helicobacter pylori: Molecular and Cellular Biology*
ISBN 1-898486-25-5 © 2001 Horizon Scientific Press, Wymondham, UK.

18

Molecular Typing of *Helicobacter pylori*

Leen-Jan van Doorn[*], Céu Figueiredo, and Wim Quint

Abstract

The genome of *Helicobacter pylori* is heterogeneous, due to a high mutation rate and a high degree of recombination. Therefore, *H. pylori* should be considered as a population of genetically related but diverse bacteria. To analyze the epidemiological and clinical implications of *H. pylori* variability, adequate molecular tools should be used for molecular typing of isolates. A variety of available methods, including RFLP, PCR-fingerprinting, sequence analysis, and probe hybridization is described in this chapter. Also, typing of virulence-associated and antibiotic resistance-related genes is discussed. Finally, some general requirements for effective molecular typing methods are provided.

Introduction

H. pylori has a global distribution and affects humans worldwide (Covacci *et al.*, 1999; Dunn *et al.*, 1997). However, not all infected individuals develop disease. Theoretically, there are several possible explanations for this discrepancy, such as differences between hosts (Azuma *et al.*, 1998), environmental factors (Malaty *et al.*, 1994), and *H. pylori* strains. Like many other microorganisms, *H. pylori* is not a uniform entity, but a population of heterogeneous bacterial variants (Covacci and Rappuoli 1998). Despite the high degree of recombination in the *H. pylori* genome (Suerbaum *et al.*, 1998) this heterogeneity appears not to be completely random, and distinct clonal groupings can still be distinguished (Achtman *et al.*, 1999). Formally, microorganisms can be classified at different levels, e.g., genera, species, types, subtypes and quasispecies, representing an increasing degree of genetic similarity among strains. For genotypes of some viruses, specific classification criteria have been defined, but such a system does not yet exist for *H. pylori*. In contrast to formal classification, typing of microorganisms serves as a practical approach to discriminate between individual isolates. Various typing methods have been developed to assess the biological heterogeneity of *H. pylori* strains.

Phenotyping involves the analysis of all constituents of the bacterium, except the genomic DNA. Several phenotyping methods, such as lipopolysaccharide (LPS) analysis (Gibson *et al.*, 1998a), protein (Costas *et al.*, 1991; Kitamoto *et al.*, 1998; Morgan *et al.*, 1991) and enzyme profiling (Go *et al.*, 1996) have been described. Biotyping comprises the analysis of the interaction of the bacterium with its (clinical) environment, such as antibiotic susceptibility testing, enzymatic activities, and the induction of a specific immune response

[*]corresponding author, email: L.J.van.Doorn@ddl.nl

Table 1. Characteristics of various genotyping methods for *Helicobacter pylori*

Method		Principle Target	Source of DNA
RAPD /REP-PCR	Random amplification of genomic fragments using primers aimed at polymorphic/repeated DNA.	Entire genomic DNA	Cultured strain
PFGE	Endonuclease treatment of genomic DNA yields fragments of variable lengths; long fragments can be visualized by PFGE.	Entire genomic DNA	Cultured strain
RFLP	Endonuclease treatment of DNA yields pattern of fragments of variable lengths. Specific fragments can be visualized by Southern blotting.	Entire genomic DNA or amplified fragments	Cultured strain or gastric biopsy specimen
AFLP	Endonuclease treatment of genomic DNA, ligation of adapter oligonucleotides and PCR amplification yields fragments of variable length.	Entire genomic DNA	Cultured strain
Type-specific PCR	Amplification of DNA fragments, using type-specific PCR primers.	Single specific genes	Cultured strain or gastric biopsy specimen
PCR-LiPA	Simultaneous PCR amplification of multiple fragments followed by specific reverse hybridization to oligonucleotide probes.	Multiple specific genes	Cultured strain or gastric biopsy specimen
Multilocus sequence typing	Phylogenetic analysis of multiple gene fragments by PCR amplification and sequencing.	Single specific genes	Cultured strain or gastric biopsy specimen

(Owen *et al.*, 1991, 1993a). Genotyping involves the analysis of the diversity of genomic DNA among *H. pylori* strains. The aim of this chapter is to present a summary of the different molecular methods for typing of *H. pylori* strains.

Molecular Typing Methods

Since phenotyping methods usually have limited discriminatory power, great efforts have been made to develop and establish the use of molecular typing methods. Some methods assess the overall heterogeneity of the entire bacterial genome, such as (1) sequence analysis of complete genomes (2) PCR-fingerprinting, using either random primers (arbitrary-primed (AP) PCR) or repetitive DNA sequence primers (REP-PCR), (3) Restriction fragment length polymorphism (RFLP) analysis of large genomic fragments, using pulsed-field gel electrophoresis (PFGE), and (4) amplification fragment length polymorphism (AFLP). These methods analyze overall genome variability and do not provide information for individual genes.

Other methods analyze only defined parts (specific genes) of the *H. pylori* genome, such as (1) multilocus sequence typing, (2) RFLP analysis of genomic DNA using Southern blot hybridization or RFLP analysis of PCR fragments, (3) type-specific PCR and (4) PCR-hybridization assays. Several aspects of the various typing methods are summarized in Table 1.

Methods to Assess the Overall Genomic Heterogeneity of *H. pylori*

Complete Genomic Sequences

H. pylori is the first organism for which two complete genomic sequences became available. The total bacterial genome comprises approximately 1.6 Mbp (Alm *et al.*, 1999; Tomb *et al.*,

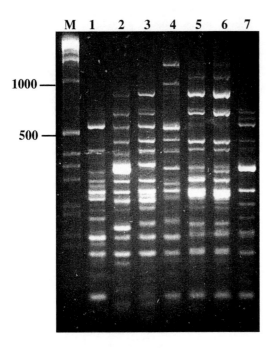

Figure 1.DNA fingerprint analysis of *H. pylori* DNA from seven cultured isolates. DNA was amplified using oligonucleotides 1254 (5'-CCGCAGCCAA –3') and 1283 (5' GCGATCCCCA –3') (van der Ende *et al.*, 1996). All samples yield a different fingerprinting pattern, except isolates 5 and 6, which were obtained from the same patient. The size of the amplimers is indicated by DNA-markers in lane M.

1997) and the majority of the genes were similar between the 2 organisms, but strain-specific sequences were also found. The genomic organization of *H. pylori* is discussed in detail in chapter 17.

Random Amplification of Polymorphic DNA (RAPD)

Since sequence analysis of the complete bacterial genome of multiple strains is not feasible for strain typing, other methods can be used to analyze the overall genomic heterogeneity. RAPD or PCR fingerprinting is based on the use of random (RAPD or arbitrary primed (AP) PCR) or repetitive (REP-PCR) primer sequences, which match different target sites that are distributed over the entire genomic DNA. Sequences between any pair of such primer target sites can be amplified by PCR and generates a large number of DNA fragments of distinct length (Akopyanz *et al.*, 1992a; Marshall *et al.*, 1996). These amplified polymorphic DNA fragments can be analyzed by gelelectrophoresis, and the electrophoretic patterns from different *H. pylori* strains compared. The amplified fragments originate from random locations distributed over the entire bacterial genome, but yields no information about specific genes.

RAPD is an effective typing method with a very high discriminatory power (Burucoa *et al.*, 1999) and therefore it is particularly suitable to analyze epidemiological relationships between individual isolates. RAPD has been used to study the epidemiology of *H. pylori* and confirmed occurrence of intrafamilial transmission (Bamford *et al.*, 1993; Miehlke *et al.*, 1999a; van der Ende *et al.*, 1996) (Figure 1). Analysis of multiple colonies from a single patient can discriminate between infection with a single or with multiple strains (Miehlke *et al.*, 1999b; Taylor *et al.*, 1995). RAPD also is useful to compare *H. pylori* strains before and after failing eradication therapy and can discriminate between recrudescence and reinfection (Fraser *et al.*, 1992; van der Hulst *et al.*, 1997; Xia *et al.*, 1995). However, attempts to associate specific RAPD patterns with clinical symptoms were largely unsuccessful, and this method appears not to be useful to identify disease-specific virulence determinants of *H. pylori* (Enroth *et al.*, 1999; Go *et al.*, 1995; Kwon *et al.*, 1998; Owen *et al.*, 1993b)

Despite its high efficacy as a typing method, it should be noted that the reproducibility of RAPD is usually poor, hampering reliable interlaboratory comparisons of RAPD patterns (Gallego and Martinez 1997; Perez *et al.*, 1998; Power 1996; van Belkum *et al.*, 1995).

Pulsed-field Gel Electrophoresis (PFGE)

Genomic DNA can be digested by restriction endonucleases such as *Not*I, *Apa*I, and *Nru*I (with only a limited number of recognition sites) resulting in large DNA fragments that can be separated by pulsed-field gel electrophoresis. Polymorphic restriction sites result in isolate-specific electrophoresis patterns, permitting discrimination between individual strains (Takami *et al.*, 1993; Takami *et al.*, 1994). PFGE has also been used to study the variability in gene order between different strains (Jiang *et al.*, 1996; Taylor *et al.*, 1992). This method has been successfully used to discriminate between clinical isolates of *H. pylori* (Hirschl *et al.*, 1994; Salama *et al.*, 1995; Wada *et al.*, 1994). However, PFGE has only limited efficacy in *H. pylori*, depending on the restriction enzyme used, and does not allow typing of every strain (Burucoa *et al.*, 1999; Wada *et al.*, 1994). This may be due to protection of restriction sites by endogenous methylases (Taylor *et al.*, 1992; Xu *et al.*, 1997). The discriminatory power of PFGE is considerably lower than that of RAPD (Burucoa *et al.*, 1999). Since the method is labor-intensive, time-consuming, and relatively poorly reproducible (van Belkum *et al.*, 1998), its use for typing of *H. pylori* in a clinical setting is limited.

Methods to Assess the Heterogeneity of Parts of the *H. pylori* Genome

Multi Locus Sequence Typing

Although sequence analysis of complete genomes is the most accurate method to distinguish genetic variants, this approach is neither feasible nor necessary. Sequencing of a representative selection of genomic fragments also permits detailed phylogenetic analysis of individual strains. This method, designated multilocus sequence typing (MLST), has already been successfully applied to *H. pylori* (Achtman *et al.*, 1999; Maiden *et al.*, 1998) (See Chapter 19)

RFLP and AFLP

Complete genomic DNA as well as PCR-amplified fragments can be digested by endonucleases and the restriction fragments analyzed by gel-electrophoresis (Akopyanz *et al.*, 1992b). If total genomic DNA from a cultured strain is used, the numerous DNA fragments yield a smear on the gel and individual bands can not be observed. To obtain a distinct banding pattern, the restriction fragments can be further analyzed by Southern blot hybridization, using a specific probe to detect individual genes of interest.

Amplification fragment length polymorphism (AFLP) is also based on endonuclease digestion of genomic DNA (Gibson *et al.*, 1998b; Savelkoul *et al.*, 1999). Restriction site-specific linkers are ligated to both ends of the restriction fragments. This permits subsequent amplification by PCR, using primers aimed at the linker sequences. Like RFLP, AFLP is particularly suitable for epidemiological purposes, but its efficacy is also determined by the choice of endonuclease, since modified restriction sites may prevent proper digestion of genomic DNA.

RFLP can not only be used on complete genomic DNA (which requires Southern blotting to visualize specific fragments), but also on amplified fragments. Since PCR amplimers are relatively small, their distinct digestion products can be directly visualized, without the need for hybridization analysis, and multiple restriction enzymes can generate a unique pattern of restriction fragments from each individual strain. In contrast to PFGE, PCR-based RFLP is not hampered by modified restriction sites that inhibit endonuclease digestion. Consequently, PCR-RFLP has a very high discriminatory power and is highly reproducible (Burucoa *et al.*, 1999). Another advantage of PCR-RFLP is that it permits direct analysis of *H. pylori* in clinical materials, such as gastric biopsies, omitting the need to culture the bacterium.

RFLP has been used extensively for epidemiological purposes e.g., to compare *H. pylori* strains from different members within the same family, from the same patients before and after eradication therapy (Stone *et al.*, 1997a) or from diverse geographic origins (Stone *et al.*, 1997b). Since PCR-RFLP is based on polymorphic restriction sites in an amplified DNA fragment, the method is less useful to differentiate between presence of single or multiple strains. Several studies have attempted to associate specific RFLP patterns with clinical parameters, such as the degree of gastritis or the presence of peptic ulcer disease, but the success of such studies has been limited (Desai *et al.*, 1994; Owen *et al.*, 1993b, 1999). Therefore, as for RAPD, the method is not suitable to identify disease-associated virulence markers.

Type-specific PCR

Type-specific PCR is based on the use of primers that allow highly specific discrimination between previously defined allelic variants. If sequences of the allelic variants are sufficiently different, type-specific PCR is usually efficient. In contrast, detection of single nucleotide polymorphisms is more difficult and requires careful primer design and optimization of the reaction conditions. Other disadvantages are that detection of each allelic variant needs separate PCRs, and it requires thorough analysis of the sequence heterogeneity of the target DNA. So far, type-specific PCR for *H. pylori* has been described for *vacA*. The distinct variants of the *vacA* s- and m-region can be separately amplified by type-specific primers, and the products are analyzed by gelelectrophoresis (Atherton *et al.*, 1995, 1999).

Hybridization-based Assays

Hybridization assays can be performed in various formats. Hybridization can be applied directly to clinical materials or cultured strains, without prior amplification by PCR. *In situ* hybridization can be used for detection of *H. pylori* in biopsy specimens (Barrett *et al.*, 1997; Park and Kim 1999), but is not suitable for molecular typing and is of little clinical use. Cultured strains can be directly analyzed by colony hybridization. This can be useful to detect mixed infections, such as *cagA+* and *cagA-* strains in a single patient (Pan *et al.*, 1997; van der Ende *et al.*, 1996).

Digested genomic DNA or amplified fragments can be analyzed by Southern blot hybridization, following gelelectrophoresis. For analyses of larger numbers of samples, microtiterplate-based assays are available, which are easier to use (Monteiro *et al.*, 1997; van

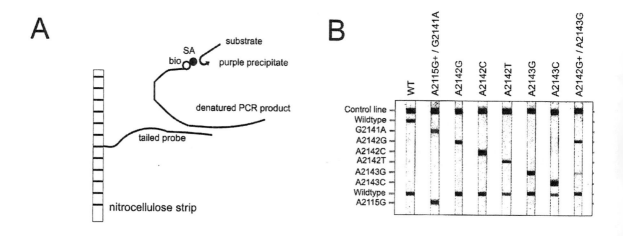

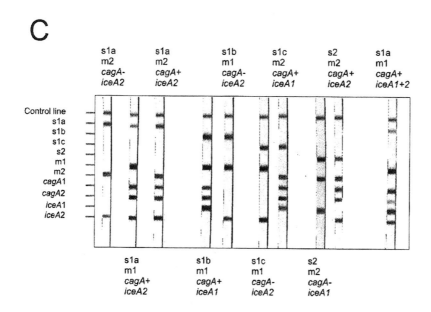

Figure 2. Typing of *H. pylori* isolates by PCR-LiPA. The principle of the reverse hybridization assay is shown in panel A. Specific oligonucleotide probes are tailed and coated onto membrane strips as parallel lines. Biotinylated PCR product is denatured and hybridized to the probes at highly stringent conditions. Hybrids are detected by a streptavidin-alkaline phosphatase conjugate and substrate, resulting in a purple precipitate. Panel B shows the LiPA for detection of specific mutations in the 23S rDNA of *H. pylori*, that are associated with macrolide resistance. Panel C shows the *H. pylori* virulence LiPA, permitting simultaneous typing of virulence-associated genes *vacA*, *cagA* and *iceA* in a single reverse hybridization step after multiplex PCR.

Doorn *et al.*, 2000). Both formats use single, individual probes, and do not easily permit testing with multiple probes.

PCR-reverse Hybridization (LiPA)

This method is based on the simultaneous amplification of multiple genomic fragments. PCR primers are not type-specific, but are aimed at conserved sequences, flanking polymorphic regions of interest. Fragments, amplified by such 'general' PCR primers, can be further analyzed by a single-step reverse hybridization line probe assay (LiPA™). This assay comprises a nylon strip, carrying oligonucleotide probes, which are immobilized as parallel lines (Figure 2). The design of the probes permits highly specific hybridization of PCR fragments under stringent conditions. Consequently, reverse hybridization allows detection of single nucleotide mismatches between probe and PCR fragment. This method is easy to use, since it requires only one PCR and a single hybridization step to obtain a multiparameter result. Large numbers of isolates can be genotyped by PCR-LiPA, making it particularly suitable for standardized epidemiological studies. Importantly, PCR-LiPA is highly sensitive for simultaneous detection of multiple strains, making it particularly useful to detect co-colonization of different genotypes, even if they only represent the minority in the patient. PCR-LiPA has been used to discriminate between the different *vacA, cagA* and *iceA* genotypes (van Doorn *et al.*, 1998a;, 1998b, 1999a), as well as the detection of 23S rRNA mutations, associated with macrolide resistance (van Doorn *et al.*, 1999b).

Clinical Use of Typing

Besides serving as research tools, microbial typing methods are useful in clinical laboratories and can be successfully applied to address various clinically important aspects of *H. pylori* infections. Most importantly, molecular typing permits detailed epidemiological studies at the level of individual patients, e.g., to investigate the route of transmission of *H. pylori*, and transmission of infection within families has been shown by RAPD and RFLP (van der Ende *et al.*, 1996). Long-term follow up of individual patients demonstrated that the same *H. pylori* strain persists over time (Kuipers *et al.*, 2000). Presence of single or multiple *H. pylori* strains in a single host has been assessed (Jorgensen *et al.*, 1996; Taylor *et al.*, 1995) and strains from various sites of infection can be compared (Prewett *et al.*, 1992). After failing eradication therapy, molecular analyses differentiated between reinfection or recrudescence by comparing genotypes of *H. pylori* strains before and after treatment (van der Hulst *et al.*, 1997). Furthermore, *H. pylori* strains from different areas can be investigated to assess the world wide or regional geographic distribution of these variants. Another important application of typing methods is to investigate the relationships between genetic variants of *H. pylori* and outcome of disease (see below "virulence-associated genes"). A great advantage of amplification-based methods such as PCR is the direct detection of low numbers of organisms in gastric biopsies, gastric juice, bile, stool and oral secretions, omitting the need for *in vitro* culture of the bacterium. This is particularly important for assessment of antibiotic susceptibility, which normally requires long-term and complicated *in vitro* culture procedures. Taken together, molecular typing methods have greatly improved our insights in the biology of *H. pylori*.

Specific H. pylori Genes

Various specific genes have been used as a specific target for molecular typing. Urease is an essential enzyme of *H. pylori* for its colonization and survival in the gastric mucosa. Therefore, the urease genes (*ureA*, *ureB*, and *ureC*) have been a popular target for genetic analyses, especially using RFLP. No relation was found between RFLP patterns and clinical parameters (Hurtado and Owen 1994; Lopez *et al.*, 1993; Owen *et al.*, 1994; Owen *et al.*, 1998; Salaun *et al.*, 1998; Stone *et al.*, 1997a). Comparison of PCR-RFLP and direct sequence analysis of PCR products revealed that PCR-RFLP of the *ureB* gene does provide an easy typing scheme to discriminate between strains, but does not reveal the true extent of genetic diversity of *H. pylori* (Tanahashi *et al.*, 2000).

Ribotyping is based on RFLP analysis of the 16S and 23S rRNA genes (Desai *et al.*, 1993). Since all bacteria contain these rRNAs as part of the ribosomes, their encoding genes have been an extremely popular target for molecular detection as well as typing of bacteria. The rDNA genes contain variable and conserved domains, and numerous sequences from a large spectrum of microorganisms have been described (Linton *et al.*, 1992; Lopez *et al.*, 1993; Owen *et al.*, 1992b; Owen *et al.*, 1992a). Analysis of rDNA genes also plays an important role in the identification of novel *Helicobacter* species (Fox and Lee 1997). Ribotyping can be performed by either PCR-RFLP or RFLP on total genomic DNA in combination with Southern blot hybridization.

Flagellin and adhesin genes have been used for RFLP analysis, but to a much lesser extent than the urease and ribosomal RNA genes (Evans *et al.*, 1995; Forbes *et al.*, 1995; Salaun *et al.*, 1998)

Virulence-associated Genes

Since not all *H. pylori* infections result in the development of disease, considerable effort has been taken to identify genetic markers for the degree of virulence of different strains. This has resulted in the identification of several genes, designated as virulence-associated genes, indicating that these genes or one of their specific allelic variants, is often present in *H. pylori* strains, isolated from patients with disease, but is mostly absent in strains from healthy individuals. Thus, the term virulence-associated genes is largely based on clinical-epidemiological observations.

vacA, Encoding a Vacuolating Toxin

vacA encodes a vacuolating toxin which is excreted by *H. pylori* and damages epithelial cells (Cover 1996; Leunk 1995). The gene is present in all strains, and comprises 2 variable parts (Atherton *et al.*, 1995). The s-region (encoding the signal peptide) is located at the 5' end of the gene and exists as an s1 or s2 allele. Within type s1 several subtypes (s1a, s1b, and s1c) can be distinguished (van Doorn *et al.*, 1998b). The m-region (middle) occurs as an m1 or m2 allele. The mosaic combination of these distinct s- and m-region allelic types determines the production of the cytotoxin and is thereby associated with pathogenicity of the bacterium (Atherton *et al.*, 1995). Presence of *vacA* s1 strains appears to be strongly correlated with the presence of peptic ulcers and gastric carcinoma (Atherton *et al.*, 1995; van Doorn *et al.*, 1998c, 1999c). Interestingly, *vacA* m1 type strains have been associated with greater gastric epithelial damage than m2 strains (Atherton 1997).

A recent study revealed that *vacA* genotypes of *H. pylori* have a particular geographic distribution (van Doorn *et al.*, 1999c). Patients from East Asia frequently contain subtype *vacA* s1c strains, whereas these are rare in other parts of the world. In fact, most non-Asian patients carrying s1c strains were born in East-Asia or have Asian ancestors, indicating a restricted prevalence of such strains. In Europe, a distribution gradient was observed. Most strains from Northern, Central, and Eastern Europe are of subtype s1a. In France and Italy, s1a and s1b were equally present, whereas in Portugal and Spain, most strains were of subtype s1b. Also, most South American strains were of subtype s1b. Similar results have been obtained by analysis of *cagA*, showing that East-Asian strains contain a different *cagA* subtype, as compared to European strains (van der Ende *et al.*, 1998; van Doorn *et al.*, 1999a). These findings strongly indicate comigration and coevolution of *H. pylori* and humans since ancient times.

cagA, a Marker for the Pathogenicity Island

cagA (cytotoxin-associated gene) is considered as a marker for the presence of the pathogenicity (*cag*) island of about 35 kbp (Censini *et al.*, 1996). This island comprises more than 25 genes, encoding proteins that play an important role in the virulence of *H. pylori*, e.g., by induction of cytokine expression by the host. Based on the presence or absence of this pathogenicity island, virulent (type I) and less virulent (type II) *H. pylori* strains can be distinguished (Covacci and Rappuoli 1998). There is a strong epidemiological association between presence of *cagA* and the *vacA* s1 genotype. The function of CagA remains unknown, but infection with *cagA+* strains (i.e., containing the pathogenicity island) is associated with an increased the risk for the development of atrophic gastritis and gastric cancer as compared to *cagA-* strains (Blaser *et al.*, 1995; Kuipers *et al.*, 1995). *cagA* contains several repeat elements, and several subtypes can be recognized (Evans and Evans 1997), which, like *vacA* genotypes, have a particular geographic distribution (van der Ende *et al.*, 1998; van Doorn *et al.*, 1999a). Several PCR primer sets have been described for detection of *cagA*, but the efficacy of each of these sets should be locally evaluated (Evans *et al.*, 1998; van Doorn *et al.*, 1999a).

iceA

Recently, a novel gene has been discovered, designated *iceA* (induced by contact with epithelium). There are 2 main allelic variants of the gene, i.e., *iceA1* and *iceA2* (Peek *et al.*, 1998). The function of *iceA1* is not yet clear, but there is significant homology to a type II restriction endonuclease from *Neisseria lactamica*. The expression of *iceA1* is upregulated upon contact between *H. pylori* and human epithelial cells and is possibly associated with peptic ulcer disease (Peek *et al.*, 1998; van Doorn *et al.*, 1998c).

babA, Associated with Binding to Bloodgroup Antigens

Since *H. pylori* produces antigens, mimicking the human Lewis A and B antigens, the binding of *H. pylori* to blood group antigens was further investigated. A group of outer membrane protcins was identified, designated *bab* (Ilver *et al.*, 1998). The *babA* gene exists in 2 distinct allelic forms. *babA1* comprises an almost complete open reading frame, but lacks the necessary initiation codon, and is therefore inactive. In contrast, *babA2* strains contain a 10 bp insert at the 5'end of the gene, containing an in-frame ATG start codon, permitting expression of the ORF. Presence of *babA2* correlates with bacterial adhesion (Gerhard *et al.*, 1999). Apart from the 10 bp insert, the sequences of *babA1* and *babA2* are completely identical. Moreover, the 5' end of another gene (designated OMP9) is completely identical to *babA*. Therefore, specific typing of the *babA* genes is not easy, and requires carefully designed primers.

Interestingly, infection with of *babA2*, *vacA* s1, and *cagA+* strains appears to be strongly correlated to the prevalence of peptic ulcers and adenocarcinoma (Gerhard *et al.*, 1999), indicating the clinical relevance of these *H. pylori* genotypes.

Molecular Analysis of Antibiotic Resistance

Macrolides

Macrolides can bind to specific regions of the 23S rRNA, and thereby inhibit ribosome functioning. Point mutations in the peptidyltransferase domain (domain V) of the 23S rRNA can change its conformation and prevent binding of macrolides, and result in resistance. So far, point mutations have been found at positions 2115, 2141, 2142, and 2143 of either of the two copies of the *H. pylori* 23S rDNA (Debets-Ossenkopp *et al.*, 1998; Hultén *et al.*, 1997; Occhialini *et al.*, 1997; Versalovic *et al.*, 1997). Therefore, macrolide resistance can easily be determined by molecular tools, such as sequencing, Ligase Chain Reaction (LCR) (Stone *et al.*, 1996, 1997c) RFLP (Debets-Ossenkopp *et al.*, 1996), specific probe hybridization in microtiter plates (Occhialini *et al.*, 1997) or by reverse hybridization LiPA (van Doorn *et al.*, 1999b). The use of PCR permits even detection of these mutations directly in gastric biopsies. Since the prevalence of macrolide-resistant strains appears to be increasing in various parts of the world, these molecular methods will become increasingly important.

Metronidazole

Resistance to metronidazole is highly prevalent worldwide. Microbiological methods to determine the susceptibility are difficult to standardize and often unreliable (Mégraud 1999). Resistance to metronidazole (a nitroimidazole) is associated to mutations in the *rdx*A gene of *H. pylori* (encoding a nitroreductase) (Goodwin *et al.*, 1998; Jenks *et al.*, 1999). However, in contrast to the point mutations in the 23S rDNA, these mutations are not clustered, and any mutation that abolishes the expression of *rdx*A, confers metronidazole resistance. Therefore, the only reliable molecular method to determine *rdx*A integrity is sequence analysis of the complete *rdx*A gene, which is not yet feasible in routine clinical laboratories.

Amoxicillin

The prevalence of amoxicillin resistance among clinical isolates is extremely low. Recently, the mechanism of resistance to amoxicillin has been revealed. A single Serine to Arginine change in the penicillin-binding protein 1A homolog (HP0597) (Kusters *et al.*, 1999) confers resistance to amoxicillin. This can be easily analyzed by various methods, such as PCR and hybridization. However, given the very low prevalence of resistance to amoxicillin, there is no immediate clinical need for such assays.

Requirements for Effective Molecular Typing Methods

Multiple molecular methods have been described for molecular typing of *H. pylori*. In order to use such methods for extensive epidemiological and clinical studies, several criteria should be fulfilled. To allow comparison of typing results between different laboratories and different experiments, the methods should be standardized and highly reproducible. The method should have sufficient discriminatory power, i.e., should be able to distinguish between *H. pylori* strains that are genetically different (Burucoa *et al.*, 1999). Especially with respect to the presence of multiple strains, the discriminatory power as well as sensitivity and specificity of the method are crucial. The typing method should have sufficient typeability (Burucoa *et al.*, 1999), i.e., it should provide an unambiguous result for each isolate and the typing schedule should be complete. Identification of specific genotypes should be based on careful phylogenetic analysis of sufficient strains from various geographic origins. For instance, initially 3 subtypes were described within the *vacA* s-region, i.e., s1a, s1b, and s2. More extensive sequence analysis revealed the existence of an additional subtype, designated s1c, and this system now seems to be complete, since virtually all *H. pylori* strains can be typed. Since *H. pylori* variants are not uniformly geographically distributed, the methods should be locally evaluated in strains from the geographic region of interest.

To investigate the epidemiological and clinical relevance of *H. pylori* genotypes, larger number of clinical specimens have to be analyzed. Therefore, rapid typing methods are required.

Summary and Conclusions

Molecular studies have revealed that the genetic diversity of *H. pylori* is high. Nevertheless, clonal groupings can still be recognized. To identify specific genetic lineages, typing methods can be employed for analysis of *H. pylori* after culture (complete genome analysis) or directly in clinical materials (partial genome analysis). Typing of *H. pylori* aims at either arbitrary discrimination between strains for epidemiological purposes or identification of specific *H. pylori* genotypes in relation to clinical parameters. Various methods are available, but they all should be rapid and must meet certain criteria, such as discriminatory power, typeability, accuracy and reproducibility. Therefore proficiency panels could play an important role.

Especially amplification methods, such as PCR, can have important implications for the laboratory organization. In addition, to permit application of typing methods in larger epidemiological and clinical studies, such methods should be locally validated.

References

Achtman,M., Azuma,T., Berg,D.E., Ito,Y., Morelli,G., Pan,Z.J., Suerbaum,S., Thompson,S.A., van der Ende,A., and van Doorn,L.J. 1999. Recombination and clonal groupings within *Helicobacter pylori* from different geographic regions. Mol. Microbiol. 32: 459-470.

Akopyanz,N., Bukanov,N.O., Westblom,T.U., and Berg,D.E. 1992b. PCR-based RFLP analysis of DNA sequence diversity in the gastric pathogen *Helicobacter pylori*. Nucleic Acids Res. 20: 6221-6225.

Akopyanz,N., Bukanov,N.O., Westblom,T.U., Kresovich,S., and Berg,D.E. 1992a. DNA diversity among clinical isolates of *Helicobacter pylori* detected by PCR-based RAPD fingerprinting. Nucleic Acids Res. 20: 5137-5142.

Alm,R.A., Ling,L.S., Moir,D.T., King,B.L., Brown,E.D., Doig,P.C., Smith,D.R., Noonan,B., Guild,B.C., deJonge,B.L., Carmel,G., Tummino,P.J., Caruso,A., Uria-Nickelsen,M., Mills,D.M., Ives,C., Gibson,R., Merberg,D., Mills,S.D., Jiang,Q., Taylor,D.E., Vovis,G.F., and Trust,T.J. 1999. Genomic-sequence comparison of two unrelated isolates of the human gastric pathogen *Helicobacter pylori*. Nature 397: 176-180.

Atherton,J.C. 1997. The clinical relevance of strain types of *Helicobacter pylori*. Gut 40: 701-703.

Atherton,J.C., Cao,P., Peek,R.M.J., Tummuru,M.K., Blaser,M.J., and Cover,T.L. 1995. Mosaicism in vacuolating cytotoxin alleles of *Helicobacter pylori*. Association of specific *vac*A types with cytotoxin production and peptic ulceration. J. Biol. Chem. 270: 17771-17777.

Atherton,J.C., Cover,T.L., Twells,R.J., Morales,M.R., Hawkey,C.J., and Blaser,M.J. 1999. Simple and accurate PCR-based system for typing vacuolating cytotoxin alleles of *Helicobacter pylori*. J. Clin. Microbiol. 37: 2979-2982.

Azuma,T., Ito,S., Sato,F., Yamazaki,Y., Miyaji,H., Ito,Y., Suto,H., Kuriyama,M., Kato,T., and Kohli,Y. 1998. The role of the HLA-DQA1 gene in resistance to atrophic gastritis and gastric adenocarcinoma induced by *Helicobacter pylori* infection. Cancer 82: 1013-1018.

Bamford,K.B., Bickley,J., Collins,J.S., Johnston,B.T., Potts,S., Boston,V., Owen,R.J., and Sloan,J.M.1993. *Helicobacter pylori*: comparison of DNA fingerprints provides evidence for intrafamilial infection. Gut 34: 1348-1350.

Barrett,D.M., Faigel,D.O., Metz,D.C., Montone,K., and Furth,E.E. 1997. *In situ* hybridization for *Helicobacter pylori* in gastric mucosal biopsy specimens: quantitative evaluation of test performance in comparison with the CLOtest and thiazine stain. J. Clin. Lab Anal. 11: 374-379.

Blaser,M.J., Perez-Perez,G.I., Kleanthous,H., Cover,T.L., Peek,R.M., Chyou,P.H., Stemmermann,G.N., and Nomura,A. 1995. Infection with *Helicobacter pylori* strains possessing *cagA* is associated with an increased risk of developing adenocarcinoma of the stomach. Cancer Res. 55: 2111-2115.

Burucoa,C., Lhomme,V., and Fauchere,J.L. 1999. Performance criteria of DNA fingerprinting methods for typing of *Helicobacter pylori* isolates: experimental results and meta-analysis. J. Clin. Microbiol. 37: 4071-4080.

Censini,S., Lange,C., Xiang,Z., Crabtree,J.E., Ghiara,P., Borodovsky,M., Rappuoli,R., and Covacci,A. 1996. *cag*, a pathogenicity island of *Helicobacter pylori*, encodes type I-specific and disease-associated virulence factors. Proc. Natl. Acad. Sci. U. S. A. 93: 14648-14653.

Costas,M., Morgan,D.D., Owen,R.J., and Morgan,D.R. 1991. Differentiation of strains of *Helicobacter pylori* by numerical analysis of 1-D SDS-PAGE protein patterns: evidence for post-treatment recrudescence. Epidemiol. Infect. 107: 607-617.

Covacci,A., and Rappuoli,R. 1998. *Helicobacter pylori*: molecular evolution of a bacterial quasi-species. Curr. Opin. Microbiol. 1: 96-102.

Covacci,A., Telford,J.L., Del Giudice,G., Parsonnet,J.E., and Rappuoli,R. 1999. *Helicobacter pylori* virulence and genetic geography. Science 284: 1328-1333.

Cover,T.L. 1996. The vacuolating cytotoxin of *Helicobacter pylori*. Mol. Microbiol. 20: 241-246.

Debets-Ossenkopp,Y.J., Brinkman,A.B., Kuipers,E.J., Vandenbroucke-Grauls,C.M., and Kusters,J.G. 1998. Explaining the bias in the 23S rRNA gene mutations associated with clarithromycine resistance in clinical isolates of *Helicobacter pylori*. Antimicrob. Agents Chemother. 42: 2749-2751.

Debets-Ossenkopp,Y.J., Sparrius,M., Kusters,J.G., Kolkman,J.J., and Vandenbroucke-Grauls,C.M. 1996. Mechanism of clarithromycin resistance in clinical isolates of *Helicobacter pylori*. FEMS Microbiol. Lett. 142: 37-42.

Desai,M., Linton,D., Owen,R.J., Cameron,H., and Stanley,J., 1993. Genetic diversity of *Helicobacter pylori* indexed with respect to clinical symptomatology, using a 16S rRNA and a species-specific DNA probe. J. Appl. Bacteriol. 75: 574-582.

Desai,M., Linton,D., Owen,R.J., and Stanley,J. 1994. Molecular typing of *Helicobacter pylori* isolates from asymptomatic, ulcer and gastritis patients by urease gene polymorphism. Epidemiol. Infect. 112: 151-160.

Dunn,B.E., Cohen,H., and Blaser,M.J. 1997. *Helicobacter pylori*. Clin. Microbiol. Rev. 10: 720-741.

Enroth,H., Nyren,O., and Engstrand,L., 1999. One stomach-one strain. Does *Helicobacter pylori* strain variation influence desease outcome? Dig. Dis. Sci. 44: 102-107.

Evans,D.G., Evans,D.J., Jr., Lampert,H.C., Graham,D.Y., 1995. Restriction fragment length polymorphism in the adhesin gene *hpaA* of *Helicobacter pylori*. Am. J. Gastroenterol. 90: 1282-1288.

Evans,D.J., Jr., Queiroz,D.M., Mendes,E.N., and Evans,D.G. 1998. Diversity in the variable region of *Helicobacter pylori cagA* gene involves more than simple repetition of a 102-nucleotide sequence. Biochem. Biophys. Res. Commun. 245: 780-784.

Evans,D.L., Jr., and Evans,D.G. 1997. Direct repeat sequences in the *cagA* gene of *Helicobacter pylori:* a ghost of a chance encounter? Mol. Microbiol. 23: 409-410.

Forbes,K.J., Fang,Z., and Pennington,T.H. 1995. Allelic variation in the *Helicobacter pylori* flagellin genes *flaA* and *flaB*: its consequences for strain typing schemes and population structure. Epidemiol. Infect. 114: 257-266.

Fox,J.G., and Lee,A. 1997. The role of *Helicobacter* species in newly recognized gastrointestinal tract diseases of animals. Lab. Anim. Sci. 47: 222-255.

Fraser,A.G., Bickley,J., Owen,R.J., and Pounder,R.E., 1992. DNA fingerprints of *Helicobacter pylori* before and after treatment with omeprazole. J. Clin. Pathol. 45: 1062-1065.

Gallego,F.J., and Martinez,I. 1997. Method to improve reliability of random-amplified polymorphic DNA markers. Biotechniques 23: 663-664.

Gerhard,M., Lehn,N., Neumayer,N., Boren,T., Rad,R., Schepp,W., Miehlke,S., Classen,M., and Prinz,C. 1999. Clinical relevance of the *Helicobacter pylori* gene for blood-group antigen-binding adhesin. Proc. Natl. Acad. Sci. USA .96: 12778-12783.

Gibson,J.R., Chart,H., and Owen,R.J. 1998a. Intra-strain variation in expression of lipopolysaccharide by *Helicobacter pylori*. Lett. Appl. Microbiol. 26: 399-403.

Gibson,J.R., Slater,E., Xerry,J., Tompkins,D.S., and Owen,R.J. 1998b. Use of an amplified-fragment length polymorphism technique to fingerprint and differentiate isolates of *Helicobacter pylori*. J. Clin. Microbiol. 36: 2580-2585.

Go,M.F., Chan,K.Y., Versalovic,J., Koeuth,T., Graham,D.Y., and Lupski,J.R. 1995. Cluster analysis of *Helicobacter pylori* genomic DNA fingerprints suggests gastroduodenal disease-specific associations. Scand. J. Gastroenterol. 30: 640-646.

Go,M.F., Kapur,V., Graham,D.Y., Musser,J.M., 1996. Population genetic analysis of *Helicobacter pylori* by multilocus enzyme electrophoresis: extensive allelic diversity and recombinational population structure. J. Bacteriol. 178: 3934-3938.

Goodwin,A., Kersulyte,D., Sisson,G., Veldhuyzen van Zanten SJ, Berg,D.E., and Hoffman,P.S. 1998. Metronidazole resistance in *Helicobacter pylori* is due to null mutations in a gene (*rdxA*) that encodes an oxygen-insensitive NADPH nitroreductase. Mol. Microbiol. 28: 383-393.

Hirschl,A.M., Richter,M., Makristathis,A., Pruckl,P.M., Willinger,B., Schutze,K., and Rotter,M.L.1994. Single and multiple strain colonization in patients with *Helicobacter pylori*-associated gastritis: detection by macrorestriction DNA analysis. J. Infect. Dis. 170: 473-475.

Hultén,K., Gibreel,A., Sköld,O., and Engstrand,L., 1997. Macrolide resistance in *Helicobacter pylori*: mechanism and stability in strains from clarithromycin-treated patients. Antimicrob. Agents Chemother. 41: 2550-2553.

Hurtado,A., and Owen,R.J., 1994. Identification of mixed genotypes in *Helicobacter pylori* from gastric biopsy tissue by analysis of urease gene polymorphisms. FEMS Immunol. Med. Microbiol. 8: 307-313.

Ilver,D., Arnqvist,A., Ogren,J., Frick,I.M., Kersulyte,D., Incecik,E.T., Berg,D.E., Covacci,A., Engstrand,L., and Boren,T. 1998. *Helicobacter pylori* adhesin binding fucosylated histo-blood group antigens revealed by retagging. Science 279: 373-377.

Jenks,P.J., Ferrero,R.L., and Labigne,A. 1999. The role of the *rdx*A gene in the evolution of metronidazole resistance in *Helicobacter pylori*. J. Antimicrob. Chemother. 43: 753-758.

Jiang,Q., Hiratsuka,K., and Taylor,D.E. 1996. Variability of gene order in different *Helicobacter pylori* strains contributes to genome diversity. Mol. Microbiol. 20: 833-842.

Jorgensen,M., Daskalopoulos,G., Warburton,V., Mitchell,H.M., and Hazell,S.L. 1996. Multiple strain colonization and metronidazole resistance in *Helicobacter pylori*-infected patients: identification from sequential and multiple biopsy specimens. J. Infect. Dis. 174: 631-635.

Kitamoto,N., Nakamoto,H., Katai,A., Takahara,N., Nakata,H., Tamaki,H., and Tanaka,T. 1998. Heterogeneity of protein profiles of *Helicobacter pylori* isolated from individual patients. Helicobacter. 3: 152-162.

Kuipers,E.J., Israel,D.A., Kusters,J.G., Gerrits,M.M., Weel,J., van der Ende,A., van der Hulst,R.W., Wirth,H.P., Hook-Nikanne,J., Thompson,S.A., and Blaser,M.J., 2000. Quasispecies development of *Helicobacter pylori* observed in paired isolates obtained years apart from the same host. J. Infect. Dis. 181: 273-282.

Kuipers,E.J., Perez-Perez,G.I., Meuwissen,S.G., and Blaser,M.J. 1995. *Helicobacter pylori* and atrophic gastritis: importance of the *cagA* status. J. Natl. Cancer Inst. 87: 1777-1780.

Kusters,J.G., Schuijffel,D.F., Gerrits,M.M., and van Zwet,A.A. 1999. A single amino acid change in PBP-1A causes amoxicillin resistance in *Helicobacter pylori*. Gut 45: A5, abstract 01-01.

Kwon,D.H., El-Zaatari,F.A., Woo,J.S., Perng,C.L., Graham,D.Y., and Go,M.F. 1998. REP-PCR fragments as biomarkers for differentiating gastroduodenal disease-specific *Helicobacter pylori* strains. Dig. Dis. Sci. 43: 980-987.

Leunk,R.D. 1995. Production of a cytotoxin by *Helicobacter pylori*. Rev. Infect. Dis. 13: S686.

Linton,D., Moreno,M., Owen,R.J., and Stanley,J. 1992. 16S *rrn* gene copy number in *Helicobacter pylori* and its application to molecular typing. J. Appl. Bacteriol. 73: 501-506.

Lopez,C.R., Owen,R.J., and Desai,M. 1993. Differentiation between isolates of *Helicobacter pylori* by PCR-RFLP analysis of urease A and B genes and comparison with ribosomal RNA gene patterns. FEMS Microbiol. Lett. 110: 37-43.

Maiden,M.C.J., Bygraves,J.A., Feil,E., Morelli,G., Russell,J.E., Urwin,R., Zhang,Q., Zhou,J., Zurth,K., Caugant,D.A., Feavers,I.M., Achtman,M., and Spratt,B.G. 1998. Multilocus sequence typing: a portable approach to the identification of clones within populations of pathogenic microorganisms. Proc. Natl. Acad. Sci. USA. 95: 3140-3145.

Malaty,H.M., Engstrand,L., Pedersen,N.L., and Graham,D.Y., 1994. *Helicobacter pylori* infection: genetic and environmental influences. A study of twins. Ann. Intern. Med. 120: 982-986.

Marshall,D.G., Coleman,D.C., Sullivan,D.J., Xia,H., O'Morain,C.A., and Smyth,C.J. 1996. Genomic DNA fingerprinting of clinical isolates of *Helicobacter pylori* using short oligonucleotide probes containing repetitive sequences. J. Appl. Bacteriol. 81: 509-517.

Mégraud,F. 1999. Epidemiology and mechanism of antibiotic resistance in *Helicobacter pylori*. Gastroenterology 115: 1278-1282.

Miehlke,S., Genta,R.M., Graham,D.Y., and Go,M.F. 1999b. Molecular relationships of *Helicobacter pylori* strains in a family with gastroduodenal disease. Am. J. Gastroenterol. 94: 364-368.

Miehlke,S., Thomas,R., Guiterrez,O., Graham,D.Y., and Go,M.F. 1999a. DNA fingerprinting of single colonies of *Helicobacter pylori* from gastric cancer patients suggests infection with a single predominant strain. J. Clin. Microbiol. 37: 245-247.

Monteiro,L., Cabrita,J., and Megraud,F. 1997. Evaluation of performances of three DNA enzyme immunoassays for detection of *Helicobacter pylori* PCR products from biopsy specimens. J. Clin. Microbiol. 35: 2931-2936.

Morgan,D.R., Costas,M., Owen,R.J., and Williams,E.A. 1991. Characterization of strains of *Helicobacter pylori*: one-dimensional SDS- PAGE as a molecular epidemiologic tool. Rev. Infect. Dis. 13 Suppl 8:S709-13: S709-S713.

Occhialini,A., Urdaci,M., Doucet-Populaire,F., Bébear,C.M., Lamouliatte,H., and Megraud,F. 1997. Macrolide resistance in *Helicobacter pylori*: rapid detection of point mutations and assays of macrolide binding to ribosomes. Antimicrob. Agents Chemother. 41: 2724-2728.

Owen,R.J., Bell,G.D., Desai,M., Moreno,M., Gant,P.W., Jones,P.H., Linton,D., 1993a. Biotype and molecular fingerprints of metronidazole-resistant strains of *Helicobacter pylori* from antral gastric mucosa. J. Med. Microbiol. 38: 6-12.

Owen,R.J., Bickley,J., Hurtado,A., Fraser,A., Pounder,R.E., 1994. Comparison of PCR-based restriction length polymorphism analysis of urease genes with rRNA gene profiling for monitoring *Helicobacter pylori* infections in patients on triple therapy. J. Clin. Microbiol. 32: 1203-1210.

Owen,R.J., Bickley,J., Lastovica,A., Dunn,J.P., Borman,P., Hunton,C., 1992a. Ribosomal RNA gene patterns of *Helicobacter pylori* from surgical patients with healed and recurrent peptic ulcers. Epidemiol. Infect. 108: 39-50.

Owen,R.J., Bickley,J., Moreno,M., Costas,M., Morgan,D.R., 1991. Biotype and macromolecular profiles of cytotoxin-producing strains of *Helicobacter pylori* from antral gastric mucosa. FEMS Microbiol. Lett. 63: 199-204.

Owen,R.J., Desai,M., Figura,N., Bayeli,P.F., Di Gregorio,L., Russi,M., Musmanno,R.A., 1993b. Comparisons between degree of histological gastritis and DNA fingerprints, cytotoxicity and adhesivity of *Helicobacter pylori* from different gastric sites. Eur. J. Epidemiol. 9: 315-321.

Owen,R.J., Hunton,C., Bickley,J., Moreno,M., Linton,D., 1992b. Ribosomal RNA gene restriction patterns of *Helicobacter pylori*: analysis and appraisal of *Hae*III digests as a molecular typing system. Epidemiol. Infect. 109: 35-47.

Owen,R.J., Slater,E.R., Gibson,J., Lorenz,E., Tompkins,D.S., 1999. Effect of clarithromycin and omeprazole therapy on the diversity and stability of genotypes of *Helicobacter pylori* from duodenal ulcer patients. Microb. Drug Resist. 5: 141-146.

Owen,R.J., Slater,E.R., Xerry,J., Peters,T.M., Teare,E.L., Grant,A., 1998. Development of a scheme for genotyping *Helicobacter pylori* based on allelic variation in urease subunit genes. J. Clin. Microbiol. 36: 3710-3712.

Pan,Z.J., van der Hulst,R.W., Feller,M., Xiao,S.D., Tytgat,G.N., Dankert,J., van der Ende,A., 1997. Equally high prevalences of infection with *cagA*-positive *Helicobacter pylori* in Chinese patients with peptic ulcer disease and those with chronic gastritis-associated dyspepsia. J. Clin. Microbiol. 35: 1344-1347.

Park,C.S., Kim,J., 1999. Rapid and easy detection of *Helicobacter pylori* by in situ hybridization. J. Korean Med. Sci. 14: 15-20.

Peek,R.M.J., Thompson,S.A., Donahue,J.P., Tham,K.T., Atherton,J.C., Blaser,M.J., Miller,G.G., 1998. Adherence to gastric epithelial cells induces expression of a *Helicobacter pylori* gene, *iceA*, that is associated with clinical outcome. Proc. Amer. Assn. Phys. 110: 531-544.

Perez,T., Albornoz,J., Dominguez,A., 1998. An evaluation of RAPD fragment reproducibility and nature. Mol. Ecol. 7: 1347-1357.

Power,E.G., 1996. RAPD typing in microbiology—a technical review. J. Hosp. Infect. 34: 247-265.

Prewett,E.J., Bickley,J., Owen,R.J., Pounder,R.E., 1992. DNA patterns of *Helicobacter pylori* isolated from gastric antrum, body, and duodenum. Gastroenterology 102: 829-833.

Salama,S.M., Jiang,Q., Chang,N., Sherbaniuk,R.W., Taylor,D.E., 1995. Characterization of chromosomal DNA profiles from *Helicobacter pylori* strains isolated from sequential gastric biopsy specimens. J. Clin. Microbiol. 33: 2496-2497.

Salaun,L., Audibert,C., Le Lay,G., Burucoa,C., Fauchere,J.L., Picard,B., 1998. Panmictic structure of *Helicobacter pylori* demonstrated by the comparative study of six genetic markers. FEMS Microbiol. Lett. 161: 231-239.

Savelkoul,P.H., Aarts,H.J., de Haas,J., Dijkshoorn,L., Duim,B., Otsen,M., Rademaker,J.L., Schouls,L., Lenstra,J.A., 1999. Amplified-fragment length polymorphism analysis: the state of an art. J. Clin. Microbiol. 37: 3083-3091.

Stone,G.G., Shortridge,D., Flamm,R.K., Beyer,J., Ghoneim,A.T., and Tanaka,S.K. 1997b. PCR-RFLP typing of *ureC* from *Helicobacter pylori* isolated from gastric biopsies during a European multi-country clinical trial. J. Antimicrob. Chemother. 40: 251-256.

Stone,G.G., Shortridge,D., Flamm,R.K., Beyer,J., Stamler,D., and Tanaka,S.K. 1997a. PCR-RFLP typing of *ureC* from *Helicobacter pylori* isolated in Argentina from gastric biopsies before and after treatment with clarithromycin. Epidemiol. Infect. 118: 119-124.

Stone,G.G., Shortridge,D., Flamm,R.K., Versalovic,J., Beyer,J., Idler,K., Zulawinski,L., and Tanaka,S.K. 1996. Identification of a 23S rRNA gene mutation in clarithromycin-resistant *Helicobacter pylori.* Helicobacter. 1: 227-228.

Stone,G.G., Shortridge,D., Versalovic,J., Beyer,J., Flamm,R.K., Graham,D.Y., Ghoneim,A.T., and Tanaka,S.K. 1997c. A PCR-oligonucleotide ligation assay to determine the prevalence of 23S rRNA gene mutations in clarithromycin-resistant *Helicobacter pylori.* Antimicrob. Agents Chemother. 41: 712-714.

Suerbaum,S., Smith,J.M., Bapumia,K., Morelli,G., Smith,N.H., Kunstmann,E., Dyrek,I., Achtman,M., 1998. Free recombination within *Helicobacter pylori.* Proc. Natl. Acad. Sci. U. S. A. 95, 12619-12624.

Takami,S., Hayashi,T., Akashi,H., Shimoyama,T., and Tamura,T. 1994. Genetic heterogeneity of *Helicobacter pylori* by pulse-field gel electrophoresis and re-evaluation of DNA homology. Eur. J. Gastroenterol. Hepatol. 6 Suppl 1: S53-S56.

Takami,S., Hayashi,T., Tonokatsu,Y., Shimoyama,T., and Tamura,T. 1993. Chromosomal heterogeneity of *Helicobacter pylori* isolates by pulsed-field gel electrophoresis. Zentralbl. Bakteriol. 280: 120-127.

Tanahashi,T., Kita,M., Kodama,K., Sawai,N., Yamaoka,Y., Mitsufuji,S., Katoh,F., and Imanishi,J. 2000. Comparison of PCR-restriction fragment length polymorphism analysis and PCR-direct sequencing methods for differentiating *Helicobacter pylori ureB* gene variants. J. Clin. Microbiol. 38: 165-169.

Taylor,D.E., Eaton,M., Chang,N., and Salama,S.M. 1992. Construction of a *Helicobacter pylori* genome map and demonstration of diversity at the genome level. J. Bacteriol. 174: 6800-6806.

Taylor,N.S., Fox,J.G., Akopyantz,N.S., Berg,D.E., Thompson,N., Shames,B., Yan,L.,

Fontham,E., Janney,F., Hunter,F.M., *et al.* 1995. Long-term colonization with single and multiple strains of *Helicobacter pylori* assessed by DNA fingerprinting. J. Clin. Microbiol. 33: 918-923.

Tomb,J.F., White,O., Kerlavage,A.R., Clayton,R.A., Sutton,G.G., Fleischmann,R.D., Ketchum,K.A., Klenk,H.P., Gill,S., Dougherty,B.A., Nelson,K., Quackenbush,J., Zhou,L., Kirkness,E.F., Peterson,S., Loftus,B., Richardson,D., Dodson,R., Khalak,H.G., Glodek,A., McKenney,K., Fitzegerald,L.M., Lee,N., Adams,M.D., Hickey,E.K., Berg,D.E., Gocayne,J.D., Utterback,T.R., Peterson,J.D., Kelley,J.M., Cotton,M.D., Weidman,J.M., Fujii,C., Bowman,C., Wattey,L., Wallin,E., Hayes,W.S., Borodovsky,M., Karp,P.D., Smith,H.O., Fraser,C.M., and Venter,J.C. 1997. The complete genome sequence of the gastric pathogen *Helicobacter pylori*. Nature 388: 539-547.

van Belkum,A., Kluytmans,J., van Leeuwen,W., Bax,R., Quint,W., Peters,E., Fluit,A., Vandenbroucke-Grauls,C., van den,B.A., and Koeleman,H. 1995. Multicenter evaluation of arbitrarily primed PCR for typing of *Staphylococcus aureus* strains. J. Clin. Microbiol. 33: 1537-1547.

van Belkum,A., van Leeuwen,W., Kaufmann,M.E., Cookson,B., Forey,F., Etienne,J., Goering,R., Tenover,F., Steward,C., O'Brien,F., Grubb,W., Tassios,P., Legakis,N., Morvan,A., El Solh,N., de Ryck,R., Struelens,M., Salmenlinna,S., Vuopio-Varkila,J., Kooistra,M., Talens,A., Witte,W., and Verbrugh,H. 1998. Assessment of resolution and intercenter reproducibility of results of genotyping *Staphylococcus aureus* by pulsed-field gel electrophoresis of *Sma*I macrorestriction fragments: a multicenter study. J. Clin. Microbiol. 36: 1653-1659.

van der Ende,A., Pan,Z.J., Bart,A., van der Hulst,R.W., Feller,M., Xiao,S.D., Tytgat,G.N.J., and Dankert,J. 1998. *cagA*-positive *Helicobacter pylori* populations in China and the Netherlands are distinct. Infect. Immun. 66: 1822-1826.

van der Ende,A., Rauws,E.A., Feller,M., Mulder,C.J., Tytgat,G.N., and Dankert,J. 1996. Heterogeneous *Helicobacter pylori* isolates from members of a family with a history of peptic ulcer disease. Gastroenterology 111: 638-647.

van der Hulst,R.W., Rauws,E.A., Koycu,B., Keller,J.J., ten Kate,F.J., Dankert,J., Tytgat,G.N., and van der Ende,A. 1997. *Helicobacter pylori* reinfection is virtually absent after successful eradication. J. Infect. Dis. 176: 196-200.

van Doorn,L.J., Debets-Ossenkopp,Y.J., Marais,A., Sanna,R., Megraud,F., Kusters,J.G., Quint,W.G.V., 1999b. Rapid detection, by PCR and reverse hybridization, of mutations in the *Helicobacter pylori* 23S rRNA gene, associated with macrolide resistance. Antimicrob. Agents Chemother. 43: 1779-1782.

van Doorn,L.J., Figueiredo,C., Mégraud,F., Pena,A.S., Midolo,P., Queiroz,D.M., Carneiro,F., Vandenborght,B., Pégado,M.G.F., Sanna,R., de Boer,W.A., Schneeberger,P., Correa,P., Ng,E.K., Atherton,J.C., Blaser,M.J., Quint,W.G.V., 1999c. Geographic distribution of *vacA* allelic types of *Helicobacter pylori*. Gastroenterology 116: 823-830.

van Doorn,L.J., Figueiredo,C., Rossau,R., Jannes,G., van Asbroeck,M., Sousa,J.C., Carneiro,F., Quint,W., 1998a. Typing of the *Helicobacter pylori vacA* gene and detection of the *cagA* gene by PCR and reverse hybridization. J. Clin. Microbiol. 36: 1271-1276.

van Doorn,L.J., Figueiredo,C., Sanna,R., Pena,A.S., Midolo,P., Ng,E.K., Atherton,J.C., Blaser,M.J., Quint,W., 1998b. Expanding allelic diversity of *Helicobacter pylori vacA*. J. Clin. Microbiol. 36: 2597-2603.

van Doorn,L.J., Figueiredo,C., Sanna,R., Blaser,M.J., Quint,W.G.V., 1999a. Distinct variants of *Helicobacter pylori cagA* are associated with *vacA* subtypes. J. Clin. Microbiol. 37: 2306-2311.

van Doorn,L.J., Figueiredo,C., Sanna,R., Plaisier,A., Schneeberger,P., de Boer,W.A., Quint,W., 1998c. Clinical relevance of the *cagA, vacA*, and *iceA* status of *Helicobacter pylori*. Gastroenterology 115: 58-66.

van Doorn,L.J., Henskens,Y.M.C., Nouhan,N., Verschuuren-van Haperen,A., Vreede,R.W., Herbrink,P., Ponjee,G.A.E., van Krimpen,C., Blankenburg,R., Scherpenisse,J., Quint,W.G.V., 2000. The efficacy of laboratory diagnosis of *Helicobacter pylori* is related to bacterial density and *vacA, cagA*, and *iceA* status. J. Clin. Microbiol. 38: 13-17.

Versalovic,J., Osato,M.S., Spakovsky,K., Dore,M.P., Reddy,R., Stone,G.G., Shortridge,D., Flamm,R.K., Tanaka,S.K., Graham,D.Y., 1997. Point mutations in the 23S rRNA gene of *Helicobacter pylori* associated with different levels of clarithromycin resistance. J. Antimicrob. Chemother. 40: 283-286.

Wada,S., Matsuda,M., Kikuchi,M., Kodama,T., Takei,I., Ogawa,S., Takahashi,S., Shingaki,M., Itoh,T., 1994. Genome DNA analysis and genotyping of clinical isolates of *Helicobacter pylori*. Cytobios 80: 109-116.

Xia,H.X., Windle,H.J., Marshall,D.G., Smyth,C.J., Keane,C.T., O'Morain,C.A., 1995. Recrudescence of *Helicobacter pylori* after apparently successful eradication: novel application of randomly amplified polymorphic DNA fingerprinting. Gut 37: 30-34.

Xu,Q., Peek,R.M.J., Miller,G.G., Blaser,M.J., 1997. The *Helicobacter pylori* genome is modified at CATG by the product of *hpyIM*. J. Bacteriol. 179: 6807-6815.

From: *Helicobacter pylori: Molecular and Cellular Biology*
ISBN 1-898486-25-5 © 2001 Horizon Scientific Press, Wymondham, UK.

19

Population Structure of *Helicobacter pylori* and Other Pathogenic Bacterial Species

Mark Achtman[*]

Abstract

This articles reviews recent data on population genetics of *Helicobacter pylori* within the framework of more extensive knowledge about other bacterial species. DNA sequences from different natural isolates of *H. pylori* differ frequently at synonymous sites that encode the same amino acids. Unlike other bacterial species, the polymorphic synonymous sites in *H. pylori* are frequently shuffled among different strains due to recombination. Thus, the population structure of *H. pylori* is panmictic. Founder effects or geographical specialization have resulted in distinct non-overlapping populations in different continents.

Bacterial Population Structures

It is only very recently that the first data on the population structure of *Helicobacter pylori* have been published. These data are reviewed here within the context of what has been learned in more extensive investigations of other bacterial species. Bacterial species vary in their population structure. Three types of population structure, designated clonal, panmictic and epidemic, were described by Maynard Smith *et al.* (1993).

Clonal Bacterial Populations

All isolates from a clonal species are derived from a single ancestral cell and differ from that ancestral cell only by intragenomic changes (single nucleotide point mutations, deletions, insertions, rearrangements) that accumulate slowly with time (Figure 1, left). The neutral mutation theory (Kimura, 1991) is applicable to such bacteria. Selectively neutral sequence polymorphisms at synonymous sites are transient and will be either eradicated or fixed within the population by random drift after a time period that correlates with the effective population size. Periodic selection and selective sweeps (Majewski and Cohan, 1999) also purge bacterial populations of sequence polymorphisms. Geographical separation can result in different polymorphisms and fixed mutations in distinct areas. Geographical spread is accompanied by bottlenecks and founder effects that eliminate sequence variation within a population and can speed fixation of individual polymorphisms (Morelli *et al.*, 1997).

[*]email: achtman@molgen.mpg.de

The lack of sequence divergence in a species or population indicates that insufficient time for divergence to accumulate has elapsed since the existence of its last common ancestor. Based on the lack of sequence diversity, it has been calculated that the last common ancestor of *Mycobacterium tuberculosis* existed 12,000 years ago (Sreevatsan *et al.*, 1997) and that *Yersinia pestis* evolved from *Yersinia pseudotuberculosis* 2,000 - 15,000 years ago (Achtman *et al.*, 1999a). These bacteria are not only clonal but also show little sequence diversity because they are so young. Similar arguments may also apply to *Bacillus anthracis* which shows very little sequence diversity (Keim *et al.*, 1997).

Other bacteria are clonal although they have existed long enough to accumulate sequence diversity. *Salmonella enterica* infects mammals and reptiles and is thought to be derived from an ancestor that existed before mammals and reptiles evolved about 120 million years ago (Ochman and Wilson, 1987). Sequences of many genes from *S. enterica* yield similar phylogenetic trees (Selander *et al.*, 1996), indicating clonal descent.

Panmictic Populations

Clonality is disrupted by horizontal genetic exchange (Figure 1, right). Recombination by transduction, conjugation or transformation disrupts phylogenetic trees because it introduces sequences whose evolutionary history differs from that of other genes in the population. Although *Escherichia coli* is fairly clonal, most of its genes have been imported from unrelated species during its long evolutionary history (Lawrence and Ochman, 1997). Its chromosomal map still resembles that of the related species *S. enterica* serovar *typhimurium*, but is pocked with numerous genes and segments that were imported after these species separated. Due to subsequent recombination within *E. coli*, 50 times more sequence polymorphisms have resulted from recombination than from mutation (Guttman and Dykhuizen, 1994). Similar ratios of recombination to mutation have also been calculated for *Neisseria meningitidis* (Feil *et al.*, 1999) and *Streptococcus pneumoniae* (Feil *et al.*, 2000). Extremely frequent recombination can disrupt evolutionary relationships sufficiently that the population structure becomes panmictic (Figure 1, right). For example, most strains of *Neisseria gonorrhoeae* have combinations of particular alleles that have arisen by transformation rather than by the sequential accumulation of mutations (O'Rourke and Spratt, 1994).

Multiple Population Structures Within Species

Currently unknown mechanisms can result in various distinct population structures within a single species. For example, the HAU strains of *N. gonorrhoeae* are clonal (Gutjahr *et al.*, 1997) although the species is panmictic. Similarly, within *N. meningitidis*, most serogroup A isolates and several other hyper-virulent clonal groupings are characterized by an "epidemic" population structure even though the general population structure of this species is panmictic (Maynard Smith *et al.*, 1993; Maiden *et al.*, 1998). Epidemic populations result when the spread and multiplication of the descendents of a single cell are so efficient that these bacteria transiently maintain a clonal structure despite frequent recombination. For serogroup A *N. meningitidis*, this seems to reflect frequent geographical spread and the resulting purification due to sequential bottlenecks (Morelli *et al.*, 1997).

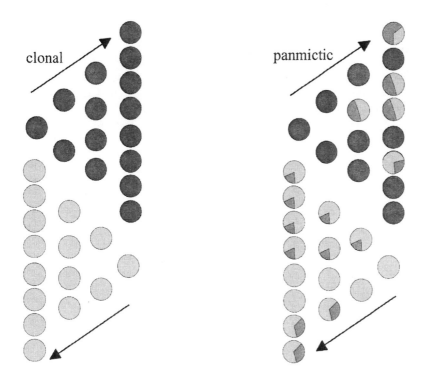

Figure 1. Clonal versus panmictic bacteria. Both models show a dark grey and a light grey lineage of bacteria that coexist in the same environment, e.g. on a mucosal surface. The clonal lineages (left) coexist without DNA exchange while the panmictic lineages (right) import sequences from the other lineage at random. The arrows indicate descent from ancestral cells.

Sequence Diversity Between Strains

General Considerations

Sequence polymorphism between strains is not uniform throughout the genome and depends on the location of the nucleotides that are investigated. Intergenic regions and non-expressed DNA can be highly polymorphic due to the absence of selection for function. These sequences are also subject to frequent deletion and replacement events (Klee *et al.*, 2000) and are less suitable for a comparison of population genetic structures of different species than are conserved sequences that are important for cell growth. Possibly somewhat unexpectedly, virulence genes are also frequently affected by horizontal genetic transfer (Li *et al.*, 1995), and correlate poorly with phylogenetic descent of other genes. rRNA genes are strongly conserved because changing the secondary structure of rRNA that interacts with the ribosomal proteins is deleterious and because gene conversion between the multiple copies can eliminate mutations. In practice, rRNA species are too conserved within species to be particularly useful for elucidating population genetic structures.

Conserved genes encoding important housekeeping enzymes are intermediate between these extremes. They contain sufficient informative sequence polymorphisms for statistical analyses and are present in all isolates. For gene products that are important for cell growth,

Table 1. Sources of sequence data used in Figure 2

Species (Nr. of isolates)	Locus	Size (bp)	Alleles	Source
S. aureus (155)	arc	456	20	1
	aro	456	30	
	glp	465	18	
	gmk	429	18	
	pta	474	19	
	tpi	402	22	
	yqi	516	27	
E. coli (see Alleles)	celC	348	13	2
	crr	507	14	2
	aceK	1722	11	3
	icd	1164	12	3
	mdh	849	15	3
	pabB	1008	11	3
	phoA	1413	8	3
	putP	1467	8	3
	sppA	972	11	3
	zwf	891	11	3
S. pneumoniae (621)	aroE	405	33	1
	ddl	441	60	
	gdh	460	39	
	gki	483	46	
	recP	450	30	
	spi	474	40	
	xpt	486	47	
S. enterica (see Alleles)	aceK	1720	16	4
	gapA	882	18	
	gnd	1162	70	
	mdh	849	26	
	putP	1467	15	
N. meningitidis (1079)	abcZ	433	91	5
	adk	465	59	
	aroE	490	106	
	fumC	465	104	
	gdh	501	82	
	pdhC	480	100	
	pgm	450	98	
H. pylori (20)	atpA	627	20	6
	efp	410	20	
	mutY	420	20	
	ppa	396	20	
	trpC	456	20	
	ureI	585	20	
	yphC	510	20	

Blanks indicate the same information as in the next non-blank preceding entry. Sources of data: 1. http://mlst.zoo.ox.ac.uk with permission from Brian Spratt; 2. Hall and Sharp, 1992; 3. a compilation of data from various sources by Thomas S. Whittam (DuBose *et al.*, 1988; Nelson and Selander, 1992; Boyd *et al.*, 1994; Guttman and Dykhuizen, 1994; Pupo *et al.*, 1997; Wang *et al.*, 1997; Nelson *et al.*, 1997); 4. sequences in GenBank (see Nelson *et al.,* 1997); 5. http://mlst.zoo.ox.ac.uk with permission from Martin Maiden; 6: Achtman *et al.*, 1999b.

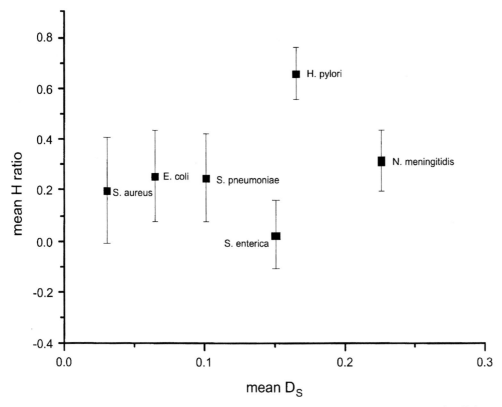

Figure 2. A comparison of the mean Homoplasy ratio (H) and mean genetic distance at synonymous sites (D_S) among different bacterial species. The sources of the data for these analyses are shown in Table 1. Mean Homoplasy ratios are the means for each species of three calculations per locus; they were calculated using HOMOPLASY (Suerbaum *et al.*, 1998). Mean D_S values were calculated using DNASP (Rozas and Rozas, 1999).

nucleotide changes that result in amino acid changes (non-synonymous polymorphisms) may be under selective pressure and are rarer than are synonymous polymorphisms. Aligned sequences of housekeeping genes from any one species are usually continuous, without any gaps, stop codons or frameshifts. In contrast, genes or parts of genes that encode immunogenic, cell-surface exposed proteins often possess more non-synonymous than synonymous polymorphisms. The sequences of genes encoding immunogenic proteins are also characterized by insertion/deletions, stop codons and frameshift mutations due to the selective pressure of the immune system (Smith *et al.*, 1995; Derrick *et al.*, 1999). Comparative sequence analyses of such genes are more likely to yield information on host-parasite interactions than on the population structure of the bacterium.

Mean Genetic Distance in Different Species

Due to gene-specific features, the analysis of the synonymous polymorphisms from multiple housekeeping genes is more suitable for measuring the genetic diversity within a species than from any single gene. Gene fragments from seven or more housekeeping genes have been sequenced from multiple strains in order to genotype *N. meningitidis* (Maiden *et al.*, 1998), *S. pneumoniae* (Enright and Spratt, 1998; Shi *et al.*, 1998), *H. pylori* (Achtman *et al.*, 1999b) and other species (Enright and Spratt, 1999) (see also http://mlst.zoo.ox.ac.uk) (Table 1). Comparative sequence data of housekeeping genes is also available from some other species (Table 1). The mean fraction of differences at synonymous sites between different alleles (D_S, mean genetic distance) (Li, 1993) is higher in *H. pylori* than in *Staphylococcus aureus*, *E. coli* or *S. pneumoniae*, roughly comparable to *S. enterica* and considerably lower than in *N. meningitidis* (Figure 2). It has been proposed that the higher sequence diversity in *H. pylori* than *E. coli* reflects different mutation rates, and possibly even an absence of mismatch repair in *H. pylori* (Wang *et al.*, 1999). However, different population structures can more readily explain the wide range of sequence diversity in different species than different quantitative levels of mutation frequency (Achtman and Suerbaum, 2000). At equilibrium, the average sequence diversity is directly related to the population size according to the neutral mutation theory. Higher mean genetic distances might reflect larger effective population sizes. The level of sequence diversity also reflects the frequency of purification events and rare purification will increase the degree of sequence diversity. *H. pylori* lacks a codon usage bias, indicating little growth competition between different isolates even during mixed infections (Lafay *et al.*, 2000). Purification will be rarer in species without growth competition than in species such as *E. coli* with a strong codon bias. Finally, sequence diversity will be considerably higher in species that frequently import genes because novel nucleotides that add to the sequence diversity are present in imported genes from related species or from geographically distinct populations that have followed independent evolutionary paths (Suerbaum *et al.*, 1998).

High Recombinant Frequency in *H. pylori*

Recombination occurs so frequently in *H. pylori* that it can occasionally be recognized simply by comparing different isolates from one patient (Kersulyte *et al.*, 1999). Multiple isolates that differ in RAPD (random amplified polymorphic DNA) pattern can arise due to the import of genes from other *H. pylori* that co-colonize the stomach. A number of methods have been devised to detect recombination from sequence data based on mosaic structures (Stephens, 1985; Sawyer, 1989; Maynard Smith, 1992) but none of these yield a quantitative measure of the frequency of recombinants in a population. A quantitative measure is provided by the recently described Homoplasy test (Maynard Smith and Smith, 1998). The Homoplasy test calculates the Homoplasy ratio which can range between 0.0 for sequences where all polymorphisms present in at least two sequences are due to the inheritance of independent mutations and 1.0 for sequences where all such polymorphisms result from recombination between alleles. The Homoplasy ratios for the *flaA*, *flaB* and *vacA* genes were about 0.9 in *H. pylori* isolated in Germany or Canada (Suerbaum *et al.*, 1998). Sequences of seven housekeeping genes from 20 isolates within a global collection (Table 1) also yielded very high values (Achtman *et al.*, 1999b). Their Homoplasy ratios are much higher than in other bacterial species (Figure 2), indicating that the frequency of recombinants is higher in *H. pylori* than in any other known bacterial species. A visual impression of the degree of clonality is also provided by compatibility matrices (Jakobsen *et al.*, 1997), which also indicated considerable recombination among the *H. pylori* sequences (Suerbaum *et al.*, 1998; Achtman *et al.*, 1999b). Further evidence for frequent recombination comes from multilocus enzyme electrophoresis of housekeeping genes (Go *et al.*, 1996) and other phylogenetic analyses of *vacA* and *flaA* (Göttke *et al.*, 2000).

Sequence polymorphisms are introduced 50 to 100 times more frequently by recombination than by mutation in *N. meningitidis* (Feil *et al.*, 1999), *S. pneumoniae* (Feil *et al.*, 2000) and *E. coli* (Guttman and Dykhuizen, 1994). For both *N. meningitidis* and *S. pneumoniae*, genes are also frequently imported from other commensal species that colonize the nasopharynx (Zhou *et al.*, 1997; Reichmann *et al.*, 1997; Linz *et al.*, 2000). The higher Homoplasy ratios in *H. pylori* than in these other species indicates that recombination occurs more frequently and/or that more recombinants accumulate. The half-life of carriage of *N. meningitidis* in the nasopharynx is only several months (Blakebrough *et al.*, 1982). Furthermore, recombinants are purified by bottlenecks during epidemic spread of serogroup A *N. meningitidis*. Distinct genotypes in parents within families (Suerbaum *et al.*, 1998; Kuo *et al.*, 1999) and frequent spread from parents to children or vice versa (see Chapter 3) provide ample opportunities for mixed infections that can lead to recombination. Recombinants may also accumulate due to the long persistence of infection (Kuipers *et al.*, 2000) and a lack of purification during spread from individual to individual.

Geographic Differences

Sequence comparisons have shown that distinct alleles of *H. pylori* genes exist in different geographic regions. This is most clear for the *cagA* gene where unrelated alleles are present in eastern Asia (China, Japan and Thailand) *versus* Africa, India, Europe or North America (van der Ende *et al.*, 1998; Achtman *et al.*, 1999b; Kersulyte *et al.*, 2000). The alleles of housekeeping genes showed geographic correlations that distinguished between strains isolated in East Asia *versus* Africa *versus* Europe or the Americas (Achtman *et al.*, 1999b). The *vacA* gene also tends to possess different alleles in these locations (van Doorn *et al.*, 1999; Letley *et al.*, 1999; Achtman *et al.*, 1999b) and also differs in India (Kersulyte *et al.*, 2000). Note that these data are somewhat difficult to interpret at the moment because the *cag* PAI did not distinguish between isolates from Europe and Africa although those bacteria differed according to housekeeping gene sequences (Achtman *et al.*, 1999b). These apparent discrepancies need to be resolved by more extensive analyses of housekeeping genes with additional strains. If those analyses confirm the limited data currently available, then the *cag* PAI (and possibly other virulence genes) might be unsuitable for recognizing certain bacterial populations.

Geographic differences in bacterial populations have not been well studied in any bacterial species. Hints of such stratification have been described for *Haemophilus influenzae* (Musser *et al.*, 1990) and certain clonal groupings within *E. coli* (Achtman and Pluschke, 1986). The fact that *H. pylori* is transmitted primarily between parents and children within families would result in only slow spread between distinct human populations. Different parts of the globe have only experienced extensive contact for the last few centuries and it is therefore possible that the current geographical stratification is a remnant of diverse strains having developed in different areas. A first test of this hypotheses has been recently published: a large global collection of *H. pylori* was tested for geographic patterns of parts of the *cag* PAI (Kersulyte *et al.*, 2000). The results showed that isolates from native South Americans were most similar to those from southern Europeans, who colonized South America several centuries ago, rather than those from East Asia.

Future Prospects

In other bacterial species, clones exist with identical alleles at numerous loci and multilocus sequence typing is based on whether bacteria possess identical or different alleles (Enright and Spratt, 1999). Such typing is not suitable for *H. pylori* where almost all isolates are unique and identical pairs are only to be expected among isolates from within families (Suerbaum *et al.*, 1998). Typing of *H. pylori* may well need to be restricted to long-term typing which focuses on larger phylogenetic differences between geographically distinct populations. One of the great advantages of multilocus sequence typing in other bacteria is that centralized, public databases of sequences are maintained. Once the loci used for long-term typing of *H. pylori* have been standardized, such a central database would be extremely useful in order to allow concurrent data collection in independent laboratories. A concerted effort could then rapidly map the global distribution of these bacteria and potentially reveal important aspects of the coevolution of bacteria and host.

Acknowledgements

I gratefully acknowledge the receipt of aligned *E. coli* sequence data from Thomas S. Whittam, Pennsylvania State University, the permission to include analyses of unpublished *S. aureus* data from Brian G. Spratt, Univ. of Oxford, and the permission to use sequence data from the MLST databases by Martin C. J. Maiden, Univ. of Oxford. I thank Sebastian Suerbaum and Silke Klee for constructive comments on how to improve this manuscript.

References

Achtman, M., Azuma, T., Berg, D. E., Ito, Y., Morelli, G., Pan, Z.-J., Suerbaum, S., Thompson, S., van der Ende, A., and van Doorn, L. J. 1999b. Recombination and clonal groupings within *Helicobacter pylori* from different geographical regions. Mol. Microbiol. 32: 459-470.

Achtman, M., and Pluschke, G. 1986. Clonal analysis of descent and virulence among selected *Escherichia coli*. Ann. Rev. Microbiol. 40: 185-210.

Achtman, M., and Suerbaum, S. 2000. Sequence variation in *Helicobacter pylori*. Trends Microbiol. 8: 57-58.

Achtman, M., Zurth, K., Morelli, G., Torrea, G., Guiyoule, A., and Carniel, E. 1999a. *Yersinia pestis*, the cause of plague, is a recently emerged clone of *Yersinia pseudotuberculosis*. Proc. Natl. Acad. Sci. USA 96: 14043-14048.

Blakebrough, I. S., Greenwood, B. M., Whittle, H. C., Bradley, A. K., and Gilles, H. M. 1982. The epidemiology of infections due to *Neisseria meningitidis* and *Neisseria lactamica* in a Northern Nigerian community. J. Infect. Dis. 146: 626-637.

Boyd, E. F., Nelson, K., Wang, F.-S., Whittam, T. S., and Selander, R. K. 1994. Molecular genetic basis of allelic polymorphism in malate dehydrogenase (*mdh*) in natural populations of *Escherichia coli* and *Salmonella enterica*. Proc. Natl. Acad. Sci. USA 91: 1280-1284.

Derrick, J. P., Urwin, R., Suker, J., Feavers, I. M., and Maiden, M. C. 1999. Structural and evolutionary inference from molecular variation in Neisseria porins. Infect. Immun. 67: 2406-2413.

DuBose, R. F., Dykhuizen, D. E., and Hartl, D. L. 1988. Genetic exchange among natural isolates of bacteria: Recombination within the *phoA* gene of *Escherichia coli*. Proc. Natl. Acad. Sci. USA 85: 7036-7040.

Enright, M. C., and Spratt, B. G. 1998. A multilocus sequence typing scheme for *Streptococcus pneumoniae*: identification of clones associated with serious invasive disease. Microbiology 144: 3049-3060.

Enright, M. C., and Spratt, B. G. 1999. Multilocus sequence typing. Trends Microbiol. 7: 482-487.

Feil, E. J., Maiden, M. C., Achtman, M., and Spratt, B. G. 1999. The relative contribution of recombination and mutation to the divergence of clones of *Neisseria meningitidis*. Mol. Biol. Evol. 16: 1496-1502.

Feil, E. J., Smith, J. M., Enright, M. C., and Spratt, B. G. 2000. Estimating recombinational parameters in *Streptococcus pneumoniae* from multilocus sequence typing data. Genetics 154: 1439-1450.

Go, M. F., Kapur, V., Graham, D. Y., and Musser, J. M. 1996. Population genetic analysis of *Helicobacter pylori* by multilocus enzyme electrophoresis: extensive allelic diversity and recombinational population structure. J. Bacteriol. 178: 3934-3938.

Göttke, M. U., Fallone, C. A., Barkun, A. N., Vogt, K., Loo, V., Trautmann, M., Tong, J. Z., Nguyen, T. N., Fainsilber, T., Hahn, H. H., Korber, J., Lowe, A., and Beech, R. N. 2000. Genetic variability determinants of *Helicobacter pylori*: Influence of clinical background and geographic origin of isolates. J. Infect. Dis. 181: 1674-1681.

Gutjahr, T. S., O'Rourke, M., Ison, C. A., and Spratt, B. G. 1997. Arginine-, hypoxanthine-, uracil-requiring isolates of *Neisseria gonorrhoeae* are a clonal lineage within a non-clonal population. Microbiology 143: 633-640.

Guttman, D. S., and Dykhuizen, D. E. 1994. Clonal divergence in *Escherichia coli* as a result of recombination, not mutation. Science 266: 1380-1383.

Hall, B. G., and Sharp, P. M. 1992. Molecular population genetics of *Escherichia coli*: DNA sequence diversity at the *celC*, *crr*, and *gutB* loci of natural isolates. Mol. Biol. Evol. 9: 654-665.

Jakobsen, I. B., Wilson, S. R., and Easteal, S. 1997. The partition matrix: exploring variable phylogenetic signals along nucleotide sequence alignments. Mol. Biol. Evol. 14: 474-484.

Keim, P., Kalif, A., Schupp, J., Hill, K., Travis, S. E., Richmond, K., Adair, D. M., Hugh-Jones, M., Kuske, C. R., and Jackson, P. 1997. Molecular evolution and diversity in *Bacillus anthracis* as detected by amplified fragment length polymorphism markers. J. Bacteriol. 179: 818-824.

Kersulyte, D., Chalkauskas, H., and Berg, D. E. 1999. Emergence of recombinant strains of *Helicobacter pylori* during human infection. Mol. Microbiol. 31: 31-43.

Kersulyte, D., Mukhopadhyay, A. K., Velapatino, B., Su, W., Pan, Z., Garcia, C., Hernandez, V., Valdez, Y., Mistry, R. S., Gilman, R. H., Yuan, Y., Gao, H., Alarcon, T., Lopez-Brea, M., Balakrish, N. G., Chowdhury, A., Datta, S., Shirai, M., Nakazawa, T., Ally, R., Segal, I., Wong, B. C., Lam, S. K., Olfat, F. O., Boren, T., Engstrand, L., Torres, O., Schneider, R., Thomas, J. E., Czinn, S., and Berg, D. E. 2000. Differences in genotypes of *Helicobacter pylori* from different human populations. J. Bacteriol. 182: 3210-3218.

Kimura, M. 1991. Recent development of the neutral theory viewed from the Wrightian tradition of theoretical population genetics. Proc. Natl. Acad. Sci. USA 88: 5969-5973.

Klee, S.R., Nassif, X., Kusecek, B., Merker, P., Beretti, J.-L., Achtman, M., and Tinsley, C.R. 2000. Molecular and biological analysis of eight genetic islands that distinguish *Neisseria meningitidis* from the closely related pathogen *Neisseria gonorrhoeae*. Infect. Immun. 68: 2082-2095.

Kuipers, E. J., Israel, D. A., Kusters, J. G., Gerrits, M. M., Weel, J., van Der, E. A., Der Hulst, R. W., Wirth, H. P., Hook-Nikanne, J., Thompson, S. A., and Blaser, M. J. 2000. Quasispecies development of *Helicobacter pylori* observed in paired isolates obtained years apart from the same host. J. Infect. Dis. 181: 273-282.

Kuo, C. H., Poon, S. K., Su, Y. C., Su, R., Chang, C. S., and Wang, W. C. 1999. Heterogeneous *Helicobacter pylori* isolates from *H. pylori*-infected couples in Taiwan. J. Infect. Dis. 180: 2064-2068.

Lafay, B., Atherton, J. C., and Sharp, P. M. 2000. Absence of translationally selected synonymous codon usage bias in *Helicobacter pylori*. Microbiology 146: 851-860.

Lawrence, J. G., and Ochman, H. 1997. Amelioration of bacterial genomes: rates of change and exchange. J. Mol. Evol. 44: 383-397.

Letley, D. P., Lastovica, A., Louw, J. A., Hawkey, C. J., and Atherton, J. C. 1999. Allelic diversity of the *Helicobacter pylori* vacuolating cytotoxin gene in South Africa: rarity of the vacA s1a genotype and natural occurrence of an s2/m1 allele. J. Clin. Microbiol. 37: 1203-1205.

Li, J., Ochman, H., Groisman, E. A., Boyd, E. F., Solomon, F., Nelson, K., and Selander, R. K. 1995. Relationship between evolutionary rate and cellular location among the Inv/Spa invasion proteins of *Salmonella enterica*. Proc. Natl. Acad. Sci. USA 92: 7252-7256.

Li, W.-H. 1993. Unbiased estimation of the rates of synonymous and nonsynonymous substitution. J. Mol. Evol. 36: 96-99.

Linz, B., Schenker, M., Zhu, P., and Achtman, M. 2000. Frequent interspecific genetic exchange between commensal neisseriae and *Neisseria meningitidis*. Mol. Microbiol. 36: 1049-1058.

Maiden, M. C. J., Bygraves, J. A., Feil, E., Morelli, G., Russell, J. E., Urwin, R., Zhang, Q., Zhou, J., Zurth, K., Caugant, D. A., Feavers, I. M., Achtman, M., and Spratt, B. G. 1998. Multilocus sequence typing: a portable approach to the identification of clones within populations of pathogenic microorganisms. Proc. Natl. Acad. Sci. USA 95: 3140-3145.

Majewski, J., and Cohan, F. M. 1999. Adapt globally, act locally. The effect of selective sweeps on bacterial sequence diversity. Genetics 152: 1459-1474.

Maynard Smith, J. 1992. Analyzing the mosaic structure of genes. J. Mol. Evol. 34: 126-129.

Maynard Smith, J., and Smith, N. H. 1998. Detecting recombination from gene trees. Mol. Biol. Evol. 15: 590-599.

Maynard Smith, J., Smith, N. H., O'Rourke, M., and Spratt, B. G. 1993. How clonal are bacteria? Proc. Natl. Acad. Sci. USA 90: 4384-4388.

Morelli, G., Malorny, B., Müller, K., Seiler, A., Wang, J., del Valle, J., and Achtman, M. 1997. Clonal descent and microevolution of *Neisseria meningitidis* during 30 years of epidemic spread. Mol. Microbiol. 25: 1047-1064.

Musser, J. M., Kroll, J. S., Granoff, D. M., Moxon, E. R., Brodeur, B. R., Campos, J., Dabernat, H., Frederiksen, W., Hamel, J., Hammond, G., Høiby, E. A., Jonsdottir, K. E., Kabeer, M., Kallings, I., Khan, W. N., Kilian, M., Knowles, K., Koornhof, H. J., Law, B., Li, K. I., Montgomery, J., Pattison, P. E., Piffaretti, J.-C., Takala, A. K., Thong, M. L., Wall, R. A., Ward, J. I., and Selander, R. K. 1990. Global genetic structure and molecular epidemiology of encapsulated *Haemophilus influenzae*. Rev. Infect. Dis. 12: 75-111.

Nelson, K., and Selander, R. K. 1992. Evolutionary genetics of the proline permease gene (*putP*) and the control region of the proline utilization operon in populations of *Salmonella* and *Escherichia coli*. J. Bacteriol. 174: 6886-6895.

Nelson, K., Wang, F. S., Boyd, E. F., and Selander, R. K. 1997. Size and sequence polymorphism in the isocitrate dehydrogenase kinase/phosphatase gene (*aceK*) and flanking regions in *Salmonella enterica* and *Escherichia coli*. Genetics 147: 1509-1520.

O'Rourke, M., and Spratt, B. G. 1994. Further evidence for the non-clonal population structure of *Neisseria gonorrhoeae*: Extensive genetic diversity within isolates of the same electrophoretic type. J. Gen. Microbiol. 140: 1285-1290.

Ochman, H., and Wilson, A. C. 1987. Evolution in bacteria: Evidence for a universal substitution rate in cellular genomes. J. Mol. Evol. 26: 74-86.

Pupo, G. M., Karaolis, D. K. R., Lan, R. T., and Reeves, P. R. 1997. Evolutionary relationships among pathogenic and nonpathogenic *Escherichia coli* strains inferred from multilocus enzyme electrophoresis and *mdh* sequence studies. Infect. Immun. 65: 2685-2692.

Reichmann, P., König, A., Liñares, J., Alcaide, F., Tenover, F. C., McDougal, L., Swidinski, S., and Hakenbeck, R. 1997. A global gene pool for high-level cephalosporin resistance in commensal *Streptococcus* species and *Streptococcus pneumonieae*. J. Infect. Dis. 176: 1001-1012.

Rozas, J., and Rozas, R. 1999. DnaSP version 3.0: an integrated program for molecular population genetics and molecular evolution analysis. Bioinformatics 15: 174-175.

Sawyer, S. 1989. Statistical tests for detecting gene conversion. Mol. Biol. Evol. 6: 526-538.

Selander, R. K., Li, J., and Nelson, K. 1996. Evolutionary genetics of *Salmonella enterica*. In: *Escherichia coli* and *Salmonella*. R. Curtiss III, J.L. Ingraham, E.C.C. Lin, K.B. Low, B. Magasanik, W.S. Reznikoff, M. Riley, M. Schaechter, and H.E. Umbarger (eds.). ASM Press, Washington, DC. p. 2691-1707.

Shi, Z.-Y., Enright, M. C., Wilkinson, P., Griffiths, D., and Spratt, B. G. 1998. Identification of three major clones of multiply antibiotic-resistant *Streptococcus pneumoniae* in Taiwanese hospitals by multilocus sequence typing. J. Clin. Microbiol. 36: 3514-3519.

Smith, N. H., Maynard Smith, J., and Spratt, B. G. 1995. Sequence evolution of the *porB* gene of *Neisseria gonorrhoeae* and *Neisseria meningitidis*: Evidence of positive Darwinian selection. Mol. Biol. Evol. 12: 363-370.

Sreevatsan, S., Pan, X., Stockbauer, K., Connell, N. D., Kreiswirth, B. N., Whittam, T. S., and Musser, J. M. 1997. Restricted structural gene polymorphism in the *Mycobacterium tuberculosis* complex indicates evolutionarily recent global dissemination. Proc. Natl. Acad. Sci. USA 94: 9869-9874.

Stephens, J. C. 1985. Statistical methods of DNA sequence analysis: detection of intragenic recombination or gene conversion. Mol. Biol. Evol. 2: 539-556.

Suerbaum, S., Maynard Smith, J., Bapumia, K., Morelli, G., Smith, N. H., Kunstmann, E., Dyrek, I., and Achtman, M. 1998. Free recombination within *Helicobacter pylori*. Proc. Natl. Acad. Sci. USA 95: 12619-12624.

van der Ende, A., Pan, Z. J., Bart, A., Van der Hulst, R. W., Feller, M., Xiao, S. D., Tytgat, G. N., and Dankert, J. 1998. *cagA*-positive *Helicobacter pylori* populations in China and The Netherlands are distinct. Infect. Immun. 66: 1822-1826.

van Doorn, L. J., Figueiredo, C., Megraud, F., Pena, S., Midolo, P., Queiroz, D. M., Carneiro, F., Vanderborght, B., Pegado, M. D., Sanna, R., de Boer, W., Schneeberger, P. M., Correa, P., Ng, E. K., Atherton, J., Blaser, M. J., and Quint, W. G. 1999. Geographic distribution of vacA allelic types of *Helicobacter pylori*. Gastroenterology 116: 823-830.

Wang, F. S., Whittam, T. S., and Selander, R. K. 1997. Evolutionary genetics of the isocitrate dehydrogenase gene (*icd*) in *Escherichia coli* and *Salmonella enterica*. J. Bacteriol. 179: 6551-6559.

Wang, G., Humayun, M. Z., and Taylor, D. E. 1999. Mutation as an origin of genetic variability in *Helicobacter pylori*. Trends Microbiol. 7: 488-493.

Zhou, J. J., Bowler, L. D., and Spratt, B. G. 1997. Interspecies recombination, and phylogenetic distortions, within the glutamine synthetase and shikimate dehydrogenase genes of *Neisseria meningitidis* and commensal *Neisseria* species. Mol. Microbiol. 23: 799-812.

Index

16S ribosomal RNA 15, 17, 276, 300
23S ribosomal RNA 17, 300, 302

A

achlorhydria 143
acquisition of infection 32-37
actin nucleation 238
acute inflammatory response 65
adherence 185-199
adhesin
    receptors 144-145
    inhibition 198
adjuvants 265-266, 268
AFLP 294, 296
age of infection 33
Agrobacterium tumefaciens pilus 230-231, 234
AlpA/AlpB proteins 194, 196-198
ammonia 4, 155, 161
amoxicillin resistance 302
animal models 113-122
    non-murine rodent, *See* non-murine rodent models
    non-rodent, *See* non-rodent animal models
    murine, *See* murine animal models
    standardized scoring 120
antibiotic resistance 264, 302
antibiotic targets 287
antibody responses 264
anticanalicular antibodies 56, 58, 74, 105
antigenic mimicry 57
antrum predominant gastritis 53, 91
AP-1 68, 236
apoptosis 74, 92, 99-108, 137
Arp2/3 238
atrophic gastritis 7, 39, 53-59, 100, 105, 116, 135, 136, 141, 147
attractants 171
autoantibodies 215-216
autoimmune gastritis 53-59, 74
autoimmunity 89, 215
autoreactive T and B cells 58

B

B7-2 73
BabA adhesin 139, 187, 194-196, 267
babA 283, 301
Bacillus subtilis 172
BacTracker 180
basal body 171
Bcl-2 family 106-108
β-barrel structure 198
bicarbonate 180
biopsy urease test 163
bismuth therapy 6
blood group antigens 191
Bordetella pertussis ptl transport system 232
bottlenecks 311
Brucella suis 233

C

cag PAI 69-70, 103, 117, 139, 186, 228-231, 317
cag type IV secretion system 228-238
cagA 70, 91-92, 103, 217, 301, 317
CagA 134, 187, 267
CagA protein, translocation of 235, 237
camouflage 215
Campylobacter 15
Campylobacter jejuni 196
Campylobacter pyloridis 15
carbohydrate binding 187
carcinogen 7
caspases 106, 108
catalase 267
Caulobacter crescentus 172, 176
C-C chemokines 65-66, 86-87
CD4 86-87, 91, 119-120
CD8 86-87, 91
CD14 212
CD30 87
CD95, CD95 ligand 104-106
cell-mediated immunity 86, 93, 119
ceramide 68
c-fos 236
chemotaxis 171, 173, 179-180
chief cells 135, 213
    LPS effect on 212

cholera toxin 268
chromosome, physical organization 281-283
chronic gastritis 53-55, 65, 73, 114-115
c-Jun kinase 68, 236
CLO test, *See* biopsy urease test
clonal groupings 279, 293
clonal species 311
codon usage bias 281-282, 316
coevolution of humans and *H. pylori* 147
cohort studies 33, 35, 36
collagen 192
compatibility matrix 316
core oligosaccharide, structure 214
corkscrew phenotype 19
corpus predominant gastritis 53-54
C-ring 173
C-X-C chemokines 65-67, 70, 86-87
cystic fibrosis 189
cytokines 63-75, 86,89, 104-105

D

dinucleotide repeats 193, 285
DNA
    restriction/modification systems 281, 283
    topology 179
    vaccination 269
Dps DNA binding protein, similarity to HP-Nap 255
dysplasia 7, 39

E

electron microscopy 4, 191
Elk-1 236
ENA-78 chemokine 64-66, 70
endotoxin, *See* lipopolysaccharide
enteric helicobacter 15, 17, 19, 21-23
enterohepatic helicobacter 15
EPEC 187, 238
epidemic strains 30
epithelial cell turnover 100
epithelial renewal 137-138
ERK1/2 236
ERL chemokines 65
exoenzyme S 139
extracellular matrix 191, 212

F

Fas ligand, *See* CD 95
Fas 74
Ferret/*H. mustelae* model 267
ferritin *pfr* 255
ferritin 190
filament 171
FlaA/FlaB flagellins 173
flagella 18, 19, 172-181, 186, 194, 233, 246
    differences among species 17
    role in adhesion 181
flagellar arrangement 173
flagellar proteins 173
flagellin 171, 173
FlgE hook protein 173
FlgM anti sigma factor 176, 181
fliP 179
flurofamide 161, 180
founder effects 311
fucosyl transferases 140, 218, 286
Fur-regulated genes 255

G

G+C content 17, 228, 276, 281
galactosyltransferases 218-219
γ-glutamyl transpeptidase 18, 117
gamma interferon, *See* interferon γ
gangliotetraosyl-ceramide 139, 192
gastric
    carcinoma 6, 39-40, 53, 105
        link to *H. pylori* 7, 133-134
    epithelial cell lineages 137
    gland 100
    H,K-ATPase 55, 58, 74, 142, 216
    helicobacter 15, 17, 19-20
    metaplasia 189
    microflora 143
    units 137
gastritis 4, 17, 53-59, 100, 119-120, 216, 264
gene regulation 285
genome sequences 173, 275-288
genomic analysis 275-288
genomics to drugs strategy 287
genotyping 293-303, 315
geographic differences 30, 32, 279, 311, 317
gerbils 136
globotetraosyl-ceramide 139
glutamine synthetase 161

glycans as adhesin receptors 139-142, 146
glycosaminoglycans 191
glycosphingolipids 190
glycosylation 173
gnotobiotic piglets 180
GRO-α chemokine 64-66, 70
growth phase dependent gene regulation 179
gyrase 178

H

H. felis 3, 20, 72, 115, 117-120, 138, 171, 179-180, 186, 264
H. heilmannii 3, 19, 20, 115, 138
H. mustelae 171, 180
H. pylori
    association with anemia 39
    association with autoimmune gastritis 55-59
    benefit to the host 8
    benefits of universal elimination 7
    cultivation 17
    cure 8
    disease to infection ratio 40-41
    epidemiological similarities
      with enteric pathogens 30
    general features 17
    lack of association with heart disease 37-38
    mode of spread 30-31
    mouse-adapted 116
    occurrence in animal species 17
    prevalence 8
    public health relevance 263
    risk factors for infection 31-33
    spread in families 30, 33, 36
    vehicles of transmission 30
H. suis 20
H+/K+-ATPase, *See* gastric ATPase
H-antigen 191
Helicobacter identification 17
hemagglutination 188, 212-213
heparan sulfate 191
heptosyl transferase 218
Herpes virus 5
histidine kinases 285
histology 53-55, 65
history 1-8, 15
HLA-DR 73
Homoplasy test 316
homopolymeric tracts 285
hook 171, 173

HOP proteins, *See* outer membrane protein superfamily
HopZ protein 194
horizontal gene transfer 284
hpaA gene 190
HpaA protein 139, 173, 176
Hpn protein 162
HP-Nap protein 190, 192, 253-255
Hsp70 heat shock protein 139, 192
HspA/HspB heat shock proteins 162, 267
human vaccination trials 266
humoral immunity 86-87, 93, 119
hyperproliferation of epithelial cells 100,105, 116

I

IκB 68
iceA 301
IgA
    antibodies 264
    role in protection against infection 32
IL-1 β 142
IL-1 71, 89-90, 119
IL-2 86-87, 89
IL-4 86-91, 93
IL-5 86-87, 89, 91
IL-6 142
IL-6 90
IL-8 64-70, 89, 142, 187, 236
IL-10 71-72, 86-87, 89, 93, 119
IL-12 72-73, 86-90
IL-13 86-87
IL-18 72-73, 86-88
immunization routes 268
infection
    of adults 35-36
    of children 32-35
    sex specificity 1, 32
infectious dose 30, 35
inflammatory response 236
iNOS 142
insertion sequences 278
integrin 213
intercellular junctions 191
interferon γ 58, 72-74, 86-93, 104-105, 119, 264
interleukin 8, *See* IL-8
intestinal metaplasia 6, 7, 39, 116, 136
intrafamilial transmission 299, 317
intramucosal cysts 143
iron binding 190, 255
IS605 insertion element 228, 278, 284

J

JNK 236

K

Klebsiella aerogenes, urease of 157

L

lactosylceramide 192
LAG-3 87
laminin 192, 212
Legionella pneumophila 233
Lewis antigen
    a antigen 214-215
    b antigen 139-141, 191, 194, 214-215
    mimicry 214-215, 217
    role in adherence 217
    X, Y antigen 57, 141, 192-193, 214-218
lipid A, *See* LPS
lipopolysaccharide (LPS) 71, 89, 108, 118, 192, 207-219, 286
Listeria monocytogenes 238
live vaccine 269
LPS
    biosynthesis genes 192, 218-219
    chain length 208
    general structure 207-208
    low biological activity 209
LPS-binding proteins 211
LT heat-labile enterotoxin 266, 268
lymphoid infiltration 57, 65
lysine, abundance of 281

M

macrolide resistance 302
MALT-lymphoma 7, 38-39, 53, 91, 93-94, 118
MAP kinases 68, 236
Marshall, B. x, 4
master operon 176, 181
MCP-1 chemokine 64-65
MEK1/2 236
metronidazole resistance 302
microarray technology 141, 147, 181
microscopic observations 3

microspheres 269
MIP-1α, MIP-1β 71
mixed infections 37
modeling of *H. pylori* infection 139, 181, 186
morphology 19
mosaic genes 316
motility 172-181
mucin 180, 186, 191
mucosal chemokines 65-66
mucus layer, barrier properties of 246
multilocus
    enzyme electrophoresis (MLEE) 316
    sequence typing (MLST) 294, 296, 317
multipotent isthmal stem cell 137
muraminidase 176
murine animal models 115-120
mutation frequency 312
mutator genes 281

N

necrosis 137
Neisseria gonorrhoeae 312
Neisseria meningitidis 312, 317
neutral mutation theory 311, 316
neutrophil
    activating protein, *See* HP-Nap protein
    infiltration 65, 70
NF-κB 68-69, 72
nickel 155
nitric oxide synthase 211
NixA 161-162
NK cells 72, 87, 94
non-murine rodent models 120-121
non-rodent animal models 114-115

O

oral versus fecal transmission 31
orthologous genes 277
outer membrane 194
outer membrane protein superfamily 193-194, 198, 283, 287
oxidase 17

P

P fimbriae of *E. coli* 186
p21 activated kinase 68
p38 ERK kinase 236
panmictic species 311-313
parietal cell ablation 143
parietel cells 135
pathogenicity islands228
pathology 53-59
PCR amplification from the environment 30, 33
PCR-LiPA 294, 297-299
pepsinogen I 213
peptic ulcer 1, 5, 40-41, 53, 73, 91, 100, 105, 191
PFGE 294, 296
phosphate sinks 180
phosphatidylethanolamine 139, 192
phylogeny 16
pit cells 137
plasminogen activator type 2 211
plasticity zone 283
polyglycosyl-ceramides 188
population genetics 311-321
porins 287
procoagulant activity (PCA) 211
profilin 238
Proteus mirabilis 155
proton motive force 173
Pseudomonas aeruginosa 189

Q

quorum sensing 139

R

rab7 GTPase 250
RANTES chemokine 64-65
RAPD 294-296
reactive oxygen radicals 255
reasons for developing a vaccine 264
recombination 283, 293, 311-312
reflux esophagitis 8
reinfection after treatment 36
repellents 171
restriction fragment length polymorphism (RFLP) 293, 294
Re-Tagging 194-195

Rhesus monkey model 267
Rhodobacter sphaeroides 179
Rho-family proteins 238
ribosomal RNA, *See* 16S ribosomal RNA
ribotyping 300
rod, flagellar 176
role of antibodies in protection 265

S

Salmonella infection 32, 35, 67
Salmonella typhimurium 172
Salmonella, attenuated strains 266
secretor phenotype 191, 215
selection 313
sequence diversity 279, 280, 312, 313
seroconversion 33
serological reactivity 32-34, 36, 55, 134
seroreversion 33-34
sheath, flagellar 173
Shigella epidemiology 30, 32, 36-37
Shigella flexneri 238
shuttle mutagenesis 196
sialic acid 188-189, 213
sialylated glycoconjugates 188
sigma factors 176, 285
Sinorhizobium meliloti 179
slipped-strand mispairing 179, 193
slipped-strand repair 285
small Rho-GTPases 68
species 15
sphingomyelin-ceramide pathway 68
spiral bacteria 3
Staphylococcus saprophyticus 15
strain-specific genes 284-285
sulfatide 190, 192
supercoiling 178
surfactant protein D 193

T

T cell responses 264-265
target identifications 287
taxonomy 15-23
Th0 cells 86-87, 91
Th1 cells 72-73, 86-94, 119, 264, 268
Th2 cells 86-94, 119, 264, 268
Tir translocated intimin receptor 238

tissue tropism 185-186
TNF-α 71, 74, 86, 89-90, 92-93, 104-105, 142
TNF-β 86-87
TnMax transposons 196-197
transcription termination signals 286
transepithelial resistance, effect of VacA on 251-252
transgenic
    knock-out mice 22, 72, 74, 118-119, 265
    mouse models 136-147
transition mutations 281
translocation of CagA 139
transposon mutagenesis 196
triple-positive *H. pylori* strains 187, 191, 196
tRNA synthetases 277
tumorigenesis 137-138, 142-144
two-component regulators 285
type I/II strains 134, 228, 267, 301
type III secretion 233
type IV secretion 139, 186
typing methods 293-303
tyrosine phosphorylation 70, 237-238

U

urea breath test 164
urea 180
urease accessory genes 159-160
urease genes 300
    use in typing 164
urease 4, 18, 19, 71, 91-92, 117, 155-165, 180, 192, 266-267
    enzymology 156
    localization of 161
    necessity for colonization of stomach 19
    role in colonization 163
    use as vaccine antigcn 164
Ussing chambers 213

V

vacA 300, 317
    s/m alleles 249, 300
VacA vacuolating cytotoxin 89, 91-92, 94, 103, 134, 187, 245-253, 267
    effect on membrane trafficking 250
    effects on immune cells 251
    pepsin resistance 249
    pore-forming activity 252

vaccination
    prophylactic 287
    therapeutic 287
vaccine antigens 267
vaccines 263-269
vacuolization, mechanism of 250
Vibrio cholerae epidemiology 30
Vir proteins 69-70, 231
vitronectin 192

Y

Warren, R. x, 4, 15
WASP 238

Y

Yersinia spp. 139, 312

Z

zymogen granules 213

Other Publications of Interest

NMR in Microbiology: Theory and Applications

Eds: **Jean-Noël Barbotin and Jean-Charles Portais.**
Foreword by **Daniel Thomas**
This book describes the theory and practical applications of this increasingly important technique and is aimed specifically at microbiologists. Discover the value and potential and of this powerful technology!
2000, 500 p. ISBN 1-898486-21-2 £84.99 or $169.99

Development of Novel Antimicrobial Agents: Emerging Strategies

Editor: **Karl Lohner**
International researchers from academia and industry present this unique collection of highly acclaimed reviews covering every aspect of this important topic. Essential reading for all scientists interested in the development of antimicrobial agents.
2001, 284 p.,ISBN 1-898486-23-9 £84.99 or $169.99

Oral Bacterial Ecology: The Molecular Basis

Editors: **Howard K. Kuramitsu and Richard P. Ellen**
Aimed at researchers in the field of oral microbial pathogenesis this book reviews in depth the molecular basis of oral pathogenesis and ecology. Particular emphasis is placed on recent advances in molecular microbiology and genomics.
2000, 314 p. ISBN 1-898486-22-0 £74.99 or $149.99

Molecular Marine Microbiology

Editor: **Douglas H. Bartlett**
Highly topical reviews on current aspects of this exciting area, originally presented as a JMMB symposium. Provides a valuable insight into recent developments in this broad field and their biomedical or biotechnological relevance.
2000, 219 p. ISBN 1-898486-20-4 £59.99 or $119.99

Prions: Molecular & Cellular Biology

Edited by: **David A. Harris**
"...... exceeded my expectations in terms of breadth of scope, clarity, insightful interpretation and thoroughness ......" *Trends in Microbiology*
1999, 218 p. ISBN 1-898486-07-7 £74.99 or $129.99

Forthcoming Titles !!!

- **The Internet for Cell & Molecular Biologists**
  By *Andrea Cabibbo*
- **Environmental Molecular Microbiology**
  Editor *Paul Rochelle*
- **Genomes and Databases on the Internet**
  By *Paul Rangel*
- **Flow Cytometry for Research Scientists: Principles and Applications**
  By *Rafael Nunez*

To be kept informed on any of these titles please email **forthcoming@horizonpress.com**

Prokaryotic Nitrogen Fixation: A Model System for the Analysis of a Biological Process

Editor: **Eric W. Triplett**
Eminent researchers review in depth every aspect of this important subject. The book emphasizes the importance of nitrogen fixation as a model for the analysis of many other biological processes and will be useful to a wide range of scientists.
2000, 800p. ISBN 1-898486-19-0 £119.99 or $239.99

Peptide Nucleic Acids: Protocols and Applications

Edited by: **Peter E. Nielsen** and **Michael Egholm**
Contains state-of-the-art protocols and applications on all aspects of PNA. Concepts are explained clearly and in practical terms and each chapter contains concise background information. Written by leading experts in the field.
1999, 262p. ISBN 1-898486-16-6 £59.99 or $119.99

Probiotics: A Critical Review

Edited by: **Gerald W. Tannock**
State-of-the-art commentaries on probiotic research.
"... highly proficient look at a rapidly moving field." *Microbiology Today.*
"I recommend reading it ... " *J. Antimicrobial Chemother.*
1999, 161 p. ISBN 1-898486-15-8 £59.99 or $119.99

Intracellular Ribozyme Applications Principles and Protocols

Edited by: **John J. Rossi and Larry Couture.**
Foreword by **Thomas R. Cech.**
The definitive guide to intracellular ribozyme applications containing reviews of the most recent pharmaceutical, therapeutic, and biotechnological applications of ribozymes. The latest principles, applications, protocols, and much, much more!
1999, 289 p. ISBN 1-898486-17-4 £74.99 or $149.99

Cold Shock Response and Adaptation

Editor: **Masayori Inouye**
Highly regarded authors review the current status of cold shock research in bacteria, plants, and mammals. Eminent scientists present fascinating insights into cellular adaptation and response to cold shock in various organisms.
2000, 140 p. ISBN 1-898486-24-7 £59.99 or $119.99

New Journals !!!

- **Journal of Molecular Microbiology and Biotechnology**
  Editor-in-Chief *Milton H. Saier Jr.*
  ISSN 1464-1801   www.jmmb.net

- **Current Issues in Intestinal Microbiology**
  Editor-in-Chief *Gerald W. Tannock*
  ISSN 1466-531X   www.ciim.net

- **Current Issues in Molecular Biology**
  ISSN 1467-3037   www.cimb.org

- **Molecular Biology Today**
  ISSN 1468-5698   www.molbio.net

For further details on any of these titles please visit
www.horizonpress.com

Also of Interest !!!

- **Gene Cloning & Analysis: Current Innovations**
  Editor *Brian C. Schaefer*
- **Genetic Engineering with PCR**
  Editors *R. Horton and R.C. Tait*
- **Internet for the Molecular Biologist**
  Editors *Swindell et al*
- **An Introduction to Molecular Biology**
  By *R.C. Tait*
- **The Lactic Acid Bacteria**
  Editors *Foo et al*
- **Molecular Biology: Current Innovations and Future Trends**
  Editors *H.G. Griffin and A.M. Griffin*

For further details on any of these titles please visit
www.horizonpress.com

Order-form for all books and journals

NO. OF COPIES	TITLE	PRICE
	Add postage and handling £4 (UK), £6 (Europe) and £8/$16 (Rest of World) for each book:	
	TOTAL	

☐ I enclose a cheque/check in £/$. Amount ....
☐ Please debit my Visa/Mastercard/Diners/Amex. Amount ....
Card number ....
Expiry date ....

Name:

Address:

Order from:

Horizon Scientific Press
32 Hewitts Lane, Wymondham
Norfolk, NR18 0JA, U.K.

Tel: +44-(0)1953-601106
Fax: +44-(0)1953-603068
Email to: mail@horizonpress.com
http://www.horizonpress.com

In the USA order books from: ISBS, 5804 N.E.
Hassalo St, Portland, Oregon 97213-3644
Tel: (800) 944-6190 Fax: (503) 280-8832

H.pylori

☞ **FAX TO:**
+44-(0)1953-603068

☞ **MAIL TO:**
32 Hewitts Lane
Wymondham, Norfolk,
NR18 0JA, United Kingdom

☞ **EMAIL TO:**
orders@horizonpress.com

10% discount !!!
For all online orders at
www.horizonpress.com